DYNAMIC FOOD WEBS: MULTISPECIES ASSEMBLAGES, ECOSYSTEM DEVELOPMENT, AND ENVIRONMENTAL CHANGE

A volume in the
Academic Press | THEORETICAL ECOLOGY SERIES

Published Books in the Series

J. M. Cushing, R. F. Costantino, Brian Dennis, Robert A. Desharnais, Shandelle M. Henson, *Chaos in Ecology:* Experimental Nonlinear Dynamics, 2003.

DYNAMIC FOOD WEBS: MULTISPECIES ASSEMBLAGES, ECOSYSTEM DEVELOPMENT, AND ENVIRONMENTAL CHANGE

Peter de Ruiter
Utrecht University
Utrecht, The Netherlands

Volkmar Wolters
Justus-Liebig University
Giessen, Germany

John C. Moore
University of Northern Colorado
Greeley, Colorado, USA

Managing Editor
Kimberly Melville-Smith
University of Northern Colorado
Greeley, Colorado, USA

AMSTERDAM • BOSTON • HEIDELBERG • LONDON
NEW YORK • OXFORD • PARIS • SAN DIEGO
SAN FRANCISCO • SINGAPORE • SYDNEY • TOKYO
Academic Press is an imprint of Elsevier

Academic Press is an imprint of Elsevier.
30 Corporate Drive, Suite 400, Burlington, MA 01803, USA
525 B Street, Suite 1900, San Diego, California 92101-4495, USA
84 Theobald's Road, London WC1X 8RR, UK

This book is printed on acid-free paper. ♾

Library of Congress Cataloging-in-Publication Data
Application submitted

British Library Cataloguing in Publication Data
A catalogue record for this book is available from the British Library

ISBN 13: 978-0-12-088458-2
ISBN 10: 0-12-088458-5

For all information on all Academic Press publications
visit our Web site at www.books.elsevier.com

Printed in the United States of America
05 06 07 08 09 10 9 8 7 6 5 4 3 2 1

CONTENTS

CONTRIBUTORS

D. Albrey Arrington
Department of Biological Sciences
University of Alabama
Tuscaloosa, Alabama, USA

Kevin Attree
Ecosystems Group
International Ecotechnology Research Centre
Cranfield University
United Kingdom

Donald J. Baird
National Water Research Institute (Environment Canada)
Canadian Rivers Institute
Department of Biology
University of New Brunswick
Fredericton, New Brunswick, Canada

Carolin Banašek-Richter
Institute of Biology
Technical University of Darmstadt
Germany

Beatrix E. Beisner
Département des Sciences Biologiques
Université du Québec
Montréal, Canada

Janne Bengtsson
Section for Landscape Ecology
Department of Ecology and Crop Production Science
Swedish University of Agricultural Sciences
Uppsala, Sweden

Jonathan P. Benstead
The Ecosystems Center
Marine Biological Laboratory
Woods Hole, Massachusetts, USA

Matty P. Berg
Department of Animal Ecology
Institute of Ecological Science
Vrije Universiteit
Amsterdam, The Netherlands

Eric L. Berlow
University of California, San Diego
White Mountain Research Station
Bishop, California;
Pacific Ecoinformatics and Computational Ecology Lab
Rocky Mountain Biological Laboratory
Crested Butte, Colorado, USA

Louis-Félix Bersier
Institut de Zoologie
University of Neuchâtel
Chair of Statistics
Department of Mathematics
Swiss Federal Institute of Technology
Lausanne, Switzerland

Ottar Bjornstad
Department of Entomology
Pennsylvania State University
University Park, Pennsylvania;
Department of Biology
Technical University of Darmstadt
Germany

Kathryn V. Bracewell
Ecosystems Group
International Ecotechnology Research Centre
Cranfield University
United Kingdom

Ulrich Brose
Department of Biology
Technical University of Darmstadt
Germany;
Pacific Ecoinformatics and Computational Ecology Lab
Rocky Mountain Biological Laboratory
Crested Butte, Colorado, USA

Stephen R. Carpenter
Center for Limnology
University of Wisconsin
Madison, Wisconsin, USA

Kevin J. Cash
Prairie and Northern Wildlife Research Centre (Environment Canada)
Saskatoon, South Kensington, Canada

Marie-France Cattin
Institut de Zoologie
University of Neuchâtel
Switzerland

Joel E. Cohen
Laboratory of Populations
Rockefeller & Columbia Universities
New York, New York, USA

Steven H. Cousins
Ecosystems Group
International Ecotechnology Research Centre
Cranfield University
United Kingdom

Wyatt F. Cross
Institute of Ecology
University of Georgia
Athens, Georgia, USA

Kim Cuddington
Ohio University
Biological Sciences
Athens, Ohio, USA

Joseph M. Culp
National Water Research Institute (Environment Canada)
Canadian Rivers Institute
Department of Biology
University of New Brunswick
Fredericton, New Brunswick, Canada

André M. de Roos
Institute for Biodiversity and Ecosystems

University of Amsterdam
The Netherlands

Peter C. de Ruiter
Department of Environmental Sciences
Utrecht University
The Netherlands

Don L. De Angelis
U.S. Geological Survey
Biological Resources Division and University of Miami
Department of Biology
Coral Gables, Florida, USA

Stefan C. Dekker
Department of Environmental Sciences
Utrecht University
The Netherlands

Anthony I. Dell
Department of Tropical Biology
James Cook University
Townsville, Queensland, Australia

Amy Downing
Department of Zoology
Ohio Wesleyan University
Delaware, Ohio, USA

Barbara Drossel
Institut für Festkörperphysik
Technische Universität Darmstadt
Germany

Jennifer A. Dunne
Pacific Ecoinformatics and Computational Ecology Lab
Santa Fe Institute
Santa Fe, New Mexico, USA

Bo Ebenman
Department of Biology
Linköping University
Sweden

Sue L. Eggert
Institute of Ecology
University of Georgia
Athens, Georgia, USA

Anna Eklöf
Department of Biology
Linköping University
Sweden

Mark C. Emmerson
Department of Zoology
Ecology and Plant Sciences
University College Cork
Ireland

Jeremy W. Fox
Department of Biological Sciences
University of Calgary
Calgary, Alberta, Canada

Nancy E. Glozier
Prairie and Northern Wildlife Research Centre (Environment Canada)
Saskatoon, South Kensington, Canada

Spencer R. Hall
Department of Ecology and Evolution
University of Chicago
Chicago, Illinois, USA

Sarah Harper-Smith
Department of Biology
Seattle Pacific University
Seattle, Washington;
University of California, San Diego
White Mountain Research Station
Bishop, California, USA

Alan Hastings
Department of Environmental Science and Policy
University of California
Davis, California, USA

Florence D. Hulot
Laboratoire d'écologie
Ecole Normale Supérieure
Paris, France

Murray Humphries
Natural Resource Sciences
McGill University
St. Anne de Bellevue, Quebec, Canada

Andrew Keller
School of Life Sciences
Arizona State University
Tempe, Arizona, USA

Roland A. Knapp
Sierra Nevada Aquatic Research Laboratory
University of California
Crowley Lake, California, USA

Mariano Koen-Alonso
Northwest Atlantic Fisheries Centre
Fisheries and Oceans Canada
St. John's, Newfoundland, Canada

Giorgos D. Kokkoris
Department of Marine Sciences
Faculty of Environment
University of the Aegean
Mytilene, Lesvos Island, Greece

Michio Kondoh
Department of Environmental Solution Technology
Faculty of Science and Technology
Ryukoku University
Yokoya, Seta Oe-cho, Otsu, Japan

Bob W. Kooi
Department of Theoretical Biology
Faculty of Earth and Life Sciences
Vrije Universiteit
Amsterdam, The Netherlands

Craig A. Layman
Section of Ecology and Evolutionary Biology
Department of Wildlife and Fisheries Sciences
Texas A&M University
College Station, Texas, USA

Mathew A. Leibold
Section of Integrative Biology
University of Texas at Austin
Austin, Texas, USA

Michel Loreau
Laboratoire d'Ecologie
Ecole Normale Supérieure
Paris, France

Neo D. Martinez
Pacific Ecoinformatics
and Computational Ecology
Lab
Rocky Mountain Biological
Laboratory
Crested Butte, Colorado, USA

Kevin McCann
Department of Zoology
University of Guelph
Guelph, Ontario, Canada

Jill McGrady-Steed
Department of Ecology
Evolution, and Natural
Resources
Rutgers University, Cook
Campus
New Brunswick, New Jersey,
USA

Alan J. McKane
Department of Theoretical
Physics
University of Manchester
United Kingdom

Carlos Melian
Integrative Ecology Group
Estacio'n Biolo'gica de Don~ana
Sevilla, Spain

José M. Montoya
Complex Systems Lab
Universitat Pompeu Fabra
Barcelona;
Department of Ecology,
University of Alcalá
Alcalá de Henares, Madrid,
Spain

Wolf M. Mooij
Department of Food Web Studies
Netherlands Institute of Ecology
(NIOO-KNAW)
Centre for Limnology
Nieuwersluis, The Netherlands

John C. Moore
University of Northern Colorado
Department of Biological
Sciences
Greeley, Colorado, USA

Peter J. Morin
Department of Ecology,
Evolution, & Natural
Resources
Rutgers University
New Brunswick, New Jersey,
USA

Christian Mulder
Quantitative Ecology Unit
(QERAS)
Laboratory for Ecological
Risk-Assessment (LER)
National Institute for Public
Health and the Environment
(RIVM)
Bilthoven, The Netherlands

J.M.Olesen
Department of Ecology and
Genetics
University Aarhus
Aarhus, Denmark

Mitchell Pavao-Zuckerman
Department of Ecology and
Evolutionary Biology
University of Arizona
Tucson, Arizona, USA

Lennart Persson
Department of Ecology and Environmental Science
Umeå University
Sweden

Owen L. Petchey
Department of Animal and Plant Sciences
University of Sheffield
Alfred Denny Building
Western Bank, Sheffield, United Kingdom

Tobias Purtauf
Department of Animal Ecology
Justus-Liebig University
Giessen, Germany

Dave Raffaelli
Environment
University of York
Heslington, York, United Kingdom

Joe Rasmussen
Department of Biology
University of Lethbridge
Lethbridge, Alberta, Canada

Tamara N. Romanuk
Département des Sciences Biologiques
Université du Québec à Montréal
Montréal, Canada;
Pacific Ecoinformatics and Computational Ecology Lab
Rocky Mountain Biological Laboratory
Crested Butte, Colorado, USA

Amy D. Rosemond
Institute of Ecology
University of Georgia
Athens, Georgia, USA

John L. Sabo
School of Life Sciences
Arizona State University
Tempe, Arizona, USA

Ursula M. Scharler
Chesapeake Biological Laboratory
University of Maryland
Center for Environmental Studies
Solomons, Maryland;
Smithsonian Environmental Research Center
Edgewater, Maryland, USA

Stefan Scheu
Institute of Zoology,
Technische Universität Darmstadt
Germany

Dagmar Schröter
Center for International Development
Kennedy School of Government
Harvard University
Cambridge, Massachusetts, USA

Heikki Setälä
Department of Ecological and Environmental Sciences
University of Helsinki
Lahti, Finland

Ricard V. Solé
ICREA-Complex Systems Lab
Universitat Pompeu Fabra
Barcelona, Spain

Candan U. Soykan
School of Life Sciences
Arizona State University
Tempe, Arizona, USA

Maciej Szanser
Centre for Ecological Research PAS
Dziekanow Lesny near Warszawa
Poland

Elisa Thébault
Laboratoire d'Ecologie
Ecole Normale Supérieure
Paris, France

Theo P. Traas
Expert Centre for Substances
National Institute of Public Health and the Environment (RIVM)
Bilthoven, The Netherlands

James Umbanhowar
Department of Zoology
University of Guelph
Guelph, Ontario, Canada

A. Valido
Department of Ecology and Genetics
University of Aarhus
Aarhus, Denmark;
Department of Animal and Plant Sciences
University of Sheffield
United Kingdom

Herman A. Verhoef
Department of Animal Ecology
Institute of Ecological Science
Vrije Universiteit
Amsterdam, The Netherlands

Matthijs Vos
Department of Food Web Studies
Netherlands Institute of Ecology (NIOO-KNAW)
Centre for Limnology
Nieuwersluis, The Netherlands

J. Bruce Wallace
Institute of Ecology and Department of Entomology
University of Georgia
Athens, Georgia, USA

Philip H. Warren
Department of Animal and Plant Sciences
University of Sheffield
United Kingdom

Richard J. Williams
Pacific Ecoinformatics and Computational Ecology Lab
Rocky Mountain Biological Laboratory
Crested Butte, Colorado, USA

Kirk O. Winemiller
Section of Ecology and Evolutionary Biology
Department of Wildlife and Fisheries Sciences
Texas A&M University
College Station, Texas, USA

Volkmar Wolters
Department of Animal Ecology
Justus-Liebig-University
Giessen, Germany

Guy Woodward
Department of Zoology
Ecology and Plant Sciences
University College Cork
Ireland

J. Timothy Wootton
Ecology and Evolution
University of Chicago
Chicago, Illinois, USA

Peter Yodzis
Department of Zoology
University of Guelph
Guelph, Ontario, Canada

1.0 | TRIBUTE

Kevin McCann, Mariano Koen-Alonso, Alan Hastings, and John C. Moore

In the time since the last symposium held at Pinagree Park, ecology lost two formidable and important ecologists in Gary Polis and Peter Yodzis. Their presence was sorely missed at the most recent food web conference in Giessen, Germany, as their passion and enthusiasm for ecology served as a catalyst at any gathering. In the area of food web ecology, they are figures of major historical importance, both scientists continuously pushing and challenging the boundaries of our understanding. In this manner they were very similar. Gary and Peter were inspired by the beauty and complexity of the world around them, and both men were fearless in their attempts to begin to understand one of nature's most complicated puzzles, the food web. Additionally, both men were powerful personalities and determined to forge their own path in the history books of ecology. In other ways it would be hard to find two men so completely different. Gary Polis was a scorpion expert, and a hardcore empirical ecologist; Peter Yodzis was a theoretical physicist specializing in general relativity before becoming an ecologist. Gary brought unbridled amounts of enthusiasm to the scientific table. In doing so he was able to inspire a new generation of ecologists to challenge old ideas. Gary was a champion of field observation and the manipulative experiment. His work more than once reminded us of the complexity of nature, the oversimplifications behind our assumptions, and the power of reason by counterexample. Peter, on the other hand, championed the development of ecological theory that was founded on the clear and rigorous tools of the physicist. He loved thought experiments (the Gedanken experiments of Einstein). To him, the thought experiment distilled the essentials of a good scientist by forcing the scientist to pose a problem that was both clear and answerable upon logic alone. This is not to say that he believed the thought experiment as an end but rather saw it as a creative way of developing one of the most important tools of the scientist—intuition. In a historical sense, their differences represent the two aspects of ecology (theory and empiricism); however, here, too, they played an important role in bringing theory to empiricism and empiricism to theory. As one can see from this book, their efforts permeate all recent advances in food web ecology.

At a personal level, both men were deeply compassionate and caring toward family and friends. Again they did this in slightly different ways. Gary Polis's magnetic character and *joie de vie* warmed and engaged all those around him. From Peter Yodzis emanated an enormous warmth and gentle concern for all those lucky enough to come into his circle. They will be deeply missed as scientist and friends.

1.1 | DYNAMIC FOOD WEBS

Peter C. de Ruiter, Volkmar Wolters, and John C. Moore

MULTISPECIES ASSEMBLAGES, ECOSYSTEM DEVELOPMENT, AND ENVIRONMENTAL CHANGE

One of the most intensively studied food webs in ecological literature is that of Tuesday Lake in Michigan (USA) (Jonsson et al., 2004). The species composition in this food web was observed in two consecutive years, 1984 and 1986, while in between three planktivorous fish species were removed and one piscivorous fish species was added. This manipulation had hardly any effect on species richness (56 in 1984, 57 in 1986), but remarkably changed species composition as about 50% of the species were replaced by new incoming species. Manipulating one species and seeing effects on dozens of species reveals the importance of species interactions. It shows that species come and species go, populations fluctuate in numbers, and individuals grow and in connection with this may alter in the way they interact with other species. It shows the open, flexible, and dynamic nature of food webs.

Food webs are special descriptions of biological communities focusing on trophic interactions between consumers and resources. Food webs have become a central issue in population, community, and ecosystem ecology. The interactions within food webs are thought to influence the dynamics and persistence of many populations in fundamental ways through the availability of resources (i.e., energy/nutrients) and the mortality due to predation. Moreover, food web structure and ecosystem processes, such as the cycling of energy and nutrients, are deeply interrelated in that the trophic interactions represent transfer rates of energy and matter. Food webs therefore provide a way to analyze the interrelationships between community dynamics and stability and ecosystem functioning and how these are influenced by environmental change and disturbance.

Naturalists long ago observed how the distribution, abundance, and behavior of organisms are influenced by interactions with other species. Population dynamics of interacting predators and prey are difficult to predict, and many ecosystems are known to contain hundreds or thousands of these interactions arranged in highly complex networks of direct and indirect interactions. Motivated in part by May's (1972) theoretical study of the complexity–stability relationship, the study of food webs gained momentum in the late 1970s and early 1980s (Cohen, 1978; Pimm, 1982). A formal means of dealing with the flow of energy and matter in food webs was ushered with the advent of ecosystem ecology (Odum, 1963), and since then the food web approach has been adopted to analyze interrelationships between community structure, stability, and ecosystem processes (DeAngelis, 1992).

The first food web symposium was convened at Gatlinburg, North Carolina, in 1982 (DeAngelis et al., 1982). That symposium was dominated by theoretical studies focused on the complexity–stability relationship and empirical studies examining features of simple topological webs (ball and stick diagrams) compiled from the published literature. The ensuing decade was marked by exploration of a greater number of issues influencing the structure and dynamics of food webs (interaction strength, indirect effects, keystone species, spatial variation, and temporal variation in abiotic drivers) and a search for more detailed and accurate food web descriptions. Some ecologists questioned the utility of analyzing features of web diagrams that quite obviously contained too few taxa, grossly unequal levels of species aggregation, and feeding links with no magnitudes or spatio-temporal variation (Hall and Raffaelli, 1997).

A second food web symposium, held at Pingree Park, Colorado, in 1993 (Polis and Winemiller, 1996), emphasized dynamic predator–prey models, causes and effects of spatial and temporal variation, life history strategies, top-down and bottom-up processes, and comparisons of aquatic, terrestrial, and soil webs. Over the last decade the ecological debate became increasingly dominated by a number of new topics, such as environmental change, spatial ecology, and functional implications of biodiversity. This has changed our view on the entities, scales, and processes that have to be addressed by ecological research, and the food web approach became recognized as a most powerful tool to approach these issues. This was the point-of-departure for the third food web symposium held in November 2003 in Schloss Rauischolzhausen, Germany. This volume presents the proceedings of this symposium.

Much more than its predecessors, this symposium highlights approaches to understand the structure and functioning of food webs on the basis of detailed analyses of biological properties of individuals, populations, and compartments within communities. Much emphasis is laid on the understanding of food web structure and stability. Some contributions approach food web structure and dynamics from 'outside' environmental variability, in space as well as in time. Other contributions take the opposite approach by looking in depth to the dynamics of populations and biological attributes of individual within populations.

Approaching food web structure and dynamics from environmental characteristics (Section 2) shows that environmental heterogeneity may create sub-systems (compartments), especially at the lower trophic levels in food webs, with organisms at the higher trophic levels that act as 'integrators' across this variability in space and time and stabilize dynamics of their resources via density-dependent adaptive foraging. Such compartmentation has been observed at the level of spatial and temporal variation of resource availability; an example is provided for soil food webs, for which records of spatial and temporal variation indicate the primary energy source of soil organic matter as major driving force, with important implications for system stability (Moore and de Ruiter, 1997). This explicitly relates to MacArthur's idea (MacArthur, 1955) that community complexity should buffer against perturbations, and thereby override inherent constraints on system stability imposed by complexity (May, 1972). Another aspect of environmental variability regards the dynamics in nutrient availability governing the interplay between competition and trophic interactions and by this the dynamics of the populations at various trophic levels. Comparison of food web structures from different habitats, soil, terrestrial and aquatic, shows regular patterns in the flows with which food is transferred and processed by the trophic groups in the food webs. This approach bridged the gap between looking at descriptive properties of food web structure, such as species richness and trophic levels and looking at species composition in detail, as it reveals regularities in food web structure that are crucial to food web stability and functioning and appears less sensitive to the dynamics in species composition in food webs.

Approaching food web structure from dynamics in populations (Section 3) shows that the evolution of realistic food web structures can be explained on the basis of simple rules regarding population abundance and species occurrence. Life-history–based dynamics within populations may even influence community dynamics in extraordinary and counterintuitive ways in the way that predators promote each other's persistence when they forage on different life stages of their prey, inducing a shift in the size distribution of the prey, leading to more and larger sized individuals and increased population fecundity. But also within populations the dynamics in the behavior of individuals, such as prey switching, may affect population dynamics, as dietary shifts inhibit rapid growth by abundant prey and at the same time allow rare prey to rally. If these shifts are fast enough, food web architecture changes at the same time-scale as population dynamics. This affects food web structure and stability, and may even result in a positive complexity–stability relationship as proposed by Elton some seventy years ago (Elton, 1927). Preferential feeding by predators may result from prey properties (body size), or from spatial and temporal variability in prey availability. While dietary shifts may be the result of adaptive behavior by the predator, predators may also 'induce' defense mechanism in the prey; the dynamics of attack and defense may have strong implications for food web structure, stability, and functioning.

The analyses of biological properties of individuals within populations show a strong explanatory power of body size to population abundance scaling rules in understanding the dynamics and persistence of trophic groups in food webs (Section 4). Ratios between predator and prey body sizes generate patterns in the strengths of trophic interactions that enhance food web stability in a Scottish estuary. This finding confirms the published analysis of the mammal community of the Serengeti, in which predator–prey body size ratios are a primary factor determining predation risks (Sinclair et al., 2003). The approach of looking at body size relationships to understand food web structure provides a novel diversity–stability context for Charles Elton's original interest in trophic pyramids derived from feeding and body size constraints (Cousins, 1995).

Resource availability and use may govern the structure and functioning of food webs, in turn food web interactions are the basis of ecosystem processes and govern important pathways in the global cycling of matter, energy, and nutrients. Food web studies in this way connect the dynamics of populations to the dynamics in ecosystem processes (Section 5). The mutual effects between the dynamics of food webs and detritus influences food web structure as well as habitat quality. Variation in the availability of one environmental factor, i.e., nitrogen deposition, affect ecosystem processes like organic matter decomposition, nitrogen mineralization, and CO_2 emission through the mediating role of the soil food web. Similarly, the interplay between the availability of various, potentially limiting, nutrients and the network of trophic interactions may strongly impact on dynamics of both populations and nutrients in stream food webs. To fully understand the role of food webs in the energy cycle requires tools to translate resource availability to energy supply necessary for population functioning and persistence. Mechanisms operating within these transitions may vary among resource of the different trophic levels (e.g., primary producers, herbivores, and carnivores). Models that calculate the interplay between ecosystem processes and food web structure and functioning have hardly accounted for such dynamics and variations; hence new ways of modeling these processes are proposed.

The food web approach may contribute to the analysis and solution of the worldwide decline in environmental quality and biological diversity due to human activities (e.g., through climate change, habitat fragmentation, invasion, pollution, and overexploitation of natural resources). The consequences of species diversity and composition for ecosystem functioning and the provision of ecosystem services have been widely explored. Most of these studies, however, have focused on the effects of biodiversity change within single trophic levels (Loreau et al., 2002) (e.g., by looking at the effects of biological diversity in plant communities on processes like plant productivity) (Naeem et al., 1994). However, the trophic context of species in food webs may strongly influence the risks of species loss, and the possible consequences of species loss for ecosystem functioning (Section 6). A modeling approach shows that in multitrophic level systems, increasing diversity influences plant biomass and productivity in a non-linear manner. These model results are supported by empirical evidence showing that the consequences of species loss to ecosystem functioning depend on trophic level. And experiments on pond food webs show that the contributions of species to ecosystem processes depend on environmental factors, such as productivity, as well as on trophic position whereby higher trophic levels tend to have the largest effects. These kinds of results indicate that the effects of a particular species loss on ecosystem functioning can be inconsistent across ecosystems. In soil food webs, the role of species in soil processes depends on trophic position with functional redundancy greater within trophic groups than between trophic groups. Similarly, community invasibility does not entirely depend on factors like resource availability, but also on community structure especially when the 'receiving' food web becomes more reticulate. These model and experimental findings ask for new ways to measure functional diversity of species depending on the trophic structure of which they are part of.

In the field of environmental risk assessment, food webs provide a way to analyze the overall assemblage of direct and indirect effects of environmental stress and disturbance (Section 7). Such indirect effects may occur through the transfer and magnification of contaminants through food chains causing major effects on species at the end of the food chain, as well as through changes in the dynamics of interacting populations. Sometimes, species extinctions can be seen as the direct result of human activities, but in other cases extinctions are to be understood from effects of primary extinctions on the structure of the food web, such as the disappearance of some bird species from Barro Colorado Island. Overexploitation by fisheries is one of the most acute environmental problems in freshwater as well as in marine systems. Regarding sustainable fisheries, most ecologists are familiar with the "fishing down food webs" phenomenon (Pauly et al., 1998). The multi-species nature of fisheries makes the food web approach intuitively appealing, with fishery harvest viewed as consumption by an additional predator, complete with functional responses, adaptive foraging, etc. The food web approach in designing sustainable fishery practices can be supported by new methods to quantify food web links. An example is given in which stable isotopes may lead to new insights in the effects of fisheries in river food webs. The complex nature of effects of human activities on ecosystem properties asks for ways to communicate these effects with resource managers and policymakers. Visualization in the form of food webs has been shown to be very helpful. In this way, food web approaches are increasingly providing guidance for the assessment of ecological risks of human activities and for the sustainable management of natural resources, and are even beginning to influence policy.

The next ten years of food web research should see continued theoretical advancement accompanied by rigorous experiments and detailed empirical studies of food web modules in a variety of ecosystems. Future studies are needed to examine effects of taxonomic, temporal, and spatial scales on dynamic food web models. For example, adaptive foraging partially determines and stabilizes food web dynamics, but environmental heterogeneity at appropriate scales also can have a stabilizing effect. A challenge will be to further elaborate the intriguing idea that trophic interactions in food webs drive patterns and dynamics observed at multiple levels of biological organization. For example, individual attributes, such as body size, influence demographic parameters in addition to predator–prey interactions. Food web research might even provide new insights into the origins and evolution of organisms. As food web science continues to increase its pace of development, it surely will contribute new tools and new perspectives for the management of our natural environment.

ACKNOWLEDGMENTS

We thank the European Science Foundation (ESF) for funding, Joanne Dalton for all her efforts to have the meeting organized. We acknowledge the contributors of the Interact Group, sponsored by the ESF, and the Detritus Wozhing group, sponsored by the National Center for Ecological Analysis and Synthesis. We especially thank Peter Morin, Dave Raffaelli, and Stefan Scheu for co-organizing, and leading sessions and discussions, the participants of Food Web 2003 for their contributions, Tobias Purtauf and 'the crew' for their work prior and during the meeting, and the staff of Schloß Rauischholzhausen for their hospitality.

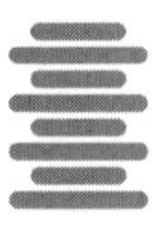

1.2 | FOOD WEB SCIENCE: MOVING ON THE PATH FROM ABSTRACTION TO PREDICTION

Kirk O. Winemiller and Craig A. Layman

This chapter explores some basic issues in food web research, evaluates major obstacles impeding empirical research, and proposes a research approach aimed at improving predictive models through descriptive and experimental studies of modules within large, complex food webs. Challenges for development of predictive models of dynamics in ecosystems are formidable; nonetheless, much progress has been made during the three decades leading up to this third workshop volume. In many respects, food web theory has outpaced the empirical research needed to evaluate models. We argue that much greater investment in descriptive and experimental studies as well as exploration of new approaches are needed to close the gap.

The most fundamental questions in food web science are "How are food webs structured?" and "How does this structure influence population dynamics and ecosystem processes?" At least four basic models of food web structure can be proposed. One model could be called the "Christmas tree" model, in which production dynamics and ecosystem processes essentially are determined by a relatively small number of structural species. Most of the species' richness in communities pertains to interstitial species that largely depend on the structural species for resources, and may be strongly influenced by predation from structural species. Hence, interstitial species are like Christmas ornaments supported by a tree composed of structural species (Figure 1A). Structural species could include conspicuous species that dominate the biomass of the system, but also could be keystone species that may be uncommon but have disproportionately large effects on the food web and ecosystem (Power et al., 1996b; Hurlbert, 1997). In many ecosystems, certain plants and herbivores clearly support most of the consumer biomass, and certain consumers strongly influence biomass and production dynamics at lower levels. This pattern may be more apparent in relatively low-diversity communities, such as shortgrass prairies and kelp forests, in which relatively few species provide most of the production, consume most of the resources, or influence most of the habitat features.

A second alternative is the "onion" model in which core and peripheral species influence each other's dynamics, with core species having a greater influence (i.e., magnitudes of pairwise species effects are not reciprocal). The core-peripheral structure is arranged in a nested hierarchy (Figure 1B). This model might pertain to high-diversity ecosystems such as tropical rainforests and coral reefs. Ecological specialization via co-evolution would result in interactions from peripheral species that may have strong effects on a few species, but weak effects on most of the community, and very weak effects on core species. In tropical rainforests, rare epiphytic plants and their co-evolved herbivores, pollinators, and seed dispersers depend upon the core assemblage of tree species, yet the converse is not true. Removal of a given pollinator species would yield a ripple effect within an interactive subset, or module, of the food web, but likely would not significantly affect core species of decomposers, plants, and animals.

A third food web structure could be called the "spider web" model in which every species affects every other species via the network of direct and indirect pathways (Figure 1C). This concept, in which everything affects everything, is explicit in network analysis (Fath and Patten, 1999), which gives rise to numerous emergent properties of networks (Ulanowicz, 1986). Signal strength, via direct or indirect propagation, may depend on proximity of nodes within the network. Propagation of indirect effects in food webs can yield counterintuitive results from press perturbations. For example, harvesting a competitor of a top predator can result in a decline rather than an increase of that predator (Yodzis, 1996; Wootton, 2001; Relyea and Yurewicz, 2002).

A fourth model of food web structure could be called the "internet" model. Following this concept, webs are networks having major and minor "hubs" in which their position within the network architecture

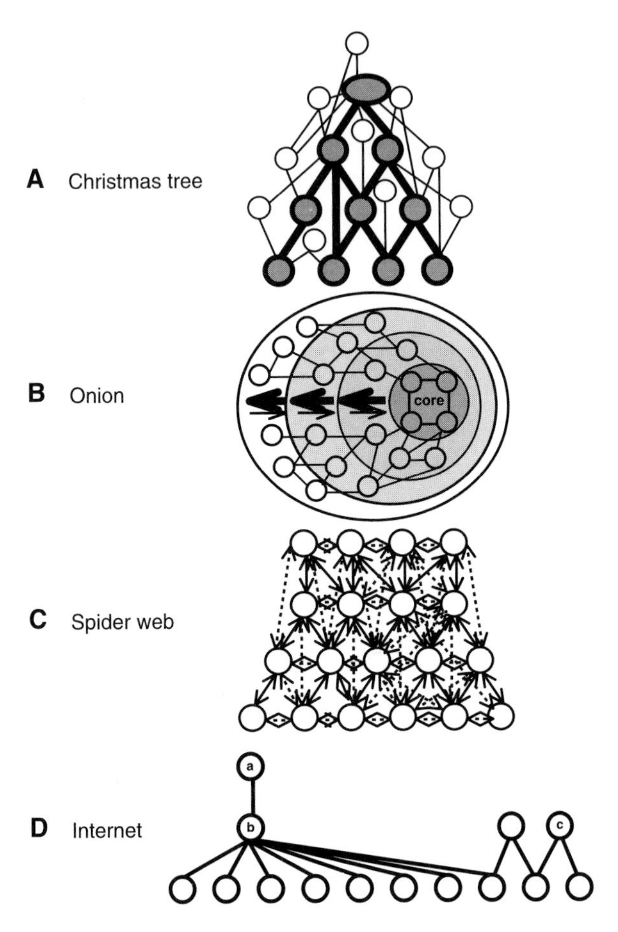

FIGURE 1 | Schematic illustrations of four models of food web structure: **A,** Christmas tree (structural and interstitial species); **B,** onion (hierarchy of core and peripheral species, the strength of effects is greater from the core outward); **C,** spider web (all species affect all others either directly or indirectly); and **D,** internet (network architecture yields disproportionate influence by hub species, which are not necessarily identified by the number of direct connections, that is, node a could actually have more influence on the system, via its control of node b, than node c).

determines the degree that a species can influence other species in the system (Figure 1D). Jordán and Scheuring (2002) reviewed the applicability of the internet model to food webs, and maintained that the density of connections to a node may be a poor indicator of the potential influence on web dynamics. For example, a highly influential species (e.g., top predator) could have only one or a few links connecting it to other species that in turn have numerous connections to other species in the system. Analysis of network features has become a popular pursuit in fields ranging from the social sciences to cell biology, but the relevance of this approach for understanding food web dynamics is uncertain (Jordán and Scheuring, 2002).

How are food webs structured? The answer will necessarily rely on accumulated evidence from a large body of empirical research. We contend that available evidence is insufficient to state, with a degree of confidence, the general circumstances that yield one or another of these alternative models. Like any scientific endeavor, research on food webs advances on four interacting fronts: description (observation), theory (model formulation), model testing (experimentation), and evaluation. Evaluation invariably leads to theory revision and the loop begins again. After several trips around this loop, a model may begin to successfully predict observations, and we gain confidence for applications to solve practical problems. Important ecological challenges already have been addressed using the food web paradigm, including biocontrol of pests, fisheries management, biodiversity conservation, management of water quality in lakes, and ecotoxicology (Crowder et al., 1996). We believe, however, that the development of food web theories (models) and their applications is greatly outpacing advances in the descriptive and experimental arenas. Although this state of affairs is not unexpected in an immature scientific discipline, it results in inefficient development of understanding. Why have empirical components lagged behind theoretical developments? We propose that unresolved issues of resolution and scale have hindered empirical research. Resolution of four basic aspects of food webs is required: (1) the food web as an operational unit, (2) components of food webs, (3) the nature of food web links, and (4) drivers of temporal and spatial variation.

Food Webs as Units

First, the spatial and temporal boundaries of a community food web are always arbitrary, and it should be emphasized that any food web is a module or subnetwork embedded within a larger system (Cohen, 1978; Moore and Hunt, 1988; Winemiller, 1990; Polis, 1991; Hall and Raffaelli, 1993; Holt, 1997; and others). Food webs are almost always defined according to habitat units nested within, and interacting with larger systems (e.g., biotia living on a single plant, water-filled tree holes, soil, lakes, streams, estuaries, forests, islands). Hence, every empirical food web is a web module. Spatial and taxonomic limits of modules are essentially arbitrary. Thus, it probably makes little sense to speak of large versus small webs, for example. Web modules vary in their degree of correspondence to habitat boundaries. Although a lake has more discrete physical boundaries than a lowland river with flood pulses and marginal wetlands, numerous links unite lake webs with surrounding terrestrial webs. Thus, broad comparative studies of food web properties necessarily deal with arbitrary units that may have little or no relationship to each other.

To illustrate this point, we examine empirical food webs from three studies, all published in the journal *Nature*, that constructed models to predict statistical features of these webs (Williams and Martinez, 2000; Garlaschelli et al., 2003; Krause et al., 2003). Leaving aside issues related to web links and environmental drivers, let us examine the number of taxa within each habitat. For statistical comparisons, these taxa were subsequently aggregated into "trophospecies" (species that presumably eat all the same resources and also are eaten by all the same consumers). The number of taxa were reported as follows: Skipwith Pond, England (35); Bridge Brook Lake, New York (75); Little Rock Lake, Wisconsin (181); Ythan Estuary, Scotland (92); Chesapeake Bay, United States (33); Coachella Valley, California (30); and Isle of St. Martin, Caribbean (44). Thus, we are led to conclude that Skipwith Pond, a small ephemeral pond in England (Warren, 1989), contains more taxa than Ythan Estuary, Scotland (92) (Hall and Raffaelli, 1991), and Chesapeake Bay (33) (Baird and Ulanowicz, 1989), one of the world's largest estuaries. These food webs were originally compiled based on different objectives and criteria. The Skipwith Pond food web reports no primary producer taxa, the Bridge Brook Lake web contains only pelagic taxa, the Chesapeake Bay web is an ecosystem model with a high degree of aggregation, and the Ythan Estuary web includes 27 bird taxa with most other groups highly aggregated. If we examine just the number of reported fish species, Skipwith Pond has none, Ythan Estuary has 17, and Chesapeake Bay is reported to have 12. In reality, Chesapeake Bay has at least 202 fish species (Hildebrand and Schroeder, 1972). These comparative studies analyzed features of Polis's (1991) highly aggregated Coachella Valley web (30 taxa) even though that author clearly cautioned against it and indicated that the web contained, among other taxa, at least 138 vertebrate, 174 vascular plant, and an estimated 2,000–3,000 insect species. The Isle of St. Martin web was reported to have 44 taxa that include 10 bird and 2 lizard species plus 8 nonvertebrate aggregations (Goldwasser and Roughgarden, 1993).

Clearly, these empirical food webs represent an odd collection of woefully incomplete descriptions of community species richness and trophic interactions, and are unlikely to provide a basis for robust predictive models. Discrepancies are due to the fact that these webs were originally compiled based on different objectives and criteria. Objective methods for defining and quantifying nested modules are badly needed. At a minimum, consistent operational definitions for units and standardized methodologies are required to make quantitative comparisons. For example, sink food webs (Cohen, 1978) can be defined based on the network of direct trophic links leading to a predator. Comparisons of different systems could be based on the sink webs associated with predators that are approximate ecological "equivalents". Alternatively, food web comparisons can be based on the collection of sink webs leading to consumers of a given taxonomic group, such as fishes (Winemiller, 1990). Source webs (tracing the network trophic links derived from a taxon positioned low in the web) provide an operational unit for food web comparisons (e.g., grasses–herbivorous insects–parasitoids) (Martinez et al., 1999), but in most cases, as links radiate upward (to higher trophic positions), they would very rapidly project outward (to adjacent habitats) in a manner that would yield major logistic challenges for empirical study.

Components of Food Webs

Our second issue is the units comprising food webs. Entities comprising food webs have been invoked to serve different objectives that are rarely compatible. Consequently, great variation is observed among food web components, ranging from species life stages to functional groups containing diverse taxa. In most empirical studies, these components have been invoked *a posteriori* rather than *a priori*. We must decide *a priori* whether we wish to examine individuals (what we catch), species populations (what we want to model), "trophospecies" (what we invoke when taxa had been aggregated), functional groups (what we think might be relevant), or trophic levels (what we once thought was relevant). Yodzis and Winemiller (1999) examined multiple criteria and algorithms for aggregating consumers into trophospecies based on detailed abundance and dietary data. Taxa revealed little overlap in resource use and the extent to which predators were shared, and almost no taxa could be grouped according to a strict definition of shared resources and predators. A similar approach was developed by Luczkovich et al. (2003) in which graph theory and the criterion of structural equivalence were used to estimate degrees of trophic equivalency among taxa. Unlike trophospecies, structurally equivalent taxa do not necessarily feed on any of the same food resources or share even a single predator, but they do play functionally similar roles in the network. We contend that species populations are the only natural food web components, because populations are evolutionary units with dynamics that are largely independent from those of heterospecific members of a guild or functional group (Ehrlich and Raven, 1969).

Food Web Links

The third issue is how we estimate food web links. Too often in the past, food web architecture was treated as binary with links either present or absent (i.e., web topology with no magnitudes or dynamics). Motivated, in part, by the seminal theoretical work of May (1973), empirical studies attempt to determine the nature and magnitude of links (i.e., interaction strength) using field experiments in which one or more species are manipulated (Paine, 1992; Menge, 1995; Wootton, 1997; Raffaelli et al., 2003). Interaction strength determines system dynamics (Paine, 1980) and stability (Yodzis, 1981a; Pimm, 1982; McCann et al., 1998), as well as the manner in which we view the basic structure of the food web (Winemiller, 1990; de Ruiter et al., 1995). Weak links are associated with greatest variation in species effects (Berlow, 1999), and food webs seem to be dominated by these weak links. For example, food webs of tropical aquatic systems are strongly dominated by weak feeding pathways as estimated from volumetric analysis of fish stomach contents (Figure 2).

Despite the critical need to understand interaction strength and the manner in which it creates food web structure and drives dynamics, many theoretical and comparative studies that relied on empirical data have not considered species abundances and have portrayed food web links simply as binary. Why has this been the case? First, it is *difficult* to inventory species in natural communities (e.g., Janzen and Hallwachs, 1994). It is *more difficult* to estimate species' relative abundances, even for conspicuous sedentary species like trees (e.g., Hubbell and Foster, 1986; Terborgh et al., 1990). It is *even more difficult* to estimate the

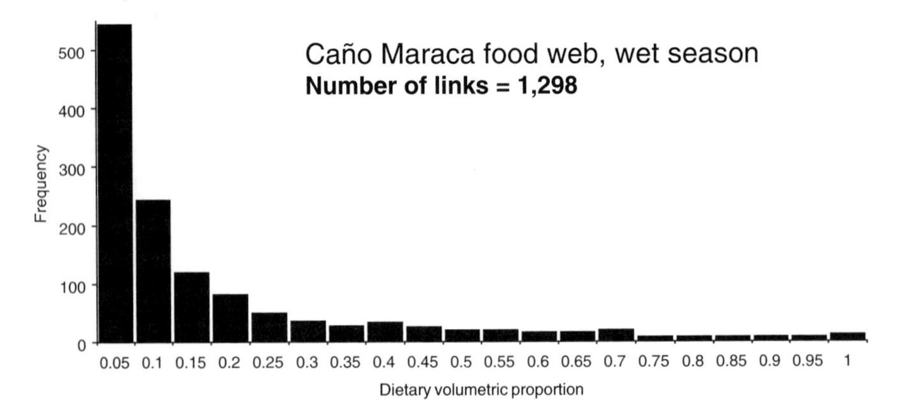

FIGURE 2 | Skewed distribution of feeding links of variable magnitudes (estimated as volumentric proportion of prey items in stomach contents) in a tropical wetland food web (Caño Maraca, Venezuela).

presence of feeding relationships (e.g., Thompson and Townsend, 1999). It is *yet more difficult* to estimate the magnitudes of feeding relationships (Winemiller, 1990; Tavares-Cromar and Williams, 1996). Finally, it is *exceedingly difficult* to estimate the strength of species interactions (Paine, 1992; Wootton, 1997).

Interaction strength can be inferred indirectly from quantitative dietary analysis, but this is extremely time consuming and requires a great degree of taxonomic and modeling expertise. The method is not viable for many consumer taxa, because most food items contained in the gut are degraded. Moreover, large samples are needed to estimate diet breadth (i.e., links) accurately and precisely and to reveal important spatial and temporal variation in feeding relationships (Winemiller, 1990). As sample size is increased from 1–20 individuals, the mean diet breadth of an omnivorous characid fish from Caño Maraca increases from 2.8–3.9, and the average number of feeding links increases from 3.7–22 (Figure 3). Similarly, sampling effort has been shown to affect food web properties associated with the number of nodes (Bersier et al.,

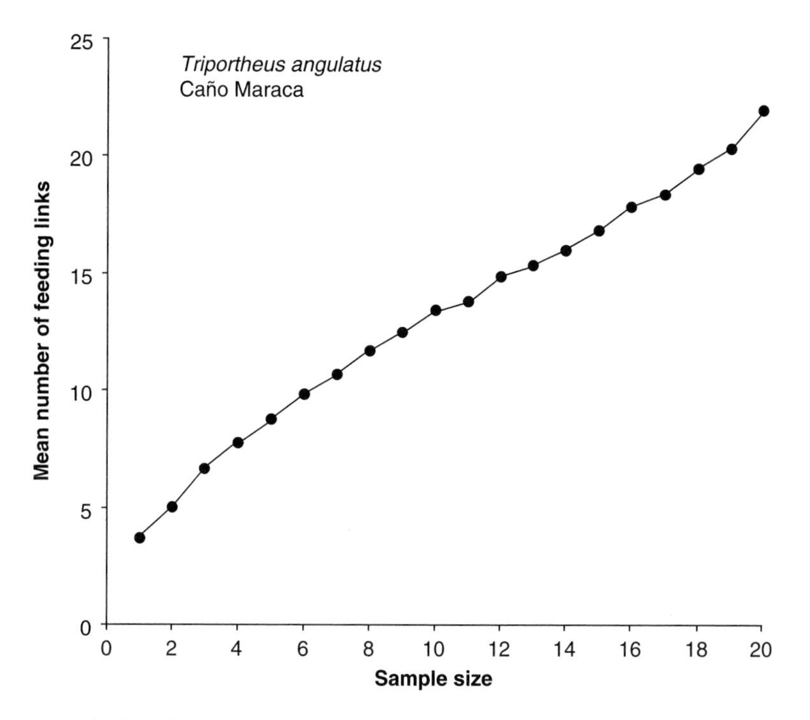

FIGURE 3 | Plot illustrating increases in mean number of feeding links (estimated from stomach contents analysis) with increasing sample size for an omnivorous characid fish, *Triportheus angulatus*, from Caño Maraca, Venezuela.

1999). Quantitative estimates of diet composition must be converted to consumption rates for use in dynamic food web models (see Koen-Alonso and Yodzis, Chapter 7.3).

Interaction strength can be directly estimated via field experiments, but this method is beset with its own set of challenges (Berlow et al., 2004). A major problem is the quantitative measure used to quantify interaction strength. Several indices have been employed (reviewed by Berlow et al., 1999), including a raw difference measure (N-D)/Y; Paine's index (N-D)/DY; community importance (N-D)/Np_y; and a dynamic index (ln(N/D))/Y_t, in which N = prey abundance with predator present, D = prey abundance with predator absent, Y = predator abundance, p = predator proportional abundance, and t = time. Different indices computed from the same set of experimental data can yield very different conclusions (Berlow et al., 1999).

Even if we could agree on a single empirical measure of interaction strength, we would still face serious challenges in estimating community dynamics with this information (Berlow et al., 2004). That is because species interactions typically are nonlinear, which implies that single estimates of interaction strength will be unlikely to assist in building dynamic community models (Abrams, 2001). According to Abrams, "Measuring interactions should mean determining the functional form of per capita growth rate functions, not trying to encapsulate those complicated relationships by a single number." Application of simple models to predict features and dynamics of complex systems would be justified if these models could, *a priori*, yield successful predictions. Clearly, considerable theoretical and empirical research remains to be done on the crucial issue of interaction strength.

An additional consideration is that food web links are usually assumed to be consumer resource; however, other kinds of species interactions, such as mutualism and other forms of facilitation, can be critical (Bruno et al., 2003; Berlow et al., 2004). Describing the functional forms of these relationships could be even more challenging. Some of the most important community interactions are not determined by resource consumption. Gilbert (1980) described ecological relationships in a food web module within a Costa Rican rainforest. This module is delimited by 36 plant species in 6 higher taxa inhabiting 3 habitat types. Each plant species has a set of generalist and specialist herbivores, pollinators, and seed dispersers, some of which are shared with other plants within the module and, in some cases, plants outside the module. In this food web, some of the most critical interactions determining species' abundances and distributions are mutualisms.

Drivers of Temporal and Spatial Variation

The fourth critical issue is the influence of environmental and life history variation on food web structure, species interactions, and population dynamics. Do food web dynamics drive species abundance patterns, or do species abundance patterns drive food web dynamics? Species' relative abundances determine functional responses, adaptive foraging, predator switching, and their effects on numerical responses. Does food web structure determine relative abundance patterns, or are other factors equally or more important?

Interaction strength varies in space and time, sometimes as a function of behavior, but sometimes as a function of environmental variation and species life histories that affect abundance patterns (Polis et al., 1996a). Species with different life histories and ecophysiological adaptations respond differently to environmental variation (Winemiller, 1989a). Species with short generation times and rapid life cycles respond faster to environmental variation (including resource availability) than species with slower life cycles that often reveal large variation in recruitment dynamics and demographic storage effects (Polis et al., 1996a; Scharler et al., Chapter 8.3). Empirical studies have demonstrated how species' abundances and web links change in response to environmental drivers. Rainfall and leaf litter deposition determine food web patterns in tree holes in tropical Australia (Kitching, 1987). Temporal dynamics in rocky intertidal webs are influenced by local disturbances (Menge and Sutherland, 1987) and coastal currents (Menge et al., 2003). Food webs of streams and rivers vary in relation to seasonal changes in photoperiod and temperature (Thompson and Townsend, 1999) and hydrology (Winemiller, 1990; Marks et al., 2000).

Theories, Tests, and Applications

So where are we now? Theory and attempts at application of theory seem to have outpaced observation and model testing. There is little agreement and consistency regarding use of operational units, methods for quantifying links, indices of interaction strength, etc. Use of confidence intervals is virtually non-existent in empirical food web research. This state of affairs is perhaps a natural consequence of an "immature" scientific discipline (i.e., abstract concepts, lack of consensus and empirical rigor). Nonetheless, society demands that ecological science address current problems. Currently, food web models have low predictive power and certainly lack the precision and accuracy of physical models that allow engineers to put a spaceship on the moon or build a sturdy suspension bridge. Food web models currently used for natural resource management are highly aggregated and employ crude quantitative estimates of production dynamics and species interactions. Output from these models can be considered educated guesses, yet, currently, we have no other options. It is unreasonable to expect individual investigators or labs to achieve predictive food web models, yet few are lobbying for empirical food web research on a grand scale. This state of affairs may be an unfortunate legacy of the IBP (International Biological Program, supported in the 1960–1970s by large sums of national and international science funding aimed at understanding major ecosystems of the planet).

Were past efforts to describe large food webs misguided? Nearly 20 years ago, the first author attempted to describe food webs of tropical streams in a standardized manner based on intensive sampling (Winemiller, 1989b, 1990, 1996). Two continuous years of field research yielded over 60,000 fish specimens and countless invertebrates. Two additional years of lab research (19,290 stomachs analyzed) produced data that supported analyses that have been ongoing for 17 years. These quantitative food webs have provided insights into how environmental variation driven by seasonal hydrology affects population dynamics and interactions. Yet, as descriptions of community food webs, these webs suffer from the same limitations that plague other webs. The many issues, both conceptual and methodological, requiring resolution in order to make meaningful comparisons of web patterns ended up being a major discussion topic (Winemiller, 1989b, 1990).

Is there a better way? We advocate a multi-faceted empirical approach for field studies as a means to advance understanding of food webs. Researchers investigating large, complex systems would be better served to investigate food web modules in a hierarchical fashion. Long-term research mindful of environmental drivers is extremely valuable in this context. Research that blends together description and experimentation will yield models that can then be tested within relevant domains (Werner, 1998). This approach obviously will require research teams with specialists that collectively provide a range of methodological and taxonomic expertise. Several groups around the world have already adopted this long-term, team research approach to investigate food webs of ecosystems ranging from estuaries (Raffaelli and Hall, 1992) to rainforests (Reagan and Waide, 1996).

We have attempted this hierarchical modular approach in our research on the Cinaruco River, a floodplain river in the Llanos region of Venezuela. Our group is describing nutrient dynamics, primary production, community structure, habitat associations, and feeding interactions in channel and aquatic floodplain habitats during various phases of the annual hydrological cycle in this diverse food web (see Layman et al., Chapter 7.4). Population abundance and distribution patterns are assessed from field surveys (Jepsen et al., 1997; Arrington and Winemiller, 2003; Hoeinghaus et al., 2003a; Layman and Winemiller, 2004), and feeding links are investigated using dietary and stable isotope analyses (Jepsen et al., 1997; Jepsen and Winemiller, 2002; Winemiller and Jepsen, 2004; see Layman et al., Chapter 7.4). We also are investigating three food web modules (Figure 4): (1) benthivorous fishes, benthic biota, detritus, and nutrients; (2) herbivorous fishes interacting with terrestrial and aquatic vegetation; and (3) piscivores and their diverse prey (see Layman et al., Chapter 7.4). Field experiments (enclosures, exclosures, and artificial habitats) have been conducted over variable spatial scales in different seasons and habitats to examine species effects on prey assemblages (Layman and Winemiller, 2004) and benthic primary production and particulate organic matter (Winemiller et al., 2006). In virtually all experiments designed to test for top-down effects, one or a small number of fish species (including large detritivores and piscivores) reveal strong and disproportionate effects in this species-rich food web (more than 260 fish species documented).

The descriptive research elements have led to creation of models that predict effects of abiotic ecosystem drivers (the most fundamental being seasonal hydrology) and aspects of species life histories (e.g., seasonal migration by a dominant benthivorous species) on food web dynamics and ecosystem processes. For example, the relative influence of top-down and bottom-up processes on benthic primary production, benthic particulate organic matter, and meiofauna diversity is a function of the seasonal cycle of hydrology, habitat volume, allochthonous nutrient inputs, migration by the dominant benthivorous fish, and changing densities of resident benthivorous fishes as a function of habitat volume. Experiments have been conducted to estimate the magnitude of treatment effects that reveal the relative influence of bottom-up (nutrient limitation and sedimentation) and top-down (grazing) effects on standing stocks of algae and fine particulate organic matter (Winemiller et al., 2006). A separate series of experiments examined effects of predators on prey fish densities and habitat use (Layman and Winemiller, 2004). The relative influence of dominant piscivores on littoral zone fish assemblages is strongly dependent on body size relationships (see Layman et al., Chapter 7.4) and habitat features which in turn are influenced by seasonal hydrology. In short, almost no aspect of this river food web could be understood without examining the direct and indirect effects of the annual hydrological cycle.

Guided by our desciptions of the overall food web, the predictive models developed for modules are being joined together based on elements of overlap. The degree to which predictions of module dynamics will agree with predictions from a model that incorporates all elements remains to be investigated. Nonetheless, it seems more rational to begin at smaller scales and work incrementally toward a model of the larger system, rather than the reverse approach. Many of the contributions in this book describe similar small-to-large approaches employing multiple lines of empirical evidence to test model predictions. Several research groups have reported results from long-term research that blends description, experimentation, and modeling—for example, temperate lakes (Carpenter and Kitchell, 1993a; Tittel et al., 2003), soils (de Ruiter et al., 1995; Moore et al., 2003), coastal systems (Menge et al., 2003; Raffaelli et al., 2003), rivers and streams (Marks et al., 2000; Nakano and Murakami, 2001; Flecker et al., 2002), ponds (Downing and Leibold, 2002), and fields (Schmitz, 2003). It is still too early to generalize about food web structure, and perhaps some systems conform to the onion model whereas others function according to the internet model, and so on. Given the disproportionate effects of a few dominant species demonstrated by field experiments in the Cinaruco River, the "Christmas tree" and "internet" models seem to be candidates for that species-rich system.

Discussion and Conclusions

Empirical food web research lags behind theoretical research. We agree with Englund and Moen's (2003) assertion "that it is vital for progress in ecology that more models are experimentally tested, and the main question is how to promote and speed up the process." They continue: "By testing a model, we mean the act of comparing model predictions with relevant empirical data." Another basic challenge identified by these authors is the critical need to determine whether or not an experimental system lies within the theoretical domain of the model being tested. In too many cases, models and tests were mismatched from the start (e.g., invalid assumptions of linear or equilibrium dynamics or inappropriate spatial scales).

Empirical food web studies must carefully consider the dynamical consequences of definitions for operational units and scale, resolution, and sample variability. Obviously, it is impossible to quantify every species and interaction in even the smallest food web modules. Even if this were possible, it is unlikely that most trophic interactions have a strong effect on system properties such as nutrient cycling and production of dominant biomass elements (e.g., the "Chrismas tree" and "internet" models). Thus, it is crucial that we determine, to the extent possible, the degree of resolution needed to make successful predictions, and then, for the sake of efficiency, not seek to achieve high levels of detail for their own sake.

We advocate a focus on a hierarchy of nested food web modules and measures of interaction strength that hold potential to yield successful predictions of population dynamics and other ecosystem features. Descriptive and experimental research should be combined in long-term studies of field sites (see also Schmitz, 2001). Such efforts require consistent funding and collaborations among scientists with different expertise. In many countries, these sorts of projects are difficult to fund and provide fewer individual rewards than short-term projects addressing specific mechanisms in small-scale ecological systems. Yet many of our most vexing ecological problems require a large-scale food web perspective. Despite the fact that a deficient empirical knowledge base is the main hurdle to scientific advancement, pressing natural resource problems require application of existing models. In many respects food web research is basic yet complicated—esoteric yet essential for natural resource management. The urgent need for application of the food web paradigm for solving natural resource problems motivates us to walk faster down the path from abstraction to prediction.

ACKNOWLEDGMENTS

A portion of this work was funded by National Science Foundation grants DEB 0107456 and DEB 0089834.

2.0 | VARIATIONS IN COMMUNITY ARCHITECTURE AS STABILIZING MECHANISMS OF FOOD WEBS

John C. Moore

The chapters in this section represent a departure from the traditional food web approaches as they seek to reconcile discordances between ecologists pursuing theoretical treatments of food webs and those making empirical observations. Theoretical treatments of food webs usually focus on those configurations of dynamic interacting species that are stable. The dynamics refer to changes in the population densities of the species. Stability has several meanings (May, 1973; McCann, 2000) but usually entails the ability of the community to recover from minor disturbances and persist in time. Many theoreticians make simplifying assumptions that remove much of the variation in structure and many of the factors that contribute to dynamics. Empiricists often cite the importance of this variation to the survival of key species or functions operating within their systems.

The chapters explore two aspects of dynamics that create variation within food webs both spatially and temporally: the changes in the strengths of interactions or flows of nutrients within food webs and the changes in the structure of the communities as defined by the constituent species and the linkages among the species. The first aspect combines the community-based perspective of modeling populations (MacArthur, 1955; May, 1972) with the ecosystem-based perspective of nutrients and energy (Odum, 1969). Banašek-Richter et al. (Chapter 2.3) note that the quantification of flows is important in general, and differences in the rates of flows affects stability and dynamics. These points were also made by Brose et al. (Chapter 2.1) in their discussion of the importance in the variability of rates nutrient uptake, and the metabolic rates of consumers on the stability and persistence of communities. Model systems dominated by ectotherm-like organisms with high metabolic rates and rapid lifecycles were able to persist longer under different conditions than counterparts dominated by endotherm-like organisms with lower metabolic rates and slow lifecycles. The implications of these studies reach beyond this section with connections to the foci of work on body size presented in Section 4, nutrients and resource dynamics in Section 5, and environmental dynamics, perturbations, and food webs in Section 7.

Just as populations and linkage strengths may vary both spatially and temporally, the proposition advanced in this section is that the underlying structure of the community changes and that linkages among populations that are distinct both spatially and are important to stability and persistence. This focus on the consequences of variation in the structure of the community represents a significant departure in how we study food webs (de Ruiter et al., 2005). From the perspective of an observer, the changes appear as a variation in community structure that might result from the seasonal addition or deletion of a mid-order consumer like an herbivore or detritivore, or by including higher order consumers or top predators that operate over larger spatial scales than their prey. In both cases the organisms and their linkages within the food web are either included in the description of a food web that represents a seasonal average or excluded altogether from the description given the low likelihood of their being present in an area in any given amount of time. From a theoretician's perspective the variation has often been treated as a nuisance to be dealt with by spatial and temporal averaging, or by deletion.

Leibold et al. (Chapter 2.2) present a series of experiments that manipulated the environmental conditions and community structure of a zooplankton and edible algae–based food web using experimental ponds. The results of this experimental approach demonstrate that food web structure and the environment interact to include qualitatively and quantitatively different dynamics. Though different, the results could be classified into two different types of oscillator—the traditional consumer-resource cycles where biomass of the consumer oscillates with the biomass of the resource, and the more recent cohort cycle construct where the demographic structure of the consumer cycles to initiate consumer-resource cycles.

McCann et al. (Chapter 2.4) investigate how predators at higher trophic positions that move from one habitat to another can operate as integrators across space and time in manners that stabilize systems. Implicit in these notions are aspects of the predator's foraging strategies, morphology relative to their prey and their life history characteristics (e.g., large long-lived predators with high vagility foraging over larger ranges than their prey in a Holling Type III manner) (Holling, 1959a). The decision to view the community structure as static rather than dynamic by either including the predator as a permanent fixture in the web or excluding such a predator altogether could have profound effects on our understanding of trophic structure and stability as the dynamic nature of the predator's movements to different habitats create variation in its presence and absence that is the core of the stabilizing influence.

McCann et al. (Chapter 2.4) provide a theoretical underpinning and framework that links these concepts. Underlying themes within the models and empirical results presented in the chapters include the notions that food webs are organized into subsystems or compartments of tightly coupled consumers and resources that are less tightly coupled by higher-order consumers (Moore and Hunt, 1988), that there is variation in the dynamics of the organisms making up the trophic structure and in the structures themselves, and that this variation provides the basis for stabilizing mechanisms. Environmental heterogeneity coupled with the variations in the life histories and metabolic efficiencies of consumers mediate the formation of subsystems or compartments. Higher order consumers link these compartments in a manner that if viewed from large enough spatial and temporal scales gives the appearance of a large inclusive web. These consumers may re-initiate or decouple themselves from one subsystem or another depending of prey densities or environmental conditions. The chapters in Section 1 represent a fitting beginning to what follows—an introduction to the concept of dynamic food webs.

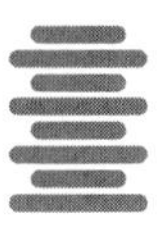

2.1 | FROM FOOD WEBS TO ECOLOGICAL NETWORKS: LINKING NON-LINEAR TROPHIC INTERACTIONS WITH NUTRIENT COMPETITION

Ulrich Brose, Eric L. Berlow, and Neo D. Martinez

Food webs traditionally specify the structure of who eats whom among species within a habitat or ecosystem. A yet-to-be realized ideal is a food web model that includes the quantity consumed and the ecological relevance of every trophic interaction among all species within a community. Far short of this ideal, current food web models provide partial yet critically important information about species' extinction risks and relative abundances by describing the dynamics of energy flow and consumer species' biomasses (Yodzis, 1998; Borrvall et al., 2000). This exclusive focus on feeding relationships or "trophic interactions" enables models of complex food webs (Brose et al., 2003; Kondoh, 2003a; Williams and Martinez, 2004a) to also incorporate exploitative and 'apparent' competition (Holt et al., 1994; Holt and Lawton, 1994) among heterotrophs. However, this focus ignores similar interactions among autotroph producers or "basal species" consuming abiotic resources (e.g., nutrients, light, etc.) along with other 'non-trophic' interactions such as many forms of facilitation and competition. In particular, the logistic growth of the food webs' producers or 'basal' species in these models typically approximates nutrient- or detritus-dependent growth of basal species each non-competitively consuming an independent pool of resources.

Explicit models of shared nutrient consumption including competition among producers for multiple resources (Tilman, 1982; Grover, 1997; Huisman and Weissing, 1999) are largely separate from trophic ecology. Still, many experimental and theoretical studies bridge this gap by elucidating how predation interacts with other non-trophic processes, such as nutrient consumption, to regulate species distribution and abundance (Menge and Sutherland, 1987; Leibold, 1996; Proulx and Mazunder, 1998; Gurevitch et al., 2000; Chase et al., 2002, and references therein). Many of these synthetic insights are based on (1) linear trophic interactions, (2) competition for one nutrient resource, (3) very simple communities (i.e., <4 spp), and (5) an equilibrium-based analytical framework. Such insights are suspect due to the discrepancy between 1-5 and frequently observed nonlinear population dynamics (Kendall et al., 1998), much greater trophic diversity and complexity (Williams and Martinez, 2000), and multiple plant nutrients in nature.

Here, we present an approach for exploring the interplay of complex trophic interactions and consumption of multiple abiotic resources among producer species using non-linear and non-equilibrium numerical simulations. This approach transforms simple food-web models into more general models of complex ecological networks and can be scaled up to systems with many more species and abiotic resources. We first describe the model and then illustrate its potential for exploring the interplay between abiotic resource competition and trophic interactions.

THE MODEL

Our model couples a nutrient-producer model (León and Tumpson, 1975; Tilman, 1982; Grover, 1997; Huisman and Weissing, 1999) with a bioenergetic model of herbivore-producer interactions (Yodzis and Innes, 1992). Both approaches have simulated non-equilibrium species dynamics that have provided important insights in this area of research that are difficult to obtain with linear, equilibrium-based models (McCann et al., 1998; Huisman and Weissing, 1999). However, these well established models had yet to be synthesized into one framework.

Our synthesis begins with a nutrient-producer model consisting of five producer species competing for five limiting resources and then expands producer-herbivore components by adding five herbivores each of which consume one of the producer species. One producer called the "guzzler" has the highest rate of consumption of the most limiting nutrient which causes it to become the competitive dominant in many nutrient-producer studies. The term "guzzler" avoids confounding a component of a very general model with results from its numerical simulation. It also avoids contradictory terminology when the conventionally termed "competitive dominant" fails to numerically dominate the community. Instead, 'competitive dominance' as used here describes results of models that apply to different species depending on the context rather than a context-independent inherent species trait. In accordance with these fundamental distinctions, the other producer species with lower resource consumption rates are called "sippers." Such species are often called 'competitively subordinate' in nutrient-producer studies.

Our model of herbivore-producer biomass dynamics is based on a widely used (McCann and Yodzis, 1994; McCann and Hastings, 1997; McCann et al., 1998) bioenergetic model of trophic interactions (Yodzis and Innes, 1992) that has been recently extended to n species (Brose et al., 2003; Williams and Martinez, 2004a). The rate of change in the biomass (B) of species i with time t is modeled as:

$$B_i'(t) = G_i(N) - x_i B_i(t) + \sum_{j=1}^{n} (x_i y_{ij} F_{ij}(B) B_i(t) - x_j y_{ji} F_{ji}(B) B_j(t) / e_{ji}) \quad (1)$$

where $G_i(N)$ describes the growth of producer species; x_i is the mass-specific metabolic rate of species i; y_{ij} is a measure of the maximum ingestion rate concerning resource j per unit metabolic rate of species i; and e_{ji} is the biomass conversion efficiency of species j consuming i (see Yodzis and Innes, 1992, for parameter details). We used a type II functional response for the flow of biomass from resource j to consumer i:

$$F_{ij}(B) = \frac{B_j(t)}{B_j(t) + B_{0ji}} \quad (2)$$

where B_{0ji} is the "half saturation density" or density of the resource at which the consumer attains half its maximal rate of consumption (Holling, 1959a, b).

To avoid the ambiguous use of "resource" (see eq. 1), we hereafter call producer resources N_l "nutrients." These are still broadly defined and may include any abiotic resource that is subject to constant turnover such as light or nitrogen. We examine the dynamics resulting from shared abiotic resources or nutrients by modifying the producer growth of Yodzis and Innes' (1992) original model using a well-studied nutrient consumption model:

$$G_i(N) = r_i MIN\left(\frac{N_1}{K_{1i}+N_1}, \ldots, \frac{N_5}{K_{5i}+N_5}\right) B_i(t) \tag{3}$$

that depends on the concentrations of five limiting nutrients N_l. This model has been widely used in plant ecological theory (León and Tumpson, 1975; Tilman, 1982; Grover, 1997; Huisman and Weissing, 1999) and experiments (Tilman, 1977; Huisman et al., 1999). In (3), r_i is the maximum growth rate of species i that is non-zero only for producer species, K_{li} is the half saturation constant for nutrient l, and MIN is the minimum operator. Therefore, $G_i(N)$ follows a Monod equation and is determined by the nutrient that is most limiting. The variation in the density of nutrient l is given by

$$N_l'(t) = D(S_l - N_l) - \sum_{i=1}^{n}\left(c_{li}\, G_i(N)\right) \tag{4}$$

where c_{li} is the content of nutrient l in species i. Nutrients are exchanged at a turnover rate D with a supply concentration of S_l, and removal depends on the current nutrient concentration in the system, N_l. Species have biomass loss rates due to metabolism but not turnover since species are not assumed to be passively drifting out of the system. We used Yodzis and Innes' (1992) empirical estimates for the bioenergetic parameters: $y_{ij} = 6$ for invertebrates, $y_{ij} = 3.9$ for ectotherm vertebrates, $y_{ij} = 0$ for producers, $e_{ji} = 0.45$ for herbivores and $B_{0ji} = 0.5$. We assumed resource-consumer body size ratios, L, of 0.1 and, in the spirit of McCann and Yodzis (1994), calculated consumer metabolic rates as

$$x_i = a x_p \left[L^{0.25}\right]^{T-1} \tag{5}$$

where x_p is the metabolic rate of primary producers (0.2), T is the consumer's trophic level and a is a constant that equals 0.54 for invertebrates and 3.48 for ectotherm vertebrates (Yodzis and Innes, 1992). This yields metabolic rates of 0.06 for invertebrate and for 0.39 ectotherm vertebrate herbivores. The parameters of the growth function are $D = 0.5$, $r_i = 1$ and $S_l = 1$. The first nutrient is the one most needed by all producer species as it has the highest content in their biomasses ($c_{1i} = 1$ and $c_{ki} = 0.5$ with $k > 1$). The half saturation densities of producer species' growth are given by:

$$K_{li} = \begin{pmatrix} K' & 0.12 & 0.12 & 0.12 & 0.12 \\ 0.06 & 0.20 & 0.16 & 0.10 & 0.02 \\ 0.10 & 0.06 & 0.20 & 0.16 & 0.06 \\ 0.16 & 0.10 & 0.06 & 0.20 & 0.09 \\ 0.20 & 0.16 & 0.10 & 0.06 & 0.09 \end{pmatrix}$$

where rows represent five nutrients and columns five producer species.

Since all producer species have similar r_i and x_i, the lowest half saturation density ($K' < 0.12$) defines producer species 1 or "guzzler" as the most voracious consumer of N_1 per unit biomass when N_1 is limited (Huisman and Weissing, 2001). The four other "sipper" producer species consume N_1 at smaller rates on a per biomass basis. We varied the strength of the nutrient uptake hierarchy by making guzzlers relatively stronger ($K' = 0.05$) or weaker ($K' = 0.11$) while ensuring that both 'strong' and 'weak' guzzlers consume N_1 more effectively than do sippers. Note that guzzling and sipping only refers to the nutrient most needed by the producers (N_1) and not to the four other nutrients ($N_{2,3,4,5}$) present in smaller concentrations in the producers' biomass. Simulations begin with randomly assigned biomass abundances ($B_i(0) =$ 0.05 to 1) and end after 2,000 time steps.

This model framework allows us to explore the following questions:

1. How do specialist herbivores alter the patterns of coexistence and relative abundances among producers that exhibit a nutrient consumption hierarchy? How does this effect vary with herbivore physiology (e.g., invertebrate vs. ectotherm vertebrate)?
2. How does the strength of the nutrient consumption hierarchy among producers influence the relative abundance and persistence of herbivores?

RESULTS

In the simpler nutrient-producer model, the modeled nutrient consumption hierarchy directly results in a competitive hierarchy where the guzzler (P1) quickly out-competes the sippers (see P2-5, Figure 1). The strong guzzler reaches abundances ($B_1 > 1.21$) close to its maximum (1.23) after 50 time steps and all sippers are extinct ($B_i < 10^{-30}$) after 1,200 time steps (Figure 1A). The weak guzzler, however, does not dominate the sippers until after 165 time steps (Figure 1B) and attains abundances ($B_1 > 1.21$) close to the strong guzzler's maximum at $t > 1150$. The weak guzzler also allows for the sippers to persist with very low biomasses

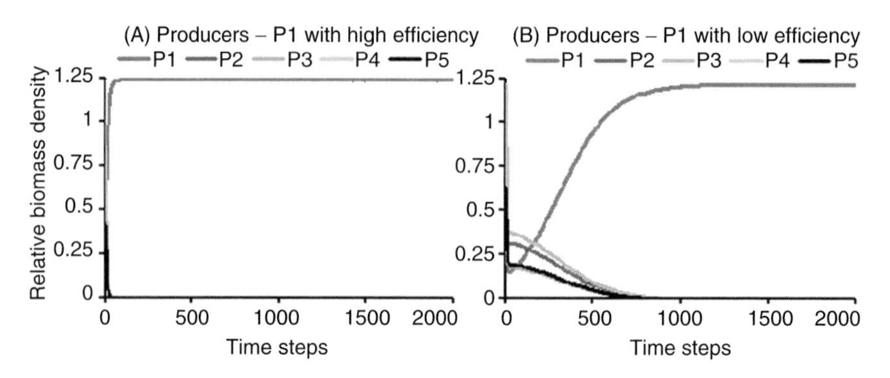

FIGURE 1 | Biomass evolution in producer communities. The guzzler (P1) is strong (**A**) or weak (**B**). (See also color insert.)

($B_i < 10^{-5}$) until the end of the simulations ($t = 2000$). Similar results were obtained in replicated runs with varying random initial abundances (data not shown).

The addition of five specialist invertebrate herbivores allows all producers to co-exist independent of the strength of the nutrient consumption hierarchy (Figure 2). In this scenario, the weak guzzler's mean abundance (0.21) is similar to the strong guzzler's (0.24), and both are

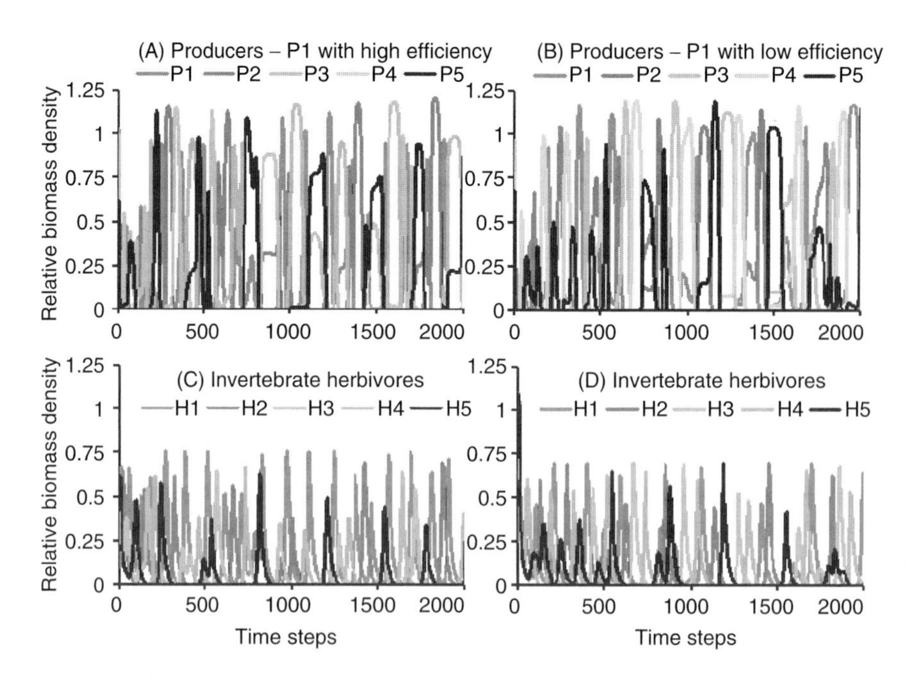

FIGURE 2 | Biomass evolution in producer-invertebrate herbivore communities. The guzzler (P1) is strong (**A, C**) or weak (**B, D**). (See also color insert.)

comparable to the sippers' mean abundances (P2-5, mean = 0.16–0.22). The abundance of the strong guzzler's herbivore (see Figure 2C, H1: mean B_i = 0.15) exceeds the other herbivores' abundances by a factor of three (see Figure 2C, H2-5: mean B_i = 0.05). In contrast, the weak guzzler's herbivore reaches lower abundances (see Figure 2D, H1: mean B_i = 0.08) and does not markedly exceed that of the other herbivores (see Figure 2D, H2-5: mean B_i = 0.06–0.07).

Consistent with the results for invertebrate herbivores, adding five specialist ectotherm vertebrate herbivores also allows all producer species to persist (Figure 3). The presence of a strong guzzler allows producers to coexist at roughly similar mean biomass densities (see Figure 3A), whereas weak guzzlers cause a sipper's biomass to dominate (see Figure 3B). In replicated runs, the dominant sipper's identity depends on initial abundances (data not shown). All herbivore species persist at low densities when the guzzler is strong, and the consumer of the strong guzzler dominates the herbivore biomass (see Figure 3C). When the guzzler is weak, sippers have lower abundance minima and their herbivores go extinct (see Figure 3D). The sipper whose herbivore goes extinct first consequently dominates the other producer species. The guzzler's herbivore, however, always persists.

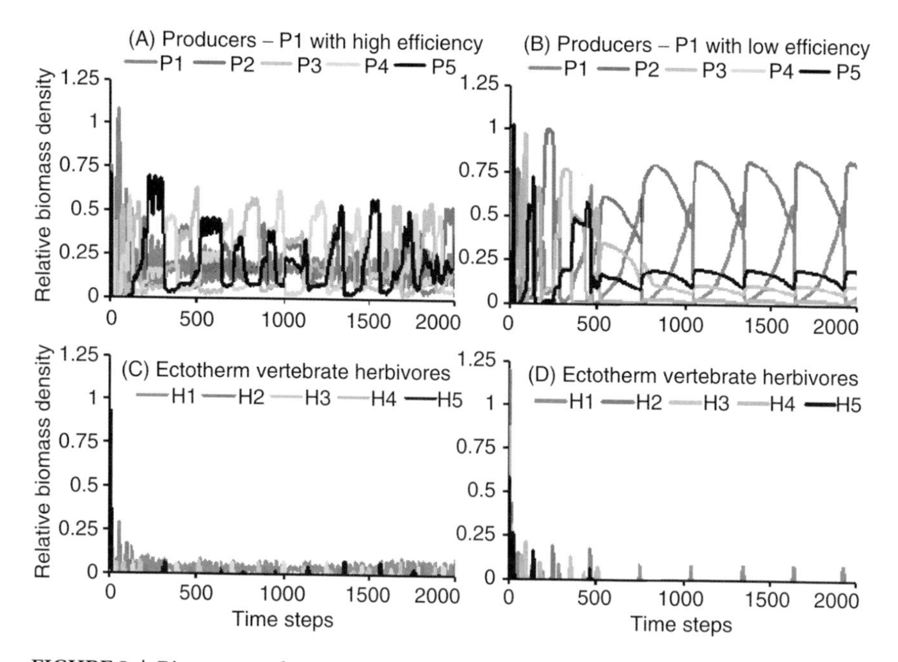

FIGURE 3 | Biomass evolution in producer-ectotherm vertebrate herbivore communities. The guzzler (P1) is strong (**A, C**) or weak (**B, D**). (See also color insert.)

DISCUSSION AND CONCLUSIONS

We have coupled a plant-nutrient competition model with a dynamic food web model. In contrast to prior studies, we simulated non-equilibrium plant-herbivore food webs with non-linear trophic interactions and producers whose growth depended on several limiting nutrients. Our results demonstrate that the herbivore physiology and strength, as well as the presence of a nutrient consumption hierarchy among producers strongly affect producer and herbivore population dynamics, relative abundances, and coexistence. Specialized herbivory allowed producer species to co-exist despite asymmetric nutrient consumption that otherwise generates competitive exclusion. This is consistent with empirical observations of predation promoting coexistence of competing prey (Gurevitch et al., 2000; Chase et al., 2002). Producers co-exist because a specialized herbivore effectively controls the abundance of the most effective nutrient consumer among the producers—the guzzler. Since there is no trade-off between nutrient uptake efficiency and vulnerability to herbivory, this effect does not rely on the assumption that the guzzler is preferentially consumed. Rather, this strong 'keystone-like' effect emerged from dynamics that quickly transformed the guzzler's higher nutrient uptake into increased abundance of its herbivore relative to other herbivores. These results suggest that changes in the producers' consumptive abilities can have "bottom-up" effects on the relative abundance distributions of higher trophic levels, which in turn regulate the strength of top-down effects. Lower nutrient supply or stronger herbivory, however, cause producer species to go extinct (data not shown), which is consistent with empirical findings (Proulx and Mazunder, 1998; Chase et al., 2002).

Our results illustrate how consumer physiology associated with the difference between ectotherm vertebrates and invertebrates might strongly affect plant-herbivore dynamics. The metabolic rates determine the herbivores' biomass loss per time step. All else being equal, the higher mass-specific metabolic rates of ectotherm vertebrates are therefore responsible for lower biomass minima. This reduces the abundance of the sippers' ectotherm vertebrate herbivores enough for the latter to go extinct, thus releasing sippers from top-down control. Our non-equilibrium approach shows that sippers regularly drop to lower abundance minima than the guzzler does, which makes extinction of the sippers' ectotherm vertebrate herbivores more likely. Initial abundances determine which sipper's herbivore goes extinct first. Extinction releases that sipper from top-down control and allows it to out-compete all other producers. In contrast, invertebrate herbivores achieve higher relative abundances, which make them less prone to extinction. In summary, herbivore physiology dramatically alters the competitive balance among producer species due to the variable likelihood of herbivore extinction.

Previous studies have demonstrated that the less consumptive producers that we call sippers may dominate producer communities in the presence of herbivores when there is a trade-off between nutrient consumption and vulnerability to herbivory (Chase et al., 2002, and references therein). This has been proposed to occur if sippers are either better defended, less palatable, or otherwise living in more enemy-free space (Grover, 1994, 1995) than are guzzler species. Our results show that none of these conditions are strictly necessary to achieve the same result.

While a prior study (Moore et al., 2004) has demonstrated coexistence of basal species or extinction of sippers, our study indicates that dominance by sippers can occur under specific parameter combinations due to variable extinction probabilities among herbivores. Surprisingly, the least effective nutrient consumer among the producer species—with the initially lowest biomass minima—may lose its herbivore first and subsequently become the competitive dominant producer species with the highest mean biomass density. Important limitations to our findings include the fact that our systems are closed to migration, which could alter our results if, for example, invertebrate consumers were able to immigrate and take over the function of the extirpated vertebrate herbivores. Other aspects in need of exploration include increased trophic complexity, such as more complex topology including more trophic levels and omnivory, as well as variation among consumers to include invertebrates, ectotherm and endotherm vertebrates in the same community.

Competition among producer species and trophic interactions have been studied in largely separate research programs. Attempts to integrate the two are typically restricted to very simple models (≤ 4 species), linear trophic interactions, and assumptions of equilibria (Chase et al., 2002). By including shared consumption of multiple nutrients among producers in a non-linear dynamic food web model, we explore the interaction of competitive and trophic systems in a more complex, non-equilibrium framework. The strength of the nutrient consumption hierarchy among producers influences the biomass distribution of herbivores rather than that of producers. “Keystone” effects, by which consumers promote coexistence of resource species and facilitate dominance by sippers that are traditionally considered “competitive subordinates,” can emerge without trade-offs between nutrient consumption ability and vulnerability to consumption. This suggests that numerical simulations of non-equilibrium systems may provide novel insights into how complex interactions between trophic and non-trophic processes regulate species coexistence and community stability. Further work is needed to understand the organizing processes of complex communities. Moving from models of food webs to more complex ecological networks that include other non-trophic interactions such as nutrient competition, facilitation, and mutualism will enable us to more rigorously analyze the complex interaction dynamics in natural systems such as keystone interactions and their context dependency (Paine, 1969, 1974; Menge et al., 1994). Following successful integration of these interactions, fitting the models to field conditions and field testing them is an exciting yet problematic next step (Martinez and Dunne, 2004; Paine, 2004).

ACKNOWLEDGMENTS

This work was supported, in part, by the German Academy of Naturalists Leopoldina by funds of the German Federal Ministry of Education and Science (BMBF-LPD 9901/8-44) and the German Research Foundation (BR 2315/1-1,2).

2.2 | FOOD WEB ARCHITECTURE AND ITS EFFECTS ON CONSUMER RESOURCE OSCILLATIONS IN EXPERIMENTAL POND ECOSYSTEMS

Mathew A. Leibold, Spencer R. Hall, Ottar Bjornstad

Until recently, much of the work on food webs has taken a very simplistic view of food web dynamics (see Hastings, 1996, for a critique). For example, work done on food web architecture has emphasized how the occurrence of species in food webs can be described and how it might be regulated but this work has not examined how patterns of dynamic behavior such as oscillations vary (but see McCann and Hastings, 1997; McCann et al., 1998; Fussmann and Heber, 2002). Similarly, comparative patterns of food web structure among communities have generally emphasized differences in their average properties but it is only recently that effects on variability have been examined (Petchey, 2000). Much insight has come from work done taking approaches that ignore dynamics and this serves as an important starting point for work on dynamics. Some major and important findings that have come from past work including both correlational and experimental methods include the following:

1. Species can have both direct and indirect interactions within and among trophic levels and 'guilds' (Levins, 1974; Neill, 1975; Dethier and Duggins, 1984; Miller and Kerfoot, 1987; Wootton, 1992)
2. Aggregate properties such as trophic structure (the distribution of biomass among trophic levels and guilds) are similarly structured by indirect effects. Two well known and reasonably well understood examples include 'trophic cascades' and 'bottom-up' effects on trophic structure (Leibold et al., 1997; Shurin et al., 2002).
3. Different food web architectures can modify the outcomes of these indirect interactions in important and sometimes dramatic ways (Leibold and Wilbur, 1992; Abrams, 1993; Hulot et al., 2000).

While we have good evidence that this last point is important, it is not always clear what the precise mechanism are responsible for having modified outcomes when food web architecture is altered. Theoretical models do indeed make predictions consistent with this general result (Phillips, 1974; Abrams, 1993; Hulot et al., 2000) but the array of predictions is almost too complex to easily evaluate using qualitative comparisons of mean values. This is especially true when one considers that mean values may not be indicative of expected equilibria in models that have non-point steady-state behavior and oscillate or show chaotic dynamics (Abrams and Roth, 1994).

An important consequence of these findings however is that many quantitative aspects of the dynamics of populations in food webs should also be altered in important ways. Recent theoretical models of population dynamics under different environmental conditions and with different food web architectures shows an impressive array of dynamics that include different types of oscillations (both periodic and complex oscillations associated with 'chaotic' dynamics) with different periods and amplitudes, as well as oscillations among different components with different phase relations (reviewed by Scheffer and Carpenter, 2003). Additional complications include the presence of different attractors (and basins of attraction for complex attractors, see Scheffer and Carpenter, 2003, Hastings, 2004). Because these dynamics are so rich in their behaviors and because they are so closely linked to the patterns of feedback in food webs, they potentially provide an important way of evaluating alternative mechanisms for the regulation of individual populations (Kendall et al., 1999) and, by extension, food webs.

Linking these dynamics to mechanisms of food web regulation however provides some very important methodological challenges. While current work on simple communities (involving a couple of species, Constantino and Desharnais, 1991) and simple food webs (McCauley et al., 1999; Nelson et al., 2001; De Roos et al., 2003b) is promising, the extension of these approaches to more complex food webs that may have 10s to 100s of resident species is daunting (see Fussmann and Heber, 2002). One possibility is that dynamics in such complex food webs can still be understood in terms of the simple subsets seen in the simpler systems. Thus, while some simple food web interactions can be thought of as simple oscillators, the behavior of more complex food webs might be understood by thinking of 'coupled-oscillators' consisting of many such simple oscillators with interacting damping and amplifying harmonics (McCann et al., 1998; Murdoch et al., 2002). The behavior of these more complex 'coupled oscillators' will be easier to study if their simpler elements maintain some aspect of their behavior in complex systems. Determining if this is likely is currently difficult because there is not yet much appropriate data.

In this study, we evaluated how the behavior of one elemental 'oscillator' in pond food webs is affected by different environmental conditions and by changes in food web architecture. We focus on the interaction of herbivorous crustacean limnetic zooplankton herbivorous grazers with 'edible' limnetic algal phytoplankton as the 'elemental oscillator' and we examine how the dynamics of this subsystem are altered by different environmental conditions involving nutrients and light, and by the diversity of the grazer assemblage. In future work we plan to evaluate how well food web dynamics models can explain the quantitative and qualitative variation we find.

We focus on the interaction between limnetic zooplankton (and especially by *Daphnia*) and edible algae because it is one of the better studied systems where population dynamics have been evaluated. The following important elements of this interaction have been identified:

1. Interactions between *Daphnia* and edible algae can be driven by 'cohort cycles' in which demographic cohorts of *Daphnia* replace each other through time (McCauley et al., 1999). One of the demographic stages (usually thought to be the juvenile stage in *Daphnia*) is capable of strongly suppressing the other stages (i.e., adults) by competing for food. However, the suppressing stage eventually matures (or dies) allowing for a pulse of reproduction (or growth) in the other stage. Thus the demographic stages cycle strongly out of phase with one another and the more competitive stage cycles strongly out of phase with the resource (the less competitive stage thus cycles in phase with the resource). The resulting period of these cycles is driven by the generation time of the grazers under the strong resource limitation assumed by the model. Estimates of this period for typical *Daphnia* systems are around 30 days.

2. Alternatively, interactions between *Daphnia* and edible algae can be driven by 'consumer resource cycles' (Lotka, 1925). Here demographic structure plays a much less important role and the various demographic stages cycle in rough synchrony, while the overall abundance of the consumer cycles with the resource with a $\frac{1}{4}$ cycle lag. The periodicity of the cycles involves multiple generations of the grazers (generally around 3–6) but the generation lengths can be quite variable and do not depend as strongly on the strong resource limitation present in the 'cohort cycles.' Reasonable estimates of this period for typical *Daphnia* systems can vary from 20–80 days depending on the mortality rate.
3. Theoretical work on the role of other possible food web components indicate that the consumer-resource cycles could be modified. For example, Kretzschmar et al. (1993), Gragnani et al. (1999), and Shertzer et al. (2002) found that adding highly grazer-resistant algae to such a system tended to strongly dampen the relatively fast cycles assumed by the consumer-resource model and found that very long period cycles (maybe around 200 days) resulted instead. Such consumer-resource cycles have also been documented in the lab in systems with rotifer grazers (Yoshida et al., 2003).
4. Stoichiometric processes have also been hypothesized to alter the stability of consumer-resource dynamics in systems such as the *Daphnia*-algae interaction when there is a mismatch between the elemental stoichiometry of the grazers, the algae, and the supply to the environment (Andersen, 1997; Muller et al.; 2001, Hall, 2004; Loladze et al., 2004). Here the effects include the presence of multiple attractors and the possibility of positive feedback in the system leading to the extinction of the grazer even in systems that initially allowed for balanced elemental budgets involving grazers and plants. The cycles, when present, are like the consumer-resource models but they cross a threshold where they diverge and lead to the extinction of the grazer.
5. Similar alternate basins of attraction can exist in models with higher trophic levels especially in the presence of macrophytes that can act as a spatial refuge for *Daphnia* (Scheffer and Carpenter, 2003).

Models with multiple grazers and multiple plants are not as well studied and we know of none that have been explicitly evaluated, but general understanding of feedback in such systems would lead to several possibilities. Among these are that additional feedback loops lead to additional types of cycles and these could lead to complex harmonic relationships with those described above but these could also be damped by diffuse interactions in webs (Fussmann and Heber, 2002). Alternatively, interactions may be so sensitive to the occurrence of specific species combinations, perhaps due to the effects of complex higher-order interactions, that the resulting dynamics are completely unrelated to those previously described.

METHODS

To see how well the limited array of grazer-algae dynamics previously described might describe dynamics of more complex food webs we conducted the following experiment:

Our experiment was conducted in 300 liter mesocosms (consisting of polyethylene cattle watering troughs (see Downing and Leibold, 2002, and Hall et al., 2004 for basic protocols)). These mesocosms had silica sand substrate on the bottom and were filled with well water. Lids made of 1 mm mesh window screening allowed us to prevent colonization by unwanted insects and crustaceans as well as other organisms larger than 1 mm in size. This screen material allowed 70% light passage. Nutrients in the form of H_2NaPO_4 and $NaNO_3$ were added to achieve target levels of either 15 μgP/L (low nutrient treatments) or 150 μgP/L (high nutrient treatments). Nutrients were subsequently added on a weekly basis at a rate of 5% per day to maintain these target levels (previous work indicated that this was necessary to counteract precipitation and accumulation of benthic and attached algae and microbes, found to be about 5% per day, Leibold and Smith unpublished data). A diverse algae inoculum (from a pooled sample of 12 nearby ponds) was then added (grazers were removed by two rounds of CO_2 narcotization and decantation) and allowed to grow for one week. Inocula of grazers were then introduced at densities of 50 individuals per species per tank. Half of the mesocosms received only *Daphnia* (thus corresponding to previous studies done on population dynamics previously described) while the other half also received *Ceriodaphnia* and *Chydorus*. Our motivation was to contrast the well studied case of *Daphnia* dynamics in the absence of competitors with dynamics that might occur in the presence of two other competitors. *Daphnia* is generally thought to be a highly efficient generalized grazer with a stoichiometry that is often relatively phosphorus limited rather than carbon or nitrogen limited (Andersen, 1997). *Ceriodaphnia* and *Chydorus* are more specialized grazers that are generally less efficient. They also differ in their stoichiometry with

Ceriodaphnia being more strongly P-limited and *Chydorus* being less P-limited than *Daphnia* (Hall et al., 2004). We hypothesized that the presence of these other species might stabilize stoichiometric dynamics and prevent the positive feedback hypothesized for *Daphnia*. A detailed analysis of the stoichiometric dynamics will be presented elsewhere (Hall et al., unpublished manuscript), however we observed very little evidence in our experiment for the hypothesis of positive feedback in any of our treatments. Our discussion focuses on food web dynamics related to consumer-resource interactions involving either cohort cycles, consumer-resource cycles, or damping by changes in algal genetic and species composition.

Three environmental conditions were imposed by the manipulation of nutrients as previously described and by the use of 90% shade cloth. The three environmental conditions were 'low nutrients, no shade cloth' (with a target level of 15 μgP/L, 70% natural light, hereafter referred to as the Low Nutrient treatment), 'high nutrients, no shade cloth' (with a target level of 150 μgP/L, 70% natural light, hereafter referred to as the High Nutrient treatment), and 'high nutrients, with shade cloth' (hereafter referred to as the Shade Treatment). After an initial period of one month during which a large algal bloom followed by the development of a substantial zooplankton community, algae and zooplankton were sampled three times every two weeks. Water was collected using a tube sampler from 14 evenly distributed points and pooled in a 20-L bucket. A 200-mL sample was taken for algal analysis and the rest was cleared of debris, sediments, and large algae using decantation methods. Zooplankton were collected using an 80-μm screen and the sample was split using a sample splitter. Half of the sample was preserved in Lugol's solution for microscopic enumeration. The other half was collected on an 80-μm steel screen and was dried and weighed on a microbalance. It was then ashed and reweighed to obtain a measure of zooplankton ash-free dry weight. The 200-mL algae sample was mixed and an approximate 50-mL subsample was filtered through a 35-μm screen from which a 3-mL subsample measured using the chlorophyll meter to obtain 'edible' algal chlorophyll. The remaining algae sample was briefly homogenized using a small electric mill to break up and mix the colonies and a 3-mL subsample measured using the chlorophyll meter to obtain total algal chlorophyll. Samples were collected for 20 sample periods over a duration of 90 days. We present data for the mesocosms that remained uncontaminated during the entire duration of the experiment and we present data only from the zooplankton mass and 'edible algae' chlorophyll in this publication. More extensive analyses will be presented elsewhere.

RESULTS

A typical set of data is shown in Figure 1. As in previous work (Murdoch et al., 2002), we found that zooplankton biomass and edible algae biomass tended to oscillate. The period of oscillation was fairly short compared to previous work, on the order of 20–25 days. Amplitudes varied but were often relatively large. To illustrate and quantify these patterns we plotted the time trajectories of the time series in phase space for each set of treatments (Figure 2). We found that the number of periods during the experiment did not vary among treatments but we found that the amplitude of both the zooplankton and of the algae did. In the low nutrient treatments the amplitude of the plant biomass was large relative to that of the herbivores for both communities with *Daphnia* alone and for *Daphnia* coexisting with the two competitors. In the shaded treatments the amplitude of zooplankton biomass was high and the amplitude of plant biomass low, and again this was true for both community types. Finally at high nutrients we observed that amplitudes of the two components were also affected but this difference depended on the community type. In the case of *Daphnia* alone the dynamics were very similar to those observed under high nutrients but in the case where *Daphnia* coexisted with the two other species, we observed large amplitudes of both zooplankton and edible algae oscillations.

We also examined phase relations and found a substantial difference for communities with *Daphnia* alone and for communities where *Daphnia* coexisted with the two other species (see Figure 3). When *Daphnia* was the sole herbivore we observed that zooplankton biomass

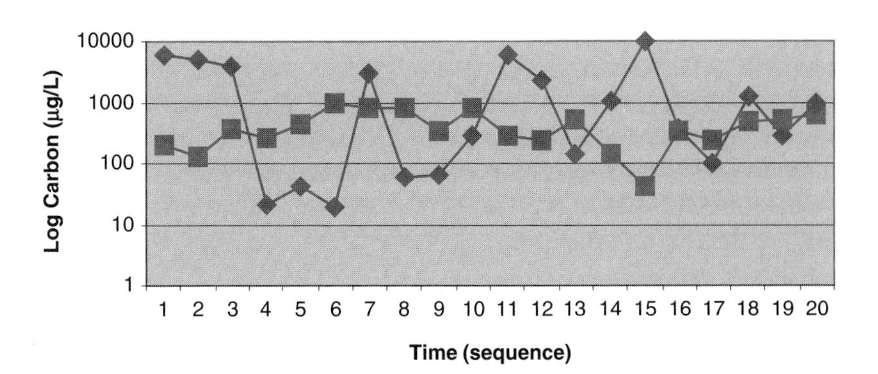

FIGURE 1 | Typical time series for zooplankton mass (square symbols) and edible algae mass (diamonds) in the experiment. Time is expressed relative to sample order and sampling dates were four or five days apart on a M, F, W, M, F, W schedule.

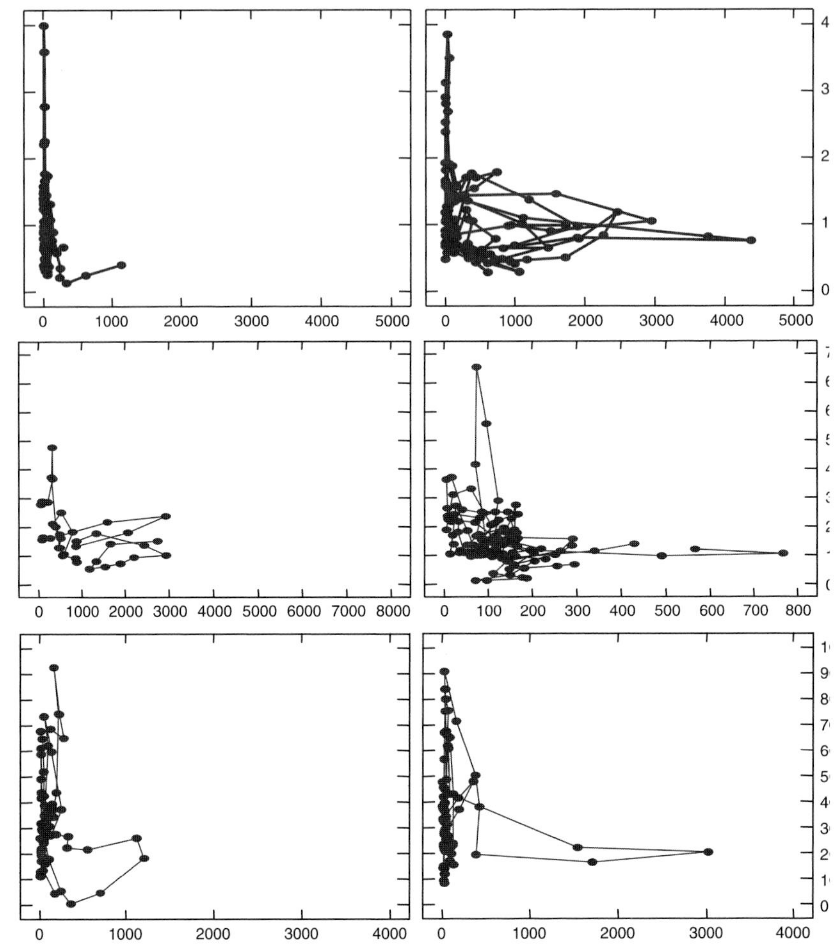

FIGURE 2 | Time series plotted in phase-space for each of the six treatments. In each cell, all the replicates are plotted jointly. Panels on the left column are for communities in which *Daphnia* was the sole herbivore, those on the right are for *Daphnia* coexisting with *Chydorus* and *Ceriodaphnia.* The top two panels are for the High Nutrient environments, those in the middle are for the Low Nutrient environments, and those at the bottom are for the Shaded environments. Note that the scales are different for each of the figure.

and edible algae were almost exactly out of phase. This is consistent with previous work and suggests that the dynamics were driven by cohort cycles. In contrast, when *Daphnia* coexisted with the two other competitors, we observed that the zooplankton biomass was lagged by $\frac{1}{4}$ cycle relative to the edible algae, consistent with the idea that the dynamics were driven by consumer-resource dynamics rather than cohort cycles.

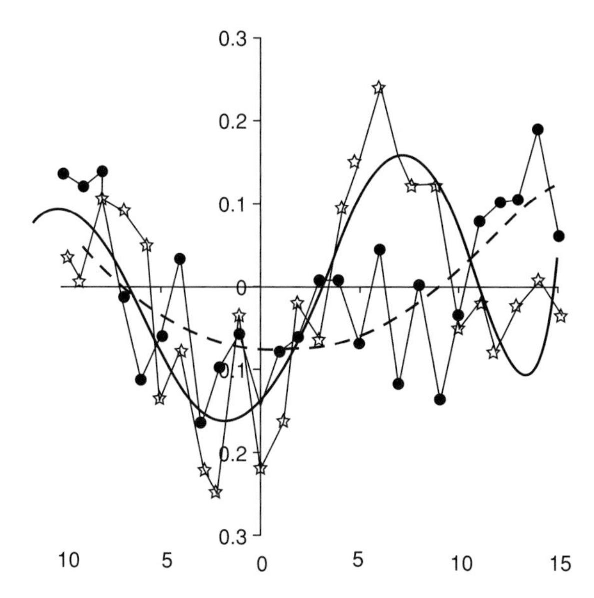

FIGURE 3 | Time series cross-correlation analysis of log-biomass of edible algae against zooplankton. Black dots show data (averaged across 4 replicates of each treatment each with 20 data points) for the cross correlation function for cases where *Daphnia* are the only grazers present, red stars indicates cases where there are three species (*Daphnia, Ceriodaphnia* and *Chydorus*). The X-axis is time lag (days), the Y-axis is the correlation coefficient. The graph shows lines that connect the data and smoothed estimators for each function (dashed black line for *Daphnia* only, solid red for three species together). (See also color insert.)

We also observed substantial populations of large 'inedible algae' (consisting especially of large colonies of filamentous Oedogonium and related species). These were present especially in the high nutrient treatments and were much less significant (though present) in the other two resource treatments. We could not evaluate whether the very long period cycles predicted in the presence of inedible algae were present given the relatively short duration of our experiments. Additionally, because we could not manipulate the composition of the algae in our experiments, we could not directly evaluate if they could damp consumer-resource interactions between the zooplankton and the edible algae (such inedible algae were present in all treatments). However the oscillation in these two other groups remained high in amplitude, even when large clumps of inedible algae developed, so that damping, if it was present, was not substantial enough to eliminate the oscillations as was found in some previous studies (Yoshida et al., 2003).

DISCUSSION AND CONCLUSIONS

A full analysis of these data will require more sophisticated methods than those we present here. In particular we hope to be able to be able to do a quantitative assessment of the various models using these data. More importantly we hope to be able to model the effects of competitors on these dynamics to assess whether our results can be linked to specific mechanisms of feedback in these food webs.

Nevertheless the simple analyses we present show several important results:

1. Cohort cycles and consumer-resource cycles, previously shown in laboratory work on *Daphnia*-algae interactions in the absence of inedible algae (McCauley et al., 1999), are both present in more complex communities.
2. The type of cycle observed can vary with food web architecture so that in our case consumer-resource cycles were observed in the presence of herbivorous competitors and cohort cycles observed in their absence. The precise reasons for this are at this point unclear.
3. The relative amplitudes of zooplankton and algae oscillations are strongly affected by environmental conditions involving nutrient supply and shading. Preliminary investigations (not discussed here) indicate that this is unrelated to stoichiometric effects but that it may rather depend on food edibility (Hall et al., unpublished manuscript).
4. Food web architecture modified dynamics in at least two ways. First by altering the incidence of cohort cycles vs consumer-resource cycles and second by altering the amplitude of the dynamics in at least some environmental conditions (in our experiment in the high nutrient treatments).

One interesting and important additional perspective is that oscillators previously described in simple communities such as consumer-resource and cohort cycles were still observable in our more complex community and that they had the same basic characteristic periodicities, amplitudes and phase relations. While naturally occurring plankton communities are frequently even more complex (e.g., often containing six or more crustacean species), our results support the perspective that these complex systems may be studied via component oscillators even though they are presumably coupled to other such oscillators (McCann et al., 1998; Hastings, 2004). While the result still seems to be quite complex it does provide evidence that these complex systems can be understood from an understanding of their component parts.

Over these time scales, our results indicated no evidence for substantially altered dynamics beyond those previously observed in the *Daphnia*-algae food chain. This indicates that interactions of zooplankton and algae in complex systems still consist of the same basic elements, in this case consumer-resource cycles and cohort cycles. Damping of the consumer-resource oscillation mediated through algal composition changes, if it was present, was not sufficiently strong to completely eliminate the consumer and cohort cycles observed here. This suggests that these dynamics might be studied as coupled oscillators that may interact with feedback due to other food web interactions. One important complication however is that we observe alternate patterns of oscillations like those observed in the simple algae-*Daphnia* (i.e., consumer-resource vs cohort cycles, McCauley et al., 1999) even though the reasons for the existence of these two patterns is not resolved satisfyingly. Understanding what determines the occurrence of these two types of cycles in naturally diverse communities will probably be a complicated question but may be an important step in understanding dynamics of planktonic food webs.

ACKNOWLEDGMENTS

We also thank L. Persson and E. Whalstrom for initial discussions that inspired this experiment. Finally, we thank A. Watterson and D. Oleyar for hugely helping with field work on this project. This work was funded by NSF grants DEB 9815799 and DEB 0235579 and is a Kellogg Biological Contribution.

2.3 | FOOD WEB STRUCTURE: FROM SCALE INVARIANCE TO SCALE DEPENDENCE, AND BACK AGAIN?

Carolin Banašek-Richter, Marie-France Cattin, and Louis-Félix Bersier

Food webs are complex and variable, and a general understanding of their structure and functioning must rely on a careful examination of their regularities. The search for scale-invariant features is of special interest in this respect (Briand and Cohen, 1984), since scale-invariance may represent basic structural constraints valuable for the discovery of underlying processes. An illuminating example of such a scale-invariant property drawn from astronomy and physics is Kepler's third law of planetary motion, which relates the revolution time t of a planet to its average distance from the sun d: t^2/d^3 is a constant value; the finding of this invariant property was a decisive step in the discovery of the underlying process, the law of universal gravitation. Most natural ecosystems are orders of magnitude more complicated and more variable than planetary motion. Their elements—species or sets of species—are themselves complex objects: they are composed of heterogeneous individuals, each involved in a wealth of interactions between themselves and their biotic and abiotic environment, and each able to adapt and to evolve. Many details have to be discarded to tackle the study of such intricate systems. Hence, food web ecologists concentrate on just one type of interaction depicting a vital aspect of ecosystems: trophic interactions describe feeding relations between species.

The burst of interest in the study of trophic interactions within communities stems from the stability-complexity debate; the finding that local stability in random systems is not a mathematical consequence of complexity (Gardner and Ashby, 1970; May, 1973; Cohen and Newman, 1984) bolstered the study of natural systems. In this framework, complexity was expressed as the product of the number of species in the community (S) and connectance (C), calculated as the quotient of the number of effective interactions (L) and the number of possible interactions (S^2). The so defined measure of complexity equals link density ($LD = L/S$). Studies of initial compilations of food webs resulted in the intriguing generalization that LD is scale invariant, meaning that this property remains constant across webs of varying size (Rejmánek and Stary, 1979; Yodzis, 1980; Cohen and Briand, 1984; Sugihara et al., 1989). This finding is in agreement with May's stability criterion (May, 1983) and was perceived as a fundamental structural constraint of food webs. However, together with criticism of the data used to assess the scaling behaviour of the link density property (Paine, 1988; Polis, 1991), a scale-dependent power law was soon proposed to provide a more accurate fit of link density to variable food web collections (Schoener, 1989; Cohen et al., 1990a; Pimm et al., 1991). Indeed, subsequently compiled collections of food webs did not uphold scale invariance for this property (Warren, 1989, 1990; Winemiller, 1990; Havens, 1992; Deb, 1995).

Link density's scaling behavior is illustrated in Figure 1 for six food web compilations on which this debate is based. Collections a and b were compiled from early literature data and appear scale-invariant. More recently assembled collections on the other hand (c – f), are clearly scale-dependent. One hypothesis to explain this discrepancy is that link density's scaling behavior is system dependent: all collections showing strong scale-dependence come from aquatic environments (Winemiller, 1990; Havens, 1992; Deb, 1995; Martimez, 1991), while the other two combine webs from various habitats (Sugihara et al., 1989; Cohen et al., 1990a). In the same vein, after factoring out the effect of web size, Bengtsson (1994a) found a similar difference between aquatic and terrestrial webs in a data set extended from collection **a** (Schoenly et al., 1991). Havens (1997) proposed biological features to explain the scaling behavior of link density in pelagic communities, namely (1) the predominance of filter-feeding predators and (2) the very high diversity of small autotrophic prey species. Most terrestrial systems however, may also be much more complex than those reported in data sets **a** and **b** (Polis and Hurd, 1996; Reagan and Waide, 1996). We studied terrestrial

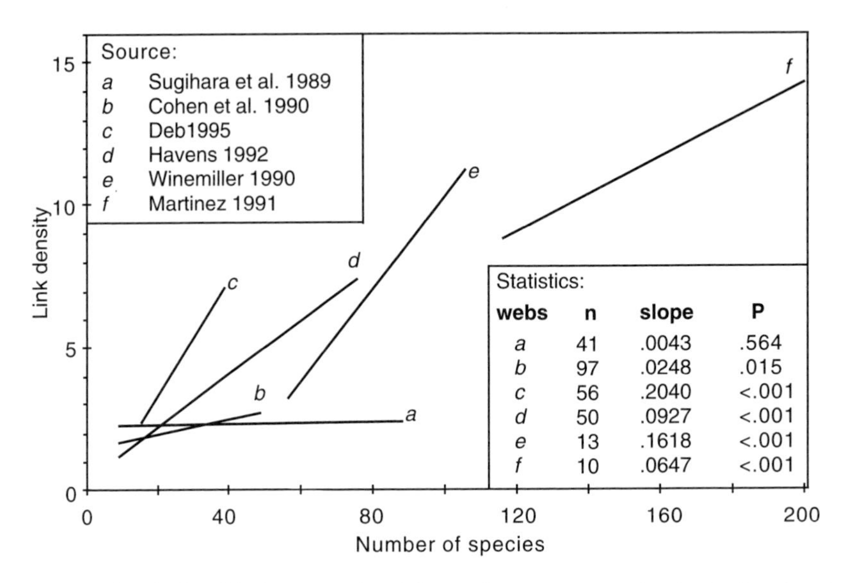

FIGURE 1 | Relationship between the link density property and web size for six collections of food webs. Lines are the results of linear regressions over the individual data points, which are omitted from the representation. Early collections of food webs drawn from literature studies not intended to represent the whole complexity of the systems (a, b) appear (approximately) scale invariant. More recent collections of webs characterized by high sampling effort (c, d, e, f) show strong scale dependence. Number of webs in each collection (n), estimates of the slope from linear regression, and *P*-values from a *t*-test are given in the inset. For collections a and b, webs with 10 species or less were omitted following Bersier and Sugihara (1997).

food webs in wet meadows, which exhibit scale-dependence for the link density property (see later discussion). Thus, intrinsic differences between systems are likely, but they cannot explain the opposing results of Figure 1 unanimously.

Another hypothesis relates these discrepancies to differences in sampling procedures and variable effort exerted toward the description of the data. Concerned specifically with the problem of sampling effort, Goldwasser and Roughgarden (1997) analyzed 21 web properties for a large, highly resolved food web (Goldwasser and Roughgarden, 1993). These authors employed different procedures to mimic increasing sampling effort, and found most properties, link density included, to be sensitive to sampling effort with high levels of sampling necessary to reach the properties' original values. A similar conclusion was reached by Martinez et al. (1999) who analyzed sampling effects in a highly resolved food web consisting of grasses and stem-borer insects. However, both former studies did not tackle the effect of sampling effort on the scaling behavior of

LD in collections of food webs. Using two models and three data sets, Bersier et al. (1999) showed that low sampling effort tends to produce the appearance of scale invariance in intrinsically scale dependent systems. This is a simple explanation reconciling the studies in Figure 1, since early collections of food webs taken from the literature (Cohen and Briand, 1984; Sugihara et al., 1989; Cohen et al., 1990a) were most often not intended to reflect the full complexity of the trophic interactions.

These sampling effects highlight the following problem inherent to the qualitative approach. The distribution of link importance in highly resolved food webs is likely to be exceedingly uneven (Goldwasser and Roughgarden, 1993). By giving the same weight to all links, binary food webs distort the true picture of their structure (Kenny and Loehle, 1991). Thus, quantitative data, which allows a more sensible approach to food web structure, is needed (May, 1983; Kenny and Loehle, 1991; Pimm et al., 1991; Cohen et al., 1993b; Bersier et al., 1999). Consequently, we must think of alternative ways of defining food web properties that take the disequitability in the distribution of link importance into account.

QUANTITATIVE LINK DENSITY

The process of formulating a quantitative counterpart for traditionally defined qualitative properties will be demonstrated in detail on the example of the link density property. The qualitative version, *LD*, is defined as the number of links (*L*) per species in the web ($LD = L/S$). For the quantitative version we base our calculations on information theory, namely the diversity index of Shannon (1948), *H*. For a system comprising *x* events, maximum diversity is attained when all events occur in equal proportion ($H_{max} = \log x$), while minimum diversity is a function of the number of cases that each event consists of. In our context, an event refers to a species and a case to a flux of biomass to or from a species (Ulanowicz and Wolff, 1991). The application of Shannon's equation to a quantitative food web matrix is visualized step by step in Figure 2: (1) In a food web matrix, species in their function as predators are conventionally listed column-wise (*j*), while the same species are arranged row-wise (*i*) in their function as prey. A matrix element b_{ij} thus expresses the amount of biomass passing from species *i* to species *j* (*j* eats *i*) per unit time and space. The total biomass output of species *k* to all its predators in the web consequently equals the sum of row *k* ($b_{k.}$). (2) The outflows from species *k* to each predator in the food web can be visualized. (3) Applying the Shannon index results in the diversity of species *k*'s biomass outflows (H_{Pk}). (4) For our purposes, the "reciprocal" of H_{Pk} is more interesting—it is understood as

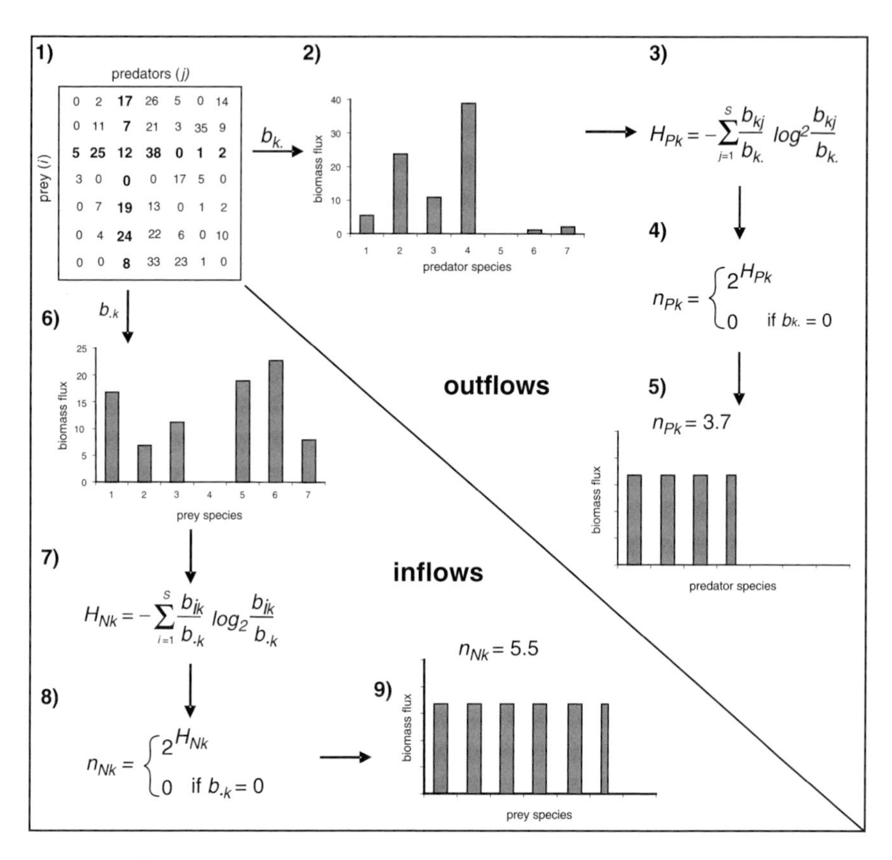

FIGURE 2 | Process of finding a quantitative definition of link density. Biomass inflow and outflow for species k are considered separately in the lower left and the upper right half of the figure, respectively. (1) Arbitrarily assembled quantitative food web matrix, highlighting the feeding interactions of species k. (2) Distribution of species k's biomass outflows; $b_{k.}$ = total biomass output from species k. (3) Shannon formula applied to species k's biomass outflows; H_{Pk} = diversity of species k's biomass outflows. (4) Reciprocal of H_{Pk}; n_{Pk} is the equivalent number of predators for species k. (5) Distribution of n_{Pk} outflows equal in magnitude, which yield the same value of H_{Pk}. (6-9) Analogous to (1) to (5), for biomass inflows to species k. $b_{.k}$ = total biomass input to species k, H_{Nk} = diversity of k's biomass inflows, n_{Nk} = equivalent number of prey for species k.

the number of predators feeding on species k in equal proportion that would generate the same diversity as H_{Pk}, and is termed the "effective number of predators" (n_{Pk}). The reciprocal of H has the desirable feature of recovering the original units, namely the number of species, but is now a real number; and (5) for the given example, n_{Pk} takes on a value of 3.7, rendering a distribution with 3.7 predators that display an equal intensity of consumption with respect to k. Thus, the diversity generated by this distribution is equivalent to that of species k's biomass output. The same approach is followed with regard to species k's biomass inflows in steps (6) to (9), yielding the effective number of prey (n_{Nk}).

A quantitative version of link density can be formulated on the basis of the previously defined indices n_{Pk} and n_{Nk}. For easier comprehension it is important to note that qualitative link density LD can be calculated either as the average number of prey computed over all species (the total number of prey divided by S), or as the average number of predators (the total number of predators divided by S). In analogy, one could formulate a quantitative link density either by averaging over all n_{Pk} values, or by averaging over all n_{Nk} values. Since it seems implausible to only consider either biomass in- or outflows, we average over both means to obtain a quantitative version of link density LD_q':

$$LD_q' = \frac{1}{2}\left(\sum_{i=1}^{s}\frac{1}{S}\cdot n_{Pi} + \sum_{j=1}^{s}\frac{1}{S}\cdot n_{Nj}\right) \tag{1}$$

This approach does not account for the fact that species vary in the amount of biomass transferred by them, and we thus refer to LD_q' as the "unweighted" quantitative link density. To include varying amounts of biomass transfer, $1/S$ in equation (1) is substituted by the quotient of each species' biomass output and total outflow over all species ($b_{i.}/b_{..}$). In other words, the effective number of predators for species i is weighted by i's contribution to total outflow. The same is done for inflow and the "weighted" quantitative link density LD_q is obtained by averaging over both equations:

$$LD_q = \frac{1}{2}\left(\sum_{i=1}^{s}\frac{b_{i\cdot}}{b_{\cdot\cdot}}\cdot n_{Pi} + \sum_{j=1}^{s}\frac{b_{\cdot j}}{b_{\cdot\cdot}}\cdot n_{Nj}\right). \tag{2}$$

For any given food web, LD_q' will always be smaller or equal to LD. This difference is an expression of the degree to which biomass flow in the system departs from a uniform distribution. The difference between LD_q' and LD_q in turn is attributable to the variation between species with respect to the partitioning of total biomass flowing in the system as expressed in unequal row and column sums for the quantitative matrix ($b_{k.}$ and $b_{.k}$ in Figure 2).

Species at high trophic levels are typically characterized by low biomass and consequently little in- and outflow, while greater biomass and more extensive biomass flux is generally a feature of species at low trophic levels. Therefore, trophodynamical constraints (Lindeman, 1942) have a determining influence on the value of LD_q.

We examined the effect of increasing sampling effort on qualitative, unweighted quantitative, and weighted quantitative link density for 10 extensively documented quantitative webs (Banašek-Richter et al., 2004). Both quantitative versions were found to be much more robust against variable sampling effort than their qualitative counterpart. This increase in accuracy is accomplished at the cost of a slight decrease in precision as compared to the qualitative link density. Conversely, the quantitative versions also proved less sensitive to differences in evenness with respect to the distribution of link magnitude. In sum, quantitative properties are not only useful as bearer of ecological information, they also represent a much more robust description of weighted matrices.

SCALING BEHAVIOR OF LINK DENSITY—RESULTS AND DISCUSSION

The scaling behavior of the conventional qualitative link density and its newly defined quantitative counterparts is compared for a collection of eight seasonal food webs from the southern shore of Lake Neuchâtel near the village of Chabrey in Switzerland (Cattin et al., 2003). These arthropod-dominated webs were collected in early summer and early fall of 2001. For each season there are two webs from a *Schoenus nigricans* dominated and two from a *Cladium mariscus* dominated vegetation zone. Of these, one web is from an area with mowing treatment, one from a control area without mowing. Species richness spans a range of 118–202 species.

The relation between link density and scale for the Chabrey collection is depicted in Figure 3. Qualitative link density (LD) is scale dependent in the sense that it experiences a significant increase with scale. Concurrently, the linear regression slopes for the two versions of quantitative link density (LD_q' and LD_q) are not significantly different from zero, thus indicating (moderate) scale-invariance for both properties. These results lead us to infer on the one hand that in a species-rich environment, consumers feed on a wider range of prey than in an environment with more limited resources (as expressed by increasing LD). On the other hand, scale-invariant LD_q' and LD_q imply that the diversity in link magnitude does not change with species richness for the food webs of the Chabrey collection. Thus, the diet of consumers (and the preys' predator list) must be more diverse only with respect to species number but not with respect to the partitioning of biomass consumed (or allocated) in large ecosystems. In fact, as systems accrue in species number, the distribution of in- and outflows must become progressively more inequitable to counteract the increase in qualitative link density.

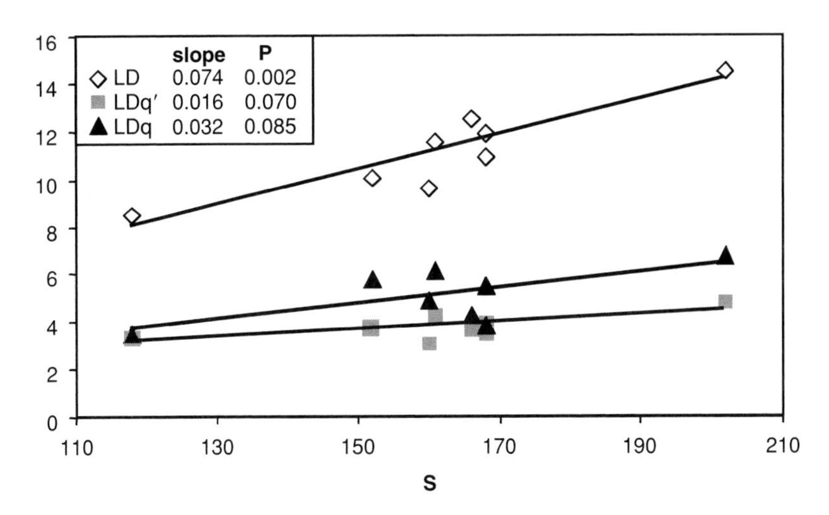

FIGURE 3 | Data and least-squares regressions (solid lines) of qualitative link density (◇), unweighted quantitative link density (■), and weighted quantitative link density (▲) for eight seasonal food webs from wet meadows on the southern shore of Lake Neuchâtel (Switzerland).

These findings raise the question if the food web collection analyzed here is structurally constrained with respect to the diversity of link magnitude and if so, what mechanisms give rise to such limitations. With regard to the search for regularities in food web structure, the result of scale-invariant (quantitative) link density would certainly be of extreme relevance with profound implications for the issue of stability and complexity.

Requisite to the derivation of new food web regularities based on the approach pursued here however is the collection and analyses of more high-quality quantitative food web data (Banasek-Richter et al., unpublished manuscript). This could serve to substantiate and supplement the obtained results, as well as to ascertain patterns in food-web structure pertaining to biomass flux which sensibly complement the results of qualitative analyses.

ACKNOWLEDGMENTS

We are grateful to R. Baltensperger, J. P. Gabriel, and C. Mermod. We are also greatly indebted to all people involved in the compilation of the Chabrey food webs, and to D. and H. M. Richter for the possibility to work undisturbed.

2.4 | THE ROLE OF SPACE, TIME, AND VARIABILITY IN FOOD WEB DYNAMICS

Kevin McCann, Joe Rasmussen, James Umbanhowar, and Murray Humphries

For some time, ecologists have sought to find a general relationship between diversity and stability (Odum, 1953; MacArthur, 1955; May, 1973; DeAngelis, 1975; Yodzis, 1981a). Today's research is beginning to turn away from this general focus and instead investigate the influence of biological structure on ecosystem stability. This change in perspective broadens the scope for stability investigations forcing us to consider the implications of structure at the population (see De Roos and Persson, Chapter 3.2 and Vos et al., Chapter 3.4), community (see Leibold et al., Chapter 2.2) and ecosystem level (see Loreau and Thébault, Chapter 6.1). The focus on structure also, importantly, says just as much about what drives instability as stability (i.e., certain structures inhibit runaway dynamics while certain structures excite such dynamics). In this chapter, we put forth ideas that bridge historical contributions to recent developments (through underlying structural assumptions) and outline an emerging perspective that ecological communities are not perfectly stable (i.e., in equilibrium) but fluctuate in response to both bottom up and top down influences. Our arguments suggest that structure created by variation in space and time is critical to food web dynamics. We further argue that ecologists should seek to understand which mechanisms inspire rapid changes in population, community and ecosystem attributes and which mechanisms contribute to the muting of such potentially violent dynamics. Finally, we comment on how factors that drive and mute variability interact to produce persistent ecological systems.

In what follows, we briefly outline historical contributions, before considering some recent perspectives in food web structure and theory that argues for space and time as critical factors in understanding food web dynamics. Ultimately, our argument is a rather simple one that postulates that variability in time and space are of extraordinary importance for food web stability. We argue that higher order predators/consumers are capable of coupling ecological systems at a range of spatial and temporal scales and in doing so they act to integrate across this variability in space and time. This *consumer integration* (via movement in space or time) either couples or decouples community dynamics and so can act to excite or muffle the noisy responses of lower level organisms. These results suggest that the scope for food web dynamics require a broad spatial and temporal perspective and ecologists must be willing to look beyond traditional scientific boundaries imposed by population, community, and ecosystem ecology.

SOME PERSPECTIVES IN FOOD WEB STRUCTURE AND THEORY

Robert MacArthur: Of Generalist Consumers and Food Webs

In 1955, Robert MacArthur put forward an intriguing argument for diversity promoting stability. Effectively, MacArthur postulated that the many different pathways in a diverse food web allowed an ecological system to buffer itself against a given perturbation. Interestingly, MacArthur's idealized food webs (similar to Winemiller's Christmas Tree webs, see Winemiller, Chapter 1.2) highlighted that he was considering a set of generalist consumers that fed across multiple prey. MacArthur argued that the different pathways effectively weakened the influence of any given perturbation by propagating the disturbance over many pathways as opposed to only one or a few pathway(s).

To fully explore MacArthur's logic it is worthwhile considering how the stability of the consumer depends on the response of a set of prey to the imposed perturbation. To see this, let us consider MacArthur's "thought experiment" under two different assumptions. First, let us assume that all prey are similarly afflicted by the imposed perturbation (*the unified response assumption*; all prey increase or decrease together). Second, let us assume that all prey respond differentially such that prey respond in an uncorrelated manner to the perturbation (*the differential response assumption*; prey essentially respond randomly). Generalist consumers foraging on prey that follow the unified response assumption are likely to be more influenced by the perturbation then generalist consumers foraging on prey that differentially respond to the same perturbation. Why? The unified prey response necessarily drives a consistent directional response by the generalist consumer. For example, if all prey respond by decreasing then the consumer responds in a consistent fashion by consuming less prey and ultimately decreasing. Under the differential prey response assumption, however, for every prey that provides less energy to the consumer there tends to exist an alternative pathway of prey that has increased it's delivery of energy to the consumer. Hence, the differential response of the prey to the perturbation tends to ultimately balance the consumers response whereas the unified prey response necessarily drives directional consumer response. We see that the differential response assumption has a very important property—it implies that the differential response imposed by the perturbation can be a very potent stabilizing effect in the webs of the type MacArthur described.

The stabilizing effect of the prey differential response can be further enhanced if the consumer has the ability to behave and switch its foraging rates in response to prey variation. Luckily, a significant foraging theory already exists that deals with the behavior of generalist consumers on prey that vary in space and time (Murdoch, 1969; Oaten and Murdoch, 1975a; Charnov, 1976). This theory suggests that the ability to behave in the face of the differential prey response ought to allow the consumer the ability to respond to the differential prey response by integrating over the different prey pathways. We believe that this result—in a sense hidden in the early explorations of MacArthur—is critical to understanding food web dynamics, composition, and stability. Additionally, this idea that variability can generate stability ties nicely into recent research developments that employ more formal mathematical arguments than the largely intuitive arguments offered above (de Ruiter et al., 1995; McCann et al., 1998; Tilman et al., 1998; McCann, 2000; Kondoh, 2003a).

Charles Elton: Of Diversity or Space?

Not many years after the seminal contribution of MacArthur (1955), Charles Elton wrote a much read book on invasions that forecast the dangers of human influence on food web composition, dynamics, and stability (Elton, 1958). In his book, Elton outlined a number of arguments for a positive diversity-stability relationship that are still discussed frequently today. His arguments can be summarized into the three following general arguments:

(E1): Simple mathematical model ecological systems and experimental model ecological systems (microcosms) tend to be violently unstable;
(E2): Island food webs are vulnerable to invasions, and;
(E3): Monocultures are vulnerable to invasion.

We assume that "vulnerable" means to be both invasable as well as significantly impacted by the invader. Although Elton's arguments are interesting and at times, persuasive, there clearly exist other factors besides diversity that vary consistently with these observations. One obvious and intriguing possibility is that each of these arguments potentially speaks to the role of space. For example, simple mathematical models and microcosms are spatially homogenous and relatively structure-free compared to real food webs. Similarly, islands food webs are spatially constrained food webs (i.e., the interactions are potentially intensified by the limited spatial extent), and monocultures, by definition, are examples of spatially simplified habitat structure. One could just as easily use Elton's arguments to suggest that space and spatial structure are of paramount importance in promoting persistent communities. In what follows, we will explore this idea more formally within the context of food webs. We will also briefly explore the potential influence of temporal structure, a close relative of spatial structure, on food web dynamics.

From Robert May to the Present

Not long after these early arguments, Robert May (1973) decided to take Elton's first hypothesis to task. His well-known work showed that in mathematical model ecoystems with interaction strengths generated from statistical universes (i.e., randomly assigned and structure-free) complexity (higher numbers of species or linkes) by itself did not drive more stability (although see Cohen and Newman, 1985a, for counterexamples). However; importantly, May (1973) and others (Gardner and Ashby, 1970) suggested a number of possible food web characteristics (e.g., interaction strength and compartmentation) that could strongly influence this result. Similarly, DeAngelis (1975) showed that self-regulation, low assimilation rates and donor control could reverse these results; while, Yodzis (1981a) generated "plausible community matrices" from a number of real food webs and found that these were more stable than May's non-structured results. Yodzis (1981a) empirically driven theoretical result showed that interaction strength was critical to the stability result.

Shortly following these findings, ecologists began a massive search for structure in real food webs (Cohen et al., 1990a). As a first pass, this research spent little effort on interaction strength (although see Paine, 1992; de Ruiter et al., 1995) and instead attempted to document static patterns in food web structure (e.g., number of predator to prey species etc.). The review of this impressive literature is beyond the scope of this chapter. However, the ideas and patterns that came out of these data and its analysis came under scrutiny in the 1990's from field ecologists who began to openly consider the problems in these early data (Winemiller, 1990; Martinez, 1991; Polis, 1991; Martinez, 1993a; Polis and Strong, 1996; Polis and Winemiller, 1996). Remarkably, many structural patterns in the early food web data still appear to exist (Martinez, 1993a) although the intense scrutiny of field ecologists revealed some potentially very important and under-explored food web structures.

In 1996, Gary Polis and Don Strong wrote a paper that crystallized some of these emerging patterns and ideas (Figure 1). Their arguments suggested that food webs were far more complex then the early data suggested. More importantly, they pointed to a number of factors that had been largely overlooked in the food web ideas of the 1970s. Specifically, they noted that food webs were replete with omnivory and generalist consumers that tended to couple different primary producers, detrital and classical grazing channels, and different habitats in space (Figure 1). Their suggestions, coined "multi-channel omnivory" (consistent with earlier terminology they could have been called multi-compartment omnivory), clearly spoke to the important role of spatial structure and generalist consumers. As previously discussed, both space and the generalist consumer are ideas that existed in the early arguments advocating a positive diversity-stability relationship (MacArthur, 1955; Elton, 1958).

The idea that spatially distinct food webs tend to be coupled in space by higher order generalist consumers suggests that food web compartments ought to exist in the lower trophic levels and become blurred at higher trophic levels. Interestingly, these suggestions have also consistently been part of the soil food web ecology literature. Here, soil ecologists have argued that distinct bacterial and fungal compartments, coupled by generalist higher trophic level consumers, are of great importance to the stability of soil food webs (Moore and Hunt, 1988; de Ruiter et al., 1995). Although the idea that spatially distinct food webs were coupled in space by higher order consumers resonated with many field ecologists (Polis, 1991), early investigations found that food webs,

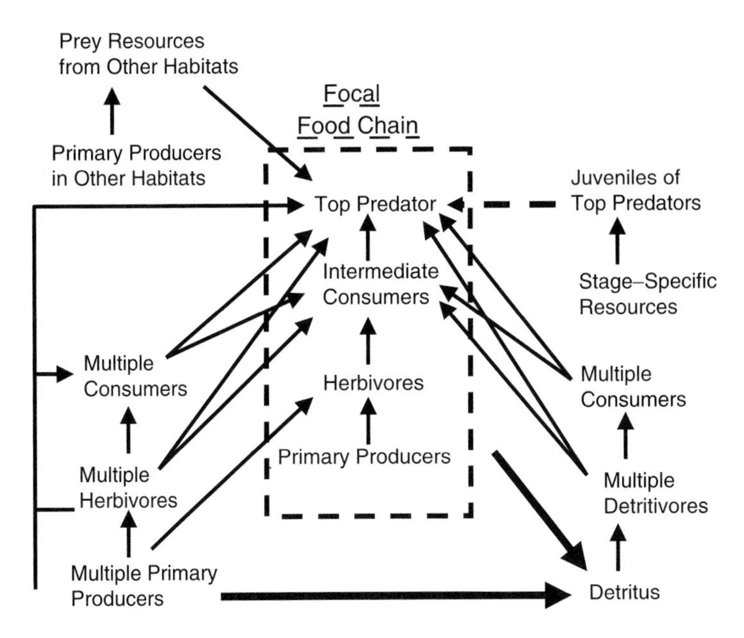

FIGURE 1 | A slightly modified version of Polis and Strong (1996) schematic representation of multi-channel omnivory highlighting factors that have been overlooked in food web ecology. Of note is the emphasis on the role of space. Specifically, they noted multiple producers existing in different habitats as well as detrital and grazing channels that tended to be coupled by higher order consumers through movement and life-history. Redrawn from Polis and Strong (1996).

in general, showed little evidence of compartmentation (Pimm and Lawton, 1980). Curiously though, Pimm (1980) found that habitat structure—at least at a large spatial scale—appeared to drive compartmentation while detrital-grazing channels were also likely candidates for real food web compartments. Despite this result and the cries of the soil ecologists (Moore and Hunt, 1988) there has been little work on compartmentation over the last 20 years (although see Raffaelli and Hall, 1992). Very recently though, Krause (2003) has developed a new compartmentation statisitic that has some clear advantages over Pimm and Lawton's (e.g., includes interaction strength). His results suggest that many food webs show evidence of compartmentation, especially when interaction strength is included in the algorithm.

In light of Krause's findings (2003) it is worth reconsidering compartmentation and its influences. Preliminary descriptions of generic food webs (Figure 2A-C) suggest that resource compartmentation may occur on a variety of different scales in food webs, from the cross-ecosystem couplings of Summerhayes (1923) to the within community coupling found in soil food webs (see Figure 2A-C).

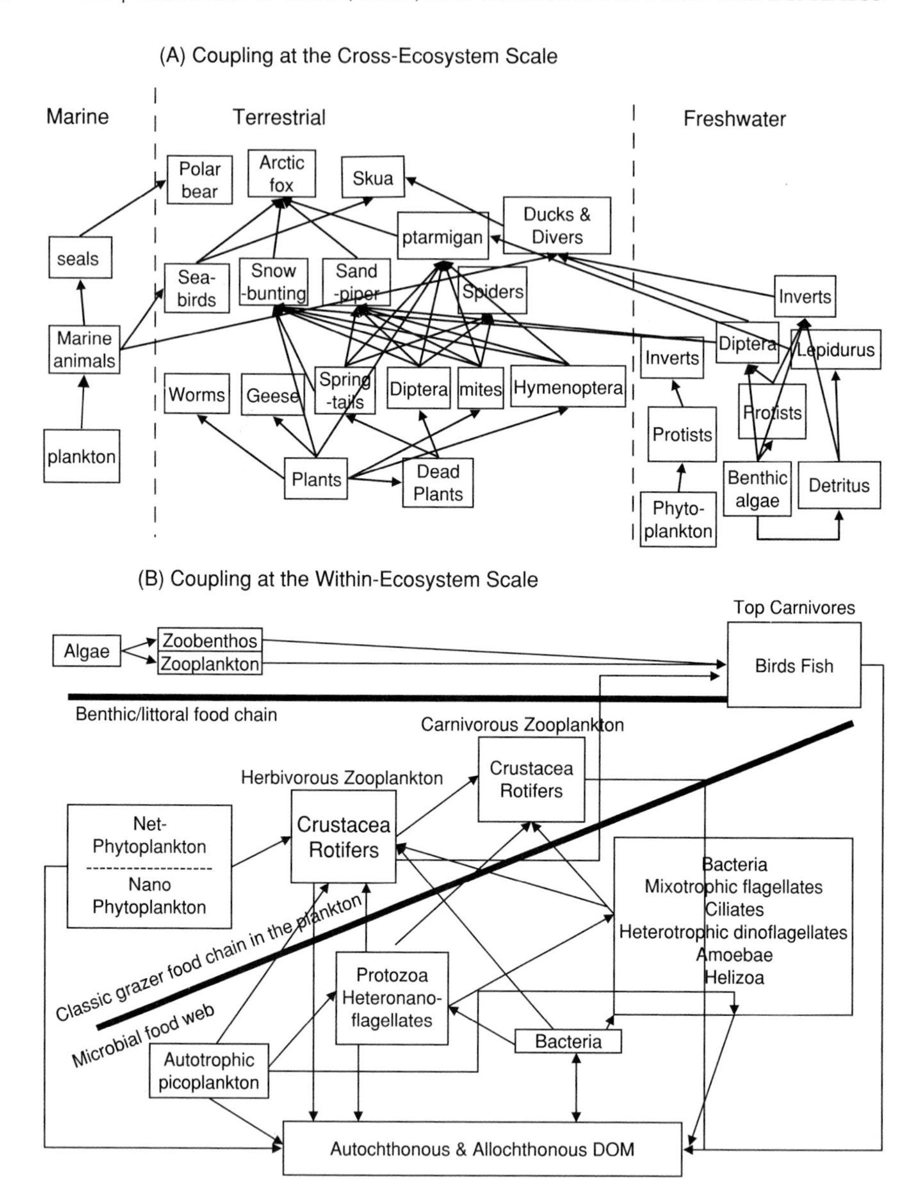

FIGURE 2 | Three potential examples of sub-system coupling or compartmentation that highlight separate sub-systems with the tendency of higher order consumers coupling these webs in space. **A,** Coupling across ecosystems in the Arctic as detailed by Summerhayes and Elton (1923), redrawn by Pimm and Lawton (1980). **B,** Coupling at the within ecosystem scale in temperate lakes (redrawn from Kalff, 2002) showing coupling between the littoral, detrital and pelagic pathways by higher order generalist consumers.

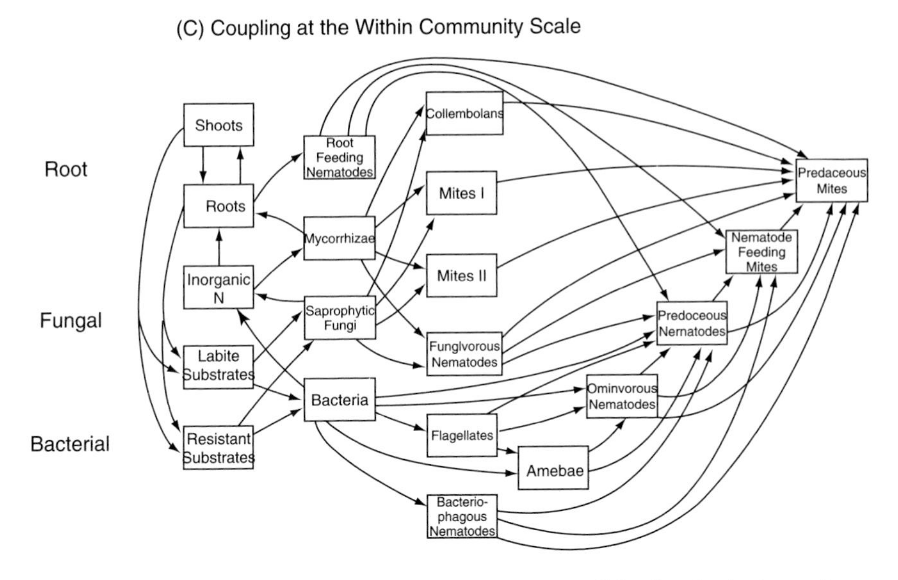

FIGURE 2, cont'd | C, Coupling with a community in soil food webs (redrawn from Moore and Hunt, 1988) depicting the fungal, root, and bacterial compartments, again with the tendency for higher order consumers coupling these pathways. Here, the fungal and bacterial channels tends to derive from recalcitrant and labile pathways, respectively.

Even less explored by food web ecologists is the potential for compartmentation to unfold along a temporal axis. One such example, though, has been put forth by Reagan (1996) for a tropical Caribbean rain forest. Here, they found that frogs and lizard create a night-day shift in foraging in this food web. Their studies of prey content at the family level, though, showed high overlap between frogs and lizards suggesting that the compartmentation may not exist despite the temporal difference in foraging. However, at a finer taxonomic resolution of the prey they found only 13% prey overlap suggesting the potential for significant compartmentation in this web. This type of temporal delineation deserves further consideration as seasonality and other periodic changes open up the possibility for large food web shifts (Winemiller, 1990). In the next section, we argue that the emerging theory suggests that generalist consumers coupling food webs in space and time, are of paramount importance to the stability of food webs. These newly emerging empirical and theoretical results clearly argue for an expanded spatial and temporal scale of investigation in food web ecology.

VARIABILITY, STABILITY, AND CONSUMER INTEGRATION

Although it has been long known to ecologists that the population dynamics of ecological systems are variable, ecologists are now beginning to fully embrace this fact (DeAngelis, 1975; Chesson and Huntly, 1997; Tilman et al., 1998). Here, we wish to argue that variability in resources in both space and time create a complex biological canvas that consumers can react to by decoupling from some consumer-resource interactions or re-initiating other consumer-resource interactions. We also suggest that if higher order consumers act to couple habitats, compartments or sub-systems in space then, importantly, consumer behavior can also similarly decouple or re-initiate interactions in habitats, compartments, or sub-systems. Intriguingly, if this decoupling occurs when densities in the sub-system are low on average then such biological structure can drive persistent food webs by allowing the sub-system a reprieve from consumptive pressures exactly when it needs it—when the sub-system is experiencing low average densities. Here, we briefly discuss these potentially potent stabilizing mechanisms as they pertain to decoupling in space and time.

Decoupling in Space

There is a lack of food web theory that considers space, although Robert Holt (2002) has developed some very interesting and important ideas. Specifically, he has suggested that space play a major role in the determination of food chain length and that immigration can act to stabilize or destabilize local communities (Holt, 2002). On a related front, a number of articles have demonstrated that preferential, or optimal foraging, in food webs can drive stability (Murdoch, 1969; Oaten and Murdoch, 1975a; McCann et al., 1998; Krivan, 2000; Post et al., 2000a; Kondoh, 2003a). This readily occurs if the functional response includes some sort of density-dependent preference and is skewed such that one resource is less preferred (this is a manifestation of the weak interaction effect (McCann, 2000)). These preference models differ from classical multi-species functional responses in that they imply that an organism, by increasing its feeding effort on one resource, decreases its ability to feed on other resources (e.g., it cannot be in all places at once). There are a number of possibilities for this trade-off but one clear case is that of a consumer integrating across multiple resources in different habitats. Thus, these model results can be interpreted in terms of their implications for a generalist forager integrating across multiple resources in space.

Preferential consumption of a generalist consumer can readily drive differential responses of its resources. This occurs because if one resource increases in density then it follows that the consumer can increase its effort on that resource. The other resources, therefore, experience less effort by the consumer and are alleviated from consumptive pressure (if they compete with the resource they may also experience competitive advantages). This behavior by the consumer readily drives an out-of-phase response of the resources (Figure 3A). This out-of-phase response, in turn, generates an emergent type III functional response of the consumer on its prey (McCann, 2000). This occurs since the consumer tends to reduce effort on a resource (i.e., decouple from that resource) when that resource is at low density simply because it profits to feed on the other resources which tend to be at higher relative densities. Hence, given variability and out-of-phase dynamics at the resource level, a generalist consumer foraging in space can be a potent stabilizer of "noisy" resource dynamics. This occurs because consumers decouple from the consumer-resource interaction when resource densities are low and re-couple the C-R interaction when the resource is at high densities. Such behavior leads to more bounded, less variable, dynamics (see Figure3A).

Importantly, it appears that such a phenomena can scale up to sub-systems (see Figure 3B). Several theoretical papers (Teng and McCann, 2004; McCann et al., 2005) have found that coupling sub-systems or compartments through higher order consumers can drive a similar result (Figure 4A-C). Here, sub-systems are stabilized the most when they are out-of-phase (see Figure 4B-C). In this case, the out-of-phase sub-system dynamics are generated by the consumer when it has a strong preference for one sub-system and a weaker preference for another. This *weak sub-system effect* is similar to the weak interaction effect in that the consumer drives the differential response of resources through time.

Although we have concentrated on top-down driven differential response, it is also possible that the differential response of resources is bottom-up driven. In such a case, if the resources vary out-of-phase and the consumer can integrate across them then again we expect the consumer to mute the noisy dynamics of the resources. Thus, the role of space and generalist consumers—those same structures emphasized by Polis and others in the 1990s—act to use variability to produce stable food webs dynamics. The higher order consumers decoupling and coupling food webs in space (see Figure 3).

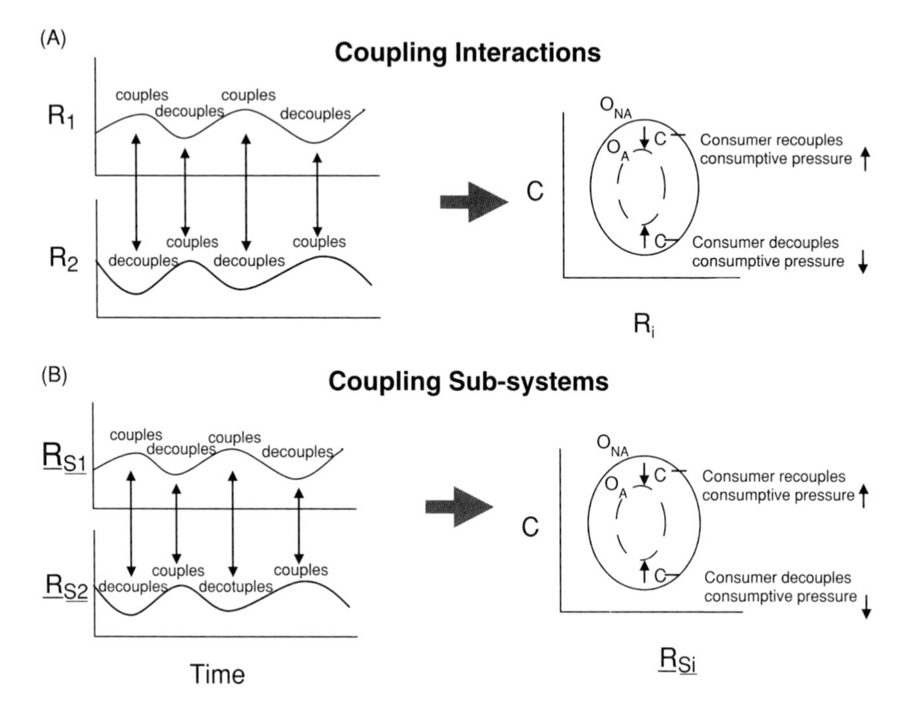

FIGURE 3 | Schematic depicting the role of out-of-phase resource dynamics with stability (well bounded oscillations, or mimina consumer and resource densities that are further from zero). **A,** Out-of-phase dynamics at the consumer-resource interaction scale (C-R_i) can lead to more bounded dynamics since the consumer couples and decouples at times that enhance stability (i.e., decrease variability). O_{NA} represents the cycle without an alternative resource and O_A represents the cycle with an alternative out-of-phase resource. **B,** Out-of-phase dynamics at the consumer-resource sub-system scale ((C-$\bar{R}_{si}$); where $\bar{R}_{si}$ represents the mean resource levels in a separate sub-system *si*) can lead to more bounded dynamics since the consumer couples and decouples the sub-systems at times that enhance stability (i.e., decrease variability). O_{NA} represents the cycle without an alternative sub-system and O_A represents the cycle with an alternative out-of-phase sub-system.

Decoupling in Time

The discussion of space as a major player in food webs dynamics suggests that other ways to decouple interactions and sub-systems can be equally as important and deserve attention. Another such area that is not well explored, but is likely to play a pivotal role in food web dynamics is the ability for consumers to decouple or couple interactions in time. McCauley et al. (1999) gave a very nice top-down example of temporal decoupling in their work on *daphnia*-phytoplankton dynamics. Here, ephippial egg production (as opposed to parthenogenic reproduction) allowed for zooplankton to decouple from phytoplankton when

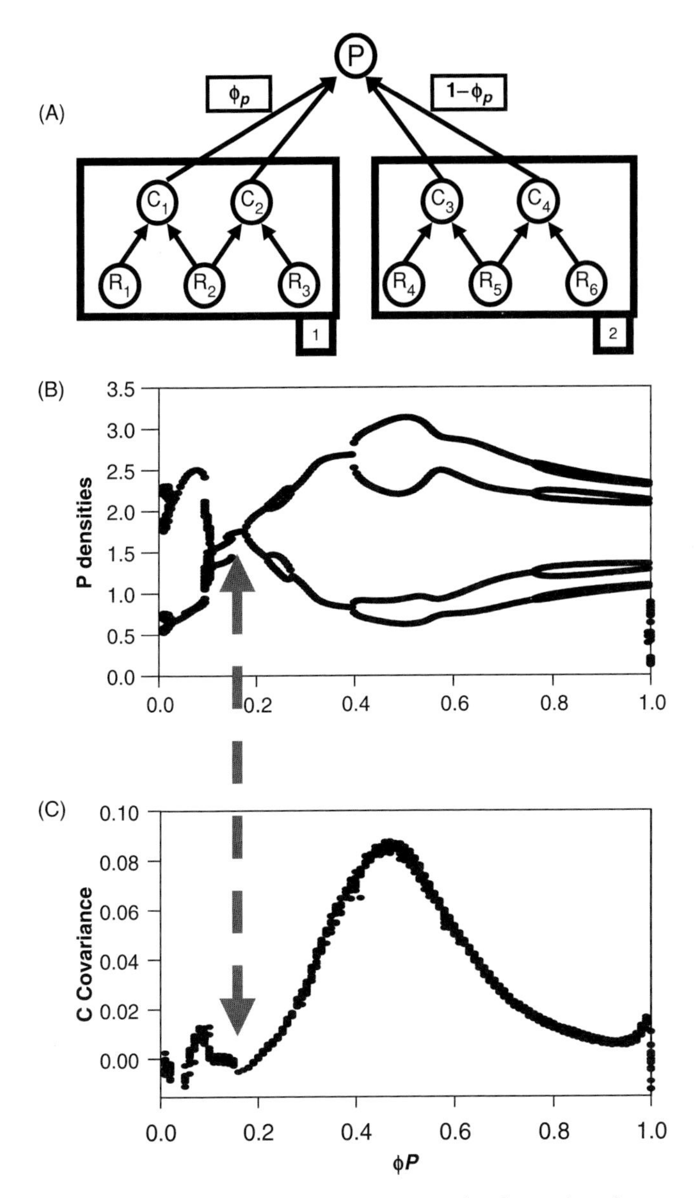

FIGURE 4 | Taken from Teng and McCann (2004). The dynamics of a top predator coupling two sub-systems. **A,** The larger food web depicting a generalist mobile top predator, P, coupling two identical sub-systems through the preference function (ϕ_p) defined in Post (2000a). **B,** Local maxima and minima of the top predators dynamics as a function of preference for sub-system 1. Variance at any value of the preference function, ϕ_p, can be visualized as the difference between the global maxima and the global minima. At intermediate sub-system preferences the whole food web attains its minimum variance (see *dashed arrow*). **C,** A measure of the covariance of the mean C's in each sub-system for each value of the preference function. The dashed arrow shows that the high stability (i.e., low variance) portions of parameter space tend to be associated with out-of-phase dynamics (i.e., lower covariance between sub-systems). The covariance was measured by randomly perturbing the solution off the attractor and measuring the covariance between the mean C's in the sub-systems.

the phytoplankton were at low densities. They found that this mechanism (a delay in the onset of zooplankton production or dispersal through time) generated less variable dynamics. A bottom-up example of temporal decoupling is provided by hibernating prey populations, which typically become inaccessible to generalist non-hibernating predators during winter, thereby shifting predation pressure to other non-hibernating prey (Humphries et al., 2004). As with space, the key requirement for a stabilizing effect of decoupling in time is differentiation in the temporal responses of one trophic level, and integration across these responses by generalists at another trophic level.

Recently, Humphries et al. (unpublished manuscript) have explored how metabolic flexibility can serve as a highly general, temporal decoupling mechanism. This approach is based on a few simple physiological principles: (1) fast metabolism permits a high attack rate on abundant resources, while a slow metabolism improves fasting endurance when resources are scarce, (2) animals differ in their capacity to vary metabolism across gradients of resource abundance (e.g., by hibernating, entering diapause, or simply by becoming less active), (3) a high degree of metabolic flexibility comes at a cost of reduced maximum attack rates. As a result of the physiological mechanisms, metabolically slow and flexible populations that can decouple from scarce resources will have stabilized population dynamics and a strong potential to coexist with metabolically fast, inflexible populations. The predicted association between metabolic rate and population variability is supported by an empirical comparison of 53 mammal species (Humphries et al., unpublished manuscript).

Viewed from a food web interaction-strength context, slow-flexible animals are capable of decoupling from the resource and maintaining a relatively weak interaction with their prey base, while the fast, inflexible species act as strong interactors. Interestingly, the existence of the slow, flexible animals can act to mute out the strong oscillatory population dynamic potentials of the fast, inflexible animals (McCann et al., 1998). This occurs because the slow, flexible consumers divert energy from the fast, inflexible competitors and in doing so limit their ability to over-consume their prey. In essence, their existence reduces the overall trophic efficiency of the competitive guild, which has the well-known effect of stabilizing the overall community dynamic (Rosenzweig and MacArthur, 1963; DeAngelis, 1975). The centrality of metabolism in trophic interactions and the directness with which variation in metabolism generates advantages and compensating disadvantages under different levels of resource abundance, suggests slow-fast metabolism may represent a basic axis of differentiation promoting food web stability.

Additional forms of decoupling in time deserve consideration. One relatively unexplored area in food web dynamics is the role stage structure and ontogentic shifts in habitat plays in food web dynamics. Here, delays in time (stage structure) and movement in space (through life history) couple in ways that may strongly influence food web composition, control, and dynamics (see De Roos and Persson, Chapter 3.2, along with Persson and De Roos, Chapter 4.2).

DISCUSSION AND CONCLUSIONS

We have argued that many food webs appear to be coupled in space by generalist higher order consumers. Further, if such couplings exist and generalism tends to increase as one goes up the trophic structure, then we expect that food webs frequently ought to be compartmented (Krause et al., 2003). We have also argued that this structure lays the foundation for a very potent stabilizing force in food webs. Higher order consumers that integrate across variable resource densities in space (either at the population or sub-system level) will act to mute lower level variability. These results beg for an empirical analysis of the covariance dynamics between different sub-systems and compartments. Examples exist that suggest such negative co-variance between major energy channels may frequently exist. In a review of detrital dynamics, Moore et al. (2004) discuss a number of examples of the variable production in space and time of detrital pathways relative to the classical grazing web. In an aquatic example, Hunt (1975) suggests that many salmonid fishes feed primarily off of aquatic invertebrates in spring and fall and then switch to terrestrial invertebrates in summer. Hunt (1975) further suggests that the production of benthic invertebrates is greatest during spring and summer while terrestrial arthropods collected as drift peak at mid-summer. Thus, the different pathways utilized by salmonids are out-of-phase with one another. Such empirical findings may aid our understanding of what inspires unified sub-system dynamics (i.e., resources increase or decrease together) as well as what inspires differential response. It may also help us identify critical levels of aggregation for food web dynamics. In other words, out-of-phase systems may identify important functional groupings for a given spatial and temporal scale.

This simple theory suggests that variability (either bottom-up or top-down driven) and higher order consumers are of tremendous influence for mediating persistent food web dynamics. Thus, any perturbation that reduces the ability of consumers to integrate over space or unifies dynamics of different species or sub-systems will excite, or destabilize, the dynamics of food webs. Within this framework, it is not so surprising that some of the most enormous impacts by invasive species have occurred when a higher order consumer invades small islands (Elton, 1958; Ebenhard, 1988; Simberloff, 2000). One expects the limited spatial extent to limit the switching potential of these larger habitat couplers with major consequences (McCann et al., 2005). Additionally, these ideas suggest that fragmented habitats ought to generate similar instabilities when the fragmentation reaches the spatial scale that eliminates the potential for coupling by higher order organisms. Finally, although we have focused largely on the implications of decoupling and coupling in space, we suggest that ecologists should also consider the role of decoupling in time on food web dynamics as nature may have also evolved to frequently solve resource depression through this mechanism.

3.0 | POPULATION DYNAMICS AND FOOD WEBS: DRIFTING AWAY FROM THE LOTKA-VOLTERRA PARADIGM

Giorgos D. Kokkoris

Food webs are networks that depict consumer-resource interactions (links) among species or trophic species (nodes). This approach has been given a mathematical representation by the use of Lotka-Volterra population and community dynamics, and may including basic biotic relationships as intraspecific competition and predation, including parasitism. The central theme of food web research is the understanding of structure, function, dynamics, and complexity. In order to be able to predict food web behavior under external and internal effects, biotic, and abiotic influences we need to be able to answer the question, "What drives food web dynamics?" (see Scharler et al., Chapter 8.3). The success in fulfilling this task will determine in part the management of our ecosystems towards sustainability (see Section 7). Unfortunately, there is little use of food web models in environmental management. Reasons for that stem from the fact that theoretical ecologists rarely care to address practical problems and managers are reluctant to use food web models to predict.

Do we really need to consider a new view of food webs and biotic communities in the dawn of the twenty-first century? Are there any important factors that have not been included in the study of natural food webs so far? Forces that act on food webs actually affect both the nodes and the links of the graphical representation of a food web. There is now enough evidence that both components of food webs need to be reconsidered and utilized approaches to be revised in future work. Populations are usually described only through reproduction and mortality. But populations consist of individuals that grow and develop and not all predator and prey individuals are identical (see De Roos and Persson, Chapter 3.2). Life history variation among species, expressed as different generation times influences population growth (see Scharler et al., Chapter 8.3). Predation pressure may induce defences to some of the individuals of the population of their prey, creating heterogeneity in the prey population (see Vos et al., Chapter 3.4). Traditional approaches ignore the dynamics resulting from the previously described aspects of heterogeneity within the populations. This missing heterogeneity may be an important determinant of the observed pattern and processes on food webs and community level properties such stability, resilience, and persistence.

One of the valid criticisms that have been developed is that interaction between species can be described by a linear function of their densities (Pimm, 1982). This is the ecological equivalent of the *Law of Mass Action* that has been inherited to younger Ecology from older Chemistry. Applied to community processes, this law states that if the individuals in populations mix homogeneously, the rate of interaction between two species is proportional to the product of the numbers of individuals in each of the species concerned. As a result, predators for instance keep consuming their prey independently of their density, which certainly cannot be true in real systems. Responding to the call for more mechanistic models, Fretwell (1977) and Oksanen et al. (1981) studied food chains representing interactions that accounted for functional and numerical response of predators. Integration of non-linear dynamics into food web models has taken place recently as a result of significant advances in computing power that is a *sine qua non* of such approaches (Drossel et al., 2004, see also Dell et al., Chapter 8.1).

Food web models that follow the dominant Lotka-Volterra paradigm use emergent food web properties such as diversity (species richness) and connectance to determine other food web characteristics (Jansen and Kokkoris, 2003). Traditional community assembly models ignore adaptation processes (Kokkoris et al., 1999). But food webs evolve and these models fail to provide clear mechanisms explaining how these characteristics and structure emerge. The ecological interactions among species in a community and the role they play in adaptation (behavioral, developmental, or evolutionary) of species traits such as body size are usually left out (Bøhn and Amundsen, 2004). Ecological interactions are the structuring links in all food webs and the patterning of them is the major factor in the stability, resilience, and persistence of biotic communities (for a review see Berlow et al., 2004).

Kondoh (see Chapter 3.3) criticizes this static representation of nature. For instance, if a trait that influences the strength of trophic interactions is controlled by adaptation, then the food web architecture should change in a way that is favored by this adaptation. These changes influence population dynamics and consequently the stability of the food web (Kondoh, 2003a). Few pioneering studies also have recently investigated how complex food webs emerge from evolutionary community assembly processes (Drossel et al., 2004; Loeuille and Loreau, 2005; McKane and Drossel, Chapter 3.1). These studies provide useful insights on the evolution of food webs and should be developed further to allow invasions (or speciation) of species that possess characteristics that may be quite different from those that already exist in the community under study.

The points previously laid out are motivated from the chapters in this section of the volume are part of the challenge in developing testable predictions from food web studies. The chapters of this section clearly justify the new approaches needed in the face of global change and extinction crisis (Lawton and May, 1995). Food web ecologists move gradually away from the dominant paradigm in the discipline and if this is combined with a willingness to address practical issues, their models will be of good service to environmental management and biodiversity conservation.

3.1 | MODELLING EVOLVING FOOD WEBS

Alan J. McKane and Barbara Drossel

In this chapter we discuss research that we have been carrying out with our co-workers over the last few years which allows us to construct model food webs. The webs are generated dynamically from ingredients that are believed to be the most important and relevant in determining the structure of food webs. However, the approach is flexible, and so the underlying model can be modified, and the effects of the modification on the web structure determined. In addition a "library" of model webs can be built up. These can then form the basis of "experiments" that would be difficult or impossible to carry out on real webs. For instance, one or more species can be deleted and the effect on the web in the short or medium term assessed, or the distribution of link strengths can be measured.

In order to be able to "grow" a food web, we need to recognize that there are two types of dynamics which are relevant in web construction. The first type gives rise to new species in the web (by speciation or immigration) and eliminates them from the web (extinction). The second type is conventional population dynamics, which describes the interaction between individuals when the number of species in the community is fixed. These two types of dynamics are coupled: we believe that population dynamics should not be defined on a static web, since the web itself changes in response to the growth or decline of species numbers, which in turn are determined by the nature of population dynamics adopted (Thompson, 1998).

In the present model we construct the web through the introduction of new species that are similar to existing species. We start from a single species, and an environment, and at every evolutionary time step we add a new species to the model. This can have originated through a speciation event or through the immigration of a related species. The new species may survive or go extinct itself—this is decided by using the equations describing the population dynamics. The model is adaptive, with the behavior of species strongly dependent on, and determined by, the fortunes and behavior of other species in the system.

There are three time scales in the model. On the longest time scale, new species are introduced. They are variants of a randomly chosen species already in the system. On the intermediate time scale, the number of species is fixed, and the dynamics is that of conventional population dynamics. On the shortest time scale, the populations of each species are fixed, but the foraging strategies may change, so that species may alter their feeding habits to take advantage of recent changes in population sizes. These aspects all add to the realism of the model, but also mean that the model is not easily amenable to analytic treatment, and all the results have been obtained through numerical simulations.

There have been several distinct approaches to the modelling of food webs. Perhaps the simplest models are purely static, and attempt to produce webs similar to those found in empirical studies by creating a large number of random webs with a given number of species and links, but subject to a constraint that is designed to mimic real webs. In the cascade model (Cohen et al., 1990a) the constraint is very simple: species are ordered, so that only those "above" a given species are allowed to prey on it. The niche model (Williams and Martinez, 2000) has a more complicated constraint that determines which species is allowed to feed on another: it involves drawing random numbers for the position of a species, the range of its predation, and the center of this range. Not surprisingly, models such as these get better at reproducing real webs the more complicated they become. Yet they all suffer from the same defect: constraints are chosen so as to produce the best fit to data, but this tells us very little about the important factors in web formation. In other words, the models encapsulate the basic web structure, but they are not rich enough to go much beyond this.

Assembly models, a second type of food web model, were popular in the 1990s. They are dynamical models and have a population dynamics built into them, just as in the model described in this article. However, the new species introduced into the web are not variants of those already present, but instead came from a "species pool." New species are added to the system, which starts with just one or two species, and remain in the system if the resulting system is stable (see Law, 1999, for a review). The disadvantage with these models, as we see it, is that the species in the pool have not co-evolved, and in fact are usually taken to be of a specific type such as an herbivore or carnivore. This means that much of the web structure itself is not truly emergent, with species finding their own place in the web. With some of the web structure predetermined in this way, it is difficult to know which properties follow from the population dynamics (for instance) and which have been put in by hand.

A third set of models can be broadly classed as evolutionary models. They have only been formulated relatively recently and only a few have appeared, although we expect many more to be put forward in the future. They are also a rather disparate group, which reflects the young age of the field. There are, for example, the networks of interacting molecular species of Jain and Krishna (2002) which are evolving, but are not models of food webs as such, and the work of Cattin et al. (2004) which only takes evolution indirectly into account, but does suggest a method of constructing food webs. Evolutionary models of food webs have been suggested by Tokita and Yasutomi (2003) and by Yoshida (2003), but these include mutualistic couplings, which produce a different kind of network. Apart from the one we will describe in detail in this article, the only published model of an evolutionary predator–prey network was developed by Lässig et al. (2001). However, so far no webs produced using this model—which could be compared with real webs or to webs produced using our model—have been published. Therefore, at the present time, there are no other models that make predictions similar to ours with which we can make a direct comparison.

A first version of our model (Caldarelli et al., 1998) contained the basic philosophy which we have previously outlined, but had a rather simple form of population dynamics. Although this model produced realistic looking webs, which had properties that were in good agreement with those of real webs, an invasion-resistant state was eventually reached. This unrealistic aspect of the dynamics was traced back to a property of the population dynamics that unfairly disadvantaged a child species as compared with the parent species. This problem was cured with the new version of the model (Drossel et al., 2001) which contained a much more elaborate form of population dynamics. All our discussions in this chapter will refer to this version of the model. Since then other papers and articles have appeared which discuss various aspects of the model (Quince et al., 2002; Drossel and McKane, 2003).

In the next section, we explain in more detail the basic features of the model and in the following section go on to describe its dynamical rules. We then summarize the properties of the webs constructed with the model and go on to discuss the results obtained when the functional response in the population dynamics equations is modified. Finally, we discuss the implications of our findings.

MODEL

In the introduction we gave a brief summary of the ideas behind our approach to modelling food webs. In this section we will give a more detailed description of the model.

In order to define the speciation process, and more generally to characterize a species, we need to introduce features which when taken together make up a species. These features represent phenotypic and behavioral characteristics—essentially these are the keywords that one would use when defining a particular species. In the model these become integers $\alpha = 1, 2 \ldots, K$. In the present version of the model these features are not correlated; they are viewed as independent numbers that do not map into any particular observed feature in real systems. Species in the model are constructed by picking L features out of the pool of K possible features. In our simulations we generally take $L = 10$ and $K = 500$, but any two numbers which allow for the existence of a sufficient number of possible species would be acceptable.

An important quantity in the model is the score of one species against another. This will tell us which species are adapted for predation on other species, and will appear in the population dynamics equations. This score will be determined by asking what scores the constituent features have against each other. So we have to begin by defining the score of one feature, α, against another, β. This will be given by a matrix $m_{\alpha\beta}$ which is antisymmetric, $m_{\beta\alpha} = -m_{\alpha\beta}$, so that if α has a positive score against β, β will have a negative score of the same magnitude against α. The $K \times K$ matrix $m_{\alpha\beta}$ is unknown, and so we take the independent elements to be randomly drawn from a Gaussian distribution with zero mean and unit variance at the start of a simulation run.

The matrix is left unchanged during a run, but a new set of elements is drawn at the beginning of every new run. Clearly, other distributions for the elements of $m_{\alpha\beta}$ could be used, but we believe the essential results of the model would be unchanged. We will come back to this point in the concluding section.

Once the scores of features against each other have been defined, we can define the score of one species i against another species j as

$$S_{ij} = \max\left\{0, \frac{1}{L} \sum_{\alpha \in i} \sum_{\beta \in j} m_{\alpha\beta}\right\} \tag{1}$$

Thus, if species i is adapted for predation on species j, $S_{ij} > 0$, otherwise $S_{ij} = 0$. The factor $1/L$ in (1) is included for normalization purposes. The external environment is represented as an additional species 0 which is assigned a set of L features randomly at the beginning of a run and which, like the $m_{\alpha\beta}$, do not change during the course of that particular run, but are randomly chosen again at the start of subsequent runs. Finally, we also need to define the overlap between two species i and j:

$$q_{ij} = \frac{1}{L} \sum_{\alpha \in i} \sum_{\beta \in j} \delta_{\alpha\beta}, \qquad (2)$$

that is the fraction of features of species i that are also possessed by species j.

DYNAMICS

Having described the basic structure of the model, we can now go on to describe the dynamics. We begin with the evolutionary dynamics defined on the longest timescale, as discussed briefly in the introduction. At each evolutionary time step we randomly choose a species. This will be the parent species. The child species is found by randomly choosing a feature of the parent species and replacing it by another feature. Thus parent and child species have an overlap of 0.9. After checking that the new species does not already exist in the system, the population of the parent species is reduced by 1, and the new species is introduced into the community with a population of 1. The population dynamics (discussed later) is applied to the new community, and new population numbers calculated. If the addition of the new species causes the population size of any of the existing species to fall below 1, then this species is removed from the system. Similarly, if the population of the newly introduced species itself falls below 1, it is removed. This is the extinction dynamics.

To describe the dynamics on the shorter time scales, an explicit population dynamics has to be introduced. We first write down a simple balance equation for N_i, the population size of species i at time t:

$$\frac{dN_i(t)}{dt} = \lambda \sum_j N_i(t)\, g_{ij}(t) - \sum_j N_j(t)\, g_{ji}(t) - d_i N_i(t) \qquad (3)$$

Here $g_{ij}(t)$ is the functional response—the rate at which one individual of species i consumes individuals of species j, d_i is the death rate of species i and λ is the ecological efficiency which represents the fraction of the resources obtained through prey consumption which are converted to predator resources.

The balance equations (3) are in many ways the natural generalisations of those appearing in ecology textbooks (Maynard Smith, 1974, Roughgarden, 1979) for systems with one prey and one predator. The first term on the right-hand side represents the growth in numbers of species i due to predation on other species, the second term the decrease in numbers due to predation by other species, and the last term the rate of death of individuals of species i, in the absence of interactions with other species. Where there is no predator–prey relationship between species i and species j, g_{ij} is zero. For species which feed on the external environment, species 0, the first term will include a non-zero rate g_{i0}. This is how external resources are input, at a constant rate R, into the system. It should be noted that the N_i are used to denote both the population of species i and the amount of resources tied up in species i.

We have deliberately chosen the form of the equations (3) to be the same for all species. We do not want to define different equations for species that are at different positions in the web, because species may change positions as the system evolves. To completely specify the model we still need to give an expression for the functional response and values for the death rates d_i and ecological efficiency λ. We will take all death rates to be equal: $d_i = d$ for all i, and choose $d = 1$ by a suitable choice of time scale. The constant λ will be taken equal to 0.1, a value accepted by many ecologists (Pimm, 1982). The choice of functional response is a more complicated affair, and later we will discuss how changes in the functional response affect the structure of the webs that are produced. We have chosen a ratio-dependent functional response, which satisfies the logical requirement that the population dynamics do not change if identical species are pooled into a single species. If there are several predators competing for a single prey species j, our functional response is

$$g_{ij}(t) = \frac{S_{ij}N_j(t)}{bN_j(t) + \sum_k \alpha_{ki}S_{kj}N_k(t)}. \tag{4}$$

In this formula, the scores have entered the definition—they characterize the strength of predation. The competition term $\alpha_{ki} = c + (1 - c)\, q_{ki}$ with $0 \leq c < 1$ includes inter-specific competition in addition to the intra-specific competition. This ensures that the greater the similarity of the predators k to i, the greater the competition. Two parameters b and c specify the functional response.

If a predator has more than one prey, a modification to $g_{ij}(t)$ is required. This is where the shortest time scale mentioned in the introduction enters the dynamics. If one of these prey species falls on hard times and its population numbers drop, the predator would naturally choose to reduce the amount of effort it puts into preying on this particular prey species, and direct its efforts elsewhere. To model this we introduce "efforts" denoted by f_{ij}. The effort f_{ij} is the fraction of available searching time which species *i* spends preying on species *j*. By definition $\Sigma_j f_{ij} = 1$ for all *i*. It seems reasonable to assume that the effort that *i* puts into preying on *j* should be proportional to its gain, that is, $f_{ij} \propto g_{ij}$ for a fixed *i* and all *j*. Using the normalization condition for the f_{ij} leads to

$$f_{ij}(t) = \frac{g_{ij}(t)}{\sum_l g_{il}(t)}. \tag{5}$$

In fact it can be shown that this choice is an evolutionarily stable strategy (Drossel et al., 2001). Including the efforts in the functional response (4) leads to the final form which was used in our simulations:

$$g_{ij}(t) = \frac{S_{ij} f_{ij}(t) N_j(t)}{bN_j(t) + \sum_k \alpha_{ki} S_{kj} f_{kj}(t) N_k(t)}. \tag{6}$$

The simulation process can now be outlined. On the shortest time scales the number of species and population numbers of all species are fixed. Let us suppose that the population numbers were adjusted some time ago, then the dynamics consists of iteration of equations (5) and (6) at fixed population numbers, until they converge on new f_{ij} and g_{ij} appropriate to these new population numbers. The population numbers can then be updated using these f_{ij} and g_{ij} and (a discrete form of) the balance equations (3). This change in population numbers occurs at the intermediate time scale. With the new population numbers, the equations (5) and (6) can once more be iterated at fixed population numbers, until they converge on new f_{ij} and g_{ij}, and so on. Eventually the population numbers will also start to converge, and a speciation is allowed to take place. Further details on the method of simulation may be found in Drossel et al. (2001).

PROPERTIES OF THE EVOLVED FOOD WEBS

Starting with one species and the external resources, we evolved model webs using computer simulations. Figure 1 shows the species number as a function of time for four different simulation runs with four different sizes of the external resources. Time is measured in units of "speciation" or "immigration" events. One time step consists of introducing a new species and subsequently applying the population dynamics until a stationary state is reached. One can see that the species number increases

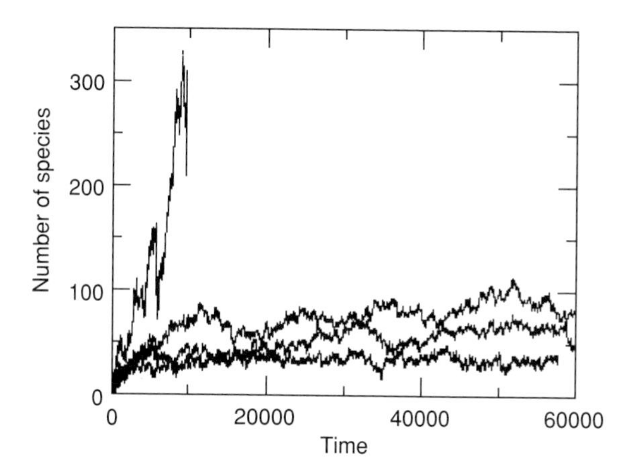

FIGURE 1 | Number of species as function of time for $c = 0.5$, $b = 0.005$, and $R = 10^4$, 10^5, 3.5×10^5, 10^6 (from bottom to top curve).

initially and eventually settles down around a stationary mean value. When there are more external resources, the size of the webs is larger. Even after long times, the composition of the food web continues to change, although the general food web structure is preserved. The food webs consist of several trophic layers, as shown in Figure 2, which gives an example of a food web obtained with our simulations. Table 1 lists several properties of the model webs, which agree well with empirical data from smaller food webs.

Figure 1 shows that not more than a few species go extinct at the same time in our model. These small "extinction avalanches," following immigration or speciation events, therefore cannot explain the large-scale extinctions seen in the fossil record. The reason is that in our model, the external resources and the model parameters are constant in time, while the large extinction events usually require a change in the external conditions. Including changes in the external conditions in our model would be a worthwhile project, which has not yet been performed.

A systematic investigation of the effect of the removal and addition of species on the stability of food webs shows that if a species is removed, this does not lead to further extinctions in approximately $\frac{2}{3}$ of the cases. We ascribe this high degree of food web stability to the features of the functional response that we have chosen. This functional response allows predators to put more effort into those prey from which they gain more energy per unit time, thereby adjusting to changes in the population sizes and the composition of the web. Species deletion stability is not correlated with complexity and the model shows extinction due to prey removal as well as extinction due to predator removal.

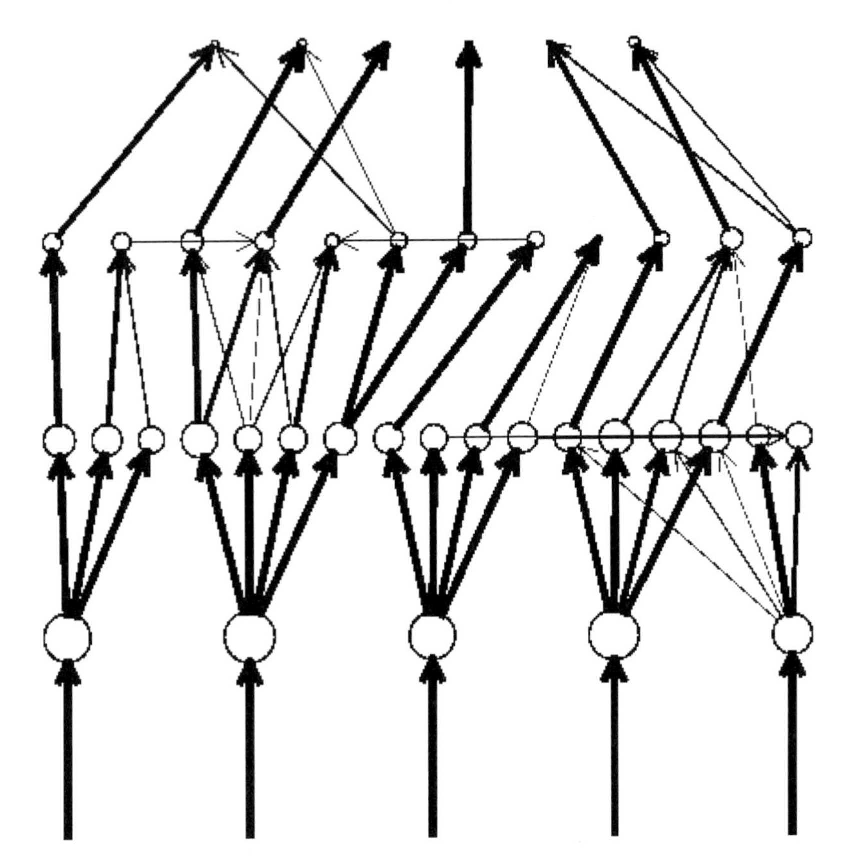

FIGURE 2 | Example of a food web resulting from the evolutionary model.

Table 1: Results of simulations of the model with $c = 0.5$ and $b = 5 \times 10^{-3}$

R	1.0×10^4	1.0×10^5	3.5×10^5	1.0×10^6
No. of species	33	57	82	270
Links per species	1.76	1.91	1.91	2.96
Av. level	1.95	2.35	2.65	3.07
Av. max. level	3.0	3.9	4.0	4.4
Basal species (%)	18	9	5	11
Intermediate species (%)	80	89	89	89
Top species (%)	2	2	6	1
Mean overlap level 1	0.32	0.34	0.31	0.27
Mean overlap level 2	0.17	0.12	0.11	0.15
Mean overlap level 3	0.19	0.09	0.09	0.12

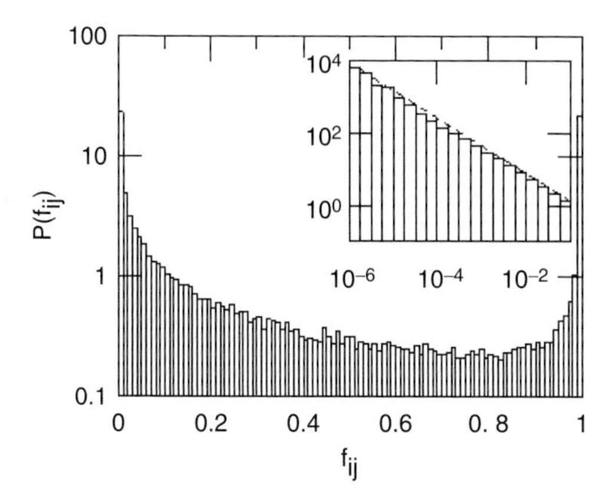

FIGURE 3 | Link strength distribution obtained from the evolutionary food web model. The inset shows the distribution for small link strength on a log-log plot. The straight line indicates a power law with an exponent −0.75.

Finally, let us discuss the distribution of link strengths obtained in our model. There has been an increasing realization that food webs have a large proportion of weak links (Paine, 1992; Tavares-Cromar and Williams, 1996; Berlow et al., 1999; Neutel et al., 2002), and that weak links tend to stabilise population dynamics (McCann et al., 1998). For this reason, we evaluated the link strength distribution in our model webs. We found that the link strength distributions were skewed towards zero and that a large fraction of links were zero. This is a highly non-trivial result, as in contrast to other work on the topic of weak links, the link strength distributions in our model are an emergent property of the system and not put in by hand. It is a strong indication that weak links are the natural outcome of long-term ecosystem evolution coupled to population dynamics. Figure 3 shows the link-strength distribution obtained with our model. The link strength was defined as the proportion of the prey in the predator's diet after the population dynamics had equilibrated. It has a large weight at weak link strength. The maximum at link strength 1 is due to the fact that many predators have a main prey from which they obtain most of their food.

RESULTS OBTAINED WITH OTHER FUNCTIONAL RESPONSES

The investigation of species deletion stability indicates that the complex structure of the model webs depends on the ability of predators to adjust their feeding behavior when changes in the food web occur. We tested this conjecture by running the computer simulations with functional responses of the general form (Arditi and Michalski, 1996).

$$g_{ij} = \frac{S_{ij} N_j}{1 + \sum_k b_{ik} N_k + \sum_l c_{il} N_l}. \tag{7}$$

If all b_{ik} and all c_{il} are zero, we have a Lotka-Volterra functional response. If the b_{ik} are nonzero, we have a Holling type II functional response, which implies saturation of consumption rates at high prey abundance. (The sum in the denominator is taken over all prey k of species i.) If also the c_{il} are nonzero, we obtain a generalized Beddington form, where the second sum is taken over all those predator species i that share a prey with j. We chose the c_{il} such that individuals belonging to the same species competed more strongly with each other than individuals belonging to different species.

Figure 4 shows a typical web obtained with any of these three models. This web consists of just one trophic level, with all species feeding on the external resources, and with many links between the species. Such a food web structure occurs even if we start our simulations with an artificially constructed complex web consisting of several trophic levels. It appears that predators on higher trophic levels cannot survive when the food web structure changes. By contrast, species with a link to the external resources cannot easily go extinct because they have a permanent resource available. It also appears that these functional responses allow too many species to feed on the same food source (here the external resources). The model misses mechanisms that prevent possible links (i.e., when $S_{ij} > 0$) from being realized and that allow predators to adjust to a modified food web structure.

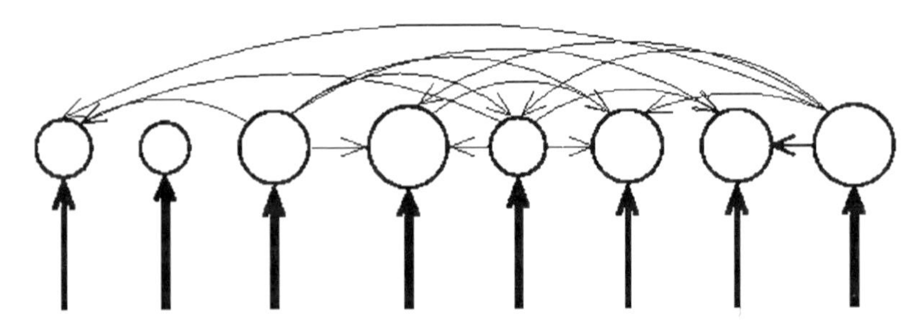

FIGURE 4 | Example of a food web resulting from an evolutionary model with Lotka-Volterrra population dynamics. The arrow direction indicates the flow of resources, and the arrow thickness is a measure of link strength. Links are only drawn if a species obtains more than 1 percent of its food through that link. The radius of the circle increases logarithmically with population size.

We introduced such mechanisms by replacing the interactions S_{ij} with adjusted interactions $S_{ij}' = S_j^{\max}(1 - (S_j^{\max} - S_{ij})/\delta)$ (Caldarelli et al., 1998), with δ being a small parameter and $S_j^{\max}$ being the largest interaction against j. Negative S_{ij}' are set to zero. Figure 5 shows a food web obtained from Lotka-Volterra dynamics with this artificial constraint.

Table 2 shows the mean number of species, of links per species, and the mean occupation numbers of the trophic levels for this model, for $R = 1 \times 10^4$, $\lambda = 0.1$, $c = 1.0$ and $\delta = 0.2$. For comparison, the results for the original Lotka-Volterra model with the parameters for $R = 1 \times 10^4$, $\lambda = 0.1$, and $c = 3.0$ are also shown. Including direct inter-specific competition or using Holling and Beddington forms with the same type of adjusted interactions gave similar web structures.

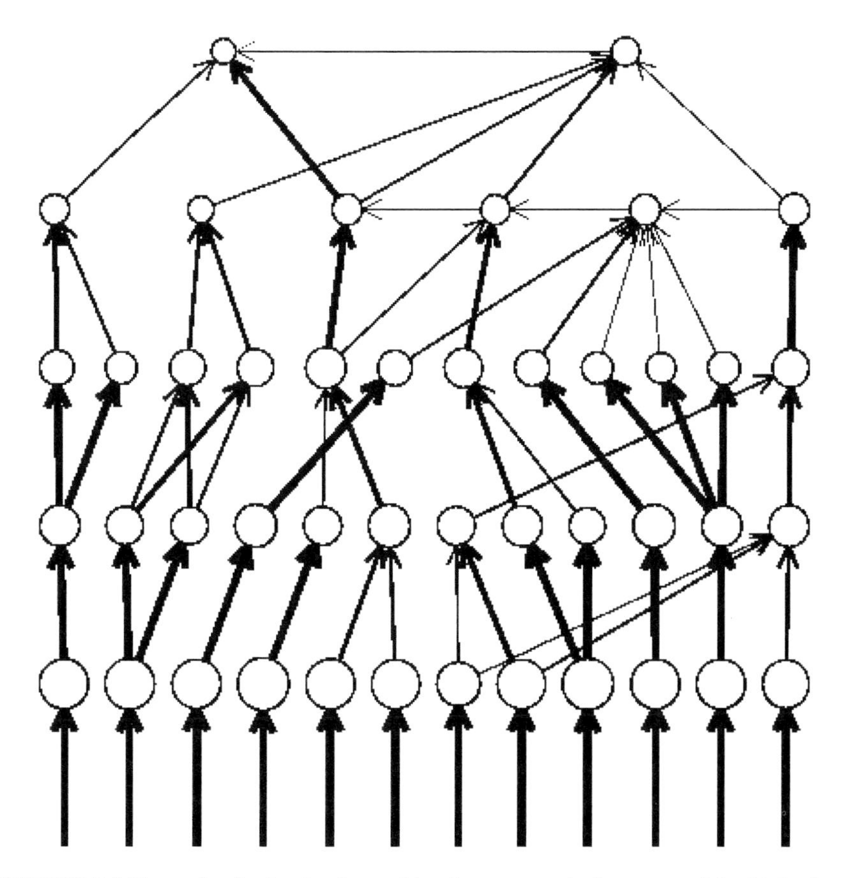

FIGURE 5 | Example of a food web resulting from an evolutionary model with Lotka-Volterra population dynamics and adjusted interactions ($\delta = 0.2$). The same conventions apply as for Figure 4 except that vertical position is now proportional to the average path length from the species to the environment weighted by diet proportions.

Table 2: Food web statistics for the two Lotka-Volterra models without and with adjusted interactions. The results are averaged over ten different simulations (lasting 200000 iterations) and over the last 20000 iterations of each simulation. The quantities in brackets give standard deviations over the ten runs for the number of species and links per species. Only links between non-environment species that constituted greater than 1% of the predator's diet were counted in the calculation of the links per species.

			Trophic level					
Model	Number of species	Links per species	1	2	3	4	5	6
			Number of species					
			Frequency of occupation					
Lotka-Volterra without adjusted interactions	40.6 (2.5)	2.3 (0.1)	40.5	1.6				
			1.0	0.061				
Lotka-Volterra with adjusted interactions	69.8 (20.4)	1.4 (0.1)	20.9	22.1	19.1	7.5	1.4	1.0
			1.0	1.0	1.0	0.987	0.14	0.003

This shows that the ability of predators to concentrate on the prey that they are best suited to exploit, rather than on all possible prey, is essential for the production of realistic food web structures. The precise form of the functional response is not important, as long as the functional response that is chosen allows for predator flexibility. We have also investigated the link strength distribution in the models with modified interactions and found that they have a large weight at weak links, just as for the model with the ratio-dependent functional response (Drossel et al., 2004).

DISCUSSION AND CONCLUSIONS

We have shown that a food web model that includes long-term changes in the food web structure, as well as population dynamics, is capable of generating large model webs that share many features with real food webs. Among these features are the existence of several trophic levels, a high proportion of intermediate species, a high stability under species deletion, a large proportion of weak or muted links, and a flexible response of predators to changes in the prey populations. An important requirement on the population dynamics of the model is that it must allow for predator flexibility. Only if this condition is implemented when the model is constructed, is a complex food web structure obtained which persists in time.

Since all the web properties are emergent, all measurements made by using the model webs are tests of the model assumptions themselves, and not of any additional input made by the modeller. This should be contrasted with the situation found in other approaches. For instance, in some assembly models the modeller may specify the nature of an introduced species and so effectively determine its level in the web. In other models the number of links per species may be specified. In our approach, the position of species in the web, as well as whether they thrive or go into decline, is a mixture of random factors and the deterministic population dynamics. Since we average over many different simulation runs, we expect random factors to average out, so that measured average quantities follow directly from the nature of the model and the choice of parameters.

At the heart of the approach is the belief that aspects such as the precise way in which species are defined or the specific nature of the distribution of the random elements of the feature matrix $m_{\alpha\beta}$ are not going to have a significant impact on the nature of the food web. This is because, for example, it is expected that as long as there are very large numbers of distinct species that may be formed, the precise way that these are defined is not going to influence web topology. On the other hand, we would expect that those aspects of the population dynamics that reflect significant biological mechanisms: predator feeding rates, strong competition—especially between similar species, adaptive foraging, and so on, would be important in determining the structure of the webs. This is the intuition that we have developed by working with this model, but of course this needs to be tested. From our comments in the preceding paragraph, it seems clear that we can directly test these claims by systematically varying each basic aspect of the model and determining whether the measured web characteristics change significantly as a result of this variation. This is a lengthy investigation, which we have not yet completed. We have carried out a reasonably systematic investigation of the result of changes in the nature of the population dynamics. We are in the process of testing the robustness of the web structure to changes in the specification of the species, features, feature matrix, etc. Preliminary results indicate that, as expected, many of these details will be unimportant in determining food web properties.

There are several advantages in the approach to the modelling of food webs that we have outlined here. The continual monitoring of the results of changing the basic assumptions of the model should reveal which aspects have to be specially tuned to achieve good agreement with data, and which aspects are insensitive to the precise choice of these basic assumptions. This in turn should indicate what are the important processes involved in generating food webs. In addition, the model produces webs that can be observed changing dynamically, and so the nature of the processes on short time scales can be studied directly. Moreover, as pointed out in the first paragraph of this article, experiments can be carried out on the model webs produced to assess the consequences of changes made to the web structure. Perhaps the greatest strength of generating food webs in the way we have described here, is that any property of webs hinted at by empirical studies can be quickly and easily investigated using the webs generated by the model. We look forward to carrying out such investigations, prompted by suggestions from empirical colleagues, in the future.

ACKNOWLEDGMENTS

We thank Chris Quince for providing Figures 2, 3, and 5, and for useful discussions.

3.2 | THE INFLUENCE OF INDIVIDUAL GROWTH AND DEVELOPMENT ON THE STRUCTURE OF ECOLOGICAL COMMUNITIES

André M. de Roos and Lennart Persson

Following Elton's (1927) seminal ideas a large body of theory about the dynamics of ecological communities is based on a conceptualization, in which species constitute trophic units, engaged in a reticulate network of consumer-resource interactions. Most theory on community dynamics is moreover based on the assumption that the interaction between two species can be described by a linear function of their densities. The resulting Lotka-Volterra class of community models have been studied extensively, especially focusing on questions regarding the relation between properties of the species network (extent, connectivity, overall interaction strength) and the stability of the ecological community (May, 1973; Pimm, 1982; Hall and Raffaelli, 1993; McCann, 2000). On the other hand, the food chain models studied by Fretwell (1977) and Oksanen et al. (1981) represent foraging interactions between species more explicitly and mechanistically than Lotka-Volterra models, as both functional and numerical response of consumer species are accounted for. In contrast with Lotka-Volterra models, these functions are typically non-linear. The models developed by Yodzis and Innes (1992) extend this approach by relating parameters of the functional and numerical response to individual-level traits of consumers, such as their characteristic body size.

Even though they do not explicitly represent individuals at all, these models of ecological communities can be argued to account for two processes at the level of the individual organism: reproduction and mortality. After reproduction and mortality the most prominent process in an individual's life history is growth and development. In virtually all species there is a delay between the birth of an organism and the onset of its reproduction, representing the prime aspect of individual development. Moreover, the duration of this delay may vary with environmental conditions, most notably food density, as maturation tends to require reaching a threshold body size (Kooijman, 2000) and growth rates are generally food dependent. In addition, the growth in body size that is apparent in the majority of species (Werner, 1988) generally increases individual foraging rates and decreases predation risks. Few studies have considered how these consequences of individual growth and development, which obviously affect the central processes in a food web, translate into effects on the structure and dynamics of ecological communities (Pimm and Rice, 1987). Studies on the influence of stage structure for population dynamics address to some extent the issue of individual growth and development. These studies have often focused on consumer-resource interactions (see Murdoch et al., 2003, for a sythesis), in which the stage structure of the consumer population is accounted for, or on insect host-parasitoid interactions with stage structure in host and parasitoids (Murdoch et al., 1987; Godfray and Hassell, 1989; Briggs, 1993). Almost all of these studies deal with life stages of fixed duration and primarily focus on the likelihood of population cycles to occur. They consequently do not address the question whether and how individual growth and development affects the structure of ecological communities.

In this paper we present a preliminary overview of possible consequences for ecological communities that may arise from individual growth and development. The consequences reported here arise because of the dependence of growth on ambient food densities and hence indirectly on consumer density. We show that the interplay of food-dependent growth in a consumer species with size-dependent mortality imposed by a predator on this consumer leads to emergent Allee effects for the predator, such that under identical environmental conditions community states with and without the predator are both stable equilibria. In addition, we show that two predator species, foraging on different life stages of the consumers, may greatly facilitate each other's persistence. The key mechanism for these effects to occur is that the size-distribution of the consumer species changes through a subtle interplay of food-dependent growth and size-selective mortality. Predation mortality decreases the overall abundance of the consumer species, but this overall decline may be more than compensated for by the changes in size distribution, such that specific size classes may increase in abundance due to predation, even the size class suffering from the predation mortality itself.

THE MODEL

De Roos and Persson (2002) present one of the simplest models, in which the interplay between food-dependent growth and size-dependent predation mortality can be studied. The model describes the dynamics of a resource R, a consumer size-distribution $c(t,l)$ and a predator density P with the following set of equations:

$$\begin{aligned} \frac{dP}{dt} &= \left(\epsilon \frac{aB}{1 + aT_h B} - \delta \right) P \\ \frac{\partial c(t,l)}{\partial t} + \frac{\partial g(R,l)\, c(t,l)}{\partial l} &= -(\mu + m(P))\, c(t,l) \\ g(R,l_b)\, c(t,l_b) &= \int_{l_j}^{l_m} b(R,l)\, c(t,l)\, dl \\ \frac{dR}{dt} &= \rho (K - R) - \int_{l_b}^{l_m} I(R,l)\, c(t,l)\, dl \end{aligned} \tag{1}$$

Consumers are born at length l_b, mature on reaching length l_j and may reach the maximum length l_m under very high food conditions. Resource ingestion of consumers is assumed proportional to their squared length with proportionality constant I_m and follows a type II functional response to resource biomass: $I(R,l) = I_m l^2 R/(R_h + R)$ with half-saturation constant R_h. A fixed fraction of ingested food is channeled to reproduction, while the remainder is spent on growth plus maintenance. Maintenance is assumed proportional to l^3 and takes precedence over growth, which hence follows a von Bertalanffy growth law: $g(R,l) = \gamma(l_m R/(R_h + R) - l)$ with l_m and γ representing the maximum length under very high food densities and the growth rate, respectively. After reaching the maturation length l_j, individuals produce offspring at a rate $b(R,l) = r_m l^2 R/(R_h + R)$ with proportionality constant r_m.

Consumer mortality equals the sum of the background mortality μ and the predator-induced mortality $m(P)$. Predation mortality is assumed larger than 0 only for individuals within a specific size range from l_l to l_h. The variable B denotes the biomass of these vulnerable consumers, which can be computed as an integral over the consumer size distribution $c(t,l)$, weighted by the consumer length-weight relation, βl^3:

$$B = \int_{l_l}^{lh} \beta l^3 c(t,l)\, dl \tag{2}$$

In this chapter we only consider predators with a negative size-selectivity, which forage on all prey individuals below an upper size threshold l_h (and $l_l = 0$), or a positive size-selectivity, which forage on all prey individuals above a lower size threshold l_l (and $l_h = \infty$).

Predators follow simple consumer-resource dynamics, experiencing a background mortality rate δ, while foraging on consumers with length between l_l and l_h with attack rate a, handling time T_h and conversion factor ε. The intake rate of consumer biomass by a single predator individual thus equals $aB/(1+aT_hB)$. Likewise, predation mortality of consumers with length between l_l and l_h follows:

$$m(P) = \frac{aP}{1 + aT_h B} \tag{3}$$

while $m(P) = 0$ for other individuals. Resource regrowth is assumed to follow semi-chemostat dynamics (Persson et al., 1998) with maximum resource biomass density K and flow-through rate ρ.

RESULTS

Figure 1 shows the possible equilibrium states at different levels of system productivity (defined as ρK) for the food chain model, in case predators are negatively size-selective. A stable equilibrium of consumers and resource can occur for system productivity levels below $2.5\ 10^{-5}$ g.L^{-1}.day^{-1} (the invasion threshold) for the default parameter set. Below this invasion threshold low densities of predators are incapable of establishing themselves in the consumer-resource equilibrium. A stable equilibrium of predators, consumers and resource can occur for system productivity levels larger than $0.8\ 10^{-5}$ g.L^{-1}.day^{-1} (the persistence threshold) for the default parameter set. Below this persistence threshold predators are unable to persist and the consumer-resource equilibrium is the only possible steady state. In between the persistence and invasion threshold both a stable equilibrium state with and without

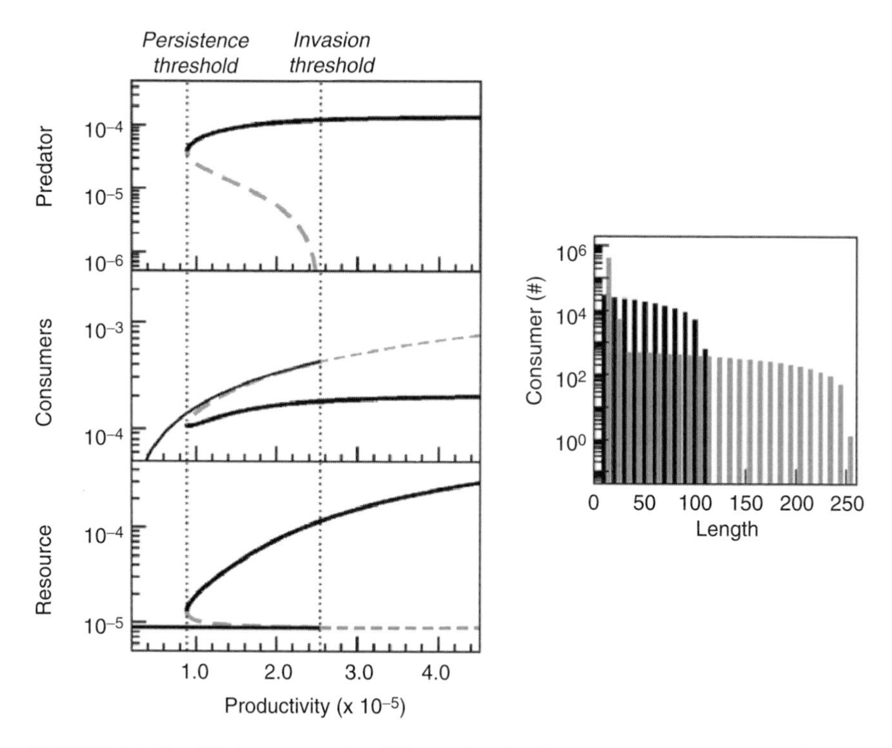

FIGURE 1 | Equilibrium states for different levels of system productivity (g.L^{-1}.day^{-1}) in case predators have a negative size-selectivity ($l_l = 0$, $l_h = 27$ mm). **Left:** Variation in predator density (top panel; individuals.L^{-1}), consumer biomass (middle panel; g.L^{-1}) and resource biomass (bottom panel; g.L^{-1}). **Thin lines:** equilibria with only consumers and basic resource. **Thick lines:** equilibria with all three trophic levels. Solid black lines represent stable equilibria, dashed grey lines unstable ones. Alternative, stable equilibria with and without predators occur between the invasion and persistence threshold (vertical dotted lines). **Right:** Consumer size distribution in an equilibrium with (grey bars) and without predators (black bars) at a system productivity of 2.0 10^{-5} g.L^{-1}.day^{-1}. Other parameters: l_b = 7 mm, l_j = 110 mm, l_m = 300 mm, I_m = 1.0 10^{-4} g.day^{-1}.mm^{-2}, R_h = 1.5 10^{-5} g.L^{-1}, r_m = 0.003 day^{-1}.mm^{-2}, γ = 0.006 day^{-1}, μ = 0.01 day^{-1}, β = 9.010^{-6} g.mm^{-3}, ρ = 0.1 day^{-1}, ε = 0.5 g^{-1}, a = 5000 L.day^{-1}, T_h = 0.1 day.g^{-1} and δ = 0.01 day^{-1}. These parameters mimic the life history characteristics of roach (*Rutilus rutilus*), feeding on *Daphnia* spp., while being predated by perch (*Perca fluviatilis*: De Roos and Persson, 2002).

predators can occur, while above the invasion threshold only an equilibrium state with predators, consumers, and resource is possible.

The reason for that the community can occur in two different states, with and without predators, is related to the difference in size distribution of the consumer population (Figure 1, right panel). In the absence of predators, this size distribution is stunted (i.e., relatively flat and constricted to a narrow size range) with only a few individuals reaching body sizes above the maturation threshold. In this state the consumer population is regulated by intraspecific competition for food, which slows down individual growth and development (from here on referred to as development regulation). As a consequence, individuals spend a long time in the juvenile state and most die before maturing. The individuals that do mature, however, have a sufficiently high fecundity to make up for the low through-stage survival as a juvenile. When predators are present the size-distribution extends over a much wider size range with the largest consumers reaching sizes well over two times the maturation size. In addition, the size distribution shows a distinct peak at small body sizes. Juvenile growth and development is rapid owing to the high resource densities in this equilibrium (see Figure 1, left-bottom panel). The total density of consumers is significantly lower when predators are present, but this decline is due to a disproportionate decrease in the density of large juveniles. The densities of small juveniles and large adult consumers are higher in the presence of predators. For specific size-classes the change in consumer size-distribution imposed by the predator more than compensates for the decline in overall consumer abundance. Most interestingly, the density of small consumers that are vulnerable to predation is higher in the presence of predators than in their absence. Predators thus exert a net positive effect on their own food density. This positive feedback of the predators comes about by a subtle interplay between the size-selective mortality they impose and the food-dependent growth of the consumers. The size-selective predation mortality decreases intraspecific competition among consumers, especially among the larger juveniles, which are much less abundant in the presence of predators. As a consequence, individual growth is rapid and many individuals reach maturation and grow to sizes well above the maturation size. Due to a combination of high resource levels and large body sizes these adults have a very high fecundity and produce large numbers of offspring. In the presence of predators total population fecundity is hence significantly larger. This leads to an increase in the density of small consumers, on which the predators can forage and impose a sufficiently high mortality to limit their recruitment to larger size classes.

Figure 2 shows that bistability between stable equilibria with and without predators is also possible when predators are positively size-selective. A stable consumer-resource equilibrium occurs for system productivities below an invasion threshold (2.5 10^{-5} g.L^{-1}.day^{-1}) that is higher than the persistence threshold (2.1$10^{-5}$ g.L^{-1}.day^{-1}), above which a stable equilibrium state with predators, consumers, and resource can occur. Clearly, for positively size-selective predators the range of system productivities with bistability between the persistence and invasion threshold is smaller than for negatively size-selective predators. The bistability between stable

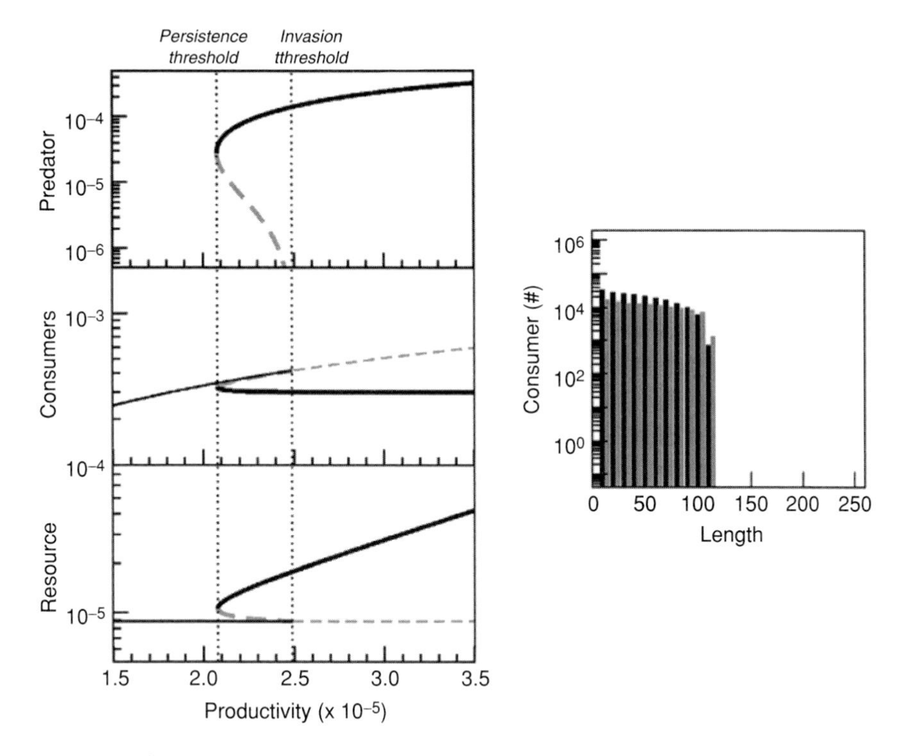

FIGURE 2 | As Figure 1, but for predators with a positive size-selectivity ($l_l = 109$ mm, $l_h = \infty$). The consumer size distributions shown in the right panel are for a system productivity of 2.3 10^{-5} g.L^{-1}.day^{-1}.

equilibrium states with and without predators is again connected to a change in consumer size-distribution, but this change is not the same as for negatively size-selective predators. In the presence of positively size-selective predators adult consumers reach sizes up to 1.5 times the maturation size, which is significantly larger than in the absence of predators. Survival of adult consumers is, however, low due to the high predation mortality. As a consequence, in the presence of predators the consumer size-distribution has a long, but very thin tail (too thin to be visible when expressed as numbers of individuals as in Figure 2) and only the density of adults with body sizes just above the maturation size is significantly higher, while the density of individuals in all juvenile size classes is smaller. Predators that are positively size-selective thus exert a positive net effect on their own food density (large juveniles and adult consumers), but negatively affect the density of all smaller-sized consumers.

The occurrence of a stable equilibrium state with and without predators in the food chain with size-structured consumers for otherwise identical conditions has been dubbed emergent Allee effect (De Roos et al., 2003a). None of the usual mechanisms that give rise to an Allee effect, such as sexual reproduction, a complex mating system, or social foraging behavior (Courchamp et al., 1999; Stephens and Sutherland, 1999) are incorporated in the food chain model, which only accounts for purely exploitative foraging by both consumers and predators. Nonetheless, in the bistability range predators experience a reduced population growth rate at low predator densities and are hence incapable of establishing themselves in the consumer-resource community. At higher densities, however, predators are capable of building up a population and attain a stable equilibrium state. This Allee effect emerges from the subtle interplay of the size-dependent mortality and the food-dependent growth experienced by the consumers. Alternative stable equilibria do not occur if predators forage on all consumer individuals or only on the intermediately sized consumers (De Roos et al., 2003a). Also, the Allee effects only emerge if in the absence of predators the prey population is regulated by juvenile development in such a way that a reduction in density of the competitively dominant, juvenile prey leads to an increase in adult recruitment rate. The relationship between adult recruitment rate and the dominant, juvenile prey density should hence be hump shaped. Both a high potential fecundity and low background mortality, which is generally the case for populations of zooplankton species like *Daphnia* and various planctivorous fish species, will promote the occurrence of emergent Allee effect (De Roos et al., 2003a). The emergent Allee effect is therefore a consequence not of the predator's life history, but of that of its prey. It crucially depends on the fact that individual growth and development change with changing resource conditions: if growth is independent of food density, the body size of an individual would be a function of its age and this relationship would be the same under all conditions. A fixed size-age relationship prevents the shifts in consumer size-distribution that cause the emergent Allee effect. Age-structured variants of the food chain model consequently do not show the Allee effect (De Roos and Persson, 2002).

Due to the emergent Allee effect, size-selective predators may exhibit catastrophic collapses of their population (i.e., rapid declines from high population densities to extinction levels), when conditions change in such a way that the persistence threshold is crossed. Decreasing levels of system productivity and increasing levels of predator mortality (De Roos and Persson, 2002) exemplify shifts that will give rise to such collapses. Because of the bistability between an equilibrium state with and without predators, these collapses are not reversible unless conditions are changed beyond the invasion threshold.

Apart from the catastrophic population collapses and the possibility of multiple stable equilibrium states under otherwise identical conditions, the results shown in Figure 1 are indicative for further community consequences of individual growth and development. Negatively size-selective predators induce changes in consumer size-distribution such that the density of both small and large consumers is increased (see Figure 1; right panel). Hence, these predators not only positively affect their own food conditions, as previously discussed, but also increase food availability for other predator species that forage on large juveniles and adults of the same consumer species. Negatively size-selective predators may therefore facilitate the persistence of positively size-selective predators on the same prey. Positively size-selective predators, on the other hand, exert a decreasing influence on the density of all smaller sized consumers (see Figure 2; right panel) and hence only exert a negative, competitive influence. Van Kooten et al. (2004) study this emergent facilitation between competing predators in a simpler food chain model, which only accounts for two size classes of consumers. In this setting it is shown that a negatively size-selective predator species can act as a facilitator, such that in its presence a positively size-selective competitor can endure mortality levels that are up to an order of magnitude larger than can be tolerated in its absence. Due to food-dependent consumer growth the persistence of a positively size-selective predator may thus crucially depend on the presence of a negatively size-selective forager on the same prey species.

Figure 3 shows the possible equilibrium states for the food chain model with a positively size-selective predator species for different values of the minimum prey size l_l that is vulnerable to predation. Thus, predators forage on all prey with a size larger than l_l. At low values of l_l, when almost all consumers are vulnerable to predation, predators exert a strong top-down control, leading to low consumer and high resource densities. With an increasing threshold value l_l this top-down control relaxes, consumer density increases and resource densities decrease as a result. Since predators take only larger sized consumers they derive a higher energy yield from consuming a single prey individual and thus exploit their prey population more prudently. Predator abundance hence increases as well with an increase in the threshold value l_l. These results exemplify a situation in which a relaxation of top-down control actually increases the density of the two adjacent trophic levels involved. This increase in predator and consumer density and the decrease in resource density continue until the threshold value l_l is slightly larger than the size l_j at which consumers start to allocate to reproduction. Here a similar collapse of the equilibrium state with predators is observed as when varying the system productivity (*cf.* Figure 1). The collapse occurs because predators that forage on adults cannot exert sufficient top-down control any more and the consumer population

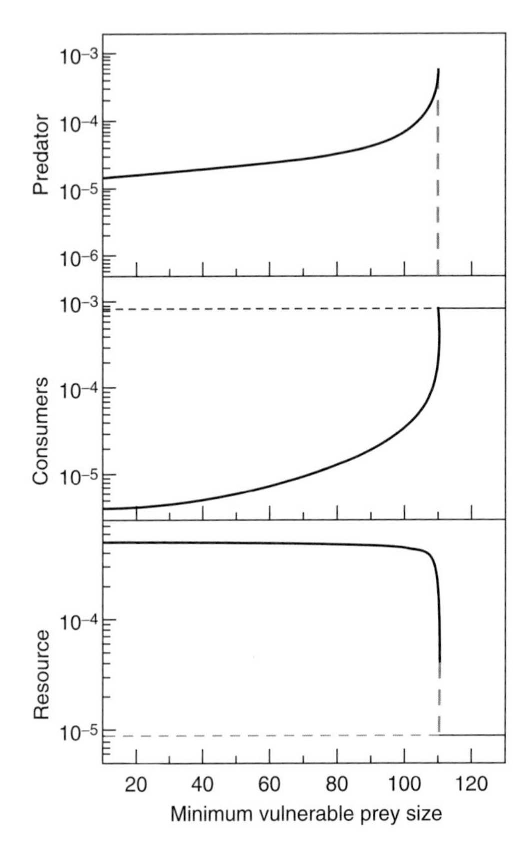

FIGURE 3 | Equilibrium states for different values of the minimum vulnerable prey size l_l in case predators have a positive size-selectivity ($l_h = \infty$). Line styles and parameters as in Figure 1, but for the system productivity which equals 5.0 10^{-5} g.L^{-1}.day^{-1}.

becomes regulated entirely by intraspecific competition for food. As shown in Figures 1 and 2, consumers that are not subjected to any predation mortality become development regulated and only reach sizes just above the maturation size. Predators that only forage on adult consumers with a larger size experience a zero food density in the consumer-resource equilibrium state and hence cannot sustain themselves. The consequence of these results is that predators, which depend for their persistence on adult consumers, will never be able to persist on their own, irrespective of the system productivity or the background mortality they experience. For their persistence other predator species have to be present, which forage on smaller, juvenile stages of the consumer and change its size-distribution such that the population includes also substantial numbers of adult consumers. The key element is that a sufficiently high juvenile mortality is required for the occurrence of substantial densities of adult consumers.

DISCUSSION AND CONCLUSIONS

That predators can have significant effects on the size distribution of their prey has been known since the work of Brooks and Dodson (1965). A number of studies (Tonn et al., 1992; Brönmark et al., 1995) have moreover shown that planctivorous fish populations indeed become stunted in the absence of predators, spanning a smaller body size range than in their presence. The analysis of the food chain model with size-structured consumers reveals the possible community consequences of such size-selective predation in combination with food-dependent individual growth and development. Traditional food web models have not revealed these consequences, as these models only account for reproduction and death of individual organisms.

Two classes of community consequences were discussed, which can be referred to as emergent Allee effects and emergent facilitation. Since the emergent Allee effect may lead to catastrophic collapses and a lack of re-establishment of the predator population in the resulting consumer-resource equilibrium state, it has been argued to offer a plausible explanation for the recent collapses of the cod populations in the Northwest Atlantic (De Roos and Persson, 2002). Drastic ecosystem changes, especially involving cod's major prey species in the area, capelin, have been observed in the Northwest Atlantic after collapse of the cod stocks in the early 1990s (Carscadden et al., 2001). The food chain model with negatively size-selective predators predicts that after the collapse of the cod population (i) the total capelin biomass should be higher, while (ii) its size distribution should be more stunted, and (iii) reproducing capelin should be smaller. These predictions are in line with the observed changes in the capelin population after the cod collapse (Carscadden et al., 2001). If applicable, the model might explain why the cod populations in the Northwest Atlantic have failed to recover from their collapse even after imposing a ban on commercial cod fisheries in the beginning of the 1990s.

Through emergent facilitation, predators that forage on larger juvenile and adult consumers may be able to persist under a significantly wider range of conditions in the presence of predators foraging on smaller juveniles. Adult-specialized predators, which exclusively forage on adult prey, can even not persist at all in the absence of predator species that forage on juvenile prey (juvenile-specialized predators). These results reveal the very different roles that juvenile- and adult-specialized predators play in structuring the ecological community they are part of. Juvenile-specialized predators are predicted to have a much larger influence on the size-distribution of their prey, open up possibilities for other predator species and thus promote the diversity of ecological communities. Persistence of adult-specialized predators, on the other hand, requires the presence of other predator species, which makes them dependent on ecological diversity. On the basis of these insights it might also be hypothesized that predators, which are specialists on the adults of a single prey species, will not exist or at least be rare in comparison with specialists on juveniles of a single prey species. We postulate that adult-specialized predators will tend to be generalists, living from a range of different prey species, while specialist predators will either forage on all stages of their prey or primarily on juveniles.

Through changes in size distribution and emergent facilitation in particular, individual growth and development may thus lead to a cascading dependence of consumers within a trophic level on each other's presence, as is shown in this chapter. In addition, since juvenile stages of predators often forage on the same resource as their future prey, these facilitating effects may indirectly also affect species at higher trophic levels. Many consequences of food-dependent individual growth and development on the structure of ecological communities therefore remain to be uncovered in future studies.

ACKNOWLEDGMENTS

This research was funded by the Netherlands Organization for Scientific Research, the Swedish Research Council, and the Swedish Research Council for Environment, Agricultural Sciences and Spatial Planning.

3.3 | LINKING FLEXIBLE FOOD WEB STRUCTURE TO POPULATION STABILITY: A THEORETICAL CONSIDERATION ON ADAPTIVE FOOD WEBS

Michio Kondoh

In nature a large number of species are connected by trophic interactions, resulting in a complex food web network (Warren, 1989; Winemiller, 1990; Martinez, 1991; Polis, 1991). Ecological theory, on the other hand, often derived from dynamic mathematical models suggesting that populations are less stable in more complex food webs (May, 1972; Pimm, 1991). According to theory, a population is more likely to become extinct if it is a component in a web with more species (May, 1972; Gilpin, 1975; Chen and Cohen, 2001a), denser trophic links (May, 1972; Gilpin, 1975; Chen and Cohen, 2001a), longer chain length (Pimm and Lawton, 1977), or more omnivory links (Pimm and Lawton, 1978). If, as theory predicts, complexity destabilizes populations, complex food webs should not persist and their occurrence should be infrequent in nature. The apparent contradiction between observation and theory gives rise to the challenging question of how populations persist in complex food webs (DeAngelis, 1975; Lawlor, 1978; Yodzis, 1981a; de Ruiter et al., 1995; McCann et al., 1998; Haydon, 2000; Chen and Cohen, 2001a; Neutel et al., 2002; Kondoh, 2003a, b).

The majority of theoretical studies encompassing the complexity-stability issue have assumed, either explicitly or implicitly, a static food web linkage pattern (May, 1972; DeAngelis, 1975; Gilpin, 1975; Pimm and Lawton, 1977; Lawlor, 1978; Pimm and Lawton, 1978; Yodzis, 1981a; de Ruiter et al., 1995; Haydon, 2000; Neutel et al., 2002). In these studies a food web model, from which population dynamics are evaluated, is usually constructed with given architectural characteristics such as connectance, link distributions, and interaction strengths. The theoretical contribution of this "static architecture" approach in food web studies is unquestionable. It provides a clear insight into how food web architecture affects population dynamics and has catalyzed fruitful discussion on complexity-stability issues. At the same time, this approach has constrained our consideration of various ecological questions to these limited cases where the basic architecture of food webs does not change over time (Paine, 1988). This is despite the well-recognized fact that a food web changes its linkage pattern over time and that food web complexity can be timescale dependent (Winemiller, 1990).

One of the forces that drive structural changes of food webs is adaptation, a distinguishing characteristic of living organisms. Adaptation occurs, at least, at two different biological levels—population and individual. A population consists of individuals with varying genotypes whose relative abundances change over time according to their relative contributions to the next generation. This results in phenotypic shifts at population level. At individual level, organisms learn through their experience and modify their behavior (Hughes, 1990). If a trait that influences the strength of trophic interactions is controlled by adaptation, then the food web architecture should change in a way that is favored by the adaptation. Such adaptation-driving changes in food web structure potentially influence population dynamics (Holling, 1959a; Abrams, 1982, 1984, 1992) and thus complexity-stability relationship (Pelletier, 2000; Krivan, 2002; Kondoh, 2003a, b).

An example of an adaptive behavior that modifies trophic interactions and food web architecture is predator diet choice (Emlen, 1966; MacArthur and Pianka, 1966). A predator often only consumes diets with higher quality or quantity from a set of available nutritionally-substitutable diets (Stephens and Krebs, 1986). Food web models with even relatively simple structure suggest that food web flexibility arising from foraging adaptation has a major effect on population dynamics and community structure (Tansky, 1978; Teramoto et al., 1979; Holt, 1983; Sih, 1984; MacNamara and Houston, 1987; Wilson and Yoshimura, 1994; Abrams and Matsuda, 1996; Sutherland, 1996; Holt and Polis, 1997; McCann and Hastings, 1997; Krivan, 2000; Kondoh, 2003a). Adaptive diet shift not only inhibits rapid growth of prey, but it also allows minor prey to rally without predation pressure and thus not fall to levels too close to zero. This prevents apparent competition (Holt, 1977) leading to species extinctions (Tansky, 1978; Teramoto et al., 1979; Abrams and Matsuda, 1996), and enhances coexistence of intraguild prey and predator (Holt and Polis, 1997; McCann and Hastings, 1997; Krivan, 2000) or competing prey species (Wilson and Yoshimura, 1994; McCann and Hastings, 1997).

In this study, I use a food web model that incorporates both population dynamics and adaptive dynamics to show how adaptive diet choice alters the previously-held static view of the complexity-stability relationship. In the model presented in this study, food web architectural properties such as connectance, link distribution, and interaction strength are not given *a priori*. Instead, they emerge as a consequence of foraging adaptation, which determines actual diets that are used from a given potential diet range. The complexity–stability relationship is determined by (i) examining how population stability changes as organisms' capability of foraging adaptation and basic food web architecture (which defines potential prey range) changes; (ii) by identifying key structural properties that characterize adaptive food webs and show how these structural properties are influenced by changing adaptation capability or potential connection; and (iii) by taking these results together to predict the emerging relationships between the food web's structural properties and population stability.

MODEL

Consider a food web comprising of B basal species and (N -B) non-basal species. The former are self-reproducing and have no resource species within the web, while the latter species rely energetically on other species. Each species is characterized by a species-specific body size, V_i (i = 1 to N), which determines its trophic role (Warren and Lawton, 1987; Cohen et al., 1993a) and physiological characteristics (references in Yodzis and Innes, 1992). A body size is chosen from the lognormal distribution (m = 0.5630 and SD = 2.2152) representing aquatic invertebrates (Cohen et al., 1993a). The smallest B species are assumed to be basal species. Species are numbered in size order from the smallest species 1 to the largest species N.

Trophic links are potential links that may be activated or inactivated according to an organism's diet choice. The potential trophic roles of two directly connected species are determined by their relative body sizes: the larger species consumes the smaller one (note that this generates cascade-model-type food webs, Cohen et al., 1990a). There is no cannibalism. To confirm that every non-basal species has at least one potential resource species, a species with smaller body sizes is randomly chosen from a web as a resource for each non-basal species. As I consider a web with only one basal species ($B = 1$) in the present chapter, this assumption prevents a basal species without a consumer. In addition, with this assumption, all species are included in a single food web and separated food webs are never generated. Each of the remaining $[N(N-1)/2 - (N-B)]$ pairs is connected with connection probability C. Thus, expected potential connectance (L / S^2) of a food web with N species and connection probability C is given by $[C\{N(N-1)/2 - (N-B)\} + (N-B)] / N^2$.

The dynamics of biomass of species i ($1 \ldots N$), X_i, is described by:

$$\frac{dX_i}{dt} = X_i\left[R_i\left(1 - \frac{X_i}{K_i}\right) - T_i + E_i - F_i\right] \tag{1}$$

where R_i is set to a maximum production rate, J_i (> 0), for basal species ($i = 1$ to B) and to 0 for non-basal species ($i = (B+1)$ to N); K_i is the basal species's carrying capacity; T_i represents the biomass loss by respiration; E_i and F_i are biomass gain and loss due to consumption, respectively, and are given by:

$$E_i = J_i \sum_{j \in \text{sp. i's resource}} f_{ij}\, a_{ij}\, X_j \Bigg/ \left(\sum_{j \in \text{sp. i's resource}} a_{ij}\, X_j + H_i\right) \tag{2}$$

and

$$F_i = \sum_{j \in \text{sp. i's consumer}} \left\{ I_j J_j f_{ji} a_{ji} X_j \Bigg/ \left(\sum_{k \in \text{sp. j's resource}} a_{jk} X_k + H_j\right)\right\}, \tag{3}$$

respectively (Type II functional response, Holling, 1959a). J_i is the maximum production rate; I_j is the inverse of the metabolization rate (> 1); f_{ij} (≤ 1) is the foraging efficiency of species i on resource species j; a_{ij} is the foraging effort of species i allocated to resource j; H_i is the half-saturation value. The parameters are set to the following values: $I_i = 1/0.85$, $T_i = 0.005\ V_i^{-0.25}$, $J_i = 0.097\ V_i^{-0.25}$, and represent carnivorous invertebrates (Yodzis and Innes, 1992). For simplicity, I present the result for this restricted setting, although the assumption on assimilation efficiency, metabolic rate or body size distribution, which affect the parameterization, could affect the dynamics (Yodzis and Innes, 1992).

I assume that predator species can increase the predation rate by allocating their foraging effort to a prey species, but this is associated with a decrease in their foraging effort to the other prey species. To represent this situation, I assume that a non-basal species has a fixed amount of foraging effort $\left(\sum_{j \in \text{sp. i's resource}} a_{ij} = const. = 1\right)$, which is distributed among potential prey species.

Further, I assume that a fraction F of non-basal species are adaptive foragers. I assumed that an adaptive forager takes a simple rule to maximize the energy consumption per unit time. Adaptive foragers increase its foraging effort to resource j if resource j's profitability ($f_{ij}X_j$) is higher than average profitability, while decreasing the effort if the profitability is lower than the average. The dynamics of foraging effort of consumer species i to resource species j is given by:

$$\frac{da_{ij}}{dt} = G_i a_{ij}\left(f_{ij}X_j - \sum_{k \in \text{sp. i's resource}} a_{ik} f_{ik} X_k\right), \tag{4}$$

where G_i the adaptation rate of consumer i., which is, for simplicity, set to a constant value, $G = 0.25$. The major result does not change if G is sufficiently high.

Species whose abundance becomes sufficiently small ($< 10^{-13}$) were removed and represent species extinction. The following initial values are used: $0 < f_{ij} < 1$, $a_{ij}(0) = 1/(\textit{potential number of sp. i's resource species})$, $0 < X_i(0) < 1$. The probability that the initial abundance falls below the extinction threshold is negligibly low.

THE PARAMETER-DEPENDENCE OF POPULATION STABILITY IN ADAPTIVE FOOD WEBS

To evaluate the effect of foraging adaptation and food web characteristics on population stability, I created 10,000 food web models with varying adaptation levels (F) and parameters representing food web structure (N and C), and evaluated population persistence in these webs. Population persistence (P_p), defined as a probability that a species randomly chosen from the N-species community does not become extinct within a given time period, is given by [$\ln P_p(C, N) = \ln P_c(C, N) / N$], where P_c is a probability that no species become extinct within a given time period ($T = 10^5$) and is directly measured by the simulations.

Adaptation has a major impact on the effect of connection probability (C) on population persistence (Figure 1). In the absence of adaptive foragers ($F = 0$), the effect of increasing C on population persistence depends on

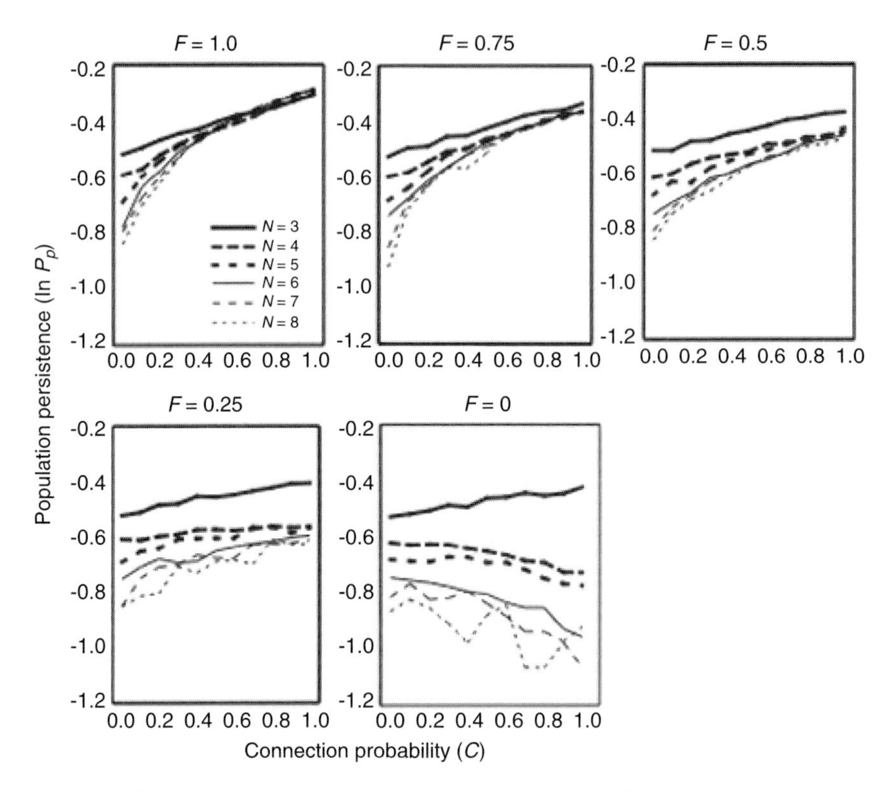

FIGURE 1 | The relationships between connection probability, (C) and population persistence with varying fractions of adaptive foragers, F (0, 0.25, 0.5, 0.75. 1) and species richness, N (3-8). Increasing adaptive foragers fraction inverses the negative relationships between C and population persistence into positive ones. $K_i = 1$. $B = 1$.

species richness (N). When $N = 3$, increasing C enhances population persistence. When $N > 3$, increasing C always decreases population persistence. It is notable that the negative relationship is partly affected by the straightforward effect that a web with low C has more basal species without consumer and thus less likely to lose a species. In the presence of adaptive foragers ($F \geq 0.25$), in contrast, population stability always increases with increasing C. The positive relationship is clearer as the fraction of adaptive foragers, F, is higher.

EMERGING ARCHITECTURE OF ADAPTIVE FOOD WEBS

As adaptive foragers may not utilize all potential diets, some potential links may not be activated in a food web consisting of adaptive foragers.

Indeed, a snapshot of adaptive food webs shows that only a small fraction of potential links is activated while most links are silent (Figures 2 and 3). The predicted pattern that connectance over a short time-scale (hereafter termed "short-term connectance") is low is interesting because natural food webs are often characterized by many weak links and a few strong links (Winemiller, 1990; Raffaelli and Hall, 1996; McCann et al., 1998). The correspondence between natural food webs and the adaptive food web model indicates that natural food web structure might be shaped by foraging adaptation (Matsuda and Namba, 1991).

How short-term connectance changes with changing potential connectance (C) depends on the fraction of adaptive foragers, F (see Figure 2). In the absence of foraging adaptation, the short-term connectance is identical to C as all links are activated. In the presence of adaptive foragers, although short-term connectance increases with increasing C, it saturates at high levels. This implies that the fraction of silent links increases with increasing C or increasing F.

The low short-time connectance does not mean that most trophic links are not activated over a long time scale, because the activated links can change their position within a web over time (see Figure 3). Such food web reconstruction would be caused by, at least, two mechanisms. One is intrinsic food web dynamics. A change in population abundance catalyzes the reconstruction of food web architecture, and this drives further changes in population abundances. This feedback between population abundance and interaction strength may lead to fluctuating dynamics of

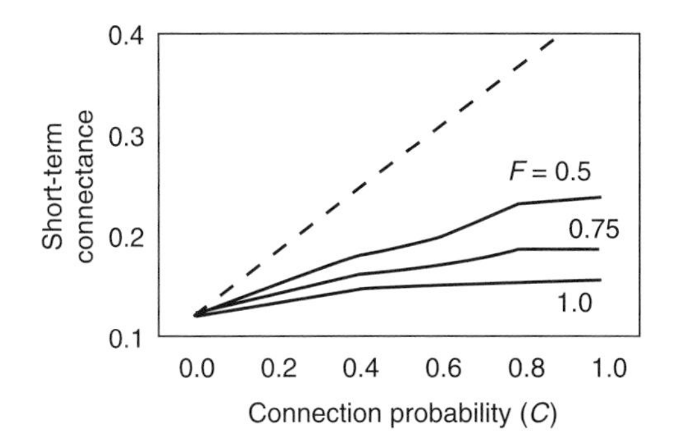

FIGURE 2 | The relationship between connection probability (C) and short-term connectance in varying fractions of adaptive foragers (F = 0.5, 0.75, 1.0). The broken line represents the connectance when all potential links are activated (F = 0). Standard deviations in short-term connectance increase with increasing C and decreasing F. For F = 1, it is 0, 0.0168, 0.0213, 0.0261, 0.0258 and 0.0311 for C =0, 0.2, 0.4, 0.6, 0.8 and 1, respectively; for F = 0.5, it is 0, 0.0274, 0.0353, 0.0533, 0.0600 and 0.0545 for C =0, 0.2, 0.4, 0.6, 0.8 and 1. $K_i = 1$.

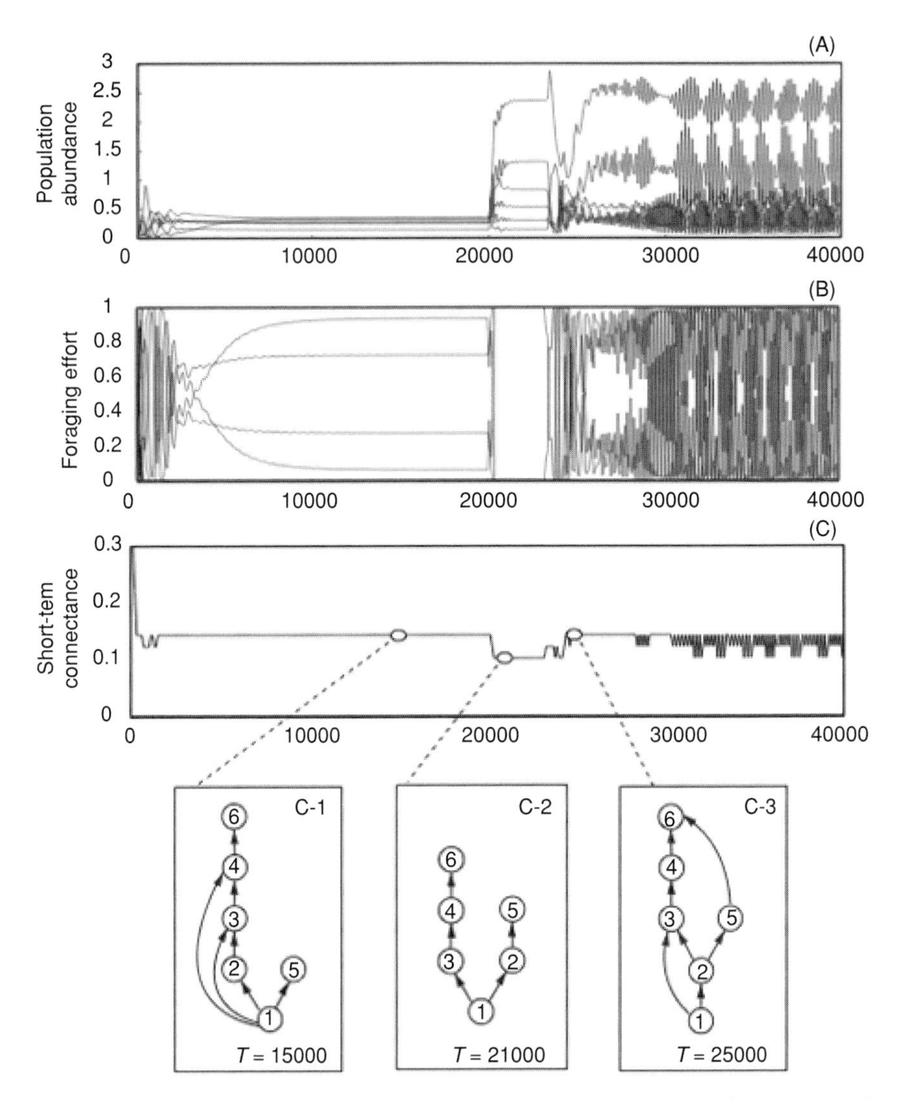

FIGURE 3 | An example of (A) population dynamics, (B) adaptive dynamics, and (C) structural changes of an adaptive food web with $N = 6$ and $C = 1$. The carrying capacity, K_1, changes from 0.5 to 1.0 at T = 20000, representing an environmental disturbance. The disturbance catalyzes fluctuation in population dynamics (A) and adaptive dynamics (B), leading to dynamics reconstruction of food web architecture. The short-term connectance (C) remains relatively constant over time. Panels (C1, 2, and 3) show the snapshots of food web linkage pattern at $T = 15000$, 21000 and 25000. Other parameters are: $F = 1$, $K_i = 1$. Trophic links with $Aij > 0.01$ are defined as "active" in C.

population and interaction strength, resulting in continuous reconstruction of food web architecture. The other is an extrinsic disturbance. The parameters determining population dynamics or adaptive dynamics, such as intrinsic growth rates, carrying capacity and foraging efficiency, are not constant, but are likely to change over time due to environmental fluctuations. Such parameter changes would alter the population dynamics or adaptive dynamics, and thus activate silent links.

In adaptive food webs, where the linkage pattern is continuously reconstructed, the fraction of active links depends on the time scale of observation (Figure 4). Consider an idealized adaptive food web with neither a species addition nor extinction; assume that any active trophic links are detectable by an infinitely short observation (this is, of course, not the case in reality). If the observation time is extremely short, the fraction of active connection would be low, since, as previously mentioned, most links are silent in this time scale. However, with increasing observation time, the probability that a link is activated during the observation period increases as population fluctuates, caused by intrinsic dynamics or environmental changes, may activate links that have been silent. Consequently, with an increasing time scale, the connectance estimated from the cumulative active links increases until it reaches an asymptote,

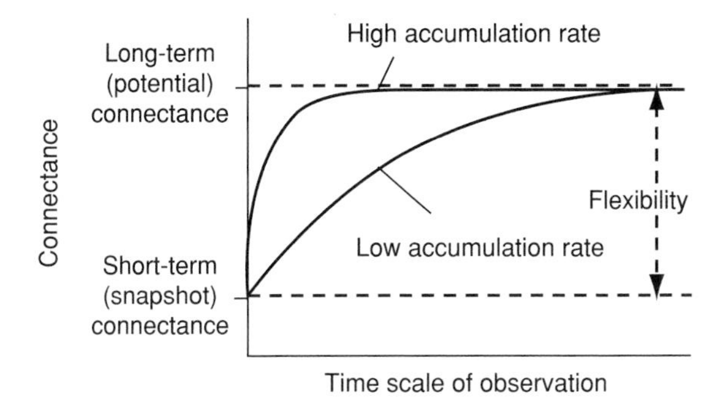

FIGURE 4 | Idealized connectance curve, showing how the connectance deriving from the cumulative number of trophic links increases with an increasing observation time period. Connectance, L, is given as an increasing function of observation time, t, with an upper limit of potential connectance. The function is characterized by the following four indices: (i) Potential connectance $(\lim_{t\to+\infty} L(t))$, representing connectance including any possible trophic links. This is obtained by infinitely long observation. (ii) Shot-term connectance $(\lim_{t\to+0} L(t))$, which is an instantaneous connectance measured by a snapshot observation. (iii) Flexibility $(1-\lim_{t\to+0} L(t)/\lim_{t\to+\infty} L(t))$, defined as the ratio of short-term connectance to potential connectance. (iv) Accumulation rate, defined as a rate at which the accumulation reaches the potential connectance. The two lines show connectance accumulation curve with a high and low accumulation rate.

which represents the potential connectance. Note that by evaluating a web with all species persisting I excluded the possibility that the connectance might change for the temporal variation in species richness.

LINKING FLEXIBLE FOOD WEB STRUCTURE TO POPULATION STABILITY

The connectance-stability relationship of adaptive food webs is different from that of static food webs in two ways. First, the relationship can be time-scale dependent, as the connectance of an adaptive food web depends on time scale. Second, the connectance-stability relationship can change depending on the factor that causes the inter-web variance of connectance, which is potential connectance or adaptation level. I start by summarizing the predicted relationships between connectance over various time scales (short-term, long-term, and intermediate-term connectance) and population stability for adaptive food webs.

First, changing C has opposite effects on population persistence in the presence and absence of foraging adaptation. Noting that potential connectance is identical to long-term connectance, this suggests that the relationship between long-term connectance and population persistence depends on adaptation level (Figure 5). In the absence of adaptation, increasing C decreases population persistence, suggesting a negative relationship between long-term connectance and population persistence. In contrast, if adaptation level is high, a positive relationship emerges. In adaptive food webs, populations are more stable in food webs with species that have, and can discriminate between, wider prey ranges on average.

Secondly, the relationship between short-term connectance and population persistence depends on which factor creates the variance in short-term connectance (see Figure 5). If the variance in short-term connectance between food webs is caused by a variance in adaptation level, the relationship should be negative. This is because lowering adaptation level increases short-term connectance while lowering population persistence. If the variance in short-term connectance is due to a variance in potential connectance (C), the relationship depends on the adaptation level. In the absence of adaptation, short-term connectance is identical to potential connectance and has a negative effect on population persistence. Therefore, the relationship is predicted to be negative. In the presence of adaptation, high short-term connectance indicates high potential connectance and the relationship would be positive, as increasing potential connectance stabilizes populations in this case.

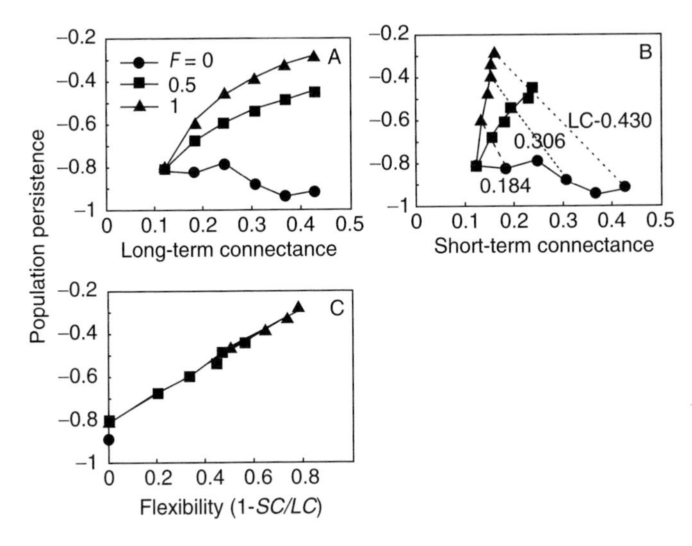

FIGURE 5 | Emergent relationships between food web properties (**A,** long-term connectance; **B,** short-term connectance; and **C,** flexibility) and population persistence. Each point represents the resultant values of a food web property and population persistence for varying connection probability (C) between 0.0 and 1.0 with steps 0.2. The different simbols represent no (F = 0; circle), intermediate (0.5; square), and high fraction (1.0; triangle) of adaptive predators. In panel B, the rigid lines connect the points with the same adaptive level at three levels, showing the connectance-stability relationship resulting from a variance in potential connectance; the three dotted lines connect the points with the same potential (long-term) connectance at three levels (0.184, 0.306, 0.430), showing the connectance-stability relationship arising from a variance in adaptive level. Parameters are: $N = 7$, $K_i = 1$.

Finally, the relationship between intermediate-scale connectance and population persistence is unclear. This is because it is unclear how adaptation or C affects the frequency of food web reconstruction, which, in turn, influences the accumulation rate (see Figure 4). However, it may be possible to relate intermediate-scale connectance to population variability. In the presence of adaptive foragers, population fluctuation catalyzes the reconstruction of food web structure and thus increases connectance accumulation rate. This is likely to increase connectance on an intermediate time scale. If potential connectance and adaptation level are kept constant, temporal variability of population dynamics is positively correlated with intermediate-term connectance. This suggests a negative relationship between intermediate-scale connectance and population variability.

In summary, a connectance-stability relationship cannot be determined without identifying the time scale of the connectance and the causal mechanism of the variance in the focal connectance. This implies a difficulty in detecting a consistent relationship between food web connectance and population stability. This is consequential because connectance is a static property and therefore cannot capture the dynamic feature of an adaptive food web. A possible way to detect a consistent relationship between food web structure and population persistence is not to use such a static index, but to use a more dynamic index that incorporates how the structure of the focal food web changes over time.

An example of such dynamic indices is food web flexibility, which is defined as a fraction of inactive links in all potential links (i.e., 1 – [short-term connectance]/[long-term connectance]). In the presence of adaptation a consumer does not use some potential resources in a short time scale. Food web flexibility can be regarded as a fraction of such an unused and alternative diet to which a consumer can change its energy source when the focal resource becomes less available. The model analysis suggests that the flexibility-stability relationship is consistently positive and not altered by which parameter's variance creates the connectance variance between food webs (see Figure 5). The positive flexibility-stability relationship clearly shows that what is essential for the population persistence in an adaptive food web is not the static structure of who eats whom at a particular moment, but the dynamic structure of how food web architecture can change over time.

DISCUSSION AND CONCLUSIONS

The dynamic food web models show that the adaptive nature of predator's diet choice can have major impact on food web properties. If adaptation is sufficiently quick, food web architecture changes in the time scale comparable with that of population dynamics, providing food web architecture with flexibility. This makes food web connectance time-scale dependent and enhances population persistence. As a result, there emerge the diverse connectance-stability relationships depending on the time scale in which the connectance is measured and the causal mechanism of the inter-web variance in the connectance. The generality of these patterns is however still an open question. Various simplifying assumptions might have affected the model outcomes or general conclusions, although the major result is not altered by using food web topology of random (Kondoh, 2003a, b), cascade (Kondoh, 2003a, b), and niche models (Kondoh, in press, but see Brose et al., 2003), using trophic interactions of Holling's Type I and II (Kondoh, 2003b, unpublished data) functional responses, and different fractions of species with a positive intrinsic growth rate (Kondoh, 2003a, b).

Understanding of population dynamics and complexity-stability relationship of food webs require more care to be paid to dynamic changes in food web architecture. However, we know too little about how food web architecture changes in nature, while we need information as to the relative time scale of web reconstruction to population dynamics is the essential determinant of the complexity–stability relationship. We need future studies, which focus more on dynamic property of food webs to show how strengths of trophic interactions change over time or how macroscopic food web characteristic such as connectance depends on the time scale of observation. In addition, it should be noted that the present model is simplified, where all flexible trophic links change with the same rate, G. However, in reality trophic links are likely to fluctuate with varying rates (i.e., there is intra-web variance in the time scale of food web reconstruction). Behavioral flexibility at individual level might lead to higher adaptive rate than evolution at population level; a shorter generation time would again lead to a higher adaptive rate. This implies that an empirical food web data may include both short-term connectance and long-term connectance, and we cannot define the relative time scale of trophic link dynamics to population dynamics at the whole web level. It is important to extend the present theory to include such variability within a web.

3.4 | INDUCIBLE DEFENSES IN FOOD WEBS

Matthijs Vos, Bob W. Kooi, Don L. DeAngelis, and Wolf M. Mooij

A variety of complementary approaches is required to understand the structure and functioning of complex food webs. One approach is to study the effects of important ecological mechanisms within relatively simple food web modules (McCann et al., 1998; Polis, 1998; Persson et al., 2001; Vos et al., 2001; Vos et al., 2002; Vos et al., 2004a; Vos et al., 2004b). Once the isolated effects of such mechanisms are known, one can study their combined effects (Abrams and Vos, 2003) and relative importance, and test whether the results can be extrapolated to more complex communities (DeAngelis et al., 1989a; Werner and Peacor, 2003; Vos et al., 2004b).

Food webs are structures of populations in a given location organized according to their predator–prey interactions. Interaction strengths and, therefore, prey defenses, are generally recognized as important ecological factors affecting food webs (Leibold, 1989; Power, 1992; Polis and Strong, 1996). Despite this, surprisingly little light has been shed on the food web-level consequences of inducible defenses. Inducible defenses occur in many taxa in both terrestrial and aquatic food webs (Karban and Baldwin, 1997; Kats and Dill, 1998; Tollrian and Harvell, 1999; Vos et al., 2004b). They include refuge use, reduced activity, adaptive life history changes, the production of toxins, synomones and extrafloral nectar and the formation of colonies, helmets, thorns, or spines (Vos et al., 2004a; Vos et al., 2004b). Here we briefly review the predicted effects of induced defenses on trophic structure and two aspects of stability, 'local' stability and persistence, as well as presenting novel results on a third, resilience. We consider differences between aquatic and terrestrial systems and discuss how inducible defenses cause flexible food web links and heterogeneous food web nodes. We focus on an important class of inducible defenses, those that are reversible (i.e., allow both induction and relaxation of defenses). Reversible defenses include refuge use and other anti-predator behaviors, colony formation and the production of toxins, synomones, and extrafloral nectar. Well-known irreversible induced defenses against visually hunting fish include a reduced age and size at maturity in *Daphnia* zooplankters (Vos et al., 2002, and references therein).

The stability of complex food webs has fascinated ecologists for decades (McCann, 2000). The wide range of ways in which the term 'stability' has been used in ecology led Grimm and Wissel (1997) to conclude that "the general term stability is so ambiguous as to be useless." However, it is possible to give precise definitions to three aspects of stability: local stability, resilience, and persistence (Pimm, 1982; DeAngelis et al., 1989b; DeAngelis, 1992a). The first, local stability, is the tendency for a system to return to steady-state, following a perturbation. Local stability is equivalent to the linearized equations having negative (real parts of the) eigenvalues. Resilience can be defined as the rate at which the system returns to steady-state (i.e., the inverse of the return time). Persistence is the property of the food web maintaining all species at positive densities bounded from zero, despite perturbations or internal factors that cause populations to undergo variations in numbers. We will use minimum population densities as a measure for persistence. When these stay further away from zero, populations have a smaller risk of extinction under demographic stochasticity (McCann et al., 1998).

Here theoretical results for the effects of inducible defenses on trophic structure and the three aspects of stability are reviewed. This is done, in part, using bifurcation analysis, a type of analysis that is applied to nonlinear dynamic systems described by a set of ordinary differential (continuous time) or difference equations (discrete time). In Kooi (2003) the application of this analysis technique for small-scale food web models is reviewed. Here we apply it to predator–prey interactions in which the prey population has inducible defenses that depend on predation pressure.

Our bi- and tri-trophic food chain models (Vos et al., 2004a; Vos et al., 2004b) are extensions of the classical Rosenzweig-MacArthur model (Rosenzweig and MacArthur, 1963; Rosenzweig, 1971; see also Kretzschmar et al., 1993; Abrams and Walters, 1996; Abrams and Vos, 2003). Each prey population consists of two subpopulations: one undefended and the other defended. Changes in the proportion of defended individuals are modelled as flows between these subpopulations depending on predator density. Predators consume both prey types, but have a reduced predation rate on defended prey. We think of our bitrophic example system as consisting of algal prey (denoted as plants) and herbivorous rotifers. It is represented by the differential equations:

$$\frac{dP_1}{dt} = P_1\left(r_1\left(1 - \frac{P_1}{k} - \frac{P_2}{k}\right) - s_{11}\right) - \frac{v_{11}H_1P_1}{1 + (v_{11}h_{11}P_1 + v_{21}h_{21}P_2)} - i_1P_1\left(1 - \left(1 + \left(\frac{H_1}{g_1}\right)^{b_1}\right)^{-1}\right) + i_1P_2\left(1 + \left(\frac{H_1}{g_1}\right)^{b_1}\right)^{-1} \quad \text{(1a)}$$

$$\frac{dP_2}{dt} = P_2\left(r_2\left(1 - \frac{P_2}{k} - \frac{P_1}{k}\right) s_{21}\right) - \frac{v_{21}H_1P_2}{1 + (v_{11}h_{11}P_1 + v_{21}h_{21}P_2)} + i_1P_1\left(1 - \left(1 + \left(\frac{H_1}{g_1}\right)^{b_1}\right)^{-1}\right) - i_1P_2\left(1 + \left(\frac{H_1}{g_1}\right)^{b_1}\right)^{-1} \quad \text{(1b)}$$

$$\frac{dH_1}{dt} = H_1\left(\frac{c_{11}v_{11}P_1 + c_{21}v_{21}P_2}{1 + (v_{11}h_{11}P_1 + v_{21}h_{21}P_2)} - s_{12}\right) \quad \text{(1c)}$$

Plant growth is logistic, with intrinsic rates of increase r_i and carrying capacity k in undefended (P_1) and defended plants (P_2). The trophic interaction is a Holling type II functional response, where handling times h_{21} may be increased and attack rates v_{21} may be decreased on defended algae. Algae may show induced colony formation in the presence of herbivorous rotifers and cladocerans (Verschoor et al., 2004b) and this defense is costly in terms of the increased sedimentation rate s_{21} of colonies (Van Donk et al., 1999; Vos et al., 2004a; Vos et al., 2004b). Modelling the costs of defenses in terms of reduced growth or an increased natural mortality or sedimentation rate is largely equivalent. The third and fourth terms in equations (1a) and (1b) show that the rates of defense induction and decay may depend on herbivore density H_1 in a sigmoidal fashion (Vos et al., 2004a). Parameter g_1 is the herbivore density at which half the maximum defense induction rate is reached. Induction substracts from the undefended part of the plant population and adds to the defended part. Similarly, decay of defenses substracts from the defended part and adds to the undefended part of the plant population. These rates combine to affect the fraction of individuals in the population that is defended at each moment in time. In permanent defense and no defense scenarios only a single prey type is present (e.g., an algal strain that is always single-celled or colonial, see (Verschoor et al., 2004b) and the induction/decay terms are lacking from equations (1a), (1b), and (1c) (Vos et al; 2004a; Vos et al., 2004b).

TROPHIC STRUCTURE

The classical hypothesis of exploitation ecosystems (EEH) by Oksanen et al. (1981) is based on a food chain model that assumes complete top-down control of plants by herbivores in bitrophic systems and of herbivores by carnivores in tritrophic systems. The EEH model predicts enrichment to increase only herbivore density in bitrophic systems, and only plants and carnivores in tritrophic systems. However, empirical data on aquatic, terrestrial, and microbial food chains often show all trophic levels increasing simultaneously in biomass in response to increasing primary productivity (Akçakaya et al., 1995; Leibold, 1996; Brett and Goldman, 1997; Kaunzinger and Morin, 1998; Chase et al., 2000b).

The discrepancy between EEH predictions and the outcome of field studies requires the identification of ecological factors that are not present in the Oksanen et al. (1981) model, but are important in governing trophic level biomass responses in nature. Vos et al. (2004b) identified inducible defenses as a potential mechanism and modified the Oksanen et al. (1981) model to include undefended and defended types at each prey level. Using the inducible defense model for bitrophic systems (i.e., equations 1a, 1b, and 1c), it was shown, in contrast to the EEH model, that herbivore-induced shifts in the fraction of defended plants caused the total plant biomass to increase in response to enrichment (Vos et al., 2004b). Numerical results extended this conclusion to show that inducible defenses may result in all trophic levels increasing in response to increased primary productivity in tritrophic systems as well (Vos et al., 2004b). Interestingly, Kaunzinger and Morin (1998) observed all trophic levels to increase under enrichment in bitrophic and tritrophic microbial food chains. This result remained unexplained in that paper, as EEH type behavior was expected in that experiment. Vos et al. (2004b) have hypothesized that consumer-induced changes may have caused this pattern, as morphological changes are known to occur in the genera *Serratia* and *Colpidium* that were used by Kaunzinger and Morin (see Vos et al., 2004b, and references therein).

Future studies face the challenge of explaining the combined effects of important ecological mechanisms on complex communities. Abrams and Vos (2003) investigated the effects of adaptive processes (including induced defenses) and density dependence in a full factorial analysis, and showed that trophic level responses and the (in)determinacy of theoretical predictions are greatly affected by interactive effects of these different mechanisms.

LOCAL STABILITY

It is a major question whether inducible defenses stabilize or destabilize the interaction between plants, herbivores, and higher trophic levels. Early empirical studies suggested that induced defenses might cause population cycles in herbivores (Haukioja, 1980; Seldal et al., 1994). However, such a destabilizing effect was not predicted in the pioneering model work by Edelstein-Keshet and Rausher (1989). These authors analyzed a model in which herbivore densities were coupled to the level of defenses in plants. In that model the dynamics of herbivores were not coupled to plant densities. Although such a simplification may not be realistic for many plant herbivore systems, the results are interesting in that inducible defenses almost invariably resulted in a stable steady-state. Only when herbivores exhibited an Allee effect could oscillations be driven by the dynamics of inducible defenses (Edelstein-Keshet and Rausher, 1989). A similar model by Lundberg et al. (1994) typically showed oscillations that damped towards a stable equilibrium. Such stability occurred despite the fact that the relaxation of defenses introduced a delay in this model. Ramos-Jiliberto (2003) also showed a stabilizing effect of predator-induced reductions in prey vulnerability. This model did not include time delays in the onset or lowering of defenses.

Interestingly, Underwood (1999) used a simulation model loosely based on the analytical work of Edelstein-Keshet and Rausher (1989) and showed that population cycles may occur when a substantial time lag occurs between herbivore damage and an increase in the level of resistance to herbivory. Underwood (1999) concluded that inducible defenses may drive population cycles in herbivores. While this confirms that oscillations may occur in models that incorporate inducible defenses, the potential conclusion that inducible defenses drive such oscillations requires that the inducible defense scenario be compared with positive and negative 'control' scenarios in which defenses are permanent or absent. Such a comparison with fixed defense strategies was not made in the previous analyses.

Vos et al. (2004a) used a modification of classical bitrophic and tritrophic food chain models (Rosenzweig, 1971; Oksanen et al., 1981; Kretzschmar et al., 1993) and showed that both stable equilibria and cycles may exist in models with inducible, permanent, or no defenses. The major point of Vos et al. (2004a) was that inducible defenses may prevent a paradox of enrichment (i.e., enrichment causing the system to become unstable and oscillate) in both bitrophic and tritrophic food chains, although such enrichment-driven instability existed in the cases in which defenses were permanent or absent (Vos et al., 2004a). It was shown analytically that the absence of a paradox of enrichment depended on differences in handling times and/or conversion efficiencies on defended and undefended prey. Vos et al. (2004a) did not analyze the effects of time lags. It would be interesting to extend their analysis of (equations 1a, 1b, and 1c) to see whether such time lags would cause the paradox of enrichment to return. Abrams and Walters (1996) analyzed a bitrophic model that differed in two ways from the one by Vos et al. (2004a); prey with induced defenses were assumed to be completely invulnerable, and the induction and decay of defenses did not depend on predator densities. That study also showed that a paradox of enrichment may be absent when defenses are inducible.

PERSISTENCE

When the densities of all populations in a food web are bounded from zero, the system is called persistent. Population persistence may be approached from the point of view of the minimum predator population densities reached when the prey have different types of defenses. Underwood's study (1999) showed that a sufficient strength of induced resistance in plants may cause herbivores to go extinct. In some systems herbivore mortality may increase considerably on plants with more damage (West, 1985), but usually induced defenses do not make plants or other prey species invulnerable (see Jeschke and Tollrian, 2000; Vos et al., 2001; Vos et al., 2004a; Vos et al., 2004b).

Vos et al. (2004a) analyzed a bitrophic plant-herbivore model (i.e., equations 1) and showed that minimum herbivore densities were lower in the inducible defense and permanent defense scenarios than in the no defense scenario. In contrast, minimum densities were highest (at all trophic levels) in the tritrophic inducible defense scenario. This result points at potential differences between simple and more complex communities. This prediction resulted from a numerical analysis that was tuned to a specific planktonic system. Thus the result is not general, but could indicate the potential consequences of defensive strategies for population persistence in other systems.

Different types of inducible defenses, such as refuge use through diel vertical migration (DVM), or predator-induced life history changes, may differ substantially in their effects on persistence. Vos et al. (2002) showed that a behavioral defense, DVM, decelerated a *Daphnia* population decline under heavy fish predation much more strongly than a reduced size at first reproduction, which decreases the visibility of daphnids for visually hunting fish. Staying down in the safe and dark hypolimnion was predicted to have an even stronger protective effect (Vos et al., 2002). Daphnids that stay in this refuge avoid the predation suffered by migrating daphnids around dusk and dawn. Previous models focused almost exclusively on the effects of DVM versus remaining in the epilimnion. Staying down was ignored in the theoretical literature, but has in fact been observed within a variety of *Daphnia* clones and species, both in the laboratory and in stratified lakes (Vos et al., 2002, and references therein). This model study suggests that staying down as an induced defense under peak predation pressure may prevent a population decline altogether, thus greatly enhancing persistence in the prey population.

RESILIENCE

Here we investigate resilience in bitrophic systems with inducible, permanent, and no defenses. Our approach is to contrast resilience in these defense scenarios along a gradient of primary productivities, as measured by the carrying capacity. Many studies have shown that nutrient enrichment may increase the resilience of a system (DeAngelis, 1980, 1992a, and references therein), although DeAngelis et al. (1989a) have warned that resilience may not always increase with nutrient inputs. The general idea is that the rate of recovery of a system is often inversely related to the turnover time of the limiting nutrient in the system (DeAngelis, 1992a). Increased nutrient inputs should shorten recovery times, that is, increase resilience.

Since defenses tend to decrease the flux of nutrients through a food chain, we hypothesize that resilience will increase more slowly under enrichment when defenses play an important role in the system. As an introduction to our results, we first elucidate how several aspects of the dynamics of food webs are connected to changes in eigenvalues of the models that describe these webs.

Eigenvalues, Bifurcation Analysis, and Resilience

A food web can be described by a set of differential equations. To determine the behavior close to steady-state, the Jacobian matrix is obtained (Case, 2000), in which the eigenvalues, λ_i, determine the dynamic behavior after a perturbation from the steady-state. The steady-state is only stable when each of the eigenvalues has a negative real part. Thus there is a clear link between local stability and the sign of real parts of eigenvalues (Case, 2000). When the steady-state is locally stable, the magnitude of the dominant eigenvalue (i.e., the eigenvalue with the largest real part) determines how fast the system returns to the steady-state following a perturbation. This makes the dominant eigenvalue a direct measure for resilience.

It is important to note that the initial (transient) behavior of a perturbed system can be irregular. This depends on the type and direction of the perturbation. Such transient irregularities do not play a role when the sizes of applied perturbations for each population are proportional to the eigenvector associated with the dominant eigenvalue. Such perturbations can easily be applied in theoretical studies, but not in experiments in the laboratory or the field. A comparison of experimental and theoretical results is facilitated most when a system is given the necessary time to reach steady-state behavior (but note that transients may be long).

Bifurcation analysis is aimed at detecting situations where the qualitative system behavior changes under parameter variation. Generally such transitions occur when the real parts of one or more eigenvalues are zero. Numerical bifurcation analysis packages calculate these transitions by continuation (Kuznetsov and Levitin, 1997; Kuznetsov, 1998; Kooi et al., 2002; Kooi, 2003). Bifurcations separate regions in parameter space with qualitatively different behavior (e.g., stable steady-state, limit cycles, or chaos).

Here we show the effects of different defense strategies on resilience using the simplest possible food web module as an example: a bitrophic predator–prey interaction (equations 1a, 1b, and 1c). This implies that we focus on carrying capacities where both species can coexist (i.e., beyond the transcritical bifurcation, where a predator can invade a prey population). We discuss only a system in stable steady-state. Therefore, with increasing the carrying capacity we do not cross the Hopf bifurcation, where the real part of the dominant complex eigenvalue becomes positive and the system starts to oscillate. We will see that there is another transition that occurs in the stable region between the transcritical and Hopf bifurcation, that is essential for our analysis of resilience. It is not a bifurcation in its own right, but it marks the transition from a stable node (the dominant eigenvalue is real and negative) to a stable spiral (the dominant eigenvalue is complex with negative real part). Following a perturbation, a system shows a monotonic asymptotic return to a stable node, while it shows damped oscillations in the case of a stable spiral.

Figure 1 shows the effects of the carrying capacity, k, on (i) herbivore density (top panel), (ii) total plant density (i.e., the sum of defended and undefended plants), (iii) the fraction of defended plants, and (iv) resilience (bottom panel). Resilience is defined as $-\text{Re}\,\lambda_1$ (i.e., minus real part of the dominant eigenvalue). Inducible defenses, permanent defenses, and no defenses in plants are indicated by solid, dashed, and dotted lines respectively (Figure 1). The results for trophic structure are more fully presented and discussed in Vos et al. (2004b).

Resilience when Defenses are Permanent or Absent

At low carrying capacity only the plant population can exist (Figure 1). The herbivore population can attain a positive density (invade) above a

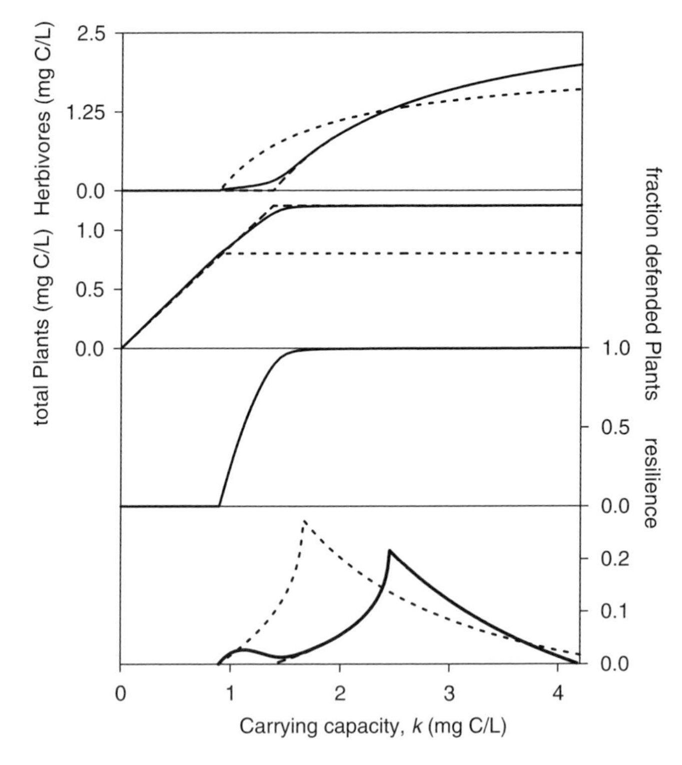

FIGURE 1 | Trophic level abundances and resilience in bitrophic systems with inducible defenses *(solid lines)*, permanent defenses *(dashed lines)*, or no defenses *(dotted lines)*. The top two panels show the total densities of herbivores and plants, along a gradient of plant carrying capacities (k), a measure for primary productivity. The panel below displays the fraction of defended plants in the inducible defense scenario. Resilience is a unimodal function of the carrying capacity, as shown in the bottom panel. All parameter values are as in Vos et al. (2004b).

carrying capacity value called the transcritical bifurcation point (see Figure 1, top panel). A higher carrying capacity is required for herbivore invasion of a defended plant system. Increasing the carrying capacity even further the Hopf bifurcation is reached, where the steady-state becomes unstable.

Resilience is zero at the transcritical bifurcation and is zero again at the Hopf bifurcation, since in both points the real part of the dominant eigenvalue is zero (see Figure 1, bottom panel). Close to the transcritical bifurcation point both eigenvalues are real and the steady-state is a stable node. Here the system's asymptotic behavior after a perturbation is a monotonic return to the steady-state. Close to the Hopf bifurcation both eigenvalues form a conjugate pair and the steady-state is a stable spiral. Here the system's asymptotic behavior after a perturbation is a damped oscillation. At some intermediate point both eigenvalues are real and equal, which marks the transition from the stable node to the stable spiral. In Hirsch and Smale (1974) this point is called a focus. At this point resilience reaches a peak value, as can clearly be seen in Figure 1 (bottom panel). In Neubert and Caswell (1997) this phenomenon was also observed but not related to the occurrence of a focus. Further increases in k cause the resilience to decrease sharply. This is in contrast with the general idea that resilience is likely to increase with nutrient inputs. DeAngelis et al. (1989a) studied a slightly different model and found also that resilience increased monotonically beyond the transcritical bifurcation. This earlier study used a Holling type III functional response, and the parameterization allowed no oscillatory behavior. These model characteristics probably caused the transition between a stable node and a stable spiral to be absent. Our results show that resilience may initially increase with enrichment, at a faster rate, and to a higher level, when defenses are absent (see Figure 1, bottom panel). This is in accordance with our expectations. The sharp decrease in resilience beyond the transition from a stable node to a stable spiral was a surprising result. However, it is not difficult to see that such a decrease has to occur in any system that can be destabilized by enrichment, since resilience is by definition zero at the Hopf bifurcation.

Resilience when Defenses are Inducible

Now we focus on the case where the plant population consists of undefended and defended individuals simultaneously. There is again a region of carrying capacities where this food chain possesses a stable steady-state. At the lower k values close to the transcritical bifurcation resilience resembles that of the no defense scenario (see Figure 1, bottom panel, solid line). However, resilience quickly starts to resemble that of the permanent defense scenario as the carrying capacity increases. This is understandable, as the fraction of defended plants increases very rapidly to almost one, with increases in k (see Figure 1, second panel from the bottom). With further increases in k the inducible defense case has a resilience that cannot be discriminated from the permanent defense case. In this example, resilience reaches its maximum value at almost the same carrying capacity in the inducible and permanent defense scenarios. In other systems with inducible defenses, where the fraction of defended prey might increase much more slowly with consumer density, resilience could peak at a lower value of k than in the permanent defense scenario. Resilience in systems with inducible defenses is strongly dependent on the tuning of defense induction and decay to consumer densities.

In the inducible defense model scenario previously discussed, parameter values allowed a Hopf bifurcation to exist, implying that resilience has to decrease towards zero, just as in the scenarios where defenses were absent or permanent. However, for this inducible defense model scenario a region in parameter space exists where the system cannot be destabilised (Vos et al., 2004a). Then a focus point does not need to exist and this allows a monotonic increase in resilience under enrichment, as was observed in the study by DeAngelis et al. (1989a), where a Holling type III functional response stabilized the system.

DIFFERENCES BETWEEN AQUATIC AND TERRESTRIAL SYSTEMS

Although the examples previously discussed are based on an aquatic system, the model also applies to terrestrial systems. This is not to say that there are no differences between aquatic and terrestrial systems in the way induced defenses act on consumers. In aquatic systems gape limitation is very important, as many consumer species have no means to tear their prey to pieces (Hairston and Hairston, 1993; Brönmark et al., 1999). Algae, zooplankters, and fish are mostly ingested as whole individuals. Consequently, many inducible defenses in aquatic systems involve morphological changes such as colony- and spine-formation and deepened body shapes, that hinder handling and ingestion by gape-limited consumers.

In contrast with algae, terrestrial plants are usually not consumed as whole individuals. Plants may tolerate or compensate for tissue losses due to herbivory. In addition, they may employ a variety of defenses. Direct defenses include the production of toxins, while indirect defenses may involve the emission of volatiles that are attractive to natural enemies of insect herbivores (Vet and Dicke, 1992; Dicke and Vet, 1999; Vos et al., 2001). Feeding damage inflicted by insects may also increase the production of extrafloral nectar that provides a sugar source to carnivores and insect parasitoids (Wäckers et al., 2001; Wäckers and Bonifay, 2004). This type of positive effect between the first and third trophic level seems less common, (or is less studied) in aquatic food webs.

Our theoretical work indicates that it is important, both in aquatic and terrestrial systems, whether induced defenses have a stronger effect on handling times or attack rates, as these have differential effects on functional responses and community dynamics (Jeschke and Tollrian, 2000; Vos et al., 2001; Vos et al., 2002; Vos et al., 2004a; Vos et al., 2004b). The duration of delays in the onset and decay of defenses, and consequently their effects on stability, may differ strongly between fast species such as aquatic algae and much slower terrestrial species such as trees.

HETEROGENEOUS FOOD WEB NODES AND FLEXIBLE LINKS

Inducible defenses modulate interaction strengths between predators and prey, because the induction and decay of defenses respond to consumer densities. Prey may 'estimate' predation risk through the concentration of predator-released kairomones and through chemical cues from crushed conspecific prey (Hagen et al., 2002; Stabell et al., 2003). Predator-induced changes occur in many prey species. Kats and Dill have reviewed responses to predator odors in more than 200 animal prey species (Kats and Dill, 1998; also see Tollrian and Harvell, 1999). Similarly, Karban and Baldwin (1997) have reviewed induced resistance in more than 100 plant species. Such induced responses to consumer densities have probably been fine-tuned by natural selection and may be a structuring force on the distribution of interaction strengths in natural food webs.

We use Figure 2 to exemplify how induction-mediated variability and heterogeneity in prey species modulate interaction strengths in a food web. Our simplified food web example consists of a fish top predator (*Perca*) that predates on an invertebrate predator (*Chaoborus*) and on waterfleas (*Daphnia*) that consume algae (*Scenedesmus*). In this system both daphnids and algae may employ morphological defenses, while *Chaoborus* and *Daphnia* may both use a refuge against fish predation. When consumer densities are low, all prey species are undefended and per capita feeding rates are high (Figure 2A). When consumer densities increase, defenses are induced. Part of the *Chaoborus* and *Daphnia*

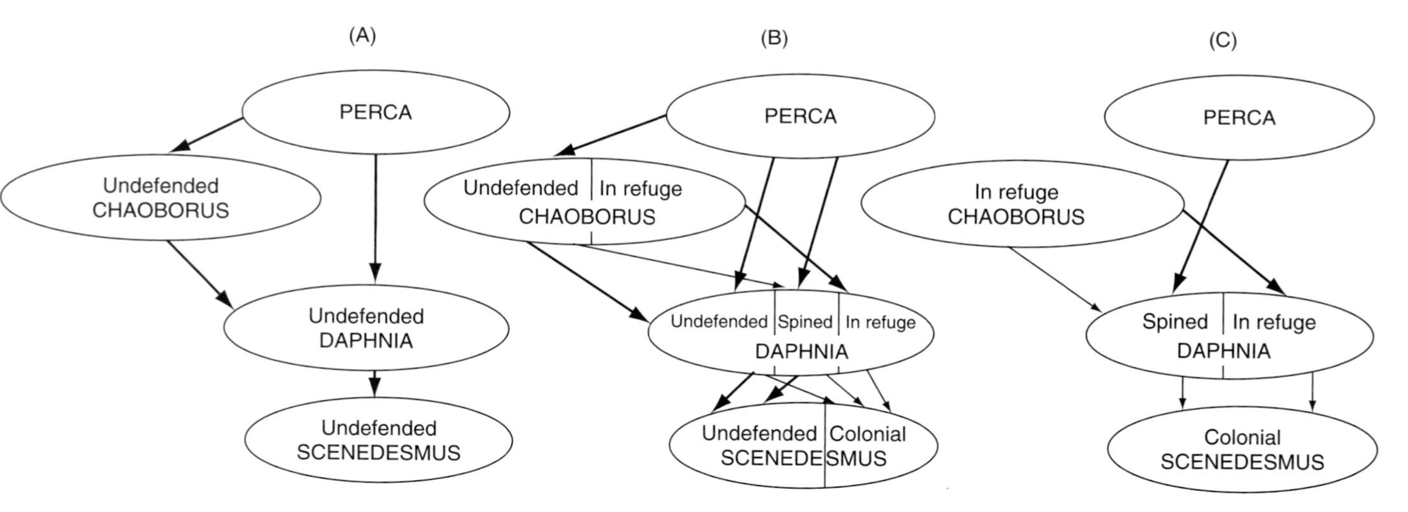

FIGURE 2 | Trophic interactions in a simplified food web, when prey are (A) undefended, (B) have induced defenses, or (C) are completely defended. Defenses and intraspecific heterogeneity in prey vulnerability determine which links are present and how strong they are. Thickness of arrows represents the strength of per capita feeding interactions. Inducible defenses are predator-density dependent. This causes variability in the fraction of defended individuals in prey food web nodes, when predator abundance changes. Species in this food web example are *Perca* (zooplanktivorous fish), *Chaoborus* (invertebrate predator), *Daphnia* (herbivorous zooplankton), and *Scenedesmus* (green alga). In this example, both *Chaoborous* and *Daphnia* can make use of a refuge from fish predation. Induced morphological defenses include spine formation in daphnids and colony formation in algae.

populations now occupy the spatial refuge and part of the algae and *Daphnia* develop morphological defenses. Individual consumers have different feeding rates on the different 'defense classes' within each prey population (see Figure 2B). In the extreme case that all prey are morphologically defended or in the refuge, some feeding links may disappear altogether (see Figure 2C). This might reduce consumer densities to the extent that the system returns to the situation in scenario (B).

Here we have focused on the differential effects of three defense strategies on trophic structure and several aspects of community stability. Changes in traits of a species, as occur in the case of inducible defenses, may have both direct and indirect effects on the traits and densities of other species in the food web. Determining the relative magnitude of direct and indirect effects remains a fascinating issue in food web theory. Abrams (1995), Werner and Peacor (2003), Bolker et al. (2003), and Peacor and Werner (2004) provide excellent overviews of the main ideas and results on direct and indirect interactions in relation to variable traits.

DISCUSSION AND CONCLUSIONS

Heterogeneous Nodes, Flexible Links, and Food Web Dynamics

Faced with the brimming complexity of natural food webs (Polis, 1998), it is tempting to ignore variation within populations. Lumping individuals, or even different species, into single food web nodes simplifies and homogenizes our view of food webs. However, the work presented here suggests that heterogeneity, as caused by induced defenses in prey species, has major effects on the functioning of food webs. Inducible defenses occur in many species in both aquatic and terrestrial systems and theoretical work indicates they have major effects on important food web properties such as trophic structure, local stability, persistence, and resilience. All of the theoretical results presented here can be tested in experimentally assembled plankton communities, using strains and species that have inducible, permanent or no defenses. Preliminary experimental results show highly replicable differences in the dynamics of such communities and strong effects of different defense strategies on persistence and stability (van der Stap, Vos and Mooij, unpublished manuscript; Verschoor et al., 2004). Such experiments may be a valuable addition to the pulse and press perturbation experiments (Abrams, 1995) that have dominated the empirical field during the past decades.

4.0 | WEARING ELTON'S WELLINGTONS: WHY BODY SIZE STILL MATTERS IN FOOD WEBS

Philip H. Warren

> Size has a remarkably great influence on the organization of animal communities. We have already seen how animals form food chains in which the species become progressively larger in size or, in the case of parasites, smaller in size. A little consideration will show that size is the main reason underlying the existence of these food-chains, and that it explains many of the phenomena connected with the food cycle.
>
> ELTON (1927) p. 59

In the laboratory where I did my PhD we had a pair of wellington boots that once belonged to Charles Elton. While the boots weren't exactly objects of reverence, there was something rather special going about sampling food webs wearing Elton's wellingtons! His classic book, *Animal Ecology* (1927), intrigued me as I labored to try and describe, quantify, and understand the structure and dynamics of even a relatively simple food web (Warren, 1989). It didn't answer all the difficult questions about complexity, stability, and interaction strength, but it spoke with encouraging confidence about the importance of simple ideas about size and abundance as organizing principles, cutting across the complexity of animal communities. Elton's thinking was clearly deeply rooted in observation and the details of natural history, but through these he sought patterns, and through them common themes in community organization. One of the factors he saw as central to animal communities, and food chains in particular, was body size. His two central points (Elton, 1927, p. 60 and p. 69) were that:

> *. . . the size of prey of carnivorous animals is limited in the upward direction by its strength and ability to catch the prey, and in the downward direction by the feasibility of getting enough of the smaller food to satisfy its needs, the latter factor being also strongly influenced by the numbers as well as the size of the food.*

and

> *. . . the smaller an animal the commoner it is on the whole.*

> *To put the matter more definitely, the animals at the base of a food-chain are relatively abundant, while those at the end are relatively few in numbers, and there is a progressive decrease between the two extremes.*

It was clear to Elton, as it is to any natural historian who gets involved in the business of looking at what organisms feed on, that size was an accessible metric, common to all animals, which provided a potential link between natural history, the economics of foraging, organization of feeding links, and relative abundance. It was not a complete story, and Elton was not always entirely correct in the mechanisms he suggested (Liebold and Wootton, 2001), but the bold insight that size was a common cause, or correlate, of many phenomena involved in structuring food webs was an important one. This importance is reflected in the inclusion, three quarters of a century later, of a set of chapters on body size effects in food webs, in this symposium on the state of the art in food web ecology. But why is size such an enduring theme?

Current interest in the role of body size in food webs is certainly not just an historical hangover. In fact the importance of size as an organizing factor in food webs lay largely unexplored for a long time after Elton's suggestions were made. Ideas about trophic organization soon began to embrace the concepts of energy flow and community stability. The former, developing out of the work of Lindeman (Lindeman, 1942), leading to an emphasis on describing food webs in terms of energy flow among functional groupings of organisms, or among trophic levels and laying the groundwork for the International Biological Program and a 'systems' approach to understanding communities (McIntosh, 1985). Elton, writing later (Elton and Miller, 1954, p. 464), accepted the role of energy as a fundamental ecological currency, linking the physiological and demographic elements of a system, but he did not see it as displacing population biology in understanding communities.

Another significant consequence of the emphasis on energy flow and trophic level was the subtle transition from Elton's pyramid of numbers, to the description of communities in terms of pyramids of biomass. Although the latter is often seen as the equivalent of the former, this is not strictly the case (Cousins, 1985, 1987, 1996). The pyramid of numbers is a representation of numerical abundance of organisms by size classes which, because in general "*. . . smaller animals are usually preyed upon by larger animals*" (Elton, 1927, p. 70), tends to result in a correlation between position in the pyramid (size) and distance from the producers in the system (measured as the number of trophic links). Elton saw the food cycle as comprised of many food chains, interconnected by species belonging to more than one chain. However, it is the individual chains that are structured by size (successive increases in size for predators, and decreases for parasites). Organisms occurring within one size class in a community can be from different trophic positions in the various food chains to which they belong. The pyramid of biomass rearranges things: here the classificatory variable is 'trophic level' (the number of trophic links energy has passed through from producer level), and the response (magnitude) of each trophic level is measured as the sum of body sizes of all individuals at that level (biomass). Although they are in fact rather different, their superficial similarity seems likely to have contributed to the subsequent relative neglect of the pyramid of numbers in trophic ecology.

However, it was just that, an idea that was neglected rather than superseded. It is probably more accurate to view the pyramid of numbers as the ancestor (albeit with something of a generation gap!) of the recent, and burgeoning, interest in the relationship between body size and population abundance (Damuth, 1981; Peters, 1983; Peters and Raelson, 1984; Lawton, 1990; Cotgrave, 1993; Blackburn and Gaston, 1999; Gaston and Blackburn, 2000). Investigations of this topic include studies of relationships compiled for populations of different species across many communities (Damuth, 1981; Peters, 1983; Robinson and Redford, 1986; Nee et al., 1991; Silva and Downing, 1995), and those for assemblages, or whole communities at a single location (Marquet et al., 1990; Strayer, 1994; Cyr et al., 1997a; Leaper and Raffaelli, 1999). The observations of Elton, and others, have been given quantitative description, cross system comparision, and scrutiny for evidence of the correct mechanisms to explain them. At the same time aquatic ecologists, facilitated at least initially by the ability to automate the enumeration and size measurement of particles from pelagic ecosystems, have been describing communities in terms of biomass spectra: the total biomass of organisms across body size classes (Sheldon and Parsons, 1967; Sheldon et al., 1972; Schwinghamer, 1981; Sprules and Munawar, 1986). Although the two literatures have proceeded rather independently, the approaches are clearly related (see Cohen and Carpenter, Chapter 4.1, and references therein).

This growing interest in size-abundance relationships has proceeded largely independently of other strands of food web ecology. Although some of the work on possible mechanisms to explain biomass spectra has utilized theory involving trophic structure, with size dependent feeding (Kerr, 1974; Dickie et al., 1987; Duplisea and Kerr, 1995), in general the body size-abundance relationship has not been approached as a food web problem. This may be partly because the bulk of the relationships are often for taxonomically restricted collections of species across many communities, rather than entire single communities. However, energy limitation—essentially the argument that similar amounts of energy can support many small, or a few large, individuals—whilst much debated (see Blackburn and Gaston, 1999; Gaston and Blackburn, 2000, for reviews) provides one mechanism with some trophic relevance (Jennings and Mackinson, 2003). Much remains to be done to explore the relationship between size-abundance patterns and trophic structure, but the analyses of Cohen et al. (2003), Reuman and Cohen (2004b), and Cohen and Carpenter (see Chapter 4.1) provide a determined attempt on the problem, using detailed data from a single well described food web, which draw these elements together, and provide both bold analyses of Elton's variables, and a detailed evaluation of the methods used to extract information from them. Such analyses—of species from a single web, rather than just collating data on individual species from very different communities—allow investigations of mechanisms in ways that would not otherwise be possible (Leaper and Raffaelli, 1999). Most importantly, by more precisely defining what a food web looks like, such analyses can reveal new patterns against which theory can be tried (Brown and Gillooly, 2003), and provoke new questions (Reuman and Cohen, 2004b).

This type of 'top-down' approach to food webs, dissecting apart patterns in whole webs, is productively complemented by 'bottom-up' approaches, in which body size is also providing a core component of theories seeking to predict food web structure, or function, as a product of the rules that determine individual feeding interactions.

Elton clearly saw much of the structure in food cycles as deriving from size limits to trophic interactions (Elton, 1927, p. 61), in all the intriguing detail of their natural history. However, only more recently have models of the basic structural properties of food webs been interpreted in body-size terms (Warren and Lawton, 1987; Cohen, 1989; Cohen et al., 1993a), leading in turn to more explicit niche based models of the arrangement of trophic links in webs (Warren, 1996; Williams and Martinez, 2000; Cattin et al., 2004). These suggest that many aspects of food web structure could be a consequence of simple niche constraints on individual resource-consumer pairs. There are many possible niche dimensions, but size is important because it is a continuous dimension, which imposes broadly consistent constraints across many species. Although models of webs in terms of binary links (connectance webs) have their limits for understanding community dynamics (Paine, 1980, 1988; Polis, 1991; Winemiller and Polis, 1996), it is worth trying to understand them for a number of reasons. First, recent application of techniques from network analysis and topology have indicated that there may be useful things to be gleaned about whole web behavior, such as consequences of extinction, even from connectance webs (Solé and Montoya, 2001; Dunne et al., 2002a; Williams et al., 2002). Second, the mechanisms responsible for constraining the distribution of links in a web is unlikely to be entirely distinct from those determining the strength, energy flow, and variability of these links. As Elton (1927, p. 60) observed:

> *The food of every carnivorous animal lies therefore between certain size limits. . . . There is an optimum size of food which is the one usually eaten, and the limits actually possible are not usually realised in practice.*

The frequency and efficiency of feeding interactions will vary within the limits of the possible. Explanations of this variation lie in the province of foraging theory, but size often provides an important axis, through its simultaneous relationships with key variables such as energy content of prey, energy demand of predators, ease of attack and handling prey, foraging range, and population density (Peters, 1983). Food web theory has made little explicit use of foraging theory (though see Hastings and Conrad, 1979; Cousins, 1985; Stenseth, 1985; Abrams, 1996), but this view of niche constraints suggests that, as well as determining which links are possible, size may be important in influencing the relative strengths of the interactions that do occur.

A number of recent studies have developed the idea of using the allometries between size and these variables to provide a biologically consistent framework for otherwise difficult process of choosing relative parameter values of consumer-resource models (Yodzis and Innes, 1992; McCann and Yodzis, 1994; Jonsson and Ebenman, 1998; Emmerson and Raffaelli, 2004a). Emmerson et al. (see Chapter 4.3) develop this idea by showing that, for a Lotka-Volterra system, via allometries of size and feeding rate, biomass, and abundance, estimates of per capita interaction strengths can be derived. The linkage strength patterns of two webs parameterized in this way result in webs that are much more likely to be locally stable than randomly permuted analogues: an intriguing outcome (with echoes of Yodzis' (1981) results for 'plausible' web models), suggesting that there are strong non-random features of web organization driven by the biology of the component organisms. The extent to which these are causal of, caused by, or coincidental with, stability in natural systems remains an important problem.

These approaches to estimation of interaction strength operate at the level of the population; the complexities of dealing with an entire food web are kept under control by assuming homogeneity (all individuals in a population are the same) and fixed properties (characteristics of individuals do not alter behaviorally, change in response to population densities, or evolve, over time) (Abrams, 1996). Incorporating such complexities into large food web models is a daunting task, but it is clearly important to investigate the effects of heterogeneities and the dynamics of foraging behavior to judge the extent to which they might affect our understanding of the behavior of food webs in models that ignore them (Abrams, 1996). Size provides a natural route to tackling this problem: many of the cross-species scalings that affect organisms and their interactions carry over to intraspecific variation; indeed size effects will often override species identity—a late instar damselfly larva feeds voraciously on the early instar nymphs of waterboatmen (Corixidae), and on adult copepods, because they are of similar size and movement characteristics, although as an adult the waterboatman would be far too large to be eaten, and the nauplii of the copepod probably too small. This variation in the likelihood or intensity of an interaction across species' life-stages, is well known—Hardy's (1924) well known food web showing ontogenetic diet changes in the herring was used as an illustration in Elton's book (1927)—and its effects on patterns in web structure have been documented (Warren 1989; Woodward and Hildrew 2002). However, it can also generate complex dynamics (see De Roos et al., 2003b, for a review), and it is aspects of these that Persson and De Roos explore here (see Chapter 4.2).

Using one consumer–one resource models that track the feeding, growth, and reproduction of individuals within populations, and in which the processes driving the model are all occurring at the individual level, Persson and De Roos illustrate both that size-dependent scaling of foraging (attack rate) in a seasonally reproducing consumer, has a strong tendency to generate non-equilibrium dynamics (see also

Persson et al., 1998) and that size-dependent cannibalism may, or may not, be stabilizing, depending on the exact size threshold for cannibalistic interactions, and the scaling of attack rate with size. The biological constraints being modelled here are features we might expect to be quite common, and there are examples from natural and experimental systems of the types of dynamics predicted (Persson et al., 1998; see Persson and De Roos, Chapter 4.2). It is clear that there are new levels of dynamical complexity revealed by allowing size-dependent interactions among individuals, the challenge of examining how these effects scale to multispecies food webs, and how this affects an understanding largely derived from the dynamics of models based on fixed species properties, remains.

So why is size important for understanding food webs? Woodward et al. (see Chapter 4.4) step back from the detail and provide an overview of the different levels and processes through which the influence of organism size can operate. They identify a range of aspects of food web structure which are related to size—some touched on here, others not—and some speculations about future directions. In many ways, the breadth of issues touched on in that chapter answers the question. Here I add just a few additional thoughts about what makes size an interesting topic for food web ecology.

First, size is a property that blurs the distinction between species and individuals; thinking in size terms reorientates our thinking away from trophic levels and, to some extent, away from species. In the most extreme view this leaves us with a food web composed of individuals of certain sizes interacting with other individuals on the basis of size, concentrating energy into packages (bodies) of particular sizes and thus making it accessible to others of sizes different again. This sort of view lends itself to individual-based approaches to modelling such systems, which might provide a useful complement to approaches based on population models, or trophic levels.

Second, size is a property of individuals (whether we choose to work with it at that, or on a more aggregated level), and individuals lie closer to the level at which natural selection is operating than do populations. This may not matter directly for predicting ecological dynamics, but it is important for developing an understanding of the causal pathways by which the size structures that produce these dynamics come about (Lundberg and Persson, 1993). These questions may be asked at scales from the evolution of optimal diet choice, though to the factors governing the (generally positively skewed) frequency distribution of species body sizes in a community. Are the latter driven by processes in the food web, or are food webs just samples from distributions produced by size-related speciation and extinction rates at larger scales (Gaston and Blackburn, 2000)? If the evolution of body size, and individuals' feeding strategies are understood, then can food webs be understood largely as emergent properties of these?

Third, size affects not just how an organism interacts with others, but also how it interacts with its environment: how far it ranges, how much environmental space it experiences (Williamson and Lawton, 1991), and how it is affected by environmental variation (Peters, 1983). This raises a number of questions about the mechanisms that might drive relationships between web structure and the spatial structure and variability in the environment; these ideas have been explored rather little (but see Jennings and Warr, 2003, for an interesting example of the latter).

Lastly, perhaps the most prosaic point: data on body size is (relatively) straightforward to collect. Whilst an apparently rather trivial observation to finish with, it is worth noting that despite this, in quantitative terms, what we know about body size relations in whole food webs from direct observation comes from a very limited number of data sets, with still fewer of these having abundance or biomass information as well. It is ironic to note that, in his book, Elton (1927, p. 60) observed:

> *We have very little information as to the exact relative sizes of enemies and their prey, but future work will no doubt show that the relation is fairly regular throughout all communities.*

I would hazard a guess that Elton did not imagine quite how far into the future this work would be. Information has accumulated gradually, mostly as a by-product of studies in which measurement of consumer and resource size was not the main focus of the work, and only recently have these these scattered resources become subject to more systematic compilation and analysis to look specifically at the form of predator–prey size relationships, and their implications for food webs (Peters, 1983; Vezina, 1985; Warren and Lawton, 1987; Cohen et al., 1993a). If we are to develop the ideas discussed in the chapters in this section, we need to give thought to methods for the systematic collection, compilation, and dissemination of 'rich' food web data, including size, abundance, and temporal change. Recent datasets (Leaper and Raffaelli, 1999; Woodward and Hildrew, 2002a; Cohen et al., 2003; Cohen et al., 2005) make clear the possibilities. This is work firmly in the tradition of Elton's "*scientific natural history*" with the tantilizing prospect of a framework that links foraging ecology, population biology, and energy flow. It's not time to put Elton's wellingtons away just yet.

ACKNOWLEDGMENTS

I am grateful to ESF for financial support. I am also grateful to Owen Petchey for helpful comments on this chapter.

4.1 | SPECIES' AVERAGE BODY MASS AND NUMERICAL ABUNDANCE IN A COMMUNITY FOOD WEB: STATISTICAL QUESTIONS IN ESTIMATING THE RELATIONSHIP

Joel E. Cohen and Stephen R. Carpenter

The quantitative patterns and mechanisms of the relationship between body mass and abundance have been examined in many ecological settings (Colinvaux, 1978; Damuth, 1981; Peters, 1983; Griffiths, 1992; Brown, 1995a, p. 94; Griffiths, 1998; Blackburn and Gaston, 1999; Leaper and Raffaelli, 1999; Kerr and Dickie, 2001; Russo et al., 2003).

The purpose of this chapter is to examine in detail some of the statistical foundations of estimating the quantitative relationship between average body mass and abundance. The empirical example analyzed here is the pelagic food web (hereafter, simply "web") of Tuesday Lake, Michigan (Carpenter and Kitchell, 1993a). The data, provided by Stephen R. Carpenter, are given in full by Jonsson et al. (2005).

Blackburn and Gaston (1999, p. 182) referred to the study of Damuth (1981) as "the single most influential study of the interspecific abundance-body size relationship." Blackburn and Gaston plotted Damuth's 467 data points on 307 species of mammalian terrestrial primary consumers in a graph in which the ordinate (vertical axis) was log N ($\log_{10}$ number of individuals per km^2) and the abscissa (horizontal axis) was log M ($\log_{10}$ body mass, not further specified as to average, adult, or maximal). According to Blackburn and Gaston, the ordinary least squares regression of the linear model

$$\log(N) = \beta_1 \log(M) + \gamma_1 \tag{1}$$

gave an estimated slope of $\beta_1 = -0.75$. Other data yielded other estimates of the slope (Peters, 1983; Peters and Wassenberg, 1983; Peters and Raelson, 1984; Griffiths, 1992; Cyr et al., 1997a; Cyr et al., 1997b; Griffiths, 1998; Leaper and Raffaelli, 1999). (Throughout the appendices of Peters (1983), the columns headed "Independent variable" should be headed "Dependent variable.")

Enquist et al. (1998, p. 164), Cohen et al. (2003, p. 1784), and most plant ecologists plotted log M on the ordinate and log N on the abscissa. Many approximated the data by ordinary least squares regression of the linear model

$$\log(M) = \beta_2 \log(N) + \gamma_2 \tag{2}$$

If all data fell exactly on a straight line, then both equations would be exact descriptions, the slope coefficients would be related by $\beta_1 = 1/\beta_2$, and the choice between Eqn. 1 and Eqn. 2 would not matter. But $\beta_1 = 1/\beta_2$ need not hold when there is random variation in the relation between body mass and abundance. It is not clear initially whether Eqn. 1 or Eqn. 2 is a more useful description of patterns in a community of plant and animal species with different modes of growth, some determinate and some indeterminate.

The results show that the two regressions may appear contradictory unless careful attention is paid to how well the data satisfy the assumptions of each linear regression model. Most of the assumptions underlying ordinary least squares regression of the linear model Eqn. 1 cannot be rejected by the Tuesday Lake data. An important limitation of this conclusion is that the assumption of negligible error variance in the measurement of log M cannot be tested with the available data. Most of the assumptions underlying Eqn. 2 can be rejected by the Tuesday Lake data. This information is sufficient to guide the choice of a linear regression model. However, the direction of causation cannot be determined on statistical grounds from static observational data.

MATERIALS AND METHODS

Data

Tuesday Lake is a mildly acidic lake in Michigan (89°32' W, 46°13' N) (Carpenter and Kitchell, 1993a) with surface area 0.9 ha. The fish populations were not exploited and the drainage basin was not developed. Data collected in 1984 and again in 1986 consisted of a list of species, and for each species, its predator species and its prey species (for the body sizes and life stages present in the lake in each year), its average body mass M (kg fresh weight per individual), and its numerical abundance N (individuals/m^3 in the non-littoral epilimnion, where the trophic interactions take place). In 1985, the three planktivorous fish species were removed and replaced by a single piscivorous fish species. In 1984, the 56 biological species consisted of 31 phytoplankton species, 22 zooplankton species, and 3 fish species. In 1986, the 57 species consisted of 35 phytoplankton species, 21 zooplankton species, and one fish species. Jonsson et al. (2005) gave details of sampling methods, the raw data, and significant limitations of the data, including the delicate estimation of the density of zooplankton species. Our analyses used biological species rather than trophic species. Two or more biological species are "lumped" into a single trophic species if the two or more biological species have an identical set of consumer species and an identical set of resource species. There were 27 trophic species in 1984 and 26 trophic species in 1986 (Jonsson et al., 2005). Cohen et al. (2003) and Reuman and Cohen (2004a, 2004b) analyzed other aspects of the data.

Definitions

The biomass B (kg/m^3) of a species is its average body mass M times its numerical abundance N. Log throughout means $\log_{10}$.

The rank(M) of a species equals 1 if that species has the largest value of M, equals 2 if that species has the second largest value of M, and so on; the rank is the order from largest to smallest average body mass. In 1984, the 56 body masses were described by 55 unique values; in 1986, the

57 body masses were described by 55 unique values. The species with tied body masses retained the ordering of the listing of species in the data appendices of Jonsson et al. (2005). Rank(N) is similarly defined by rank ordering the species from largest to smallest values of N and by retaining the ordering of Jonsson et al. (2005) in case of ties. The rank (with values 1, 2, 3, ...) increases as the numerical value of the variable decreases.

Theory

In the absence of measurement error or random fluctuation, numerical abundance N depends allometrically on average body mass M if and only if biomass B depends allometrically on M. For if $b > 0$ is the exponent in an allometric relation $N \propto M^{-b}$, then $B = MN \propto M^{1-b}$ and conversely. Thus the study of the relation between the numerical abundance and the body mass of biological species is related to the study of the biomass spectrum, as has been widely recognized (Platt and Denman, 1977, 1978), but if the biomass spectrum uses size classes rather than biological species as the unit of analysis, then the relation depends on the number of biological species within each size class. Likewise, N depends allometrically on M if and only if M depends allometrically on N, for $N \propto M^{-b}$ if and only if $M \propto N^{-1/b}$, so biomass B also depends allometrically on numerical abundance N according to $B = MN \propto N^{1-1/b}$.

The two cases of special interest are b = 3/4 and b = 1. If b = 3/4 (as Damuth, 1981 suggested), then $N \propto M^{-3/4}$ so $B = MN \propto M^{1/4}$, and the biomass of a species increases with average body mass. If b = 1, then $N \propto M^{-1}$ and B is the same for every species, regardless of M or of N or of rank(M) or of rank(N). In this case, c = MN implies that log N = log c – log M and log M = log c – log N, that is, all species should fall on a line with slope –1 in the plane with coordinates log M and log N, regardless of which coordinate is the abscissa and which is the ordinate.

Methods

Computations were done using the statistics toolbox and other functions of Matlab, version 6.5.0.180913a (http://www.mathworks.com/, Release 13). Linear regressions were done using 'regress.' All regressions used all species, whether they were connected to the main web or isolated (i.e., not connected to most other species).

Five principal assumptions must be satisfied to justify the probability values and confidence intervals generated by ordinary least squares linear regression analysis of the model $y=a+bx+\varepsilon$. In a sample of size n, for any data point (x_i, y_i), with $i = 1, \ldots, n$, the residual r_i is defined as the difference in the vertical direction $r_i = y_i - (a+bx_i)$ between the observed ordinate y_i and the predicted ordinate $y_{pred,i} = a+bx_i$ (pred = predicted) given by the linear model, where a and b are the least-squares estimates of the regression coefficients. The five assumptions are *linearity* of the average (conditional expectation) of y as a function of x, *normality* of the residuals ε (with unknown variance), *homoscedasticity* (i.e., the variance of the residuals is independent of x), serial *independence* of the residuals with increasing x, and *no error in the value of x*. A diagnostic Matlab function 'regressiontest' was written to assess the validity of the first four of these assumptions for any pair x, y of data vectors of equal length.

This function evaluated *linearity* in two ways, by classical hypothesis testing and by using the Akaike information criterion. The classic test of nonlinearity was performed by fitting a quadratic equation $y=a+bz+cz^2+\varepsilon$, where $z=x-\text{mean}(x)$ was a centered translation of x used to reduce the collinearity between x and x^2. The null hypothesis of linearity was rejected if the confidence interval for the parameter c did not include 0.

To use the Akaike information criterion, the corrected Akaike information criterion AIC_c (Burnham and Anderson, 2002, pp. 63, 66) was computed once for the residuals from the linear model and then again separately for the residuals from the quadratic model, using in each case the formula:

$$AIC_c = n\log\left(\sum_{i=1}^{n} r_i^2 / n\right) + 2Kn/(n-K-1). \qquad (3)$$

For the linear model, K=3. For the quadratic model, K=4. Intuitively speaking, Eqn. 3 says that, in its application to ordinary least squares regression, AIC_c is a rescaling of the mean squared residual, and the smaller AIC_c the better the model fits. Then

$$\Delta = |\, AIC_c\,(\text{linear model}) - AIC_c\,(\text{quadratic model})\,| \qquad (4)$$

was computed to compare the goodness of fit of the linear and quadratic models for the same set of (x, y) data. Although statistical significance could not be assigned to any value of Δ, a value of 10 or more is interpreted to mean that the model with the higher AIC_c is essentially without empirical support (Burnham and Anderson, 2002, e.g., p. 226).

Whether the residuals from the regression line were *approximately normal* with unknown variance was examined in three ways. First, for visual inspection, the Matlab function 'qqplot(x)' plotted the quantiles of the residuals r_i as a function of the quantiles of the normal distribution. The more nearly the residuals were normally distributed, the more nearly the quantile-quantile plot approximated a straight line. Second and third, the Matlab functions 'jbtest' performed the Jarque-Bera test and 'lillietest' performed the Lilliefors test of normality with unknown mean and variance. Neither test was impressively sensitive to deviations from normality. For example, at the 0.01 level of significance, 'lillietest' failed to reject normality as a model for the first 358 natural numbers 1, 2, . . . , 358, but did reject normality as a model for the first 359 natural numbers. Similarly, at the 0.01 level of significance, 'jbtest' failed to reject normality as a model for the first 147 natural numbers but did reject normality as a model for the first 148 natural numbers. For samples as small as the 56 species in 1984 or the 57 species in 1986, these tests would not easily detect minor deviations from normality.

If the residuals were *homoscedastic*, there should be no trend in the residuals, or in the absolute value of the residuals, as a function of either the independent variable x or of the predicted value of the dependent variable y_{pred}. Homoscedasticity was tested by fitting linear and quadratic regressions of r_i against $y_{pred,i}$. Homoscedasticity was further tested by fitting linear and quadratic regressions of the absolute value of the residuals $|r_i|$ against $y_{pred,i}$. The null hypothesis of homoscedasticity was rejected if any of the confidence intervals (from either the linear or the quadratic models) of the coefficients of the linear or quadratic terms did not include 0.

The *independence* of successive residuals was tested by sorting the residuals in the order determined by increasing y_{pred} and then comparing computed values of the Durbin-Watson statistic for the residuals against tabulated critical values (Stuart and Ord, 1991, p. 1077 for the formula, pp. 1245–1246 for critical values). The computation of the Durbin-Watson statistic was programmed from scratch. Each numerical result was identical to that obtained from the publicly available Matlab function dwatson.m, accessed from http://econpapers.hhs.se/software/bocbocode/t850802.htm on January 15, 2004.

Testing the assumption of no error in x would require replicate measurements of *M* and *N* for all species. Unfortunately such data are not available for Tuesday Lake. See the discussion and conclusion subsection on *Errors in measurement.*

RESULTS

Regression Coefficients of Allometric Relations

Scatter plots of log N as a function of log M in 1984 (Figure 1A) and 1986 (Figure 1B) suggested that a linear relation is plausible. When log N was regressed as a linear function of log M (following Damuth, 1981) separately in 1984 and 1986, the point estimates of the slope coefficient were –0.83 and –0.74 (Table 1). The 99% confidence intervals for the slopes included –0.75 and excluded –1. Scatter plots of log B as a function of log M in 1984 (Figure 2A) and 1986 (Figure 2B) suggested no very clear, but perhaps an increasing, relation. When log B was regressed as a linear function of log M separately in 1984 and 1986, the point estimates of the slope coefficient were 0.17 and 0.26. Since log B = log M + log N, it follows mathematically, and is observed numerically, that $\beta_1 + 1 = \mu_1$ and that $\beta_2 + 1 = \nu_1$, using the notation of Table 1. Both 99% confidence intervals for the slopes included 0.25 and excluded 0. The regression coefficients of log B as a linear function of log rank(M) were significantly negative

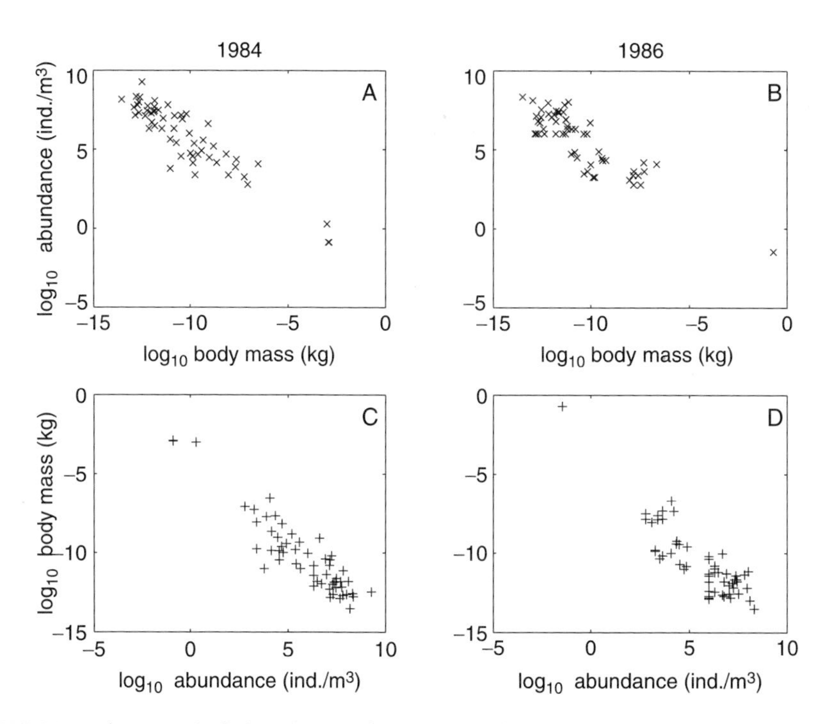

FIGURE 1 | Numerical abundance of species as a function of average body mass in Tuesday Lake, Michigan, in 1984 (**A**) and 1986 (**B**) and average body mass as a function of numerical abundance in 1984 (**C**) and 1986 (**D**).

Table 1: Slope coefficients of linear regressions between log numerical abundance log N and log average body mass log M of all species in Tuesday Lake in 1984 (56 species) and 1986 (57 species), and regressions of log biomass log B as a function of log M, log N, log rank(M), and log rank(N). The 99% confidence intervals for the slopes are in parentheses. The putative p value is the nominal statistical significance with which the data reject the null hypothesis that the slope is zero.

x	y	year	slope param-eter	point estimate (99% confidence interval)	r^2	putative p
log M	log N	1984	β_1	−0.8271 (−0.96, −0.70)	0.84	<0.001
log N	log M	1984	β_2	−1.0178 (−1.18, −0.86)	0.84	<0.001
log M	log N	1986	β_1	−0.7397 (−0.89, −0.59)	0.75	<0.001
log N	log M	1986	β_2	−1.0149 (−1.23, −0.80)	0.75	<0.001
log M	log B	1984	μ_1	0.1729 (0.04, 0.30)	0.19	<0.001
log N	log B	1984	ν_1	−0.0178 (−0.18, 0.14)	0.002	0.77
log M	log B	1986	μ_1	0.2603 (0.11, 0.41)	0.27	<0.001
log N	log B	1986	ν_1	−0.0149 (−0.23, 0.20)	0.001	0.85
log rank (M)	log B	1984	μ_2	−1.0762 (−1.89, −0.26)	0.19	<0.001
log rank (N)	log B	1984	ν_2	−0.2750 (−1.17, 0.62)	0.01	0.42
log rank (M)	log B	1986	μ_2	−1.4506 (−2.31, −0.59)	0.27	<0.001
log rank (N)	log B	1986	ν_2	−0.5807 (−1.57, 0.40)	0.04	0.12

($p < 0.001$). These statistical results are all consistent with $N \propto M^{-3/4}$ and inconsistent with $N \propto M^{-1}$.

When the independent variable is changed from average body mass to numerical abundance, scatter plots of log M as a function of log N in 1984 (Figure 1C) and 1986 (Figure 1D) likewise suggested that a linear relation is plausible. However, when log M was regressed as a linear function of log N separately in 1984 and 1986, the point estimates of the slope coefficient were −1.02 and −1.01 (Table 1). The 99% confidence intervals for the slopes excluded −1.33 = 1/(−0.75) and included −1. Scatter plots of log B as a function of log N in 1984 (Figure 2C) and 1986 (Figure 2D) suggested no clear, but perhaps a decreasing, relation. When log B was regressed as a linear function of log N separately in 1984 and 1986, the point estimates of the slope coefficient were −0.02 and −0.01. Both 99% confidence intervals for the slopes included 0 and

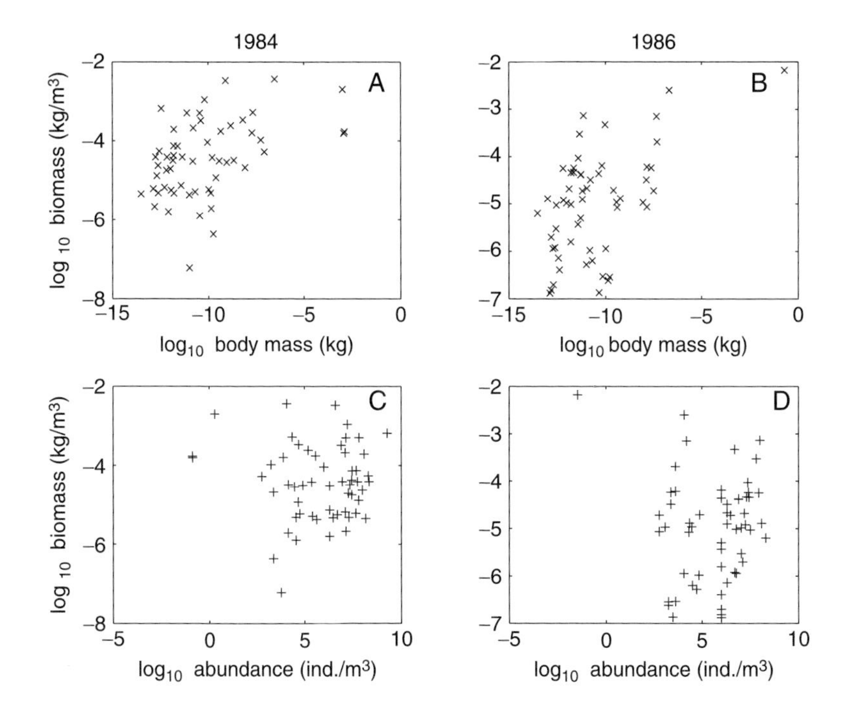

FIGURE 2 | Biomass of species as a function of average body mass in Tuesday Lake, Michigan, in 1984 (**A**) and 1986 (**B**) and biomass as a function of numerical abundance in 1984 (**C**) and 1986 (**D**).

excluded –1/4 and –1/3. The regression coefficients of log B as a linear function of log rank(N) were not significantly different from 0 ($p > 0.12$). These statistical results are all consistent with $N \propto M^{-1}$ and inconsistent with $N \propto M^{-3/4}$

The 99% confidence interval (–1.18, –0.86) for β_2, the 1984 slope of Eqn. 2, implies a confidence interval (1/–0.86, 1/–1.18) = (–1.16, –0.85) for the slope coefficient in a linear equation of the form of Eqn. 1 (though not for the least squares fit of Eqn. 1), and this interval intersects the 99% confidence interval (–0.96, –0.70) for β_1, the slope coefficient in Eqn. 1, in the range (–0.96, –0.85), which is incompatible with Damuth's estimate of slope of –3/4 and with constant biomass across species (slope –1). Similarly, for 1986, the 99% confidence interval (–1.23, –0.80) for the 1986 slope of Eqn. 2 implies a confidence interval (1/–0.80, 1/–1.23) = (–1.25, –0.81) which intersects the 99% confidence interval (–0.89, –0.59) for the slope coefficient in Eqn. 1 in the range (–0.89, –0.81), which is incompatible with Damuth's estimate of slope of –3/4 and with constant biomass across species (slope –1). If one hopes that the slope coefficients are the same in 1984 and 1986, then the intersecting confidence intervals for the slope coefficient in Eqn. 1 in 1984 of (–0.96, –0.85) and in 1986 of (–0.89, –0.81) themselves intersect only in the range (–0.89, –0.85).

To resolve the contradiction between the apparent conclusions in the first two paragraphs of this section of Results regarding the slopes –3/4 and –1, we investigated whether the data conformed to the assumptions that justified the computed p values and confidence intervals.

Conformity of Data to Regression Assumptions

Quantile-quantile plots of the residuals vs. a standard normal distribution were examined visually. Discrepancies from a straight line were apparent in some cases. However, the Jarque-Bera and Lilliefors tests failed to reject the null hypothesis of composite normality for the distribution of the residuals even at the 0.05 level of significance in any of the regressions. In none of the regressions was there significant evidence even at the 0.05 level of significance for a linear trend or for serial correlation in the residuals. In these respects, the data did not deviate detectably from the assumptions underlying the linear regression model.

In other respects, the data rejected some assumptions underlying the linear regression model (see Table 2). Quadratic regressions showed that log *M* and log *B* were significantly nonlinear functions of log *N*: the coefficient of the quadratic term differed from 0 according to the classical test of statistical significance with $0.01 < p < 0.05$ in 1984 and $p < 0.01$ in 1986 (Table 2, column 4). The strong rejection of linearity for both models in 1986 was exactly confirmed by the Akaike information criterion: in these two regressions, and in no others in Table 2 (column 5), AIC_c for the linear model exceeded AIC_c for the quadratic model by more than 10. In short, classical hypothesis testing (with $p < 0.01$) and the Akaike information criterion (with $\Delta > 10$) agreed perfectly in identifying marked nonlinearities. When the evidence against linearity was weaker, as in 1984, the classic test and the Akaike information criterion agreed well though imperfectly. Excepting only the value of Δ in 1986 for the regression of log *B* as a function of log rank(*N*), shown in the last line of Table 2, all of the values of Δ between 2.27 and 10 identified models that the classic test rejected as nonlinear with $0.01 < p < 0.05$, and vice versa.

As the models in which log *M* and log *B* were quadratic functions of log *N* were markedly better in 1986 than the models in which log M and log *B* were linear functions of log *N*, it is of biological interest to report the estimated 99% confidence intervals of the coefficients of the quadratic models for 1986. Let $z = \log N - \text{mean}(\log N)$ be the centered log numerical abundance of each species. Then in 1986:

Table 2: Quantitative tests of assumptions underlying putative p values in Table 1. Cells left blank represent no statistically significant violation of the assumptions of linear regression due to the property in the column heading. Cells with + represent a statistically significant violation with probability between 0.01 and 0.05. Cells with ++ represent a statistically significant violation with probability < 0.01. Columns are omitted for the Jarque-Bera and Lilliefors tests of composite normality, for the presence of linear trend in the residuals, and for the presence of serial correlation in the residuals because none of these tests was significant at the 0.05 level. In columns 6 and 7, AIC_c is defined in Eqn. 3. In column 5, Δ is defined in Eqn. 4.

column 1	2	3	4	5	6	7	8	9	10
x	y	year	classic non-linearity	Δ	AIC_c linear model	AIC_c quadratic model	residuals quadratic	absolute residuals linear	absolute residuals quadratic
log *M*	log *N*	1984		2.26	–9.96	–7.71			
log *N*	log *M*	1984	+	2.28	1.65	–0.63	+		
log *M*	log *N*	1986		2.02	–2.77	–0.75			
log *N*	log *M*	1986	++	10.38	15.26	4.88	++	+	
log *M*	log *B*	1984		2.26	–9.96	–7.71			
log *N*	log *B*	1984	+	2.28	1.65	–0.63	+		
log M	log *B*	1986		2.02	–2.77	–0.75			
log *N*	log *B*	1986	++	10.38	15.26	4.88	++	+	
log rank(*M*)	log *B*	1984		2.09	–9.89	–7.80			
log rank(*N*)	log *B*	1984		1.65	1.05	2.70		+	
log rank(*M*)	log *B*	1986		0.42	–2.57	–2.99			
log rank(*N*)	log *B*	1986		2.32	12.78	15.10		++	+

$$\log M = (-11.29, -10.46) + (-1.07, -0.62)z + (+0.02, +0.15)z^2, \quad (5)$$

$$\log B = (-5.70, -4.87) + (-0.07, +0.38)z + (+0.02, +0.15)z^2. \quad (6)$$

Because $\log B = \log M + \log N$, it is expected mathematically (and observed numerically here) that the confidence intervals (and point parameter estimates) of the coefficients of the quadratic terms in Eqn. 5 and Eqn. 6 are identical, and that the confidence intervals (and point parameter estimates) of the coefficients of the linear terms in Eqn. 5 and Eqn. 6 differ by exactly 1. It is of more biological interest to compare the interval (–1.07, –0.62) of the linear coefficient in Eqn. 5 with the interval (–1.23, –0.80) of the linear coefficient of the corresponding *linear* model in Table 1, line 4: while both intervals include a slope estimate of –1, the latter interval excludes both –2/3 and –3/4 while the former interval (which incorporates the influence of the quadratic term) includes both –2/3 and –3/4. Similarly, the interval (–0.07, +0.38) of the linear coefficient in Eqn. 6 includes both 1/3 and 1/4, while the interval (–0.23, –0.20) of the linear coefficient of the corresponding *linear* model in Table 1, line 8, excludes both 1/3 and 1/4.

The residuals of the regressions of log *M* and log *B* as linear functions of log *N* had significant quadratic dependence on the predicted value of the linear regression, and therefore on log *N* (Table 2, column 8). In 1984 and 1986, linear regressions of log *B* on log rank(*N*) had absolute residuals that had a statistically significant linear dependence (column 9), and in 1986 a significant quadratic dependence (column 10), on the predicted value of the linear regression, and therefore on log rank(*N*). All 6 regressions in which the independent variable (x) was log *N* or log rank(*N*) displayed statistically significant deviations from the assumptions of linear regression, whereas none of the 6 regressions in which the independent variable (x) was log *M* or log rank(*M*) displayed any statistically significant deviation.

Artificial Example

The following artificial example emphasizes that the linear regression of y on x may satisfy the underlying assumptions while the linear regression of x on y may not. Suppose the random variable X took the values 1, 2, 3, or 4 each with probability 1/4, and suppose the random variable Y took the values $1 + 10X + \varepsilon$, where each realization of ε was an independent normal random variable with mean 0 and variance 1. By construction, (X, Y) satisfied perfectly the assumptions of the linear regression model: normality, linearity, homoscedasticity, independence, and exact knowledge of X. Figure 3A illustrates the distribution of (X, Y) with 100 values of ε, and therefore 100 values of Y, for each value of X. The quantile-quantile plot (see Figure 3B), the distribution of the (signed) residuals (see Figure 3C), and the distribution of the absolute residuals (see Figure 3D), as well as all of the quantitative statistics, were compatible with the assumptions of linear regression.

When X and Y were exchanged, the graph of the function (Figure 4A) approximated a step function. While the marginal distribution of the residuals was very nearly normal (see Figure 4B), neither the residuals nor the absolute residuals were serially independent for increasing values of the linear prediction (see Figure 4 C,D). The Durbin-Watson statistic easily detected these departures from serial independence.

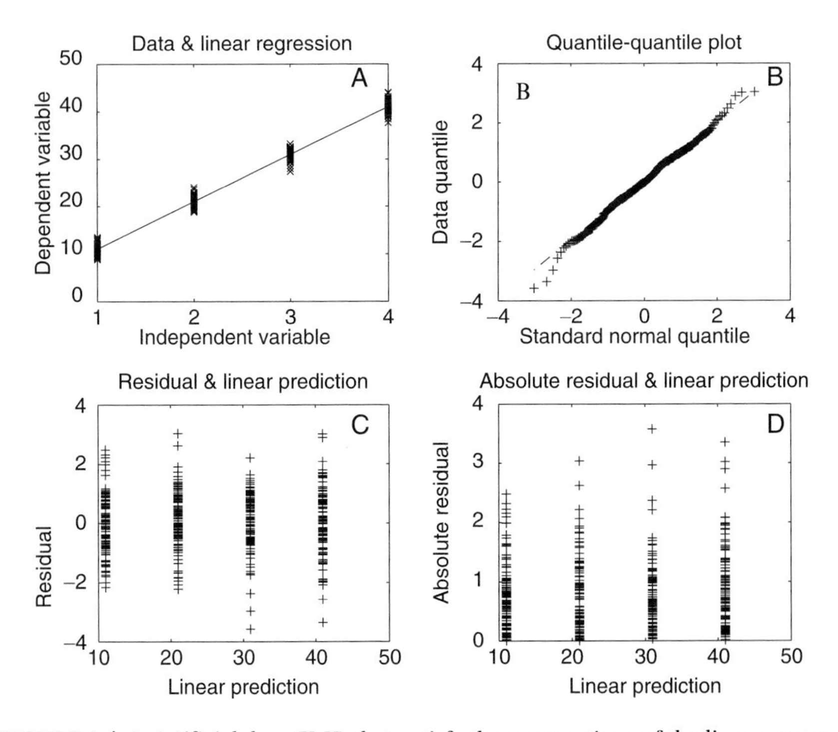

FIGURE 3 | A, Artificial data (X, Y) that satisfy the assumptions of the linear regression model regarding normality, linearity, homoscedasticity, independence and no error variance in x (see text for details). **B,** Quantile-quantile plot. **C,** Residuals as a function of the linear prediction. **D,** Absolute residuals as a function of the linear prediction.

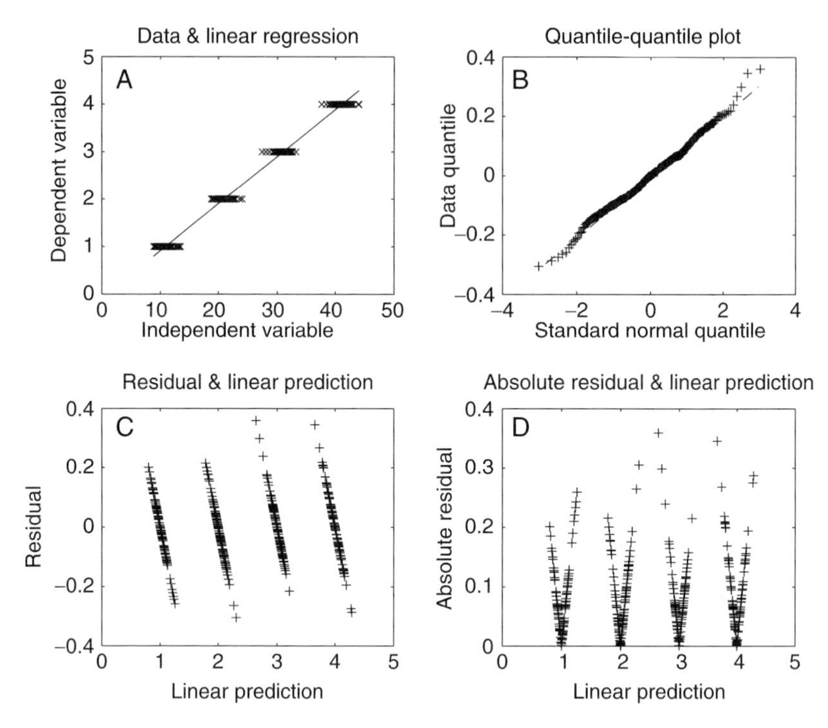

FIGURE 4 | **A,** The same artificial data as in Figure 3 but with axes exchanged: (Y, X) data do not satisfy linearity, homoscedasticity or independence. **B,** Quantile-quantile plot. **C,** Residuals as a function of the linear prediction. **D,** Absolute residuals as a function of the linear prediction.

DISCUSSION AND CONCLUSIONS

Principle Findings

As many have pointed out before, studies of the connection between body size and numerical abundance should carefully investigate whether the probability statements associated with linear regression are justified by the properties of the data being analyzed (Strathmann, 1990; Blackburn and Gaston, 1997, 1998, 1999; Russo et al., 2003; and references they cite). The present study may be the first to examine carefully both choices of independent variable, body size and numerical abundance, for a single set of data. These data include essentially all plant and animal species in a single pelagic community with a known food web. In linear regressions of the Tuesday Lake data, it is more defensible to use $\log M$ or $\log \text{rank}(M)$ as an independent variable and $\log N$ or $\log B$ as the dependent variable than it is to use $\log N$ or $\log \text{rank}(N)$ as an independent variable and $\log M$ or $\log B$ as the dependent variable. On the other hand, using log N as the independent variable seems to be more sensitive at picking up a nonlinear relationship, if that nonlinearity is real. The generality of these conclusions for other webs remains to be determined.

Ecological Significance

Choosing an appropriate statistical model matters ecologically. For example, in Table 1, in the models where the underlying statistical assumptions were not rejected by the data, the 99% confidence intervals for the allometric exponent β_1 did not include −1, but rather lay above −1. Equivalently, the 99% confidence intervals for the allometric exponent μ_1 lay above 0. That is, in Tuesday Lake in 1984 and 1986, the bigger the average individual body mass of a biological species, the larger the biomass of that species, on the average.

The important conclusion that the biomass of biological species in Tuesday Lake increased with the average individual body mass of biological species was not evident in Table 1 when the independent variable was log N. However, in these cases the confidence intervals of the slopes were statistically unjustified. If unjustified models were to be used, an important conclusion about the upward trend of biomass with increasing average individual body mass of biological species in the Tuesday Lake pelagic food web would be lost.

Variation Among Studies

Among the many studies of the relation between numerical abundance and body size, Leaper and Raffaelli (1999, their Table 2) tabulated 31 estimates of the exponent b in the relation $N \propto M^{-b}$, in addition to two estimates of their own for the Ythan estuary. Kerr and Dickie (2001) reviewed many studies of biomass spectra, which are related through the frequency distribution of species according to body size. These studies differ in numerous significant respects, including: the units of observation, the universe of units of observation, the measurement of body size, and the choice of independent variable (body size or abundance). We consider each of these four differences in turn.

First, the units of observation have been diverse. Examples include: functional groups, such as detritus, phytoplankton, and macroalgae (Leaper and Raffaelli, 1999, p.192); taxonomic groups resolved as nearly as possible to biological species (Leaper and Raffaelli, 1999, p. 192, and the present study); size classes without regard to taxonomic identification (Jennings and Mackinson, 2003); and narrowly defined size classes within a mixture of species and broad taxonomic units, for example, heterotrophic bacteria, *Prochlorococcus, Synechococcus,* and ultra- and nano-plankton (Rinaldo et al., 2002). Regressions of numerical abundance against body size using taxonomic units of observation are necessarily affected by the distribution of body size within the taxonomic units and may be biased by failure to sample some life history stages adequately.

Such regressions may differ from regressions of numerical abundance against body size using size classes as the units of observation. Leaper and Raffaelli (1999) showed that improving the taxonomic resolution among organisms of small body size substantially decreased the estimate of b in the Ythan estuary from 1.03 to 0.63, and observed that it was difficult to compare studies with differing degrees of taxonomic resolution. Russo et al. (2003) compared estimates of b obtained by using individual biological species, ecological guilds (frugivore-omnivore, granivore, insectivore, raptor), and groupings based on phylogenetic relatedness.

Second, the units of observation have been selected from different universes. Leaper and Raffaelli (1999, p. 192) usefully classify studies as direct observations of a local community; compilations of community data; global compilations; and compilations of data on one broad taxon, such as birds or terrestrial herbivorous mammals. Russo et al. (2003) analyzed direct observations of sympatric assemblages of tropical birds in three local rain forest habitats. The Tuesday Lake data were direct observations of phytoplankton, zooplankton, and fishes, excluding microbes and benthic and littoral organisms (Jonsson et al., 2005).

Third, how body size is measured has two aspects: how is the size of a single individual measured, and how is a composite indicator of size for the units of observation derived from the measurements for individuals? The size of a single individual may be measured by wet weight (this study), volume (Rinaldo et al., 2002), dry weight (Leaper and Raffaelli, 1999), length (converted to mass by an allometric relation between length and mass, a conversion that introduces additional error), or as a vector of weights of each component chemical element (e.g., by carbon and phosphorus content, Sterner and Elser, 2002). The composite indicator of size may be the average (this study, Enquist et al., 1998, p. 164), the "mean adult body-size" (Leaper and Raffaelli, 1999), or the maximum body size. Only the average body size has the convenient property that the species' biomass is the product of the numerical abundance times the average body mass. On the other hand, the average body size depends on the body size distribution within the unit of observation.

The other indicators of population body size, such as adult or maximum size, may not be well defined for species of indeterminate growth.

Fourth, which variable (body size or abundance) is viewed as independent may depend on the observer and the unit of observation. As previously noted, most plant ecologists have plotted log *M* on the ordinate and log *N* on the abscissa. Damuth (1981), Blackburn and Gaston (1999), Leaper and Raffaelli (1999) and most animal ecologists have reversed the assignment of variables to axes. If there were no errors of measurement in log *M* and log *N* and no stochastic variations from an exact allometric relation between *M* and *N*, the choice would be immaterial. In the presence of deviations from an allometric relation and error variance in measurements, the choice could substantially affect the estimated coefficients of the allometric relation.

Errors in Measurement

Given replicate measurements of *M* and N for all species, an explicit allometric model that allows for measurement errors in both variables *M* and *N* is:

$$\log M = \mu + \upsilon,$$
$$\log N = \eta + \omega,$$
$$0 = \beta_0 + \beta_1\mu + \beta_2\eta + \varepsilon,$$

where υ, ω, and ε are independent, normally distributed errors with mean zero, μ is the true but unobservable log *M* and η is the true but unobservable log *N*. One could estimate the regression coefficients β and the variances of υ, ω, and ε using Error-in-Variables regression (Clutton-Brock, 1967; Reilly and Patino-Leal, 1981). The estimates of the β_is will be sensitive to the variances var(υ) and var(ω). Alternative statistical models such as the general structural relation, reduced major axis regression and ordinary least squares, and further references, are discussed by Griffiths (1992, 1998) and Russo et al. (2003, p. 272), among others.

For pelagic ecosystems, measurements of *M* tend to have lower observational error than measurements of *N*. For subsets of communities similar to Tuesday Lake, S.R.C. has estimated the measurement error in *M* and *N* using replicates and found the error variance in *N* to be larger than the error variance in *M*, frequently by as much as two orders of magnitude. This difference in error variance results in part from the sampling design, which could allocate replicates to increase the precision of either *M* or *N*. For patchy or mobile species, it can be laborious to increase the precision of *N* by sampling more intensively. Estimating *N* requires handling many individual organisms. With only the limited extra effort of measuring each individual's *M*, it is possible to build up a large number of replicates to increase the precision of measurement of *M*. Thus attempts to improve the precision of measurement of *N* may yield even further improvements in the precision of measurement of *M*.

It is tempting to speculate that the difference between animal and plant ecologists in which variable they choose as independent and plot on the abscissa may result from differences in the difficulty of measuring *M* versus *N* for different types of organisms. Trees (and most terrestrial plants) stand still to be counted but it is often not easy to estimate their mass. For example, considerable error may be involved in estimating tree mass from trunk diameter. On the other hand, the numerical abundance of animals is often difficult to measure precisely but measurements of body mass may be relatively easy to replicate, especially for species with determinate growth. Where the direction of causality is not obvious, perhaps animal and plant ecologists are both inclined to put the more precise measurement on the abscissa.

Causality

A linear regression model is not informative about causality because the direction and quantitative form of causation cannot be determined on statistical grounds from static observational data. The statistically defensible regression model in Eqn. 1 suggests but does not prove that the average individual body mass of a species determined the species' numerical abundance in Tuesday Lake in 1984 and 1986.

Damuth (1981) suggested that *M* determines *N* for energetic reasons. If *M* is fixed for each unit of observation (e.g., species, other taxonomic unit, or size class); and if the energy required to support each individual is aM^b; and if the energy available to support each unit of observation is a constant *E*; then the number *N* of individuals that can be supported by the available energy in each unit of observation is $E/(aM^b)$ (i.e., $N \propto M^{-b}$). The fixed amount of energy and the scaling of metabolic rate with body mass establish a relationship between *M* and *N* and could equally well argue for *M* as a function of *N*. Damuth's crucial assumption, for the present discussion, is to regard *M* as fixed and *N* as variable. This and other mechanisms are reviewed in greater detail by Blackburn and Gaston (1999).

An alternative causal scheme is that the abundance of a species is fixed by underlying birth and death rates, and that the average body size adjusts up or down according to the number of individuals who make a living from a fixed supply of food or light or other essential input. Causation could run from numerical abundance to body mass in a nonlinear relation with a nonstandard error structure. Such an alternative causal model is not the simplest model, given our data, but it is also not logically impossible when species interact in a community web, as in Tuesday Lake.

The assumption that M is fixed for each unit of observation can be attacked at the level of the individual, at the level of the population, and at the level of the community.

At the level of the individual, density-dependent body growth is well known in plants, animals and other organisms. A large number of seedlings or young-of-year may, when resources are limited, result in reduced growth for most or all individuals. The mechanistic basis of the determination of individual body size is little understood (Hafen and Stocker, 2003).

At the level of the population (excluding those units of observation defined by size alone, such as a size class, for which size is necessarily fixed or bounded), a familiar feature of the demography of age-structured populations is that rapidly growing populations have a higher proportion of individuals in young age classes and a lower proportion of individuals in old age classes than stationary or declining age-structured populations with exactly the same life table (that is, holding the life table constant and varying only the intrinsic rate of natural increase or Malthusian parameter usually denoted by r, not to be confused with the residual r). In such a case, if body mass increases with age, then a more rapidly growing population (with a younger age structure) will have smaller average body mass than a stationary or declining population, even if the life tables, fertility schedules, and age-specific growth schedules are identical in the populations being compared. Roff (1986, p. 317) argued that "although an increase in body size increases fecundity and tends to increase [the Malthusian parameter] r, the concomitant increase in development time decreases r"; the observed distribution of body sizes in a population results from the balance between these countervailing selective forces and has, in some instances, been demonstrated to be under genetic control.

Changes in the size distribution within a population influence, not only the average body size, but also the average energy consumption per unit of mass if younger, faster-growing individuals have higher respiration rates per unit of mass than older, slower-growing individuals, as indicated by Riisgård (1998). If the average energy consumption per unit of mass influences a species' numerical abundance (as in the energetic model of Damuth), then population age-structure or size-structure influences both M and N and may affect the estimated allometric exponent b.

At the level of the community, trophic, competitive and mutualistic interactions may affect the average body size of a population. For example, predators (including human harvesters) may preferentially remove vulnerable small individuals or large trophy individuals. In Tuesday Lake, all but a handful of species were linked by the food web and were therefore subject to influence by resource species or consumer species. Even the half dozen isolated species were subject to competition (for light or nutrients, for example) that could affect average body size.

Some ecologists view a species' average body mass as determined on an evolutionary time scale and its numerical abundance as responding on an ecological time scale. However, a fuller view suggests that a species' average body mass is determined on at least three time scales: ontogenetic (by the way an individual's genes guide its development in interaction with its proximate environment), ecological (by how individuals interact with other individuals of the same and different species and with the abiotic environment), and evolutionary (by how individuals' selective advantages, resulting from their ontogeny and ecological interactions, contribute to the heritable characteristics of the next generation). The balance among ontogenetic, ecological, and evolutionary influences on a species' average body mass varies among species and, for a given species, according to the biotic and abiotic surroundings of the species.

Body size (individually and on the average) and numerical abundance interact dynamically, on multiple time scales, within and between species or other units of observation, and through biotic interaction with the abiotic environment. The Tuesday Lake data demonstrate, once again, that statistical care is required in assessing even the simplest pattern generated by this dynamic process.

ACKNOWLEDGMENTS

We thank T. M. Blackburn, James H. Brown, Adam E. Cohen, Heinrich zu Dohna, Kevin Gaston, Simon Jennings, Carlos Melián, Daniel C. Reuman, Sabrina E. Russo, and two anonymous reviewers for excellent suggestions, comments, and corrections, and the U.S. National Science Foundation for support from grant DEB9981552 to Rockefeller University. J.E.C. thanks Mr. and Mrs. William T. Golden for hospitality during this work and Kathe Rogerson for help. S.R.C. is grateful for the support of N.S.F. and the A.W. Mellon Foundation.

4.2 | BODY SIZE SCALINGS AND THE DYNAMICS OF ECOLOGICAL SYSTEMS

Lennart Persson and André M. de Roos

A striking feature of ecological communities is the variation in body size that can be observed among organisms (Gaston and Lawton, 1988; Werner, 1988; Brown, 1995a; Cohen et al., 2003). Not surprisingly, this variation in body size has been the focus of many studies considering the structure of food webs. For example, size relationships between predators and their prey has been a central aspect in the investigation of patterns of distribution of species and body masses in ecological systems (Cohen et al., 1993a; Woodward and Hildrew, 2002a). Predator–prey size ratios have formed the basis for food web models such as the cascade model (Cohen et al., 1990b; Chen and Cohen, 2001b) and are also used to estimate interactions strengths in food webs (Emmerson and Raffaelli, 2004b; see Emmerson, et al., Chapter 4.3). Additionally, individual body mass has been a key characteristic in the analysis of food web patterns in terms of numerical and biomass abundances at different trophic positions (Cohen et al., 2003; see Cohen and Carpenter, Chapter 4.1). Finally, body size is the key variable in the analyses of trophic structure in ecological systems based on size spectra (Kerr and Dickie, 2001).

Characteristic of these studies is that they have focused on equilibrium conditions and the (local) stability of food webs as a function of small perturbations away from equilibrium including the implications of transient dynamics following perturbations (Chen and Cohen, 2001b). Thus, these studies essentially have not considered situations systems that are intrinsically non equilibrium. Yodzis and Innes (1992) studied the dynamical properties of a consumer-resource system where energy intake and metabolic costs scaled with body size (among other things) to investigate the ecological scope of ectotherms versus endotherms. In many studies, this approach has been used to analyse the effects of perturbations and/or dynamics of marine food webs (Yodzis, 1998, Koen-Alonso and Yodzis, Chapter 7.3). Although forming an important step forward in our mechanistic understanding of the implications of body size relationships on population and community dynamics, Yodzis and Innes (1992) examined only variation among species. Size variation within populations, however, has been shown to be a major feature of most ecological communities. For the overwhelmingly majority of the earth's taxa, individuals grow over a substantial period of their lives including organisms undergoing restructuring of overall body architecture (i.e., metamorphosis) (Werner and Gilliam, 1984; Sebens, 1987; Ebenman 1988; Werner, 1988). Furthermore, the rate by which individuals grow is also generally dependent on their energy intake and hence food availability. The exceptions to this rule, except for unicellular organisms, occur among birds and some mammals where the size difference between a juvenile (just independent of its parents) and an adult is less than an order of magnitude.

The purpose of this chapter is to give a baseline overview of the implications of intraspecific body size scaling for the dynamics of ecological communities. We use two trophic configurations to illustrate how different size scalings have profound effects on population dynamics, with strong feedbacks on overall trophic dynamics. The first example involves a consumer feeding on a resource where the size scaling of foraging rate turn has profound effects on the dynamics of the system and where the dynamics of the system for most parameter values is characterized by non equilibrium dynamics driven by cohort competition. The second example involves cannibalism where the scaling of the lower size boundary for which cannibal-victim interactions take place has a major effect on dynamics, and where ecological systems may shift between different trophic structures over time. We argue that a mechanistic consideration of species interactions, including an explicit consideration of population dynamics, is essential in analysis of implications of body size scaling for population, community, and ecosystem dynamics and structure.

MODELLING APPROACH

The modelling approach we use—physiologically structured population models—is based (in contrast to Lotka-Volterra type of models that dominate food web research) based on a two-state level description:

an *i* (individual) state and a *p* (population) state (Metz and Diekmann, 1986; Metz et al., 1988; De Roos et al., 1990; De Roos, 1997). In addition, an *E* (environmental) state is defined which represents both the resources and the predators of an individual. The *i*-state represents the state of the individual in terms of a collection of characteristic physiological traits such as size, age and energy reserves, and the *p*-state is the frequency distribution over the space of possible *i*-states. The model formulation process consists of deriving a mathematical description of how individual performance (growth, survival, reproduction) depends on the physiological characteristics of the individual and the condition of the environment (*i*-state description). This involves deriving size-dependent functions for food gathering (see later discussion) and metabolism, starvation mortality, energy channelling between somatic and gonad growth. Handling the population level (*p*-state) dynamics is subsequently just a matter of bookkeeping of all individuals in different states, and, importantly, no further model assumptions are made at this level. The core of these models is thus the individual state. In the modelling examples we present here, all processes are assumed to be continuous except reproduction in the consumer which is assumed to be a pulsed event at the start of each growing season. Hence, the model structure is a mixture of continuous and discrete processes (Persson et al., 1998; Claessen et al., 2000). The assumption of pulsed recruitment is easily relaxed to handle completely continuous systems (cf. De Roos and Persson, 2002). The formulation of equations for *i*- and *p*-state processes and the interaction between individuals and the environment is beyond the scope of this chapter, and the reader is referred to previous papers (Persson et al., 1998; Claessen et al., 2000; De Roos and Persson, 2001).

SIZE SCALING OF CONSUMER-RESOURCE DYNAMICS

At the individual level, implications of size scaling have been considered in a number of studies (Peters, 1983; Calder, 1984; Werner, 1988; Lundberg and Persson, 1993; Werner, 1994; Kooijman, 2000). In particular, the relationship between the size scaling of foraging rate relative to the size scaling of metabolic demands has formed the basis for deducing the competitive ability of organisms as a function of their size (Werner, 1994). Other studies also have considered the effects of digestive capacity (Lundberg and Persson, 1993). The size scaling of metabolic demands has generally been found to be described by a power function with an exponent that varies between 0.7 and 0.8 using body mass as the base (cf. West et al., 1997; Gillooly et al., 2001). Mechanistic derivations for the size scaling of foraging rate have also been advanced (Kooijman, 2000). Experimental studies suggest that the size scaling of foraging in terms of the attack rate varies considerably (size scaling exponent for body mass based encounter rate = 0.1–0.7) depending on both the foraging method of the consumer (f.e. particulate vs. filter feeders) and the environment in which prey is searched for (e.g., 2-dimensional versus 3-dimensional environments) (Wilson, 1975; Mittelbach, 1981; Sebens, 1982; Werner, 1988; Lundberg and Persson, 1993; Werner, 1994).

To study effects of different size scalings of foraging rate on population dynamics, Persson et al. (1998) derived the following size-dependent attack rate:

$$a(w) = A\left(\frac{w}{w_o}\exp\left(1-\frac{w}{w_o}\right)\right)^{\alpha} \qquad (1)$$

where A is the maximum attack rate, w is body size, w_o is the body size at which this maximum rate is achieved, and α is a size-scaling exponent (for mechanistic explanations of the form of the function see Persson et al. 1998, Figure 1). By combining the energy gained by foraging and the metabolic costs we can calculate the competitive ability of an organism as a function of its size defined as the lowest resource level that the individual can tolerate without losing mass (i.e., $\mathrm{dEnergy}_{growth}/\mathrm{dt} = 0$) (Lundberg and Persson, 1993; Persson et al., 1998, Figure 1). In what follows this minimum resource level will be referred to as the critical resource demands.

Different values of the size scaling exponent result in three different forms of the relationship between critical resource demands and body size. For low values of α, critical resource demands are monotonically increasing with body size (Figure 1, right panel). In contrast, for high values of α, critical resource demands first decreases with body size to thereafter increase again. Finally, for intermediate values of α, the relationship between critical resource demands and body size is relatively flat with small differences in competitive ability between differently size individuals. Interestingly, the three different forms of the critical resource demands functions translate into three qualitatively different forms of population dynamics. For a montonically increasing critical resource demands function, population dynamics are characterized by high amplitude cycles where strong recruiting cohorts outcompete older (larger) individuals (Persson et al., 1998). For a convex downwards critical resource demands function, also high amplitude dynamics prevail,

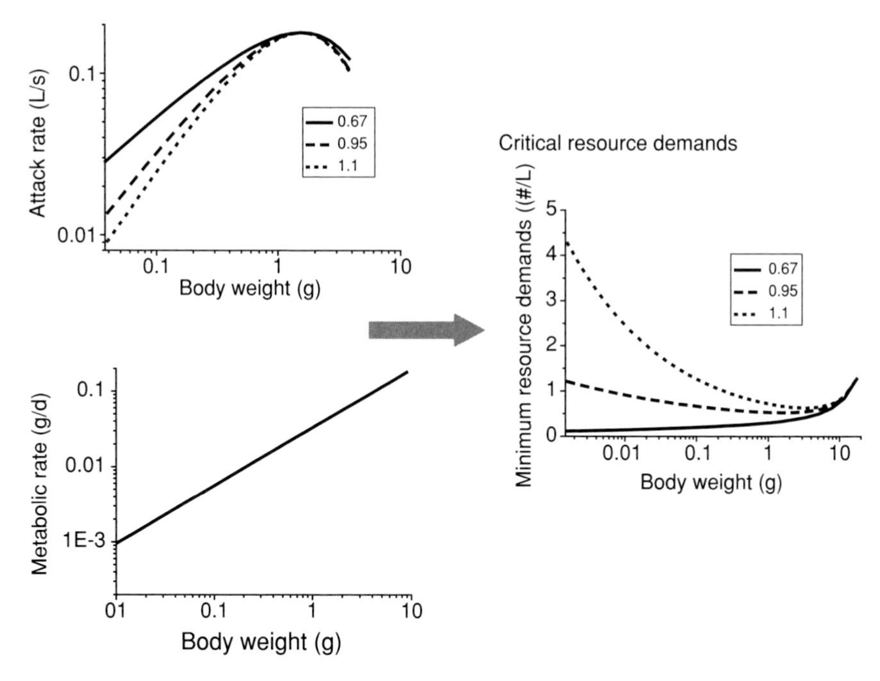

FIGURE 1 | Critical resource demands (right) as a function of body size, given different size scalings of the attack rate (upper left) and the size scaling of metabolic demands (0.75, see text) (lower left).

but in this case older juveniles prevent recruiters to enter the system because of the former's higher competitive ability. Three important conclusions can be made based on this analysis. First, population dynamics observed are heavily dependent on and can be predicted from the size scaling relationship. Second, the occurrence of fixed point dynamics is restricted to a narrow parameter range. This means that the conditions, under which a simple relationship between individual size and community and food web structure exist, are limited without the explicit consideration of population dynamics. Third, empirical data on individual size scalings suggests that recruit-driven cycles should be the dominating dynamics for exploitative consumers.

For size-structured consumer resource interactions, there also exists ample empirical evidence for cohort-driven cycles particularly for fish populations (Murdoch and McCauley, 1985; Hamrin and Persson, 1986; Townsend et al., 1990; Shiomoto et al., 1997; McCauley et al., 1999; Sanderson et al., 1999; Lammens et al., 2002). Moreover, the majority of these examples come from multispecies systems, suggesting that the signal from the consumer-resource interaction prevails in a multispecies context (see also Murdoch et al., 2002). The cohort-driven signal is hence not only visible in the resource of the consumer population (Cryer et al., 1986; Hamrin and Persson, 1986, Figure 2A), but also in the density of the resource of that resource (Shiomoto et al., 1997, Figure 2B). The effect of intraspecific size scaling thus does not only translates into a specific population dynamics but also has strong ramifications for trophic dynamics and can consequently not be ignored in food web studies.

SIZE-SCALING AND CANNIBAL-VICTIM INTERACTIONS

Physiologically structured population models are based on a mathematical description of how the individual's environment influences its performance and how the individual in turn affects its environment. As a result, density dependence is assumed to operate through an individual's environment, which can both result from the effect of the total population on a common environmental factor such as food density and through considering all other individuals of the population as a part of an individual's environment. When introducing cannibalism in a consumer-resource system, the environmental state, in addition to the basic resource, will for an individual of a specific size include its potential victims and its potential cannibalistic predators (Claessen et al., 2000).

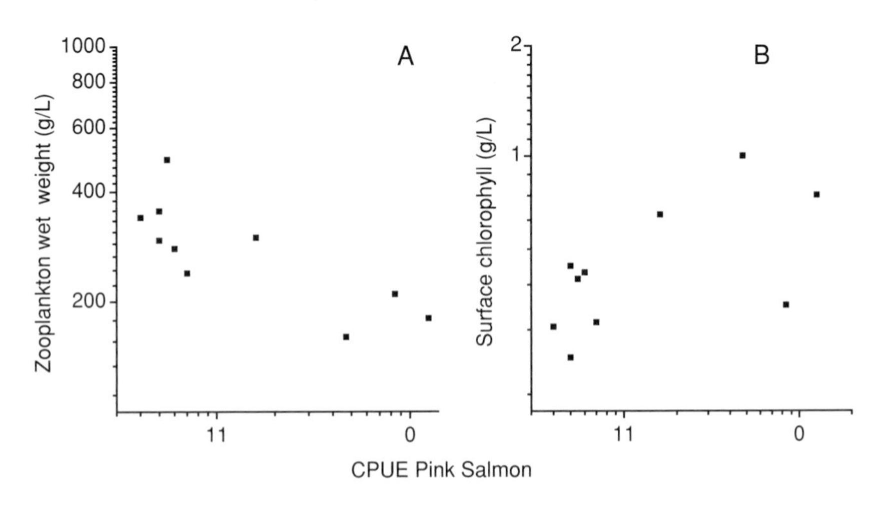

FIGURE 2 | Relationship between the abundance of pink salmon and biomass of zooplankton (**A**) and the biomass of phytoplankton and the abundance of pink salmon (**B**). (Based on Shiomoto, et al., 1997.)

To describe a cannibalistic system, the functions that determine the interactions between cannibals (c) and victims (v) have to be specified. Cannibal-victim interactions have been assumed only to take place in a so called cannibalistic window which is bounded by the lower (δ) and upper (ε) ratio of victim and cannibal body size (Figure 3A). Within the size boundaries where cannibal victim interactions take place, the cannibal attack rate ($A_c(c,v)$) is assumed to increase (βc^{σ}) along a ridge for the optimal victim size (φc) for a specific cannibal size resulting in an overall tent function (Claesssen et al., 2000):

$$A_c(c,v) = \begin{cases} \beta c^{\sigma} \dfrac{v - \delta c}{(\varphi - \delta)\, c} & \text{if } \delta c < v < \varphi c \\ \beta c^{\sigma} \dfrac{\epsilon c - v}{(\epsilon - \varphi)\, c} & \text{if } \varphi c < v < \epsilon c \\ 0 & \textit{otherwise} \end{cases}$$

Investigations of the behavior of a size-structured cannibalistic system where cannibal and victims compete for a shared resource have shown that the dynamics are heavily affected by two parameters, the overall rate by which the attack rate increases with cannibal size (β) and the victim/cannibal size ratio (δ) (Claessen et al., 2002). For the lower size boundary (δ), three major dynamical regimes can be observed (see Figure 3B). For a high size boundary, the dynamics are essentially a cohort-driven competition system where recruiting cohorts outcompete

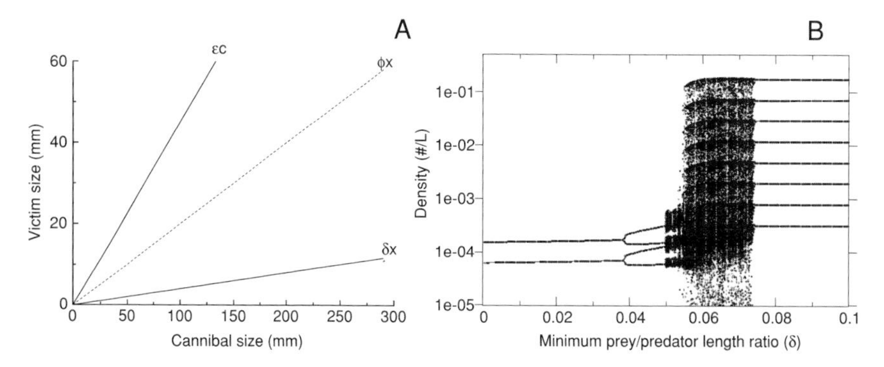

FIGURE 3 | **A,** Two-dimensional plot of the cannibalistic window showing the minimum size ratio (δ), the optimum size ratio (φ) and the maximum size ratio (ε) of victims to cannibals for which cannibalistic interactions take place. **B,** Bifurcation plot showing the effects of the lower size limit (δ) on cannibalistic population dynamics for $\beta = 200$ and $\varepsilon = 0.45$. Densities reflect all cannibals excluding young-of-the-year individuals. (See also color insert.)

their cannibals by depressing the shared resource before victims have grown large enough to be cannibalized. In contrast, for a low δ, cannibals quickly start to cannibalize recruiting victims, and thereby prevent recruits to depress the resource. This leads to fixed point dynamics, or a dynamics characterized by low amplitudes (see Figure 3B). The dynamics which emerge for intermediate values of δ are characterized by a mixture of cohort competition-driven cohort cycles and cannibal-driven dynamics alternating over time.

Overall, cannibalism may stabilize consumer resource interactions by preventing recruiting cohorts from depressing the shared resource, if the lower size ratio is sufficiently small (Claessen et al., 2004). A comparison of cannibal species with different life histories including different lower size boundaries for victim cannibal interactions, suggests that we can a priori predict the dynamics observed based on size boundaries (Persson et al., 2004). Importantly, in systems where the dynamics shift between cannibalism as the dominant factor to cohort competition as the dominant factor, the whole food web undergoes major structural shifts involving both herbivores and primary producers (Figure 4). In the case of cannibalism, we can therefore conclude that an explicit consideration of intraspecific size scaling in the case of cannibalism allows us to deduce population dynamical consequences, as was the case for consumer-resource interactions. Second, shifts in population dynamical regimes resulted in shifts in overall trophic structure, implying an

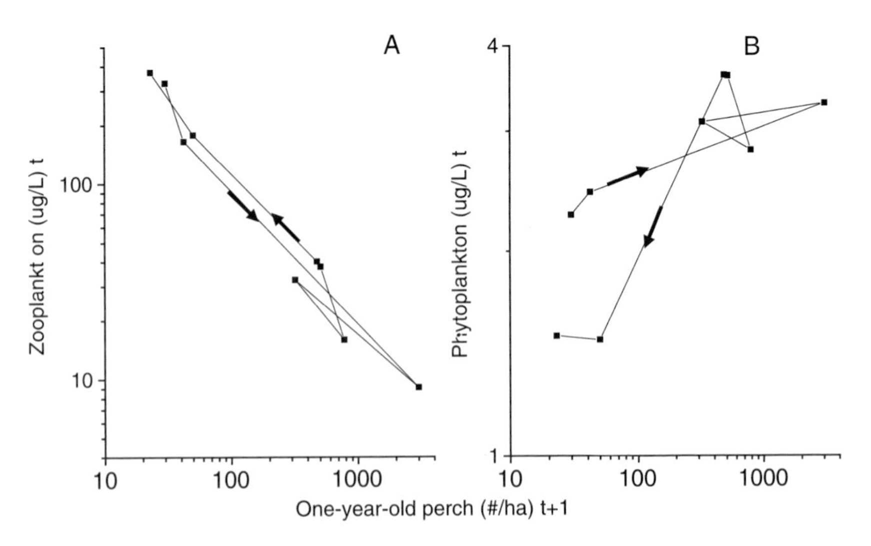

FIGURE 4 | Relationship between the biomass of zooplankton (**A**), and phytoplankton (**B**), and the abundance of one year-old perch. Based on Persson et al. (2003).

explicit consideration of population dynamics may be essential for predicting and understanding overall trophic dynamics.

DISCUSSION AND CONCLUSIONS

Our analysis shows that the size scaling of foraging capacity has important implications for population dynamics and overall trophic dynamics. For consumer-resource interactions, observed body size scalings of foraging and metabolic rate have a strong tendency to lead to nonequilibrium recruit-driven cycles. A recent review of cycles in different populations also suggests that this type of generation cycles is the most commonly observed (Murdoch et al., 2002). For cannibal-victim interactions, cannibalism may counteract the advantage of competitively superior individuals, and eventually lead to a stabilization of population dynamics (Claessen and De Roos, 2003; Claessen et al., 2004). This stabilization, however, depends critically on the lower size boundary for cannibalism; if this size boundary is too high, high amplitude dynamics with effects on overall trophic dynamics of the system can result (Persson et al., 2003). Furthermore, it has been shown that although an increase in overall cannibalistic voracity (β) in a restricted parameter range first will dampen cohort cycles, a further increase leads to high amplitude dynamics (Claessen et al., 2000; Claessen et al., 2002). For both consumer-resource interactions and cannibal-victim interactions, it is also apparent that *intraspecific* body size scalings and predator–prey body size relationships have important consequences for population and community dynamics. These effects have not been considered in the contemporary body size literature.

Our analysis of cannibal-victim size relationships shows that intensity of cannibalistic interactions depends heavily on cannibal and victim size. This result suggests that using predator–prey size ratios may be used to estimate interaction strengths (see Emmersson et al., Chapter 4.3). At least three cautionary notes, however, should be considered. First, for both consumer-resource and cannibal-victim dynamics, the size scaling parameters that have been shown to be most important for population dynamics have experimentally been found to be different for equally sized individuals of different species (Byström and Gàrcia-Berthóu, 1999; Hjelm and Persson, 2001; Persson et al., 2004). Size scaling parameters are closely connected to characteristics of the feeding apparatus (gape size, gill raker size, etc.), which suggests predator–prey body size ratios have only a limited ability to predict intensity of predator–prey interactions. Second, the extent to which estimates of interaction strengths based on Lotka-Volterra theory carries over to a size/stage-structured context is not known as we, at present, do not have a size structured food web theory equivalent to, for example, the community matrix approach of non-structured models. Third, the extent to which the dynamics and structure of ecological communities can be understood based on an equilibrium perspective is also far from obvious (Hastings, 1996; Chesson, 2000). For size-structured systems, time delays and food dependent development times have a strong tendency to result in non equilibrium dynamics. Some important elements of size-structured community theory such as the presence of emergent Allee effects and facilitation are obviously results of equilibrium analyses (De Roos et al., 2003a; see De Roos and Persson, Chapter 3.2). Nevertheless, the extent to which an equilibrium theory for size-structured communities can explain structural patterns and coexistence in these communities remains to be investigated.

Overall, we have argued for that the explicit consideration of population dynamics is a key to derive at an understanding of large scale trophic implications of size-structured interactions. The next stop is to extend this approach to consider multispecies size-structured interactions and we are presently investigating model configuration including four to five species. Modelling and empirical studies of consumer-resource systems and cannibal-victim systems, however, suggest that a focus on a limited number of strong interactors may be sufficient to predict overall trophic dynamics even in multispecies contexts. Such an approach is consistent with analyzing "community modules," where a relatively limited set of interacting entities are assumed to cover essential elements of components of ecological systems (Holt, 1997; but see Yodzis, 1998, for a different view). The extent to which we can extend the analysis to size structured multispecies community similar to that which has been done for unstructured models assuming equilibrium conditions (cf. de Ruiter et al., 1995; Neutel et al., 2002) remains to be seen. But as previously stated, the domain of an equilibrium analysis to explain patterns in size-structured food webs (as well as in unstructured webs) is an open question. Importantly, the results of the present study clearly indicate that that size-structured interactions do have major implications for the dynamics of populations, communities, and food webs.

ACKNOWLEDGMENTS

This research was funded by the Swedish Research Council and the Swedish Research Council for Environment, Agricultural Sciences and Spatial Planning to L. Persson. and the Netherlands Organization for Scientific Research to A.M. De Roos.

4.3 | BODY SIZE, INTERACTION STRENGTH, AND FOOD WEB DYNAMICS

Mark C. Emmerson, José M. Montoya, and Guy Woodward

Given recent concerns over the accelerating loss of biodiversity and its consequences for the functioning of ecosystems and the provision of sustainable and renewable resources, it seems imperative that contemporary ecologists have a good understanding of the processes that structure real ecosystems. Predicting the consequences of species loss through extinction and invasion requires that we understand which species interact with which, that is, we have a good approximation of food web structure (and other non-trophic interactions) in real, albeit highly complex systems. Focusing on feeding relationships, knowledge of food web structure would enable ecologists to understand how disturbances permeate throughout a community. The effects of such disturbances, whether they were large or small, could be mapped, understood, and perhaps predicted. However, disentangling the many interactions amongst species remains a constant challenge for modern ecologists. The most conspicuous limitation to the collection of food web data is the logistic constraints imposed on the surveying, sampling, and identification of species in whole ecosystems. Natural ecosystems are comprised of many disparate taxonomic groups, and their realistic study would require exhaustive collaborative fieldwork to resolve species identity and the interactions amongst species. There are very few, if any, large-scale studies that have explicitly set out to document the species composition and trophic structure of whole ecosystems.

Most of what is known of food web composition and structure has resulted typically from small-scale studies, which have been carried out often by postgraduate and postdoctoral researchers (Woodward and Hildrew, 2001). Well-resolved food webs have also been constructed as by-products of other research programs (e.g., Tuesday Lake, Cohen et al., 2003), or from synthetic reviews of existing data sets or studies (e.g., Ythan Estuary, Raffaelli and Hall, 1996). We are unaware of any large-scale collaborative studies dedicated to combining the study of species composition (species identity, richness, body size, and abundance), food web structure (topology of interactions), and ecosystem functioning (an exception is perhaps the International Biological Program [IBP] in the 1960s). The lack of such studies is particularly pertinent now that ecologists are being asked increasingly to address fundamental questions over the consequences of environmental change and species loss for the sustained provision of global ecosystem services. It is exactly this type of data (and access to the sites from which they were collected), which would facilitate exploration of these pressing issues.

Even if ecologists are able to fully characterise the complex network of interactions that connect species in real communities together, empirically assigning strengths to these interactions remains an elusive goal. This is because measurement of per capita interaction strengths is extremely difficult: in an ideal world this would be achieved by the experimental removal of all species pairs to determine the effects on the remainder of the community. For large, complex (i.e., real) communities this is clearly not practical, or even logistically possible. If a surrogate-correlated variable could be used as a substitute for interaction strength, in simple food web models then this would facilitate exploration of the dynamics of such complex systems. If the mechanisms underlying food web structure could be better understood, then this would enable us to make predictions regarding the consequences of species loss (and at appropriate and manageable scales experimentally test these predictions).

In this chapter we explore some community-wide patterns obtained from real ecosystems where body size, abundance, and food web structure are known. Trivariate relationships among these three properties of communities remained elusive until Cohen and collaborators explored them (Cohen et al., 2003; Jonsson et al., 2004). However, they did not assign interaction strengths to their feeding relationships, which is a fundamental step for understanding the development of community composition and food web structure (see Berlow et al., 2004, for a review). Specifically, we explore the use of allometric (i.e., body size) relationships for predicting the strength of trophic interactions; these estimations can be used to investigate the dynamics of complex real systems. We show that, perhaps surprisingly, it is possible to find stable parameterizations of these dynamical systems simply by using observations of predator and prey body size and abundance. We explore the connection between allometric relationships and the strength of species interactions and show that there is an analytical relationship between predator–prey body size ratio and interaction strength. Our goal throughout is not to simulate the dynamics of these real systems, but to gain a heuristic insight into the mechanisms that might produce the empirical patterns evident in such food webs.

BODY SIZE RATIO AND PER-CAPITA INTERACTION STRENGTHS

Throughout we make use of Lotka-Volterra (L-V) dynamics of the form

$$\frac{dX_i}{dt} = X_i(r_i + \sum_{j=1}^{n} a_{ij} X_j) \tag{1}$$

where r_i is the intrinsic growth rate of species i, X_i is the population size of species i, and a_{ij} is the per-capita effect of species j on species i. Our use of L-V dynamics is motivated by (i) their simplicity (a type I functional response); (ii) by recent studies, which provide insights into the stability properties of dynamical L-V systems using empirically obtained data (Neutel et al., 2002); and (iii) by the relative ease (compared to type II or III, predator- or ratio-dependence functional responses), with which they can be estimated from and compared with field data and experimentation (although typically only for subsets of real communities, e.g., Paine, 1992; Fagan and Hurd, 1994; de Ruiter et al., 1995; Raffaelli and Hall, 1996; Wootton, 1997; Sala and Graham, 2002).

Recently, de Ruiter et al. (1995) proposed equivalence between the dynamic Lotka-Volterra model and an individual compartmented static model proposed by Hunt et al. (1987). They showed that equivalence between the two models existed when the Lotka-Volterra model was at equilibrium, so that

$$\begin{aligned} &i)\quad a_{ik} X_i^* X_k^* = e_i F_{ki} \quad \text{for feeding by sp. } i, k < i. \\ &ii)\quad a_{ij} X_i^* X_k^* = -F_{ij} \quad \text{for feeding by sp. } i, j > i. \\ &iii)\quad r_i X_i^* + a_{ii} X_i^* X_i^* = -r_i B_i \quad \text{for death of species } i. \end{aligned} \tag{2}$$

Here the terms on the left of the equality refer to the elements of the Lotka-Volterra equation. The terms to the right of the equality refer to the elements of the Hunt model, where F_{ij} refers to the feeding rate of species j on species i at equilibrium (and is a population level effect) and e_j refers to the ecological efficiency of species j (i.e., the product of the assimilation efficiency and production efficiency). B_j refers to the biomass density of species j observed in the field. Hunt et al. (1987) used empirically obtained measures of biomass density from a soil micro-arthropod community to parameterise a static model. DeRuiter et al. (1995) showed that by rearranging (2*ii*) above it was possible to obtain the elements of the Jacobian matrix for a Lotka-Volterra system, so that

$$a_{ij} X_i^* = \frac{-F_{ij}}{X_j^*} = \frac{-F_{ij}}{B_j}. \tag{3}$$

de Ruiter et al. (1995) also assumed that there was an equivalence between the notation used to represent biomass density and population density, so that $X_i^* = B_i$. In fact, both notations have units of biomass, so we will use biomass from now on. Therefore we have

$$a_{ij} B_i = -\frac{F_{ij}}{B_j} \tag{4}$$

and hence

$$a_{ij} = -\frac{F_{ij}}{B_i B_j}. \tag{5}$$

Estimating population-feeding rates experimentally for all species in a community would be almost as difficult to achieve as direct measurement of the interaction coefficients. However, for each individual species per-capita feeding rates, f_j, can be obtained from well-established allometric relationships (Boulière, 1975, Farlow, 1976, cited in Peters, 1983, and Calder, 1984) and metabolic theory. These per-capita feeding relationships are well described by power functions of the general form $f_j = \psi W_i^{\varphi}$ and describe the amount of food that must be ingested (in terms of weight or energy) by a single consumer per unit time. Here, W_i represents the body size of species i, ψ represents a constant (with dimensions $[fW^{-\varphi}]$) and φ is a dimensionless exponent. Per-capita feeding rate is described by Peters (1983, p113) as $f_h = 0.13W_i^{0.69}$ for homeotherms and $f_p = 0.0096W_i^{0.82}$ for poikilotherms (and represents food ingested in kg per animal^{-1} d^{-1}). Calder describes the relationship as $f_v = 0.157W_i^{0.84}$ for vegetable eating mammals and $f_c = 0.234W_i^{0.72}$ for carnivorous mammals. Keeley (2003) showed $f_j \propto W^{0.73}$ for salmonid fishes (e.g., f_i *is proportional to* $W^{0.73}$). All these experimental results are supported by metabolic theory showing that whole-organism metabolic rate (E) scales with body-size in the form $E = aW^{-3/4}$ (Kleiber, 1947, 1961; Peters, 1983; Calder, 1984; West et al., 1997; Brown et al., 2000). It is reasonable to assume that the amount of food required per individual consumer (f_j) is proportional to E, and hence $f_j = bW^{-3/4}$.

To obtain the ingestion rate of consumer j on resource i (F_{ij}), we need to multiply individual ingestion rate (f_j) by consumer biomass (B_j) and consumer preference for resource i (p_{ij}), so $F_{ij} = f_j \cdot B_j \cdot p_{ij}$. The simplest and most general form of prey preference by predators is to assume that preference is in proportion to prey abundance on the set of prey of each predator, so $p_{ij}\ \alpha X_i$ (de Ruiter et al., 1995).

Just as per-capita (individual) feeding rate is described by an allometric relationship so too is population density so that $X_i = \gamma W_j^{-\varphi}$. The relationship between body size and abundance is one of the most comprehensively documented macro-ecological patterns in real systems. The relationship is typically negative, so that as the body size of a species increases, its population density declines, although empirical estimates of the exponent φ vary (see Blackburn and Gaston, 1997, for a critical review). The presence of such a relationship may indicate that the system exists in some form of state; here we assume that the relationship can be used to define equilibrium population densities.

In the few cases where population densities and mean body sizes are measured for whole local communities, the exponent φ is approximately 1. In lake communities studied during the IBP 1965–1975, where three trophic levels were considered (phytoplankton, zooplankton, and fish), the exponent varies $1.2 < \varphi < 0.77$, and the median was 0.95 (Cyr, 2000). In the Tuesday lake studied by Cohen and collaborators, $\varphi = 1$ (Cohen et al., 2003). And here again observations and theoretical predictions clearly fit. Metabolic theory predicts that $X_i \alpha W_i^{-1}$ in whole food webs (assuming three trophic levels, 10% of energy transfer between trophic levels, and average difference in body size between predators and preys is 10^3–10^4), while for a single trophic level, $X_i \alpha W_i^{-0.75}$ (see Brown and Gillooly, 2003).

Based on this metabolic theory and on the empirical relationships addressed, the relationship between biomass (B) and body size (W) is straightforward. If $B_i = X_i\, W_i$ then (*i*) for the whole food web $X_i \propto W_i^{-1}$, and therefore $B_i \propto W_i^0$, so that biomass density is constant for each species within the web; and (*ii*) for a single trophic level, if $X_i \propto W_i^{-0.75}$

and therefore $B_i \propto W_i^{0.25}$ (here the symbol $\propto$ indicates that X_i and B_i are proportional to W_i raised to the power of an exponent).

If we substitute the previous form of the ingestion rate in equation (5) we obtain the following expression:

$$a_{ij} \propto -\frac{E_j B_j}{B_i B_j} p_{ij} \tag{6}$$

Then, we can use the allometric relationships of biomass (B) and energy use (E) corresponding to whole food webs. We have assumed $p_{ij}\, \alpha X_i$, but how does X_i scale with B_i? Prey preference by a particular predator depends on the abundances of the prey of that particular predator, and not on the abundances of all the species composing the food web. The bulk of the prey of a particular predator commonly belongs to a single trophic level, so that it is more appropriate to use the scaling relationship $X_i \alpha W_i^{-0.75}$. Replacing all these allometric relationships in (6) we obtain:

$$a_{ij} \propto -\frac{W_j^{0.75} W_j^{0}}{W_i^{0} W_j^{0}} W_i^{-0.75} \rightarrow a_{ij} \propto -\left(\frac{W_j}{W_i}\right)^{0.75} \tag{7}$$

Equation (7) indicates that the predator-prey body size ratio is important for the estimation of per-capita interaction strengths and suggests that simply by combining observations of the body size of predators and prey (compositional information) with observations regarding their feeding relationships (structural information) it may be possible to explore the dynamics, and hence functioning of such systems. The idea that predator–prey size ratios are important for the functioning (sensu Pimm, 1982) of food webs is not new, Jonsson and Ebenman (1998) explored the relationship between predator–prey body size ratio and food web stability. They showed that the distribution of body sizes can influence the relative magnitudes of interaction coefficients in Lotka-Volterra systems and this, in turn, can affect the resilience of a food web, increasing the probability of local stability in linear food chain models. Emmerson and Raffaelli (2004a) used an experimental approach in which they placed predators and prey, which varied in size, into feeding arenas and quantified per-capita effects (a_{ij}) using the dynamic index (or log ratio) measure of interaction strength (Berlow et al., 1999). These experiments demonstrated that the relationship between predator prey body size ratio and interaction strength could be well described by a power function of the form $\psi(W_j/W_i)^{\varphi}$. When this relationship was combined with macroecological patterns between body size and abundance and used to parameterise interaction coefficients of a system of Lotka-Volterra equations for the Ythan food web, locally stable Jacobian matrices of this highly complex system (incorporating 88 species) could be found (see Emmerson and Raffaelli, 2004a, and later discussion for details). These studies indicate that a relationship does exist between the body sizes of predators and prey and the strength of the trophic interaction that takes place between them. Equation (7) shows that there is a theoretical basis for the allometric relationship used by Emmerson and Raffaelli (2004a) that is analytically tractable. To explore this further, we used (7) to parameterize two highly resolved food webs, the Ythan Estuary and the Broadstone Stream.

MACROECOLOGICAL PATTERNS AND FOOD WEB STABILITY

The topological properties of the Ythan Estuary food web have been studied extensively in the past and are now well known (Hall and Raffaelli, 1991; Huxham et al., 1996; Raffaelli and Hall, 1996). Here we make use of a conservative version of the food web, which was used by Emmerson and Raffaelli (2004a) and which features 88 species. This version of the food web excludes a number of species known to be consumers, but for which no food resources in the web have been identified (Emmerson and Raffaelli, 2004a). Information on the body size of these 88 species was reviewed from previously published studies, Ph.D. theses, M.Sc. theses, and unpublished data. Abundance estimates for the 88 species used in this version of the Ythan web have been made previously (Leaper and Raffaelli, 1999), but unfortunately, this comprehensive data is not available in the present study. Emmerson and Raffaelli (2004a) reviewed abundance data for this version of the previous food web for body size; however, abundance estimates are only available for a subset of the species in the food web. Rather than use site-specific relationships, here we make use of the general relationships reported in the empirical and theoretical literature (Peters, 1983; Calder, 1984; Kerr and Dickie, 2001; Rinaldo et al., 2002), our aim being to determine whether such relationships might permit parameterization of feasible and stable food webs. Broadstone Stream has had a long history of food web research (Hildrew et al., 1985; Lancaster and Robertson, 1995; Woodward and Hildrew, 2001), culminating in one of the most complete food webs for any system (Schmid-Araya et al., 2002a). The stream is naturally acid and, as a result, it has a relatively 'simple' insect-dominated community, and fish are absent. Even so, the most detailed web described to date contains 131 species, including the permanent meiofauna (species < 500 μm), and over 800 links, although there are only about 25 common macrofaunal species (Woodward and Hildrew, unpublished data). It is this macrofaunal 'subweb' that we have focussed on in the current study, as it is effectively impossible to quantify the abundance of the meiofaunal component. Benthic samples were taken on alternate months over a single year, and all (>40,000) individuals were measured. Over 4,000 predator guts were dissected, and ingested prey were identified and measured. Linear dimensions of the benthic and ingested invertebrates were converted to biomass using published regression equations (see Woodward and Hildrew, 2002a, for further details).

Importantly, for each of these webs detailed information regarding body size of each species is available. This enables us to use the generalized relationships previously detailed to parameterize Lotka-Volterra interaction coefficients (a_{ij}) for each of the food webs using equation (7). Specifically, we used the relationships $X_i^* = 3W_i^{-0.98}$ and $f_i = 0.0096W_i^{0.82}$ in equation (7). We parameterized the interaction coefficient matrix, **A**, for the Ythan Estuary and Broadstone Stream food webs respectively. For basal species intraspecific terms (a_{ii}) were set to −1 and for nonbasal species they were set to 1×10^{-6}. The general relationship between body size and density reported by Peters (1983 p. 169; $X_i^* = 3W_i^{-0.98}$), and predicted by metabolic theory ($X_i \alpha W_i^{-1}$) (Brown and Gillooly, 2003), were used to define the abundance of each species, assuming that this represented an estimate of the equilibrium population size for each species. Using the standard error of $\overline{X}_i^*$ estimated from $W_i^* = S_{\overline{X}} = S_{wx}/\sqrt{n} = 1.44/\sqrt{291}$ (Peters, 1983, p. 294) we randomly parameterized the equilibrium population size for species i using the 95% confidence limits as the upper and lower bounds for our estimate of $\overline{X}_i^*$, respectively. This estimate of population size was then used to calculate the biomass densities of each species so that, $B_i^* = X_i^* W_i$. The equilibrium biomass densities thus obtained were used to define the elements of the Jacobian matrix, $\boldsymbol{C}$, so that $\boldsymbol{c}_{ij} = \boldsymbol{a}_{ij}\boldsymbol{B}_i^*$ ($\boldsymbol{C} = \boldsymbol{BA}$, where $\boldsymbol{B}$ is an n x n diagonal matrix with the biomass densities on the diagonal and $\boldsymbol{A}$ is the interaction coefficient matrix). The terms of the Jacobian matrix, c_{ij}, represent the partial derivatives of each species population growth rate (h_i) with respect to the biomass density of each species $\boldsymbol{B}_i^*$ at equilibrium (i.e., $c_{ij} = \partial h_i / \partial B_j$ where $h_i = dB_i/dt$).

We then carried out a local stability analysis on parameterizations of each food web, to determine whether stable parameterizations could be generated using this approach. The Jacobian matrix describing the dynamics of each food web can be considered stable if the real parts of all eigenvalues are negative. Table 1 shows the percentage of stable

Table 1. Percentage of stable food web parameterizations before and after three forms of permutation: a) Elements of the Jacobian matrix were permuted; b) The vector of population biomass densities was permuted before generation of the Jacobian; c) The interaction coefficients in the matrix A were permuted before generation of the Jacobian matrix.

	% Stable	$a_{ij}B_i$	B_i	a_{ij}
Broadstone	92.30	0.10	86.10	0.30
Ythan	100.00	0.20	100.00	0.40

Broadstone Stream (92%) and Ythan Estuary (100%) food webs obtained from 1,000 parameterizations of the corresponding Jacobian matrices. To determine exactly how body size contributed to the stability of these Jacobian matrices, we disrupted the patterning of Jacobian elements determined by body size (after Yodzis, 1981a, i.e., we maintained the sign structure of the food web but permuted the off diagonal predator prey interactions, predator prey pairs were maintained, but the position of these interactions within the food web were randomized) and reanalyzed the webs for local stability. We see that this randomization destroyed the stabilizing pattern of Jacobian elements produced by body size so that only 0.1% and 0.2% of the Broadstone Stream and Ythan Estuary webs were stable, respectively, following the permutation. Size effects enter into the formulation of the Jacobian elements via the biomass densities ($B_i = \gamma W_i^{-\varphi} W_i$) and via the interaction coefficients. To determine the relative importance of these effects we carried out two further permutations: first, we randomized the vector of biomass densities before combining these with the interaction coefficients to form the Jacobian, and second, we permuted the arrangement of interaction coefficients in the matrix $\boldsymbol{A}$ as for the Jacobian matrix previously described. Table 1 shows that quantitatively disrupting the pattern of interaction coefficients has a much greater effect on food web stability than disruption of the biomass densities. Neutel et al. (2002) developed a metric describing the numerical weight of food web loops (omnivorous loops nested within food webs) using the terms of the Jacobian matrix, which they termed 'loop weight.' Loop weight was defined as the geometric mean of the absolute values of the elements of the Jacobian matrix, which correspond to the consecutive strengths of the interactions in a food web loop. They showed that for stable food webs, short loops in food webs tend to have large absolute loop weights and that there is a pattern of declining loop weight as loop length increases. We analyzed the loop weights for the Ythan Estuary and Broadstone Stream respectively for stable versions of these webs and versions of the webs where we had permuted the Jacobian, ***C***, the vector of biomass densities, ***B***, and the interaction coefficient matrix, ***A***. Our intention here was to identify how the pattern of loop weights could be disrupted and whether this would lead to the loss of stability in the food webs. For both the Broadstone Stream (Figure 1 A–D) and the Ythan Estuary (see Figure 1 E–H), it is evident that for unpermuted food web matrices the shortest loops are heaviest (i.e., the maximum loop weight for a particular food web parameterisation is always found in the shortest omnivorous loop in the food web (see Figure 1 A and E)). When the arrangement of Jacobian elements is permuted for both food webs the maximum loop weights increase in overall magnitude (by 1 to 2 orders of magnitude) (i.e., the loops become heavier and the pattern of maximum loop weights has been disrupted with longer loops having larger maximum loop weights). When the vector of biomass densities is permuted and then combined with the interaction coefficients to produce a Jacobian matrix, the resulting Jacobians tend to be stable (see Table 1). The corresponding maximum loop weights are still found in the shortest loops for each food web (see Figure 1 C and G). When the interaction coefficients alone are permuted and then combined with the vector of biomass densities to form the Jacobian there is a loss of stability and the longer loops again become heavier containing the maximum loop weights for each web parameterization (see Figure 1 D and H). It is evident from these results that the pattern of weak links in long loops confers stability on the food webs. The body size ratios presented here produce a patterning of interaction coefficients in the dynamical system that leads to an appropriate arrangement of loop weights in the food webs, and this confers dynamic stability on the webs. When the arrangement of these interaction coefficients is disrupted the corresponding increase in loop weights is sufficient to disrupt the stability of the food webs.

DISCUSSION AND CONCLUSIONS

Here we have shown that there is a theoretical relationship between the body size ratio of a predator and its prey and the per-capita interaction strength between them. These interactions can be estimated simply by observing empirical patterns in real systems and indicates that metabolic scaling may underlie the stability of whole communities and ecosystems. The approach provides us with an invaluable tool for exploring the consequences of extinctions and invasions in complex systems and allows us, importantly, to predict the dynamic conse-

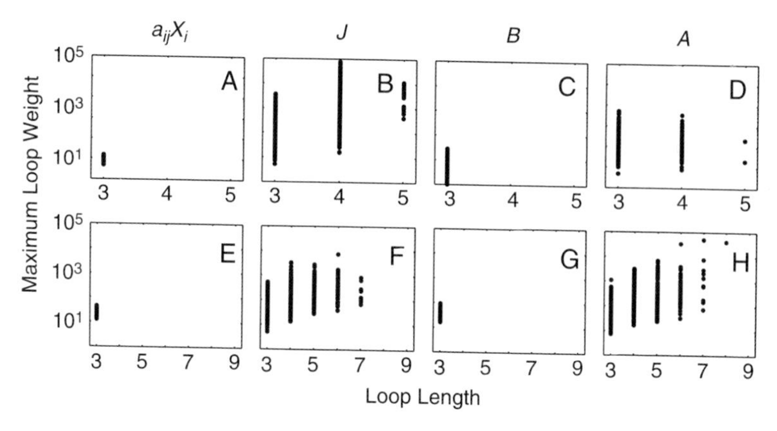

FIGURE 1 | Maximum loop weights for 1,000 parameterizations of the Broadstone Stream (**A–D**) and Ythan Estuary (**E–H**). Max loop weights are expressed as a function of the loop length for each parameterization of the respective web. Jacobian matrices were parameterized (**A & E**) and then three permutations performed: (**B & F**) each Jacobian matrix was permuted; (**C & G**) biomass densities were permuted before the Jacobian was created; and (**D & H**) the interaction coefficients were permuted before the Jacobian matrix was created. Prior to any permutations the maximum loop weights are found in the shortest loops (**A & E**); once the Jacobian matrix is permuted the maximum loop weights increase considerably and are commonly found in the longer loops (**B & F**). When the biomass densities (**C & G**) are permuted the effect on maximum loop weight is negligible and the shortest loops remain heavy; however, when only the interaction coefficients (**D & H**) are permuted before formation of the Jacobian, it is the longer loops that again become heaviest, thus disrupting the original pattern of loop weights, which confers local stability on the food web.

quences of these processes in real systems. Testing these predictions in experimentally tractable ecosystems has yet to be done and is now a pressing goal. Identifying experimental communities or ecosystems, whether terrestrial, freshwater, or marine, that can be discretely manipulated and whole system responses followed will be difficult, but surely not impossible. The approach we have outlined here has still only been tested using two empirically documented food webs. It is important to test the validity of these results using other similarly well-resolved food webs (e.g., Tuesday Lake, Cohen et al., 2003). If such consistent relationships can be falsifiably demonstrated in a greater range of systems, then perhaps a general understanding of the interplay between food web composition (species identity), structure (food web topology), and ecosystem functioning (energy flow, process rates, and stability) can realistically be achieved. We envisage that, potentially, macroecological patterns such as allometric relationships may provide a relatively simple means of uniting the previously disparate topological and dynamical approaches to food web ecology that have been employed to date.

4.4 | BODY SIZE DETERMINANTS OF THE STRUCTURE AND DYNAMICS OF ECOLOGICAL NETWORKS: SCALING FROM THE INDIVIDUAL TO THE ECOSYSTEM

Guy Woodward, Bo Ebenman, Mark C. Emmerson, José M. Montoya, J. M. Olesen, A. Valido, and Philip H. Warren

The importance of body size in trophic ecology, as a determinant of both the occurrence and strength of feeding interactions, has long been acknowledged (Elton, 1927; Peters, 1983), but it is only relatively recently that its role in structuring food chains and entire food webs has begun to be explored systematically (Cousins, 1980; Warren and Lawton, 1987; Cohen et al., 1993b; Warren, 1996; Jonsson and Ebenman, 1998; Woodward and Hildrew, 2002a, b; Cohen et al., 2003; Emmerson and Raffaelli, 2004). Many life history traits of animal species are strongly correlated with their body size (Peters, 1983; Calder, 1984; Ebenman and Persson, 1988; Gaston and Blackburn, 2000; Brown et al., 2004) and, as Nee and Lawton (1996) observed, ecologists record body sizes just as journalists report people's ages. Here we provide a brief overview of the role of body size in food webs across a range of ecological scales, from effects expressed within the individual populations that are embedded within the food web, to those that are manifested at the scale of the entire food web (Figure 1). In most food webs each species is depicted simply as a single 'node,' with little further information supplied as to the ecological characteristics of that population (Cohen et al., 2003). Each node implicitly subsumes, into a single point, not only the population size and growth rate of each species, but also its size-distribution and cohort structure (see Figure 1). These population characteristics can have powerful influences on food webs, since body size is related to so

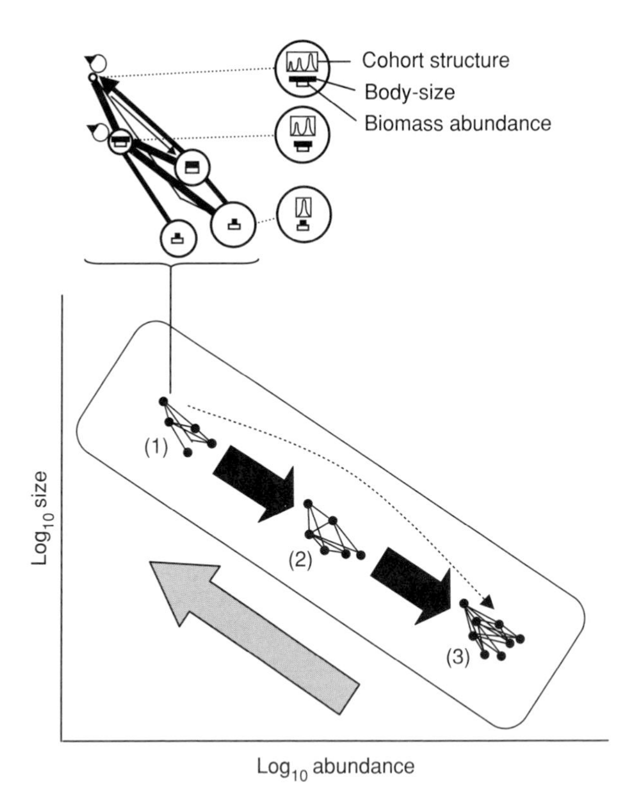

FIGURE 1 | Schematic representation of the size-disparity hypothesis of food web structure. Species may have strong interactions with other members of their subweb, and with adjacent subwebs (e.g., fish [subweb 1] and macroinvertebrates [subweb2]), but these become weakened or broken when the size-disparity between consumers and their resources is very large (e.g., fish [subweb 1] eating meiofauna [subweb 3], shown by the dotted line). The abundance and number of species in each subweb decreases with body-mass (~ trophic status). The areas of the circles in the inset subweb represent possible P/B ratios, which are typically higher for smaller taxa (e.g., Stead et al., unpublished manuscript).

many aspects of a species' ecology, including its numerical and biomass abundance, its energy demands and productivity, its longevity and associated life history traits, its spatial scale of resource use, and its competitive relationships with other members of the food web (Table 1).

BODY SIZE EFFECTS WITHIN SPECIES: ONTOGENETIC SHIFTS, COHORT DOMINANCE, AND CANNIBALISM

'Within-node' variations in population size-distributions can have important consequences for individual species which can, in turn, affect the wider food web, especially if the nodes involved are keystone species and/or top-predators (i.e., species that are often both large and strong interactors). For instance, De Roos and Persson (2002) have demonstrated how cycles of cohort dominance can have catastrophic effects on populations of piscivorous fish. Because these top predators can have strong effects in lake food webs, even to the extent that they are able to trigger regime shifts between alternative equilbria, changes in their population size-structure can have powerful cascading effects in the wider food webs (De Roos and Persson, 2002; Jones and Sayer, 2003; Scheffer and Carpenter, 2003). In other situations, the picture may be more complex and strong 'cohort' effects are unlikely to be ubiquitous, even if a food web is clearly size-structured: many organisms must reach a fixed size and/or life-stage before they can reproduce, whereas others (e.g., reptiles, many fish) can increase in size considerably following the onset of sexual maturation. Many insect species provide good examples of the former, as the adult stage is short-lived and discrete from the larval form, especially in those taxa that undergo complete metamorphosis, and the potential size range of the adult is relatively restricted. Despite the limited scope for insects to exhibit true 'cohort dominance,' many species do show clear ontogenetic shifts in their diet, resource use, and trophic position, even among larval instars, and these changes often track a seasonal cycle (Woodward and Hildrew, 2002a, b). In general, changes in the ecological niche during ontogeny, caused by size increase and metamorphosis, are common phenomena in many animal taxa (Werner, 1988; Ebenman, 1992). For instance, omnivorous lizards are often primarily insectivorous as juveniles but become increasingly herbivorous as adults (Mautz and Nagy, 1987; Valido et al., 2003). Cannibalism is driven by shifts in the relative sizes of the 'predator' and 'prey' individuals, and is often particularly prevalent at times of the year when generations

Table 1: Summary of examples of relationships between body size and ecological traits, and the potential implications for food web structure and/or dynamics

	↑ body size?	Within node effect	Between node effect	Food web effect	*e.g.*'s
Species richness	–	n/a	Smaller species have more potential competitors; more potential resources than consumers	Webs are triangular (~exponential decline in *S* with trophic status); greater potential for redundancy at lower trophic levels?	Petchey et al (2004)
Numerical abundance	–	More small than large individuals	Prey more abundant & smaller than predators; prey less prone to extinctions than predators	Trophic pyramid of abundance; reduced likelihood of population extinctions for species lower in the web	Cohen et al (2003); Schmid-Araya et al. (2002a & b); Woodward & Hildrew (2002b)
Number of size-classes (~ cohorts)	+	Increased potential for cannibalism / cohort dominance	Ontogenetic reversals of trophic status; predators have more generations than prey	Potential for intra-guild predation, self-damping & mutualistic loops increases towards top of the web	Woodward & Hildrew (2002b); De Roos & Persson (2003)
Metabolic rate	–	Small individuals have high energetic demands ∴ high	Prey need to spend more time foraging than predators;	Ecological efficiency increases with trophic status . . . but, as mini-	Cyr (2000); West et. al (2000)

		per-capita ingestion	smaller species have more encounters with predators	mum size of ectotherms < endotherms, the latter are more common at higher trophic levels.	
Nutrient immobilisation	+	Larger individuals are greater sinks for nutrients	Larger species have slower return times for nutrients & ∴ may act as important sinks	Stoichiometric imbalances between resources & consumers	Elser & Urabe (1999); Cross et al. (2003)
Size-refugia	−/+	Access to size refugia increases during ontogeny	Size constraints on feeding links	Size-delimited 'subwebs' within the community web	Huryn (1996); Woodward & Hildrew (2002a); Schmid-Araya et al. (2002a & b)
$K:r$-selected traits	+	Small individuals more vulnerable to predators & are weaker competitors	Small species more opportunistic; Large species more likely to be keystones (e.g., Chase 2000).	Top-down effects (e.g., cascades) stronger at higher trophic levels	Chase (2000)
P/B ratios	−		Lower for predators than prey	Production per unit biomass decreases with trophic status	Benke & Wallace (1997); Stead et al. (2004)
2° production	+	Larger individuals account for majority of 2° production	Larger prey species account for relatively large proportion of predator ingestion & production	Large (but often rare) species drive energy/matter flux within webs; vertical pyramid of production across trophic levels	Benke & Wallace (1997)

(Continued)

Table 1: Summary of examples of relationships between body size and ecological traits, and the potential implications for food web structure and/or dynamics—Cont'd

Diet width	+	Smaller individuals have narrower diets than larger individuals	Small predators feed on a subset of the diet of next largest predator; diet overlap is greatest among similar-sized species	Nested hierarchy of feeding niches as predator size increases towards the top of the web; upper triangularity (basis of cascade & niche models)	Cohen et al. (1993); Woodward & Hildrew (2001 & 2002b)
Trophic status	+	Large individuals more carnivorous (& cannibalistic)	Prey are smaller than predators; large omnivores more predatory than small spp.	Potential for seasonal ontogenetic reversals in trophic status; Assimilation efficiency increases towards top of the web	Jennings et al. (2001); Woodward & Hildrew (2002b); Cohen et al. (2003)
Scale of 'perception'	+	Ontogenetic shifts in perception of environment	Scale ratio between adjacent species pairs declines with increasing body size	Fractal structuring of food webs; spatial scaling determines web size	Ritchie & Olff 1999
Dispersal ability	+ (–)	Smaller individuals operate at smaller spatial scales	Predators operate over larger spatial scales than their prey	Large spp., especially top predators, may link several 'webs' together	Steinmetz et al. (2003)

overlap, such that large instars of the previous generation overlap with small instars of the new generation (i.e., inter-cohort predation). At present, although this type of size-driven variation in basic trophic structure has been documented in some community food webs (Warren, 1989; Woodward and Hildrew, 2002a, b), very few studies have examined the implications of these ontogenetic and seasonal shifts in body size distributions for the dynamics of entire webs, as most of the research to date has focussed primarily on intraguild patterns within the predator sub-web (Polis et al., 1989; Holt and Polis, 1997; Persson, 1999).

Interactions between species that depend on the size distribution within species can also generate Allee effects that can cause the collapse of top predator populations and, ultimately, this may lead to regime shifts in ecological communities (De Roos and Persson, 2002; Scheffer and Carpenter, 2003). Once a regime shift has taken place it can be difficult for the top predator to recover/re-invade. More detailed, size-structured, and individual-based models are likely to provide important insight into how such 'within node' variations propagate through food webs, especially in systems where the trophic role of keystone species is driven by their size distribution (e.g., in shallow mesotrophic lakes, Scheffer and Carpenter, 2003).

BODY SIZE AND PATTERNS IN FOOD WEB STRUCTURE

Predators are usually one to four orders of magnitude larger than their prey, in terms of body mass (Elton, 1927; Peters, 1983; Gittleman, 1985; Vezina, 1985; Cohen et al., 1993b). Although there are exceptions (in particular, host-parasitoid systems, and many pack hunters), this generalization holds true for many food webs (Cohen et al., 1993a; Cohen et al., 2003). Elton (1927) recognized that this general biological observation about feeding biology provided a link between the trophic structure of whole communities and body size patterns, and this idea has found quantitative support in more recent studies. For example, Jennings et al. (2001) have shown that $\log_2$ body-mass explained 93% of the variation in trophic level (as defined by δ15N%) among 15 fish communities in the North Sea. Cohen et al. (2003) have demonstrated clear trivariate relationships between body size, abundance, and food web structure in the Tuesday Lake food web, and analogous patterns are evident in the Broadstone Stream food web (Woodward, Speirs, and Hildrew, 2005). The generality of this link between body size and feeding interactions, and the robustness of its biological explanation, have major implications for understanding the patterns of link organization in food webs.

Any set of species, put together with the simple constraint that predators feed on a certain size range of prey, generally smaller than themselves, will exhibit non-random patterns of food web structure: specifically this results in a nested hierarchy of dietary niches within a web, such that a given predator's potential diet is effectively a subset of that of the next largest predator. For instance, in a study of predation in the highly diverse mammal community of the Serengeti, Sinclair et al. (2003) found that the diet niche of smaller carnivores was nested within that of larger carnivore species. Such hierarchical niche structure, which generates 'upper triangular' food web matrices, is a central component of the 'cascade model' of food web structure proposed by Cohen and Newman (1985), and hence body size provides one plausible biological explanation for this assumption (Warren and Lawton, 1987, Cohen, 1989). Development of this type of model has resulted in other niche-based food web models. Warren (1996) used body size explicitly as one niche dimension in a model of biological constraints on food web structure, and showed that plausible predictions of web features such as food chain length and connectance could be made with such a model. The niche model of Williams and Martinez (2000) is based on similar principles. Like the cascade model, it uses a single, general, niche dimension (which could represent body size), but in this case the strict hierarchy is relaxed so that up to half the trophic niche of a consumer's can include species with niche values higher than itself. This model successfully reproduces many of the patterns that characterize real food webs, such as the prevalence of high trophic similarity, omnivory, food chain length and numbers of feeding loops, and suggests that community niche space can be collapsed into a single dimension, potentially body size, at least when considering food web structure (Williams and Martinez, 2000).

The simple hierarchy of feeding niches, generating upper triangularity in food web matrices, will, of course, break down when the size difference between predators and prey becomes so great that feeding links at the lower end of the size range of a predator are lost (e.g., large predatory fish are unlikely to feed directly on rotifers). This has been termed the "size-disparity hypothesis" (Hildrew, 1992; Schmid-Araya et al., 2002a), and suggests that food webs might be partially compartmentalized on the basis of body size, such that although upper triangularity may exist within 'subwebs,' perhaps delimited by the unimodal humps within the community size-spectrum (e.g., microbes, meiofauna, macrofauna, megafauna), non-adjacent subwebs might have relatively few (and weak) connections between them. Very few published food webs include equally well resolved data from across the entire community size-spectrum, with taxa of <1 mm in size typically being poorly represented despite their high abundance and species richness, but there is evidence to support the size-disparity hypothesis from detailed connectance webs of twelve stream communities that include three unimodal portions of the community size-spectrum: diatoms, meiofauna, and macroinvertebrates (Schmid-Araya et al., 2002b). These webs exhibited far lower connectance (but considerably higher species richness and links) than is usually reported for other detailed webs that are restricted to more limited portions of the size spectrum. In addition, the exponent ($\beta = 1.3$) of S when plotted against L did not conform to the predictions of either the link-scaling law (Cohen and Newman, 1985, $\beta = 1$) or the constant connectance hypothesis (Warren, 1990; Martinez, 1992, $\beta = 2$), suggesting a degree of decoupling between very large and very small species.

When considering the way in which it sets the pattern of the 'trophic scaffolding' in a food web, it is convenient to regard body size as constant, both in time and also for all individuals of a species. However, this is probably rarely, if ever, true. Individuals grow, and the seasonal, size-driven changes seen within species, as described in the previous section, are often also apparent among species of both predators and prey. Such seasonal changes in a species' diet, driven by size changes in either the consumer or its prey (termed metaphoetesis; Hutchinson, 1959) have been documented in food webs for a long time (Hardy, 1924). Where body size is a primary determinant of feeding interactions across an entire web, trophic structure will track these within- and between-species size changes, resulting in a dynamic food web structure over time (Warren, 1989; Woodward and Hildrew, 2002a). The types of 'static' niche models previously outlined are still useful however, but now can be seen as describing the food web at any single point in time. Seasonal ontogenetic changes in diet as a result of changes in the relative sizes of species appear to account for many of the feeding loops seen in nature, especially within insect-dominated food webs, where size overlaps are particularly prevalent (Polis et al., 1989; Holt et al., 1997). Indeed, at least in some systems, the overriding determinant of a species' position within a web is its (variable) size, rather than its (fixed) taxonomic identity (Cousins, 1980, 1996; Woodward and Hildrew, 2002a). The exploration of these important details in a simple manner is a major challenge for ecological modelling, since such details usually introduce instabilities and high levels of uncertainty in model outcomes. The simple case of two species both acting as prey and predators of each other depending on their size class (e.g., large sea-stars feeding upon small crabs and large crabs eating small sea-stars), makes models extremely unstable (Pimm and Rice, 1987).

Even considering size effects at a simplistic level, as in the models outlined here, there is clearly a case that biological constraints on feeding imposed by size can play a significant role in determining broad patterns in food web structure. This suggests that these ideas would repay further exploration and development. For example, the use of more realistic functions to describe the food size niche of consumers, and the incorporation of actual size distributions of species in a community (e.g., right skewed, or multimodal distributions), is likely to provide important insights, yet this remains relatively uncharted territory. Clearly, there is a need to be able to mimic the strong seasonal and/or ontogenetic size-driven shifts in seen in nature and to produce 'reasonable' parameterizations of model food webs. This is essential if we are to be able to model the consequences of these commonly observed phenomena, which, in extreme cases, can result in complete reversals in trophic status between species pairs.

BODY SIZE AND SPATIAL SCALING OF FOOD WEBS

Food webs are rarely discrete entities. Boundaries between webs are blurred, both by the intergrading of habitats, and differences in habitat use and scale of movement of organisms. The magnitude of the flux of matter or individuals between webs is determined by their spatial and temporal isolation, physical transport processes, and the home range and mobility of the individual species, which are often themselves functions of body size (Polis et al., 1997; Woodward and Hildrew, 2002b). Larger species tend to have larger home ranges or territories than do smaller species (Pimm, 1982; Hall and Raffaelli, 1993) and, as a result, very large animals may be unable to persist within a single spatially-delimited web. These species may therefore serve as conduits that join several webs together: avian top predators, anadromous fishes, and migratory ungulates and cetaceans are notable examples of taxa that often operate within more than one 'food web' (Polis et al., 1997; Hall, 1999; Woodward and Hildrew, 2002b). In the same way that larger organisms may operate at scales that go beyond the area under study, there will also be smaller species that occur only in smaller patches within the area (Raffaelli and Hall, 1992). Thus, within a community food web with an arbitrary spatial definition, smaller organisms may well form localized subwebs within the area—perhaps comprising those species that are restricted to a particular microhabitat patch—which are then linked by (larger) species which operate across the larger patches, or the whole habitat. These may, in turn, be linked to other community webs at even larger scales, by organisms for which the area under study is only a part of their range (see also the chapter of Brose et al., Chapter 8.4). Body size is not the sole determinant of spatial scale of activity or population area requirement, but in many systems will be a strong correlate of both (Peters, 1983). This size-dependent scaling of food web extent has led to the suggestion that the 'ecosystem trophic module'—the area of habitat from which the energy required to support a group, or population, of the "largest species that can predate (not parasitize) the largest prey species" in a system, is drawn—provides one non-arbitrary way of defining and counting ecosystems (Cousins, 1987, 1990).

Because resources (e.g., nutrients) are nested within food (e.g., 'prey species'), and available food is nested within space (e.g., habitats), spatial scaling is implicit in real food webs and will be determined by body size. Within a habitat, different species of similar trophic position may consume food of different sizes to obtain the same resources (Ritchie and Olff, 1999). Distributions of resources, food, and habitat often appear to be self-similar, at least at ecologically meaningful scales of about three to four orders of magnitude, such that their spatial patterning can be described using fractal geometry. In fractal environments, body size is a critical determinant of the availability of food that a species perceives. For instance, the environmental grain experienced by meiofaunal species (e.g., rotifers) may span only a few cubic millimeters, whereas that experienced by some megafaunal species (e.g., cetaceans) may exceed hundreds of cubic kilometers. This spatial nestedness of consumers and resources, and their fractal nature, forms the basis of a model developed by Ritchie and Olff (1999), which predicts that species richness depends on the number of 'spatial niches' available to a guild of consumers. If these findings can be extended to encompass several guilds and trophic levels simultaneously, the approach could be used to explore the spatial structure and fractal scaling of entire food webs. For instance, the availability of physical refugia from predators may provide an analogue of the 'spatial niches' previously described, but within a food web, rather than competitive, context. It seems likely that if habitat patches are fractal, then similar scaling laws could be applied to refugia availability, especially as predator and prey body sizes are also often correlated.

Clearly, body size affects the way in which organisms use space and the extent to which movement of species results in flows of matter between different food webs (home range, dispersal patterns). Spatial food web dynamics is perhaps one of the least well-studied aspects of food web ecology, but size may provide one way in which we might start to think about general patterns in the spatial structure of food webs. For instance, an intriguing question is whether the apparent fractal nature of distributions of body size in space is due to spatial constraints per se (Schmid et al., 2000), metabolic constraints (sensu Brown et al., 2004), or some form of amalgam of the two. Essentially, the close coupling between food and habitat that is true for animals that operate at very small spatial scales (e.g., benthic meiofauna consuming biofilm on sand particles) may be weakened as body size increases and animals need to forage further afield (e.g., large predatory fish that feed in the water column and the benthos). This might suggest that fractal relationships hold at small spatial scales, but not necessarily at larger scales, where metabolic constraints may be more important: a detailed, empirical, and theoretical comparison of fractal properties and allometric relationships relating to metabolic theory has yet to be done.

ECOLOGICAL STOICHIOMETRY AND BODY SIZE

Community food webs that simply describe changes in population parameters without being constrained explicitly by thermodynamics or chemistry (i.e., via mass balance of elements) may overlook important fluxes of energy or matter, particularly where nutrient imbalances occur between consumers and resources (Sterner, 1995; Elser and Urabe, 1999). The constituent members of a food web possess an array of size-dependent species traits (e.g., longevity, Peters, 1983; metabolic rate, West et al., 2000; trophic status, Cohen et al., 2003) that can determine the rate of nutrient cycling. Shifts in the species composition and size-structure of a web can, potentially, alter biogeochemical cycles, thereby inducing feedbacks that can have powerful effects on population dynamics, energetics, and nutrient availability. Nutrient imbalances in a consumer's diet can affect its own growth rate as well as those of its predators and prey (Elser and Urabe, 1999). Since each consumer species must keep elemental ratios (e.g., C:N:P) in its body tissues within certain narrow limits, changes in the ratios among food resources will determine which species can persist within the food web (Elser and Urabe, 1999). Because species within the web operate as both sources and sinks of elements, resource dynamics (e.g., availability of P) can, potentially, be driven by the population dynamics of the consumers (e.g., low turnover rates among large predators may restrict the availability of limiting nutrients to the lower trophic levels by creating bottlenecks in nutrient fluxes). Since body size is usually positively associated with trophic status and longevity, but negatively associated with measures of abundance, growth, and turnover (Peters, 1983; West et al., 2000), the potential for stoichiometric imbalances to arise may be linked implicitly to the distribution of body sizes within a web. Recent data suggest that imbalances in C:N:P ratios between consumers and resources could be widespread (Elser et al., 2000). For instance, consumer-driven nutrient dynamics can have profound effects in lake food webs, even to the extent that imbalances in C:N:P ratios may create and maintain alternative equilibria, whereby the primary producers are dominated either by small phytoplankton or by large macrophytes (Elser and Urabe, 1999).

In addition to the within-web effects previously mentioned, large migratory species can supply important pulses of limiting nutrients to spatially-subsidised food webs, which might otherwise experience severe stoichiometric imbalances that can, potentially, extend beyond ecosystem boundaries (Ebise and Inoue, 1991; Justic et al., 1995; Humborg et al., 1997). For instance, Pacific salmon are among the largest members of many North American stream food webs and once they have died, after spawning, their carcasses represent a significant nutrient pool (Wipfli et al., 1999; Milner et al., 2000). Marine-derived N in their carcasses has been traced through the stream food web and out into the surrounding terrestrial vegetation (Hilderbrand et al., 1999). Because stoichiometric analysis integrates the community and ecosystem-based approaches to food webs, it provides a means with which to explore potential interrelations between body size, population dynamics, and nutrient cycling in apparently disparate systems, using a standardized currency (e.g., C:N:P ratios). Indeed, metabolic theory suggests that materials and energy are not necessarily fundamentally different currencies that act independently of one another (Brown et al., 2004).

In future, it would be highly desirable to integrate stoichiometric data into food webs, by recording both consumer-resource fluxes and the pools of important elements or nutrients, and C:N:P ratios are arguably the most obvious place to start. We could then start to explore the consequences for the higher trophic levels of altering resource quality and nutrient availability. Specifically, stoichiometric techniques could be employed to assess the impact of eutrophication in aquatic food webs, in which the environment and, in turn, the basal resources have elevated N and P relative to C. Similarly, the effect of removing large species that are significant nutrient sources (e.g., Sockeye salmon carcasses in young glacial streams) could be investigated experimentally. More subtle stoichiometric imbalances could also be measured: there are marked differences in the role of phosphorus and how it is apportioned within organisms that also vary among species in relation to body size. For instance, unicellular and small multicellular organisms, with high biosynthetic and metabolic rates, allocate most of their total body P for ribosomal RNA, whereas larger, slower growing vertebrates need most of their P for skeletal structures (Brown et al., 2004). Potentially, there are strong links between metabolism, body size, and stoichiometry that could pervade multiple levels of organization in ecological networks but which have yet to be explored in depth.

BODY SIZE, INTERACTION STRENGTH, AND CASCADING EXTINCTION

There has been much theoretical, and some empirical, evidence to support the 'paradigm' that complex (i.e., real) food webs can be stable, if most links are weak (see McCann, 2000, for review), in agreement with early intuitive ideas of Margalef (1968) and May (1972). However, less attention has been given to addressing how body size might be related to interaction strength, and how this might affect stability (but see de Ruiter et al., 1995; Jonsson and Ebenman, 1998; Emmerson and Raffaelli, 2004). It is becoming increasingly clear that the distribution of body size among the members of a web has important consequences for its stability (see Emmerson et al., Chapter 4.3), and for the propagation of disturbances (see Montoya et al., Chapter 7.2). For instance, disrupting the size distribution of a web can have effects on stability that are at least as strong as those produced by altering the distribution or magnitude of interaction strengths. This is especially clear when large predators disrupt community dynamics. For example, abrupt declines of the populations of harbor seals, fur seals, sea lions, and sea otters in the North Pacific Ocean appear to have resulted from increased predation by killer whales, which had previously preyed on the great whales before industrial whaling severely reduced their numbers (Springer et al., 2003). Thus the invasion of a large, strongly interacting and mobile predator to the coastal food webs might have triggered the collapses of otter and pinniped populations which, in turn, altered the interactions between sea urchins and kelp in these food webs. Some authors have suggested that, ultimately, it is predator–prey body size ratios that are the major determinant of interaction strengths, and hence, of food web dynamics (Jonsson and Ebenman, 1998; see Emmerson et al., Chapter 4.3).

If the size spectrum of the community is altered via some form of perturbation, as is often the case (e.g., overfishing, Hall, 1999; invasions of top predators; Woodward and Hildrew, 2001; acidification, Woodward and Hildrew, 2002b), this could potentially affect the stability of the entire system. The harvesting of wild populations typically has a negative effect on average adult body size, which can shift towards smaller body size classes. Emmerson et al. (Chapter 4.3) have shown that interaction strength scales with body size ratio of predator and prey raised to some power b $(aW_j/W_i)^b$. If the average adult body size of a top predator in a community is reduced (because of harvesting) then the size ratio of predator and prey will decline. Such declines in predator–prey size ratios should, in theory, reduce the strength of trophic interactions. Depending on where these interactions occur, weaker interactions are typically associated with stabilizing effects on community dynamics. Such scenarios are possible notwithstanding evolutionary shifts in prey body size, or size related shifts in prey preference towards smaller size classes of prey. The latter are quite probable in real communities and this is certainly an area worthy of future research. Chase (2000) has argued, for instance, that disturbance in running waters constrains the ability of species to exploit defensive (i.e., anti-predator) traits that weaken interaction strengths. In general, perturbations often have disproportionately strong negative effects on larger taxa, which also tend to be strong interactors (i.e., keystones and/or large predators) (Duffy, 2003; Petchey et al., 2004). It is important to bear in mind which trophic levels are being considered: removal of large, well-defended primary consumers from a web without the concurrent loss of predators might serve to increase interaction strength, and hence the likelihood of cascading interactions (Chase, 2000). However, if the large predators are also lost, this might have the opposite effect, by reducing average interaction strength (Petchey et al., 2004). In general, non-randomness of species losses might have important consequences for the risk of cascading extinction (Borrvall et al., 2000; Dunne et al., 2002; Solé and Montoya, 2002; Ebenman et al., 2004) and for the loss of functional diversity (Petchey and Gaston, 2002).

One of the outstanding questions that needs to be addressed empirically is: Are there systematic relationships between patterns in the distribution of body sizes of species in food webs and the structural/topological properties of the webs? If so, are these 'static' relationships dictated by dynamical constraints, and do certain patterns in the body sizes of prey and predators enhance stability and allow complex food webs to persist in the face of external perturbations? More specifically, this might constitute a fundamental difference between pelagic and terrestrial food webs, which differ greatly with respect to the body sizes of primary producers relative to those of the primary consumers. In pelagic ecosystems small, short-lived primary producers (phytoplankton) are consumed by larger and longer-lived primary consumers (zooplankton) while the opposite is often the case for terrestrial forest ecosystems (e.g., trees consumed by insects). Pelagic food chains are also often longer than terrestrial ones (Schoener, 1989; Cohen, 1994). Do body mass ratios between primary producers and primary consumers set up dynamical constraints on food chain length that can explain these differences in food web structure between systems?

We now know that size is inherently connected with biomass and turnover, and that it is also linked to demographic rate processes. It therefore provides one means by which processes framed in terms of population dynamics may also be interpreted in terms of ecosystem processes (matter/nutrient/energy flux), and a way in which the reverse mapping might also be made: indeed, in theory, Lotka-Volterra models can be used with biomass substituted for numbers (de Ruiter et al., 1995; Emmerson and Raffaelli, 2004; see Emmerson et al., Chapter 4.3). Integrating these different views of food webs, and testing the resultant theoretical predictions with empirical data, is one of the main challenges currently facing food web ecology.

ON THE ROLE OF BODY SIZE IN MUTUALISM WEBS

Most interaction webs explored in ecology are 'pure food webs' (Cohen et al., 1993b). Mutualistic interactions such as pollination and seed dispersal, although common in nature, are rarely considered (but see Jordano, 1987; Olesen and Jordano, 2002; Jordano et al., 2003, 2004). The role of body size in mutualistic webs is even less well-known, despite its potential importance: pollinators and seed-dispersers range in body size from the tiny 1-mg Thysanoptera to 5-ton elephants, a size spectrum that spans nine to ten orders of magnitude. The suggestion that large species are more likely to be lost from a food web than smaller species (Chase, 2000; Woodward and Hildrew, 2002b; Petchey et al., 2004) also seems to be applicable to mutualistic networks. For instance, some of the first mutualists to go extinct after the arrival of humans to the Mascarene Islands were the large dodo, the giant flightless parrot, and the giant tortoises (Quammen, 1996). In the Canary Islands, the seed-eating giant tortoises and the >1-m giant lizard *Gallotia goliath* have all gone extinct following human colonization, whereas the smaller, intermediate-sized *Gallotia* spp. have persisted (Barahona et al., 2000). Similarly, in the Americas examples of disproportionate extinctions of large-bodied species can be found among the megafounal Gomphotheres (Janzen and Martin, 1982). Human colonization appears to truncate the size spectrum of consumers. The extirpation of these large species has resulted in the loss of mutualistic partners for several hundred plant species, thus altering the structure of these mutualistic networks.

Frugivore and pollinator species use only a subset of the available plant species, because they respond differentially to the chemical composition of their food source (e.g., the ratio of sucrose/hexose in nectar; the presence of tannins and secondary metabolites in fruits), and there are also physical constraints on mutualistic links that are imposed by the size and morphology of the consumer and the resource. This size-dependent coupling can be very tight, especially where mutualistic links have coevolved for a long time, as is true, for instance, of many orchids and their specialist pollinators. The ratio between gape width and fruit size is important in Neotropical birds (Wheelwright, 1985) and Canarian frugivorous lizards (Valido and Nogales, unpublished manuscript): small animals cannot use large fruits (i.e., body size influences the linkage level (the number of links per species)). Similarly, the range of beak/corolla tube length ratios determines linkage level among long-beaked South American hummingbirds and their flowers (Lindberg and Olesen, 2001). Size, or morphological mismatches, may make a mutualistic link physically impossible or "forbidden" (Jordano et al., 2003). These forbidden links constitute different proportions of the total number of unobserved links in mutualistic webs. For instance, 24% of all possible links are absent from a web containing 16 Spanish passerine birds and 17 fleshy-fruited plant species (Jordano et al., 2004). Among pollinators, 6% of possible links between Danish syrphid flies and their weed flowers (Olesen and Myrthue, unpublished manuscript) are 'missing' because the tongue or beak is too short relative to the corolla tube length. In terms of patterns of connectedness in mutualistic networks, Jordano et al. (2003) have recently demonstrated the ubiquity of truncated linkage level distributions, whereby the most connected species were less connected than predicted by a power-law model (Barabási and Albert, 1999). One of the constraints causing truncation was suggested to be link mismatch (e.g., chemical or body size (or the size of the relevant body parts)). Thus, scattered information about mutualism webs suggests that body size can be an important determinant of the characteristics of both links (e.g., linkage level and the number of forbidden links) and nodes (e.g., population size and the likelihood of extinction) in ways that mirror many of the patterns reported in 'true' food webs, although these observations have yet to be integrated and considered more formally within the context of web theory. The need to develop an equivalent body of theory and to explore the empirical detail from a body size perspective is becoming increasingly evident in mutualistic networks: revisiting the data already collected is likely to be a rewarding first step in this direction.

DISCUSSION AND CONCLUSIONS

Body size is clearly an important determinant of a host of ecological traits which, in turn, influence the structure and dynamics of food webs and, potentially, other types of ecological networks (e.g., mutualistic webs). The effects of body size can be manifested within nodes (e.g., cannibalism), between nodes (e.g., mutual feeding loops), among groups of nodes (guilds, trophic levels) and, ultimately, at the scale of the entire food web (e.g., trivariate relationships between body mass, abundance, and web architecture). The distribution of species body sizes affects the dynamics of food webs in at least two ways. First, through its effects on the interaction strengths between predator and prey species (distribution and magnitude of interaction strengths) and on which links are feasible (link structure of the web), and second, through its effects on the intrinsic growth and mortality rates of species. Both these effects of the body size distribution will have important consequences for the response of food webs to different types of perturbations. Because many of the correlates of body size tend to act in a similar way, in terms of their influence on the food web (see Table 1), measuring body size provides a relatively simple means of encapsulating a large amount of the information 'embedded' within a web, in a single dimension. Size provides a basis on which it may be possible, with increased computing power, to structure individual-based (agent-based) models of multi-species systems, in order to address questions about the effects of population size structure on trophic interactions. It provides a framework for structuring such approaches that allows the retention of some plausible biological structure without requiring detailed individual species information. In addition, there is the intriguing possibility that metabolic theory, which underpins the quarter-power allometric relationships that are common in nature, could be integrated with food web theory to provide trophic ecologists with an over-arching theoretical framework within which to operate.

Can simple allometric scaling relationships be used to predict the structure and dynamics of seemingly complex ecological networks? There is some recent theoretical evidence to support this but, as yet, empirical data remain scarce (Emmerson and Raffaelli, 2004; Reuman and Cohen, 2004; see Emmerson et al., Chapter 4.3). If we can link metabolic rate, arguably the most fundamental biological process, with higher-order processes at the food web and ecosystem scale, we will have moved significantly closer to developing a unifying (although not necessarily all-encompassing) theory in ecology akin to the role genetic theory plays in evolutionary biology (Brown et al., 2004). If we are to attempt this, in future we will need to quantify both the nodes (body size, population abundance, and production) and the links (ingestion rates, interaction strength) in the network, and to integrate pattern and processes more closely (e.g., energy flux in trivariate systems, sensu Cohen et al., 2003; Woodward et al., 2005). Ideally, we would measure community and species-specific metabolism directly (e.g., using field and lab-based respirometers), in addition to the surrogate estimates obtained from body size measurements, wherever possible. These data can then be used as empirical tests of the emerging metabolic and size-based models (Reuman and Cohen, 2005), and to develop the next generation of more sophisticated models based on real data. As yet, no single food web study has integrated all of these approaches: the need to do so now is pressing, and the potential rewards are high.

5.0 | UNDERSTANDING THE MUTUAL RELATIONSHIPS BETWEEN THE DYNAMICS OF FOOD WEBS, RESOURCES, AND NUTRIENTS

Tobias Purtauf and Stefan Scheu

Understanding the transfer of nutrients and resources is in the heart of food web ecology since its early days (Lindeman, 1942). A formal means of dealing with the flow of energy and matter in food webs was ushered in with the advent of systems ecology (Odum, 1973), and since then the food web approach has been adopted to analyze interrelationships between community structure, stability, and ecosystem processes (DeAngelis, 1992a, b). After decades of theoretical analyses and empirical approaches, however, food web ecologists still face the problem of relating food web structure and dynamics to the dynamics of nutrients and resources. Does this imply that the food web approach actually is not suited (yet) for reflecting the spatial and temporal dynamics of the real world (see Bengtsson and Berg, Chapter 5.1)? The chapters in the section approach this challenging question by providing insights into and perspectives of this issue from very different angles. Cross et al. (see Chapter 5.4) point out the need of looking at the dynamics that biochemical nutrients and resources exert on animals in freshwater ecosystems. They show that the role of invertebrates on nutrient dynamics relates to their differential utilization or storage of limited elements in different habitats. Other contributions focus on one of what seems now the most difficult and diverse ecological systems (i.e., the soil) (Giller, 1996; Scheu and Setälä, 2002; Wardle, 2002a). Schröter and Dekker (see Chapter 5.3) relate the nutrient and resource dynamics of food webs in a geographic range of European forest soils to drivers of global change such as atmospheric nitrogen deposition and climate. These drivers affect belowground carbon fluxes via shifts in food web composition, mainly by altering microbial activity. These effects seem to be particularly strong under the conditions of nitrogen limitation. Sabo et al. (see Chapter 5.2) highlight the role of leaf litter as another external factor impacting soil food webs. Leaf litter is shown to function as both resource (nutrients, carbon) and habitat, thereby differentially affecting the structure of decomposer food webs. In particular, the predatory arthropods that mainly depend on litter as a habitat are very sensitive to litter removal. Hence, structural rather than energetic attributes of litter seem to control food web structure. Cousins et al. (see Chapter 5.5) point out the need of defining resource availability and use in terms of energy that can effectively be used to perform necessary growth and persistence for consumer populations and hence the persistence of communities as a whole. Dekker et al. (see Chapter 5.6) propose new ways of modelling food webs and nutrient cycling by including environmental variability, more precise energy parameters, and indirect effects through trophic pathways in the modelling of soil food webs. They point out that indirect effects of limited nutrients and resources (see Cross et al., Chapter 5.4) or feedbacks of soil organic matter via faeces production have to be considered in further studies to make food webs more realistic.

The important role of space and time in governing food web dynamics (see McCann et al. Chapter 2.4; Brose et al., Chapter 2.1) necessarily also affects the dynamics of nutrients and resources. Bengtsson and Berg (see Chapter 5.1) stress the lack of knowledge concerning the impact of spatio-temporal variability on the flow of energy and nutrients through food webs. By again exemplarily concentrating on the belowground community, these authors criticize the tendency for ignoring horizontal and vertical spatial variability in empirical food webs by averaging over large spatial and long time scales. Ecological systems generally show a characteristic variability over a range of spatial, temporal, and organizational scales (Levin, 1992), with local populations and communities being affected by factors simultaneously acting at different scales (Gaston and Lawton, 1990; Whittaker et al., 2001). It is thus not *a priori* clear at which spatial or temporal scale organisms perceive their environment. For example, Holt (2002) showed that food chain length is sensitive to area or distance to resources. Scaling problems (i.e., what spatial and temporal scale is analyzed in soil food web studies) arise when results are up scaled from plot analyses to larger spatial scales for addressing problems such as habitat loss or fragmentation. Dekker et al. (see Chapter 5.6) see a solution to this type of problem by integrating field experiments and by including uncertainties into food web analyses.

The chapters of this section demonstrate that understanding nutrient and resource dynamics requires a much more holistic view. Currently, however, there are still strong boundaries among different approaches to food web analysis, both within (e.g., aboveground vs. belowground in terrestrial environments) and between systems (e.g., terrestrial vs. aquatic). These differences reflect historical developments rather than being justified by scientific grounds. The multiple links between energy and element pathways above- and belowground (Scheu, 2001; Bardgett and Wardle, 2003) as well as between aquatic and terrestrial food webs (Odum, 1973; but see Rose and Polis, 1998; Thorp and DeLong, 2002) have now been convincingly demonstrated. Therefore, looking out of the perspective of dynamic food webs, above- and belowground terrestrial systems and aquatic and terrestrial systems can also be seen as independent webs linked through processes. This fuels the hope for our ability to come up with more realistic and less exclusive approaches to the dynamics of nutrients and resources in the near future.

5.1 | VARIABILITY IN SOIL FOOD WEB STRUCTURE ACROSS TIME AND SPACE

Janne Bengtsson and Matty P. Berg

In September 2003 we performed a literature search on temporal or spatial variability in soil food webs. With "spatial, temporal or variability" and "soil food web(s)" in the title we came up with zero hits. Taking soil out of the equation, we got 41 hits, mainly concerning aquatic ecosystems, from 1989 onwards (Source: Biosis previews).

Despite the fact that it has been repeatedly pointed out that information on temporal and spatial heterogeneity is crucial for understanding how soil food webs and soil biodiversity affect key ecosystem processes, surprisingly little is known on spatial and temporal variability in soil communities and food webs (Ettema and Wardle, 2002; Wardle, 2002a). Moore and de Ruiter (1991) used information on the temporal variability of functional groups to determine the main energy channels in agricultural soil food webs. Wardle (2002a) discussed how plant species and herbivores could influence spatial and temporal variability in food web structure, but also pointed out the lack of studies on soil food web variability in time and space.

Variability in food webs has most often been studied in aquatic ecosystems. Substantial spatial and temporal variability has been found at the base of the food web (Findlay et al., 1996), as well as in the whole food web (Warren, 1989). Closs and Lake (1994) found considerable temporal variability in food web structure in an intermittent stream, whereas little spatial variability between sites was found. Schoenly and Cohen (1991) analyzed a set of terrestrial and aquatic food webs for temporal variability, and found that very few species in the collection of food webs occurred on each sampling occasion, and most of the species were found only once.

Sensitivity analyses of soil food web models have shown that small changes in the biomass of particular groups, for example basal organisms or certain predators, may have a marked and disproportionate effect on soil processes, such as decomposition and mineralisation of nutrients (Hunt et al., 1987; de Ruiter et al., 1993b; Hunt and Wall, 2002). In contrast to the averaging of time and space in many models, these observations suggest that it may often be useful to take temporal and spatial variability in food web structure into account when examining carbon and nitrogen dynamics, especially when effects of variability are non-linear. This is also suggested by theoretical studies, which indicate that variability in food web structure can, depending on the circumstances, result in either mobilization or immobilization of nutrients (Zheng et al., 1999).

Soil community composition has been shown to vary over time (Bengtsson, 1994b), over space (Ettema and Wardle, 2002), and down the soil profile (Berg et al., 1998a; Berg et al., 1998b). Nevertheless, in empirical and modeling studies that aim to enhance our understanding of the role of food web composition in energy and nutrient fluxes variability in food web composition has received little attention. In this chapter we focus on the variability of soil food webs (i.e., the composition of functional groups, rather than species communities). We ask the questions (i) how variable is food web structure over time and across space, and (ii) which are the major factors that determine the variability of food web structure? We discuss our findings in the context of ecosystem functioning.

VARIABILITY IN SOIL FOOD WEBS

In soils, no clear effects of species richness on soil process rates have been reported, whereas functional redundancy at higher levels of taxonomic aggregation (i.e., at the level of functional groups) is generally small (Laakso and Setälä, 1999a; Wardle, 2002a). Variability in species composition may not be clearly related to food web variability and ecosystem functioning because of species redundancy and complementarity within functional groups (Laakso and Setälä, 1999a; Loreau et al., 2001), and because the scales of spatio-temporal changes in the abundance or biomass of many functional groups are very small compared to the scales at which ecosystem functioning is relevant to study (Anderson, 1988, 1995). Therefore, for the analyses in this paper, the focus is not on the species composition of functional groups but on the functional group composition of the food web, which are based on what is known about feeding habits (cf. Moore et al., 1988). We have not estimated the variability in trophic link strength because in the Hunt et al. (1987) model link strengths are calculated from the biomass of the food web components and we do not have independent data on trophic fluxes. For a more detailed analysis on variability in food web and species composition in our study system, we refer to Berg and Bengtsson (unpublished manuscript).

In soil communities three types of variability can be distinguished (i.e., variability over time) across space horizontally and vertically across soil horizons. We have analyzed the three types of variability for the same soil food web to see how they relate to each other and to illustrate general gaps in our knowledge on variability in food webs.

To be useful, measures of variability must enable comparisons between studies. Bengtsson (1994b) and Bengtsson et al. (1997) used Kendall's coefficient of concordance, W, which is based on ranked abundances, to examine temporal variability in soil and bird communities, respectively. However, a better and more precise option is to use measures of community or food web variability based on absolute abundances, such as the Bray-Curtis (BC) similarity (Legendre and Legendre, 1998). The ordination techniques used by Collins (2000) are powerful but not suitable for comparisons of variability between studies, as their interpretation is specific to the data set used in each analysis. Hence we expressed food web variability as BC similarity in functional group composition over time or across space. The BC index varies between 0 and 1, with high values of similarity indicating low variability.

Our analyses of temporal variability followed the same procedure as in Bengtsson (1994b), apart from the fact that Bengtsson used W, while we used BC similarity. We calculated the average similarity in biomass between samples ($n = 6$ for mean values at each point in time) taken at different time intervals and regressed similarity against the time interval between samples ($n_{time} = 14$). For horizontal variability, BC similarities were calculated for pairs of litterbags within the same horizon. These values were correlated with the distance between pairs and significance was tested with Mantel-tests. For vertical variability across horizons, we calculated the average BC similarity between the adjacent L-F and F-H horizons and the non-adjacent L-H horizons. We analyzed the biomass composition of (i) all animal functional groups (16 groups) in Figure 1 and (ii) all microarthropod functional groups (6 groups; Collembola, fungivorous Oribatida (grazers and browsers), fungivorous Prostigmata, Predaceous Acarida, and Araneae). Horizontal variability in all animal functional groups could not be calculated because microarthropods were extracted from different sets of litterbags at different locations than the nematodes and enchytraeids.

Temporal Variability in Food Web Structure

There are some long-term data on temporal variability within taxonomic groups (carabid beetles (Den Boer, 1977), springtails (Takeda, 1987; Wolters, 1998), enchytraeids (Abrahamsen and Thompson, 1979)) showing that variability in species composition in soils is often low but usually increases over time (Bengtsson, 1994b) in data encompassing around a decade. Bengtsson also examined the variability of taxonomically defined food web components although none of the studies were longer than five years. Temporal variability of functional group composition was significantly lower than for species composition (average W-values for functional groups were 0.89 and for species 0.80, lower values indicating higher variability).

Within-year variability in food web composition can be as large as between-year variability (see Figure 2: data from Huhta et al., 1967). This

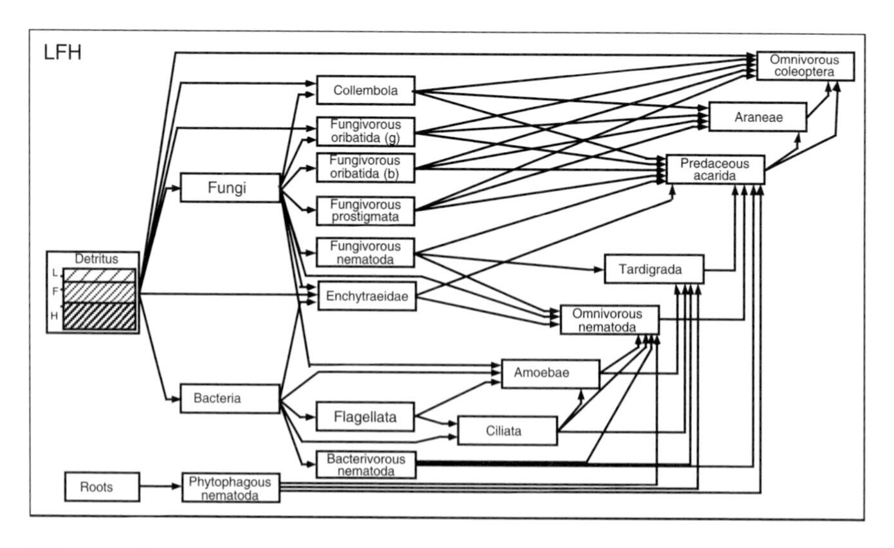

FIGURE 1 | Connectedness food web for the successive litter (L), fragmented litter (F), and humus horizon (H). Boxes represent the functional groups of organisms (sensu Moore et al., 1988), the vectors the flow of energy and nutrients between them. Between some functional groups dashed vectors were drawn as the importance of these resources for these groups are unknown (after Berg et al., 2001).

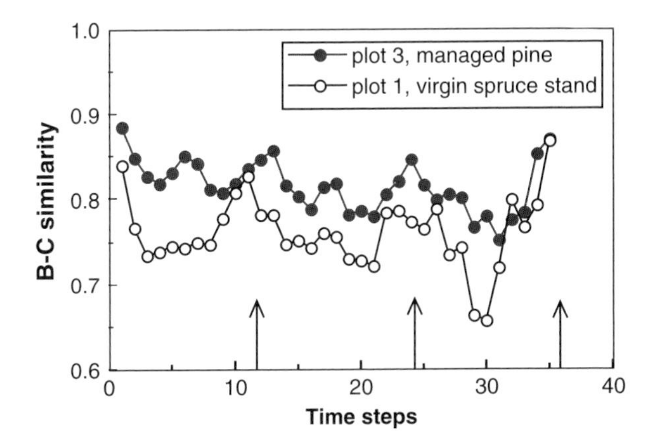

FIGURE 2 | Temporal variability (expressed as Bray-Curtis (BC) similarity, higher values indicate low variability) in animal functional group composition in two coniferous forests (data from Huhta et al., 1967, for a large part of the food web, 14-16 taxonomically defined functional groups). One time step is approximately one month; arrows indicate one-year intervals. In both plots, similarity in food web composition peaks at approximately one-year intervals, indicating strong seasonal effects on food web structure. The data points show the average similarity of samples taken at different time intervals. At t=1 the point represents the average BC of sample 1 compared to sample 2, 2 to 3, 3 to 4 and so on until sample 34 to 35. At t=35, the point represents the similarity of sample 1 to sample 35. Hence, n-values for each point decreases from 35 at t=1 to 1 for the last data point.

suggests that food webs do not have a fixed structure within a year. A consequence of this is that conclusions from analyses of food webs based on yearly averages may have to be treated with caution until we know how large an effect temporal variability has on, for example, food web stability and mineralization rates, especially in systems with fast population, nutrient, and carbon dynamics. Generation times of dominant organisms in soils, such as microbes and nematodes, are often short, in the order of some hours to two months (Anderson, 1988). We are not aware of any publications that have *explicitly* modeled the effects of a temporally varying food web structure on (e.g., mineralization rates) as compared to one based on yearly averages (some papers do that in some way, like that of de Ruiter et al., 1993b; de Ruiter et al., 1994; de Ruiter and van Faassen, 1994).

In our study, the food web in the litter horizon was significantly more variable over time than it was in the other horizons (BC similarity (95% confidence limits): L = 0.47 (0.042), F = 0.62 (0.028), and H = 0.60 (0.024)). Furthermore, the variability increased significantly with time between sampling points in the L horizon but not in the deeper F and H horizons (ANCOVA: Layer × time $F=9.33$, $p<0.005$). This difference could be due to the short duration of our study, only 2.5 years, in relation to the turnover rate of the organic matter in the F and H horizons. The temporal variability of the combined six microarthropod functional groups in the litter horizon was significantly higher than the variability calculated using all animal functional groups (Figure 3). This finding suggests that our choices of which organisms to include in our studies can influence the magnitude of food web variability that we perceive.

We also observed a significant seasonal effect on food web composition when all animal functional groups were analysed. This indicates that the composition of food webs was more variable within a year than between years, just as in the data after Huhta et al. (1967) shown in Figure 2. This seasonal effect was also observed at the level of individual species for the combined microarthropod species composition (data not shown; Berg and Bengtsson, unpublished manuscript).

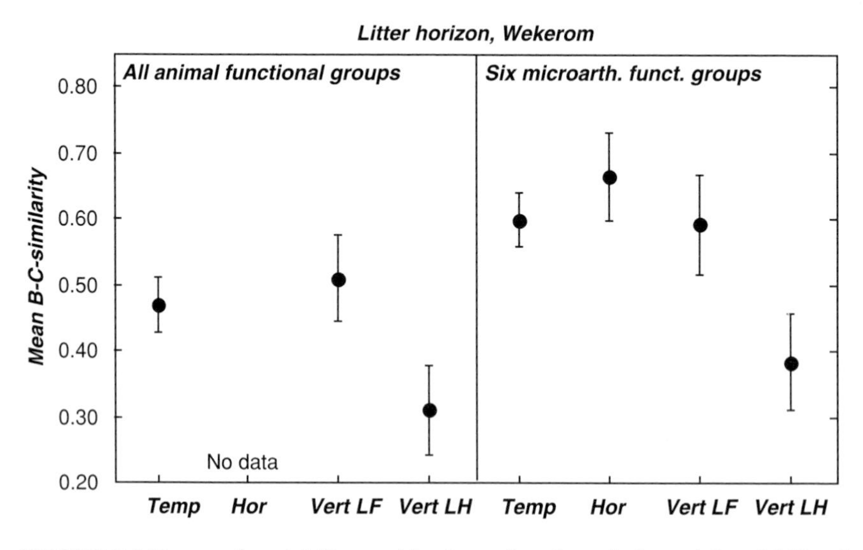

FIGURE 3 | Temporal variability, and horizontal and vertical spatial variability, in the composition of all animal functional group and of microarthropod functional groups in a Dutch pine stand over two years. Mean values of Bray-Curtis (BC) similarity for the litter horizon over time (Temp), in horizontal space (Hor), between the adjacent litter and fragmented litter horizons (Vert LF) and between the non-adjacent litter and humus horizons (Vert LH). 95% Confidence Limits (C.L.) are shown, means with non-overlapping 95% C. L. are considered to differ significantly. Lower BC-similarity values indicate higher variability.

Horizontal Variability in Food Web Structure

Horizontal variability in soil communities can be caused by both plants and animals (Saetre and Bååth, 2000; Ettema and Wardle, 2002) and environmental factors. A limitation in previous studies of spatial variability is that they have been done on individual organism groups rather than soil food webs. We found no published papers even on subsets of food webs.

We observed a relatively low horizontal variability for the six microarthropod functional groups in the pine forest (BC similarity: mean over all horizons = 0.65; Figure 3 for L-horizon). A horizontal spatial structure was seldom observed, which means that the level of variability did not depend on the distance between sampling points. The results can be explained by the facts that we studied a planted pine monoculture without understory plant cover, and in addition lacking soil engineers (sensu Lawton and Jones, 1995) because of the acid soil. On a small spatial scale engineers, like earthworms, can create variability in the environment by forming casts, redistribute organic matter, and alter soil texture. It is interesting to observe that horizontal variability resembles temporal and vertical (LF) variability (see Figure 3), despite the fact that it is calculated based on single litterbags rather than averages of six litterbags. One would have expected that variability due to sampling errors was greater and hence that BC similarity was lower for horizontal variability.

Vertical Variability in Food Web Structure

It is well known that soil process rates (Berg et al., 1982; Clays-Josserand et al., 1988; Faber, 1991) and soil organisms show a vertical spatial variability (Berg et al., 1998a; Berg et al., 1998b; Ponge, 2000). Microclimate, soil texture, and resource quantity and quality are among the factors that differ between soil horizons, and which may cause this variability. This vertical variability can be important for ecosystem processes. Briones et al. (1998) found that experimental climate warming resulted in changes in the vertical distribution of enchytraeids which in turn significantly increased nitrogen mineralization. In our study area food web variability was much higher on a scale of a few centimeters down the soil profile than both temporal and horizontal variability (see Figure 3). The variability was significantly higher between the non-adjacent L-H horizons than between the adjacent L-F and F-H horizons.

IMPLICATIONS FOR FOOD WEB STUDIES

We found that vertical spatial variability in the stratified pine forest soil was significantly greater than both temporal and horizontal variability. Thus, a high proportion of the variability in food web structure in soils can be attributed to different soil horizons. However, most empirical food web studies as well as models have disregarded this source of variability, presumably because empirical studies are labor intensive enough without sampling and accounting for different horizons, and the complexity of food web models comprised of interacting soil horizons appears overwhelming. Nevertheless, it may be appropriate to analyse the effects of vertical variability in food web structure in more detail in the future (see for instance de Ruiter et al., 1993b, and Berg et al., 2001). For example, are food webs in different layers distinct and compartmentalized to such a large extent that we should treat them as separate webs?

The largest temporal variability over two and a half years was observed in the litter horizon, which also showed the highest decomposition rate (Berg et al., 1998b) and hence largest change in organic matter quality over time (Ågren and Bosatta, 1996). This suggests that changes in organic matter quality are a major driving force for temporal dynamics of the soil food web structure, just as differences in organic matter are of importance for vertical variability. In turn, it has been reported that variability in food web structure can influence decomposition, nutrient dynamics, and primary production (Setälä and Huhta, 1991; Zheng et al., 1997; Zheng et al., 1999; Berg et al., 2001; Schröter et al., 2003). Therefore, the mutual interactions between organic matter and food web composition should be explicitly taken into account in future studies.

In our study area, we found no spatial structure in horizontal variability. In other ecosystems, for example in mixed species forests, significant spatial structure has been found (Saetre and Bååth, 2000), much of which seems to be induced by tree species (Saetre and Bååth, 2000; Bengtsson and Saetre, unpublished manuscript). On the other hand, spatially structured variability in soils may mainly occur on very small scales, within a few millimeters (Anderson, 1988). This component of variability is not captured by our, nor by most other sampling schemes. The existence of spatial structure in many ecosystems suggests that food webs could be structured into horizontal compartments in which interactions are localized. This might strongly influence analyses of food web stability (May, 1973; Hall and Raffaelli, 1993), and potentially also estimates of carbon and nitrogen dynamics.

An open question is the relation between spatial and temporal variability in soil communities. Does the amount of horizontal spatial variability influence the magnitude of vertical or temporal variability, and visa versa? There may be general relationships between variability in time and variability in space that constrain the possible combinations of the three types of variability we have examined. For example, we consider it unlikely that communities exist with a high horizontal variability but a low temporal variability. We also consider soil communities with low vertical variability to be rare. However, we are not aware of any comparative data on these issues.

We have studied a highly stratified soil without any soil engineers, like earthworms. This may be the reason for horizontal variability being low and vertical variability being high. We hypothesize that in soils in which ecosystem engineers are mixing the organic matter with mineral soil this pattern might be reversed.

Many models of soil food webs and C and N mineralization have averaged the small-scale temporal and spatial variability that we have discussed in this chapter (Hunt et al., 1987; Moore and de Ruiter, 1991; de Ruiter et al., 1995; Berg et al., 2001; Schröter et al., 2003), using generalized webs for years and sites. Such simplifications have often produced reasonable estimates of mineralization rates. However, in view of the large variability in food web composition that has been observed, it may be time to evaluate if including variability makes a difference for the outcome of these models. This could be done by comparing models with and without temporal and spatial variability in the food web. In addition, we argue that models that structurally ignore the variability that is observed on scales like months and meters may not provide as much mechanistic understanding of the relation between soil food web structure and ecosystem processes as we originally hoped it would.

It is clear that we currently do not know enough about variability in food webs. For some important soil organism groups we have scattered information on temporal variability in community composition but almost no data on spatial variability (Wardle, 2002a). We also need food web models that can take small-scale temporal and spatial variability into account. The proposition that incorporating variability in models would make a difference from averaging over years and sites, and give better estimates of mineralization can then be evaluated. Obviously, field researchers and food web modelers should work together to examine these issues.

CONCLUSIONS

On the temporal and spatial scales we have investigated, soil food webs appeared not to have a fixed quantitative structure but are dynamic. This suggests that food webs in seemingly stable and homogenous environments, as our study site, may be much more variable than implied by many food web models, and also by experimental studies in the laboratory. The assumption of compositional steady state in food web models that analyze interaction strengths and stability may be an oversimplification that obscures our understanding of the mechanisms behind stability in food webs. On the other hand, the scale at which these models apply might be considered to average over these small scales. That is, they might apply only at larger spatial scales (e.g., fields, stands, forests) and long temporal scales (e.g., decades). Whether ignoring the variability at smaller or shorter scales is a useful approximation for examining many basic and applied problems regarding food webs remains to be shown, and in most cases we consider it doubtful that it is.

ACKNOWLEDGMENTS

We thank Nico van Straalen, Herman Verhoef, and Niklas Lindberg for their constructive comments on an earlier version of the manuscript. Janne Bengtsson was funded by the Swedish Research Council, and Matty Berg was financially supported by the Royal Netherlands Academy of Arts and Sciences (Academy Fellow program) and a NATO linkage grant (Grant No. EST CLG 978832).

5.2 | FUNCTIONAL ROLES OF LEAF LITTER DETRITUS IN TERRESTRIAL FOOD WEBS

John L. Sabo, Candan U. Soykan, and Andrew Keller

Detritus is a central component of food webs in freshwater, marine, and terrestrial ecosystems (Lindeman, 1942; Vannote et al., 1980; Setälä and Huhta, 1991; Cadisch and Giller, 1997; Polis et al., 1997; Wallace et al., 1997). Dead organic matter forms the energy base for detrital loops of food webs (Fisher and Likens, 1973; Flecker, 1984; Polis and Strong, 1996; Meyer et al., 1997; de Ruiter et al., 1998; Hall et al., 2000) and links decomposers to producers directly via mutualisms with mycorrhizae (Reid and Woods, 1969; Fogel, 1980; Fogel and Hunt, 1983; St. John et al., 1983) and indirectly via nutrient mineralization associated with decomposition in soils or in the water column (Ingham et al., 1985; Setälä and Huhta, 1991; Vanni and Layne, 1997; Vanni et al., 1997; Wardle, 1999a). Detritus also couples food webs in above- and belowground compartments of terrestrial ecosystems (Wardle, 2002a) as well as between aquatic and terrestrial ecosystems (Vannote et al., 1980; Polis et al., 1997).

Detritus is both food and habitat for members of detrital- and plant-based compartments of food webs. For example, soil provides the substrate for decomposition as well as the matrix within which soil fauna interact and carry out the decomposition process (Moore et al., 2004). In streams, coarser forms of detritus—especially large senesced trees—provide a direct energy source for only very specialized detritivores (Cummins and Klug, 1979; Fowler and Whitford, 1980; Mankowski et al.,

1998), but provide substrate for producers and microbial decomposers (Aumen et al., 1983; Tank and Webster, 1998; Vadeboncoeur and Lodge, 2000). More importantly, coarse detritus such as large woody debris provides key habitat structure for detritivores (Wallace and Benke, 1984; Benke and Wallace, 1990) as well as predators of these invertebrates such as salmon in streams (Harmon et al., 1986; Fausch and Northcote, 1992; Everett and Ruiz, 1993), and small mammals and birds in riparian habitats (Steel et al., 1999).

Smaller forms of detritus have often been treated solely as a resource for detritivores. For example, leaf litter is a fundamental component of almost all forested ecosystems (Cadisch and Giller, 1997). Leaves and associated microbial and fungal decomposers are the primary source of carbon for invertebrate and vertebrate detritivores living on the soil surface and in deeper soil horizons (Anderson, 1975; Mitchell and Parkinson, 1976; Swift et al., 1979; Ponge, 1991; David et al., 2001; Hattenschwiler and Bretscher, 2001). Leaf litter food webs exemplify the inherent donor-controlled nature of detrital consumer-resource systems (Polis and Strong, 1996). The input rates and standing stocks of litter determine detritivore biomass (Blair et al., 1994), while consumption of litter may have little effect on the standing stock of litter (Zheng et al., 1997).

Like coarser forms of detritus, leaf litter can also determine the relative abundance of species not dependent on litter as an energy source. For example, ground spiders have a strong affinity for leaf litter on forest floors (Uetz, 1979; Wise, 1995; Vargas, 2000; Wagner et al., 2003). These predators may depend on the structural complexity of litter as cover from other predators (Uetz, 1979; Vargas, 2000). The depth and complexity of litter may also alter the thermal environment on the soil surface and in deeper soil horizon, which can influence the distribution and abundance of spiders (Rypstra et al., 1999) and other invertebrate species. Finally, decomposition of litter by soil flora and fauna directly influences local pools of nutrients available for plant production (Ingham et al., 1985; Setälä and Huhta, 1991), and the activities of predators can in some cases depress decomposer populations thereby inhibiting decomposition (Wise and Chen, 1999; Lawrence and Wise, 2000; Rosemond et al., 2001).

Here we use meso-scale (25 m^2) litter removals in a desert riparian floodplain habitat to investigate the influence of litter on the abundance of invertebrates occupying two adjacent trophic levels within the forest floor food web. We hypothesize that litter has stronger positive effects on the abundance of invertebrate predators (spiders) than on the abundance of detritivores (crickets and cockroaches). Litter enhances the three dimensional structure of forest floor habitat and thus, the availability of suitable cover for spiders from higher-level predators.

METHODS

Study Site

Experiments were carried out in cottonwood-willow gallery forest stands found along the upper San Pedro River in the San Pedro Riparian National Conservation Area (SPRNCA) in SE Arizona (31°37′33″, 110°10′26″; NAD27) at an elevation of ca. 1200 m. The site receives on average 36 cm of rain annually (91-year average, National Weather Service record for Tombstone, AZ) arriving mostly as summer and monsoon storms between July and September. Summer months of June–September are hot (highs > 40 °C) while winters (November-February) can be relatively cold (lows < 0 °C). Diel temperature variation during the dry season, spring and fall months, can span a similar range (10–40 °C).

The San Pedro, a tributary of the Colorado River, represents one of the last free-flowing rivers in the Desert Southwest. A prominent feature of the upper San Pedro River is extensive coverage by cottonwood-willow (CW; *Populus fremonti* and *Salix goodingii*) gallery forests and mesquite (*Prosopis* spp.) bosques. Both forest types typically occur in unconstrained valleys of low-gradient (<0.005 m/m), perennial rivers of the southwestern United States (Stromberg, 1993a, b). CW galleries develop in the river floodplains, while mesquite bosques develop adjacent to and slightly above these galleries on lowland terraces. Because soil moisture can limit growth of cottonwood and willow trees (Pope, 1984), these trees grow well in alluvial floodplain substrates maintained by channel meandering (i.e., oxbows). At these sites, CW production is not limited by water availability as the trees use subsurface water from shallow aquifers (Dawson and Ehleringer, 1991). High leaf production leads to abundant litter, varying in depth from 3 to 25 cm along the forest floor (Sabo et al., unpublished data).

The detrital component of the terrestrial food web includes a number of aboveground detritivores including, crickets (*Gryllus* sp.), wood cockroaches (*Parcoblatla* sp.), and several isopods. Leaf litter also is habitat for a guild of ground spiders (Lycosidae), an abundant group of generalist predators. These spiders seek cover in litter during the day from diurnal spider-hunting wasps (Sphecidae), but can be found in abundance on the litter surface and open forest floor by night (Sabo, personal observation). At the top of the floodplain forest-floor food web, two generalist vertebrate predators forage within the leaf litter for prey at more than one trophic level. Deer mice (*Peromyscus maniculatus*) are omnivores that consume both plant and animal resources including seeds, leaves, and invertebrates (Hoffmeister, 1986). The Sonoran grassland whiptail (*Aspidoscelis uniparens*) is an abundant generalist predator that forages incessantly in leaf litter. Both species are extremely abundant in the CW gallery forest, up to three to four times more abundant than in adjacent mesquite and desert scrub habitats that lack a comparable amount of litter (Sabo et al., unpublished data).

Experimental Design

In early May of 2002, we set up six 25-m^2 plots within each of two oxbows at Gray Hawk Nature Center (situated within the SPRNCA). In order to homogenize conditions among treatments as much as possible, the plots were designated in pairs located within 50 m of each other. One member of each pair was raked to remove all loose leaf litter (no litter treatment) while the other was left undisturbed. Litter removal was done carefully, to minimize the impact of our raking on the living vegetation in the plots. Subsequent removal experiments coupled with controls in which we raked litter but left it in place suggest that the disturbance associated with raking has little effect on the variables of interest in this chapter—ground-dwelling arthropods (Sabo et al., unpublished data).

Sampling

We sampled invertebrates using pitfall traps—small plastic cups (10 cm diameter, 10 cm in depth), half-filled with a dilute soapy-water solution (to break the surface tension of the water). Each round of sampling involved one pitfall trap placed in the center of each plot for 48 hours. We sampled a total of five times, once prior to treatment establishment and four times following litter removal/disturbance. We stored specimens collected in the pitfall traps in whirl-zip bags and 70% ethanol until they could be processed in the lab.

Lab Processing

We created digital images of all of the specimens using a Leica S6D microscope (Wetzlar, Germany) and a Spot Insight QE© digital camera (Diagnostic Instruments, Inc., Sterling Heights, MI). These digital images were then analyzed using Scion Image software (Frederick, MD), allowing us to count and measure each individual invertebrate. We grouped the invertebrates into four categories—crickets, cockroaches, lycosid spiders, and all other taxa grouped as "other." The dry mass of each taxon and whole samples were estimated using length-mass regressions (Sabo et al., 2002).

Analysis

We analyzed pre- and time-averaged post-manipulation samples using Mann-Whitney U tests with treatment (litter/no litter) as the independent variable (n=6 for the litter treatment; and n=3 for the no litter treatment). Standard repeated measure analyses (e.g., rmANOVA) were not appropriate for these data as changes in response variables across time were both strongly non-linear and non-concordant over time. Low sample size and unequal replication further prohibited more data-intensive time-series methods. For these reasons, we pooled data across time (within plots) and analyzed post-manipulation responses of taxa using these time-averaged samples. We used a non-parametric test in this case because of low sample size and violations of normality and equal variance assumptions of parametric two sample tests. Finally, three of our no-litter plots were sacrificed for an additional experiment, precluding paired analysis of the data.

RESULTS

Prior to litter removal, abundance and biomass were not significantly different between litter and no-litter plots for spiders (Mann-Whitney $U = 23.5, 23.0$; $n = 12$; $P = 0.34, 0.39$, abundance and biomass, respectively), crickets (Mann Whitney $U = 22.5, 24.0$; $n = 12$; $P = 0.47, 0.38$, abundance and biomass, respectively), or cockroaches (Mann Whitney $U = 19.0, 19.0$; $n = 12$; $P = 0.87, 0.87$, abundance and biomass, respectively). Similarly, the abundance and biomass of all taxa combined did not differ between litter and no-litter treatments (Mann-Whitney $U = 24.0, 21.0$; $n = 12$; $P = 0.34, 0.63$, abundance and biomass, respectively).

Following litter removal, temporal patterns of abundance and biomass differed visually among spiders, crickets, cockroaches and all arthropods combined (Figures 1 and 2). Spider abundance was higher in litter vs. no litter treatments during most post-manipulation census dates, whereas the relative abundance of detritivores (crickets and cockroaches) in litter and no litter treatments differed seasonally. For crickets and all taxa combined, abundance was generally higher in no litter treatments during the dry season (Julian Day 157) but this pattern reversed

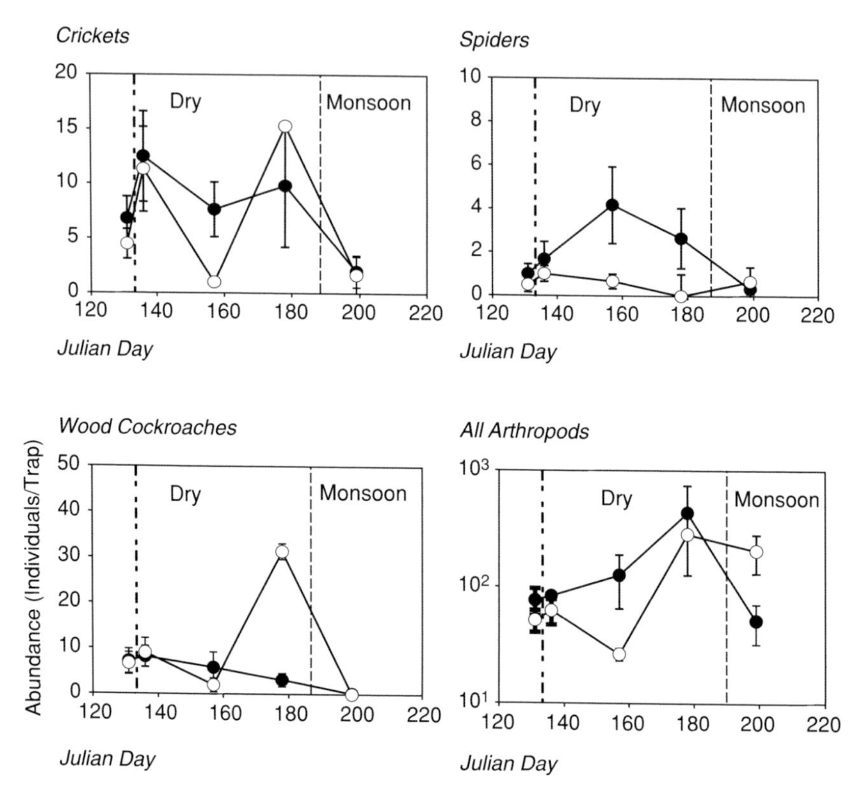

FIGURE 1 | Number (mean ± SE) of spiders, crickets, cockroaches, and all arthropods combined captured in pitfall traps during 48 hour intervals on five sampling dates between May-July 2002. Open circles represent samples from no-litter treatments; closed circles represent litter treatments.

(i.e., equal or lower in litter treatments) during the monsoon season (Julian Day 199). Both taxonomic groupings showed strong reversals in relative abundance between dry and monsoon seasons.

Ground spiders were more abundant in litter vs. no litter treatments when averaged over the four post-manipuation census periods (Figure 3; Mann-Whitney U = 16.5; n = 9; P = 0.05). Spider biomass did not differ significantly over the same period (see Figure 3) suggesting that smaller rather than larger individuals may have recruited to litter plots (Mann-Whitney U = 11.0; n = 9; P = 0.6). In contrast to the negative response of spiders to litter removal, detritivores exhibited no significant response to litter removal. Abundance and biomass were weak (see Figure 3), but not significantly higher in no litter vs litter plots for crickets (Mann-Whitney U = 7.5, 9.0; n = 9; P = 0.7, 1.0, abundance and biomass, respectively) and cockroaches (Mann-Whitney U = 4.0, 9.0; n = 9; P = 0.2, 1.0, abundance

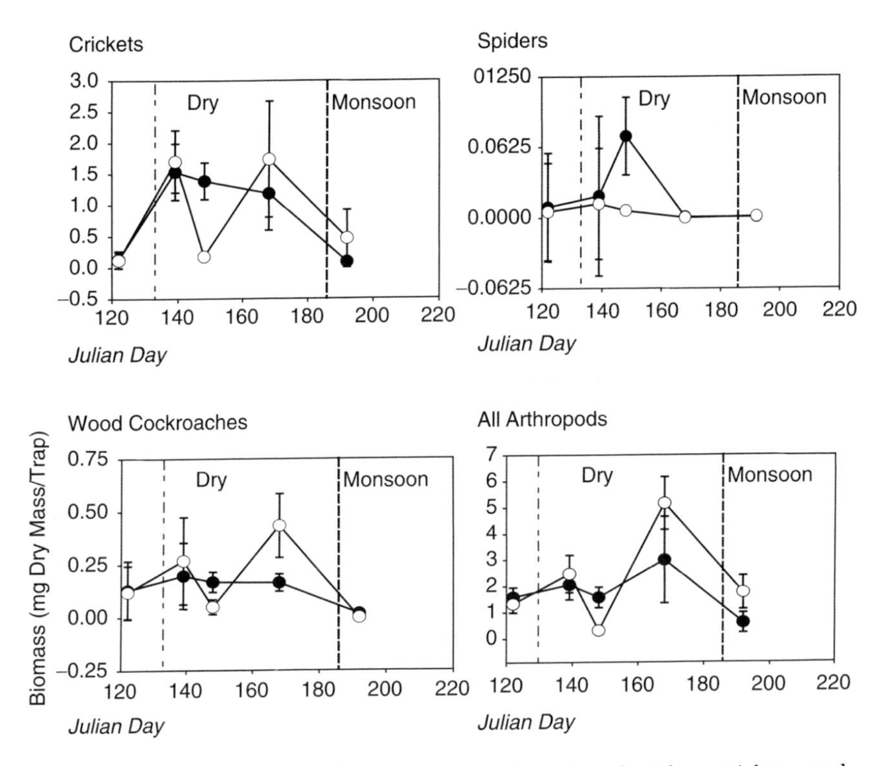

FIGURE 2 | Biomass (mean ± SE, mg dry mass/sample) of spiders, crickets, cockroaches, and all arthropods combined captured in pitfall traps during 48 hour intervals on five sampling dates between May and July 2002. Open circles represent samples from no litter treatments; closed circles represent samples from litter treatments.

and biomass, respectively). Finally, the abundance of all taxa combined did not differ significantly between treatments across the four post-manipulation samples. (Mann-Whitney $U = 9.0, 6.0$; $n = 9$; $P = 1.0, 0.44$, abundance and biomass, respectively).

DISCUSSION AND CONCLUSIONS

Leaf litter is both a resource and a structural component of habitat for terrestrial fauna inhabiting forest floors (Moore et al., 2004). In this chapter we show that one group of terrestrial predators responds more strongly than two common detritivores to the removal of leaf litter—the dominant basal resource in the system. Spiders were greater than four times less abundant in riparian plots in which litter was experimentally removed, than in control plots in which litter was left undisturbed.

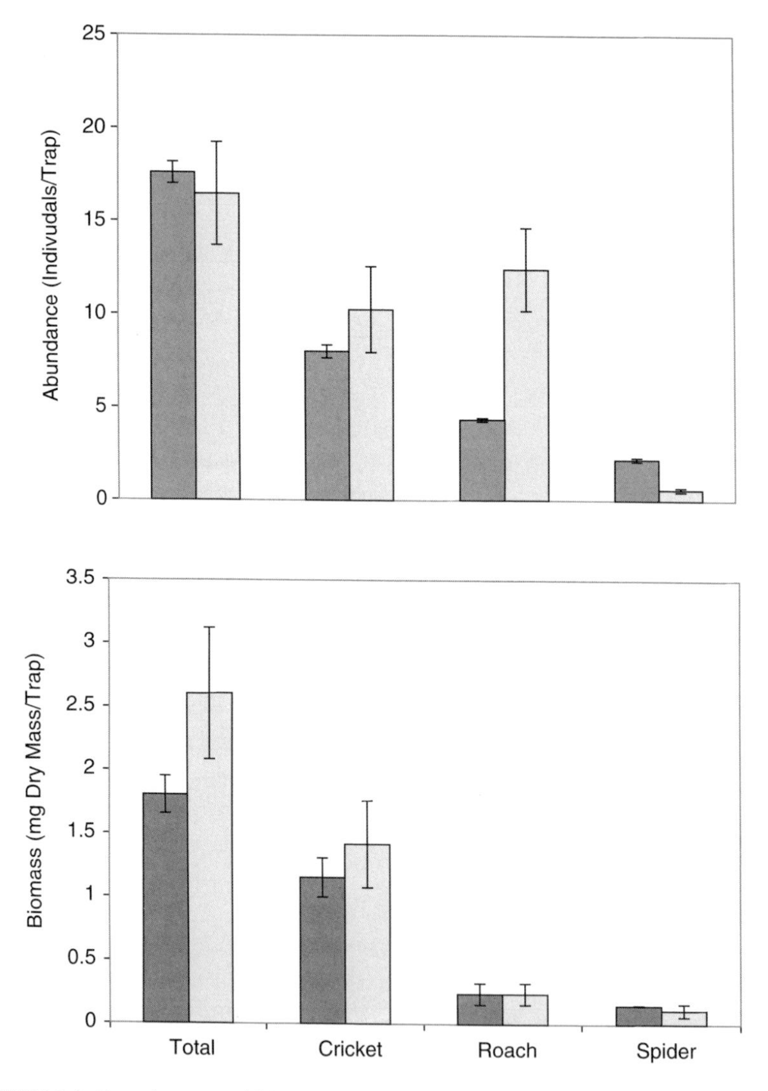

FIGURE 3 | Abundance and biomass estimates for all samples averaged across the four post-manipulation sampling dates. Means and standard errors (bars) were calculated using time-averaged abundance or biomass estimates from each replicate plot. Open bars represent no litter treatments; shaded bars represent litter treatments.

Moreover, the spider response to litter was more consistent over time (across sampling dates) than the responses of detritivores. Detritivores appeared to recruit to plots with and without litter differentially, but differences between these two treatments typically changed between dry and monsoon seasons. These results suggest spiders rely on litter as cover from predators at higher trophic levels through dry and monsoon seasons, and that detritivores respond to some combination of structural, thermal, and energetic attributes of leaf litter.

Structural Attributes of Litter and Patterns of Trophic Level Abundance

Spiders are intermediate predators in terrestrial food webs (Pacala and Roughgarden, 1984; Schoener and Spiller, 1987; Spiller and Schoener, 1988, 1994, 1995) and as such, vulnerable to predation from higher level predators including lizards, birds, and wasps (Figure 4). Spiders are also notorious intraguild predators and cannibals (Polis et al., 1989; Wise and Chen, 1999). At the San Pedro River, ground spiders are likely the prey of birds (e.g., *Sayornis nigricans*) lizards (mainly *C. uniparens*), parasitic wasps (Sphecidae, J. Sabo, personal observation), and other spiders. We hypothesize that the threat of predation—mainly by lizards and wasps by day and large mobile spiders by night—restricts spiders to microhabitats with relatively thick litter cover. Litter provides cover from these predators by decreasing their search efficiency and attack rates.

In contrast to spiders, crickets and cockroaches are omnivores, likely eating a combination of litter, living plant material and some animal

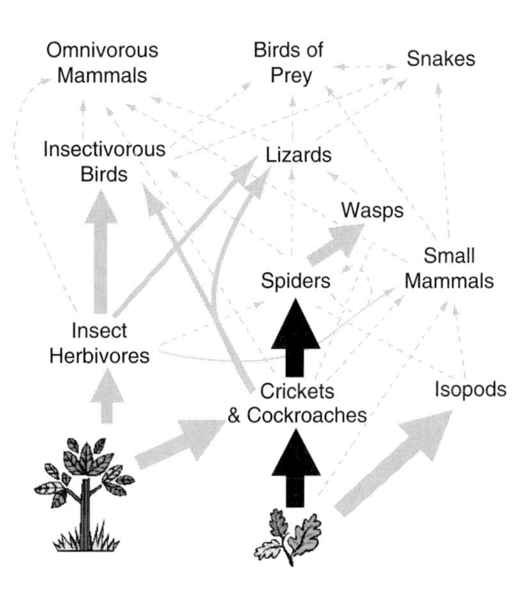

FIGURE 4 | Food web relationships of desert riparian floodplain excluding large mammals. Black solid arrows indicate energy flow pathways examined in this chapter. Gray solid arrows indicate energy flow pathways known to exist via direct observation. Gray dashed arrows indicate hypothesized energy-flow pathways.

material alive and dead. Resources are typically not limiting to detritivores in leaf litter systems (Zheng et al., 1997). In this study, we removed only large, structural leaf litter. Small particulate detritus was present on litter and no litter plots thereby providing detrital resources for crickets and cockroaches in predator-ridden and predator-free habitats. Predators of these detritivores include large spiders, parasitic wasps, lizards, and birds. Of these predators, spiders are numerically dominant and likely exert the strongest effect on cricket behavior (Lawrence and Wise, 2000).

We hypothesize that seasonal shifts in the relative abundance of detritivores in litter and no litter plots are by driven by risk of predation from spiders, microclimate differences on surface soils with and without litter cover, or some combination of both mechanisms. The relative abundance of spiders was higher in litter vs. no litter plots more consistently across seasons than the abundance of detritivores. In contrast to the consistency of spider abundance patterns, the abundance of detritivores was initially higher in litter plots (e.g., Julian Day 157) and then later in the dry season, higher in no litter plots (e.g., Julian Day 178). This change in the relative abundance of detritivores in habitats with and without litter is temporally consistent with trends in predator abundance and changes in microclimate. Specifically, we observed an overall decline in spider abundance (across treatments) and increasing nighttime low temperatures (Sabo et al., unpublished data) during this same period. These observations suggest that leaf litter may determine the relative abundance of detritivores via one of two potential mechansims—increased predation risk associated with the structural benefits litter provides to predators of detritivores or increased thermal buffering of ambient temperatures by the litter layer. Further experiments are needed to determine which of these mechanisms more strongly determine the relative abundance of detritivores.

Litter, Invertebrates and Riparian Forest Floor Food Web Structure

Desert ecosystems are notoriously devoid of large standing stocks of litter-based detritus. Water limitation and plant adaptations to alleviate water stress preclude prolific production and senescence of leaf material. In our desert riparian system, leaf litter may rival living plant material as the basis for production at higher trophic levels. Ground water, more near surface soils in near river habitats, allows more mesic species like cottonwood to maintain high levels of leaf production in an otherwise arid ecosystem. High leaf production coupled with seasonally dry surface soils and low decomposition, provide the ideal conditions for litter accumulation on the forest floor—in our system, litter depth was as high as 25 cm.

The abundance and diversity of terrestrial vertebrate species in the San Pedro Watershed is unrivaled in the continental United States, including as many as 250 species of breeding and migratory birds, 80 species of mammals, and 40 species of reptiles and amphibians. Many of these vertebrate species are found in riparian cottonwood forests and are intimately tied to cottonwood-based production both living and dead (see Figure 4). At the regional scale, high vertebrate diversity within the San Pedro Watershed likely owes to three factors: 1) the San Pedro's geographic location along the biogeographic boundary of the deserts of Chihuahua and Sonora (Brown, 1994); 2) strong gradients with elevation in climate and vegetation between the river basin and surrounding sky-island peaks (Brown, 1994); and 3) strong gradients in vegetation driven by increasing depth to ground water with increasing distance from the wetted river channel (Stromberg, 1993a, b). In addition to these three large-scale factors, we hypothesize that riparian oxbow habitats dominated by cottonwood-willow forest contribute significantly to regional diversity by providing oases with abundant water supply, buffered thermal conditions and hotspots of production relative to a more arid matrix of surrounding desert habitat.

The deep litter layer characteristic of oxbow forests at the San Pedro plays all three of these roles. Litter reduces water loss from surface soils and insulates this habitat from extreme high and low temperatures during dry season days and nights, respectively (Sabo et al., unpublished data). Further, litter provides both a resource for detritivores and a critical structural component of habitat for the invertebrate predators of these detritivores. Litter-based production of invertebrates in this system is staggering. Average activity abundance as measured here by pitfall traps, exceeds 15 individuals and 2 grams dry weigh per plot for crickets and 5 individuals and 0.05 grams dry weight per plot for spiders (see error bars in Figures 1 and 2). Visual surveys suggest that densities as high as 10 individuals per m^2 are common for both species, especially in deep litter layers (Sabo, personal observation). We argue that these two groups of invertebrates are the numerically dominant source of carbon for a variety of vertebrate species (see Figure 4) whether resident (e.g., lizards, riparian obligate birds, and omnivorous rodents) or transient visitors to these productive habitats (e.g., migratory birds and larger carnivores) and that high production of these invertebrates is maintained at least in part, by cottonwood leaf litter.

In conclusion, leaf litter removal in a riparian gallery forest led to a consistent reduction in spider abundance but seasonally variable responses of detritivores to the manipulation of this basal resource. Thus, the structural rather than the energetic attributes of litter may more strongly control the abundance and activities of detritivores in leaf litter systems by enhancing the abundance of animals at higher trophic levels.

ACKNOWLEDGMENTS

We thank Jerome Clark, Christie Deering, Paul Midura, and David Simmons for aid in processing invertebrate samples in the lab. Two anonymous reviewers provided constructive criticism that improved an earlier draft of this manuscript. This experiment was funded by a Research Incentive Grant to JLS from the College of Liberal Arts and Sciences and Department of Biology at Arizona State University.

5.3 | STABILITY AND INTERACTION STRENGTH WITHIN SOIL FOOD WEBS OF A EUROPEAN FOREST TRANSECT: THE IMPACT OF N DEPOSITION

Dagmar Schröter and Stefan C. Dekker

Food and energy production for an increasing number of people, with increasing consumption per capita, have boosted the emission of nitrogen to the atmosphere (Galloway, 2001). In the Northern Hemisphere, contemporary nitrogen deposition averages more than four times that of pre-industrial times. Cultivated lands and mixed forests even receive over sixteen times more nitrogen through deposition today than they did a century ago (Holland et al., 1999). The emission of nitrous oxides mainly stems from fossil fuel combustion, emission of ammonium mainly from domestic animal production and synthetic fertilizer use. Effects of nitrogen deposition on fauna and ground flora (as opposed to trees) have rarely been studied (Bobbink et al., 1998). Therefore we take the opportunity to look at four decomposer food webs that were studied on a European forest transect, with special focus on the effects of different levels of nitrogen deposition in the context of other environmental factors. We will examine (a) the contribution of different functional groups to the mineralization of carbon, (b) the possible effects of increased nitrogen inputs on the bacterial and the fungal subsystem and implications for the total terrestrial carbon balance, and (c) the stability of the decomposer food web as indicated by the steepness of biomass pyramids within the main energy channels and the minimum degree of intraspecific interaction strength (May, 1973).

DESCRIPTION OF CASE STUDY

The study sites form a North-South transect of European coniferous forests: N-SE (Northern Sweden, Åheden), S-SE (Southern Sweden, Skogaby), DE (Germany, Waldstein), and FR (France, Aubure; Table 1). They are part of a larger network of forest sites that were studied within the European project CANIF (Schulze, 2000). A detailed description of the sites and their history can be found in Persson et al. (2000).

The sites lie on a latitudinal gradient and are subject to different levels of nitrogen (N) deposition. The site in Northern Sweden, N-SE, differs from the others with respect to its boreal climate and the nearly total absence of N deposition (2 kg ha^{-1} a^{-1}; Table 1). S-SE and FR receive intermediate levels of nitrogen via the atmosphere (15-16 kg ha^{-1} a^{-1}), while DE is subject to high amounts of atmospheric N deposition (20 kg ha^{-1} a^{-1} wet and dry nitrogen deposition; wet deposition are means for

Table 1. Characteristics of the four study sites (from data given in Persson et al., 2000). N deposition is total wet and dry nitrogen deposition (wet deposition are means for 1993–1997; dry depositions are means for 1993–1996). Abbreviations: N-SE = Northern Sweden, Åheden; S-SE = Southern Sweden, Skogaby; DE = Germany, Waldstein; and FR = France, Aubure. *P. sylvestris* = *Pinus sylvestris*; *P. abies* = *Picea abies*; *B. pendula* = *Betula pendula*.

	N-SE	S-SE	FR	DE
N deposition (kg ha^{-1} a^{-1})	low (2)	intermediate (16)	intermediate (15)	high (20)
Latitude	64°	56°	48°	50°
Altitude asl (m)	175	95-115	1050	700
Mean annual air temperature (°C)	1.0	7.6	5.4	5.5
Mean annual precipitation (mm)	488	1237	1192	890
Stand age (a)	183	36	145	95
Most common tree species	*P. sylvestris* *P. abies* *B. pendula*	*P. abies*	*P. abies*	*P. abies*

1993–1997, dry depositions are means for 1993–1996, Table 1). The nitrogen deposition gradient studied here is not symmetrical in the sense that the intermediate levels of nitrogen input are closer to the highest level, than to the lowest. Along the three southern sites the increasing altitude towards the South counteracts the climatic gradient. S-SE, the youngest site, lies on the lowest altitude having the highest mean annual temperature and precipitation. The two more southern sites, DE and FR, lie on high altitudes resulting in lower mean annual temperatures (see Table 1).

The study sites are subject to an N depositional and a latitudinal gradient. The case study was not designed to pin down the exact effect of these factors on the decomposer food web and its function. However, it may be possible to find patterns that, apart from small scale variations, are likely to be caused by the large scale factors involved. Here, the sites are viewed primarily in order of the nitrogen depositional gradient: N-SE (low N deposition), S-SE and FR (intermediate N deposition), and DE (high N deposition).

Soil samples were collected at all four sites at four sampling dates: (i) October/November 1996, (ii) May/June 1997, (iii) September 1997, and (iv) March/April 1998. At each time, between 80 and 110 samples (soil corer: Ø 5 cm) of the organic layer (LFH) were taken at each site. Sample treatment and methods for abiotic and biotic measurements including abundance and biomass of soil biota are described in detail elsewhere (Schröter, 2001; Schröter et al., 2003).

The food web approach of Hunt et al. (1987) was used to calculate C mineralization by the belowground community (O'Neill, 1969; Hunt et al., 1987; de Ruiter et al., 1993a). A detailed description of the food web modelling approach including climatic adaptation, as well as the physiological parameters used can be found in Schröter et al. (2003). Annual average biomasses of functional groups at the different sites are given in Table 2. The food web modelling approach was used to calculate carbon mineralization rates for each functional group. Additionally, as a measure of food web stability, the minimum degree of intraspecific interaction was calculated for the entire food web as described in Neutel (2001) and Neutel et al. (2002).

The food web model was run with the biomass input data from four separate sampling occasions per site, resulting in four sets of estimates for each site. Analyses of variance revealed no significant main effect of 'sampling time' on the biomass and C mineralization rates of any functional group nor the total food web (data were adequately transformed prior to analyses whenever necessary to improve normality). Therefore the four sets of estimates per site were treated as

Table 2. Annual average biomasses (kg C ha^{-1}) of functional groups at each site. Mean values of four sampling occasions are shown, standard deviations are given in parentheses (see Table 1 for site abbreviations). The term 'panphytophagous' is used for functional groups that feed on detritus and microbes (bacteria and fungi).

	N-SE		S-SE		FR		E	
microflora								
Bacteria	24	(6)	25	(9)	61	(21)	19	(8)
Fungi	436	(103)	308	(112)	423	(146)	381	(157)
microfauna								
Testate amoebae								
panphytophagous	8	(4)	14	(7)	28	(20)	40	(28)
predaceous	2	(1)	5	(2)	6	(4)	10	(6)
Nematoda								
bacterivorous	0.74	(0.15)	0.13	(0.03)	0.32	(0.26)	0.22	(0.21)
fungivorous	0.20	(0.04)	0.05	(0.01)	0.09	(0.08)	0.06	(0.06)
omnivorous	0.10	(0.02)	0.06	(0.01)	0.08	(0.06)	0.20	(0.19)
predaceous	0.061	(0.012)	0.000	(0.000)	0.000	(0.000)	0.003	(0.003)
mesofauna								
Acari								
panphytophagous	1.2	(0.6)	1.6	(0.8)	2.0	(0.7)	3.1	(0.9)
predaceous	0.1	(0.1)	0.3	(0.1)	0.4	(0.1)	0.6	(0.2)
Collembola								
panphytophagous	2.3	(1.4)	1.4	(0.8)	6.6	(3.6)	6.1	(3.7)
predaceous	0.01	(0.02)	0.01	(0.01)	0.51	(0.34)	0.07	(0.02)
Enchytraeidae	0.6	(0.9)	1.2	(0.3)	9.5	(3.7)	8.4	(3.1)
total fauna	15	(5)	24	(9)	54	(29)	69	(34)
total food web	476	(105)	357	(122)	538	(195)	469	(197)

"replicate" estimates to calculate mean values and standard deviations per site. However, since this is not a true, independent replication no further statistical analyses were performed. The standard deviations obtained by this approach represent general, as well as temporal variations at each site.

CARBON TURNOVER AND FOOD WEB STRUCTURE

During nitrogen saturation, soil microbial communities move from being fungal dominated to being bacterial dominated (Parmelee, 1995; Tietema, 1998). Fungal-based food webs are typically a result of extreme moisture fluctuations and have a greater tendency of nutrient immobilization and slower turnover of nutrients, while bacterial based food webs indicate a more stable moisture level and fast nutrient cycling (Parmelee, 1995). Increased fertility decreased the relative abundance of fungi in a number of boreal systems (Pennanen et al., 1999), while total microbial biomass, as in our case study, remained unchanged. Therefore we expected a shift from fungal to bacterial based food webs along the latitudinal and depositional gradient. Figure 1 shows the contribution of bacteria, fungi, testate amoebae, and the rest of the fauna to total carbon mineralization at the different sites. At the low N input boreal site N-SE, the bacteria-to-fungi-ratio of C mineralization was estimated to be around 30/70 (Figure 1). In contrast to this, the sites that received intermediate or high loads of nitrogen deposition exhibited higher mineralization rates and larger contributions of testate amoebae and other fauna. The bacteria-to-fungi-ratio of C mineralization was more balanced at these sites (bacteria-to-fungi-ratio ca. 50/40, see Figure 1). We suggest that both climatic conditions and nitrogen deposition contributed to the structural difference between these food webs.

How are these structural differences reflected in the total carbon turnover of the food webs? The amount of carbon mineralized per unit carbon biomass per year can be calculated by dividing total mineralization rates by total food web biomass. At N-SE ca. 1.6 units carbon were mineralized per unit carbon biomass and year. At S-SE and FR the amount was considerably higher with 4.3 and 4.9 a^{-1} respectively. At DE, the site with the highest load of nitrogen deposition, 6.0 units carbon were mineralized per unit carbon biomass and year (total carbon mineralization ca. 800, 1500, 2600 and 2800 kg C ha^{-1} a^{-1} at N-SE, S-SE, FR and DE respectively, Schröter et al., 2003). Model calculations comparing different scenarios suggest that this pattern is due not only to

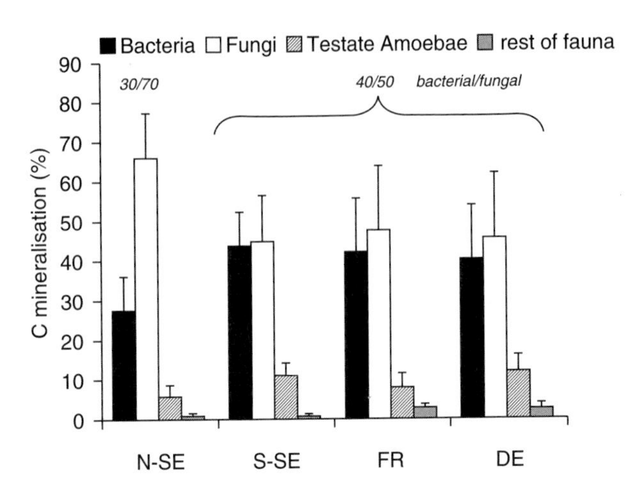

FIGURE 1 | Annual average contribution to total carbon mineralization by bacteria, fungi, testate amoebae, and the rest of the fauna (%). Mean values of four sampling occasions are shown, bars represent standard deviations (see Table 1 for site abbreviations). The ratio of bacterial to fungal contribution is given in italics.

environmental factors like climate and nitrogen availability, but also to food web structure (Schröter et al., 2003).

Besides climate, how could increased nitrogen availability via N deposition lead to enhanced mineralization rates? Along the transect highly increased metabolic activity of both, bacteria and fungi was observed that coincided with increasing N deposition (Schröter et al., 2003). The increase of metabolic activity of bacteria was more pronounced than that of fungi. In the following we look at each subsystem separately and propose mechanisms that could lead to enhanced metabolism and thus to enhanced mineralization.

In the bacterial subsystem, enhanced nitrogen availability may be directly stimulating for bacteria if degradable carbon is available as an energy source to successfully exploit the nitrogen (Paul and Clark, 1989). Triggering the bacteria to enter growth phase may then have a stimulating, strong bottom-up effect on the primary predators, the Protozoa group testate amoebae. Increased grazing by Protozoa would further enhance bacterial turnover and finally, enhance mineralization. Figure 1 shows that indeed the contributions of testate amoebae and other fauna to total carbon mineralization was slightly increased at the nitrogen rich sites with milder climate and high bacterial mineralization rates. Average annual biomass of bacteria was similar at all sites, except for an extraordinarily high value at FR (see Table 2). The annual average absolute mineralization by bacteria did not follow this pattern of bacterial biomass (absolute mineralization by bacteria in kg C ha a^{-1}: 210 at N-SE, 665 at S-SE, 1162 at FR, 1167 at DE). Therefore, the mineralization pattern points to increased bacterial turnover, especially at S-SE and DE. Increased mineralization due to increased bacterial turnover and flow of carbon through bacteria-bacterivore channels of the soil food web have also been found in a nitrogen and phosphorus fertilization study in forest clear-cuts (Forge and Simard, 2001).

In the fungal subsystem, the availability of plant derived labile carbon seems to be especially important, since many fungi are equipped to exploit organic nitrogen sources (Leake and Read, 1997). It has been suggested that increased nitrogen availability leads to increased carbon allocation to tree roots (Bauer et al., 2000; Scarascia-Mugnozza et al., 2000; Burton et al., 2001). Well-nourished trees could thus supply more energy to the mycorrhizal fungi and exude more labile carbon to the rhizosphere to the benefit of saprophytic fungi. Fungi could expend this supplementary energy to produce more degradative enzymes, thus increasing the overall availability of nutrients within the decomposer system and further enhancing turnover rates (Read, 1991; Lindahl et al., 2002). Increased fungal growth may then have a bottom-up effect on fungal feeders like Enchytraeidae and fungivorous Collembola. Enhanced fungal grazing would further increase turnover of fungal biomass and enhance mineralization. At our case study sites, Collembola and Enchytraeidae contributions to total mineralization were indeed increased especially at the two southern sites, FR and DE (Schröter et al., 2003). While the relative contribution of the fungal subsystem to total carbon mineralization was smaller at the nitrogen rich sites with milder climate than at the boreal site N-SE, the absolute contributions were higher than at the boreal site N-SE (absolute mineralization by fungi in kg C ha a^{-1}: 509 at N-SE, 686 at S-SE, 1190 at FR, 1249 at DE). Enhanced absolute mineralization rates of fungi did not follow the pattern in fungal biomass (see Table 2) and therefore point to increased fungal turnover. The link between plant derived carbon supply and the fungal subsystem was illustrated in an isolation experiment in a mature spruce forest by cutting the root-mycorrhizal connection. Disrupting the plant derived carbon supply to the soil food web had a negative effect on the fungal subsystem: the biomass of fungi and fungivorous Collembola was reduced while bacterivorous nematodes seemed to profit slightly (Siira-Pietikainen et al., 2001).

In addition to effects mediated by the plant, nitrogen availability has direct effects on fungi. For example, nitrogen content of the substrate determines the growth of both fungi and fungivorous Collembola (Hogervorst et al., 2003). In a culturing study with Scots pine litter, Hogervorst and co-workers found that elevated levels of nitrogen track through the fungal part of the soil food web. Intermediate levels of N availability enhanced fungal and fungivorous growth, while high levels had an adverse effect. Enhanced nutrient availability feeds up from both bacteria and fungi to higher trophic levels via increased food quantity and quality. In a fertilization study in sub-arctic heaths, food quality (e.g., nutrient content of the micro-organisms), had a larger effect on the microbivores than the size of the microbial biomass (Schmidt et al., 2000).

In summary, when nitrogen is limiting, and other environmental factors are favorable, addition of nitrogen via deposition may enhance mineralization through its effect on the bacterial and fungal subsystems of the decomposer food webs. The proposed mechanisms suggest short term effects in correspondence with the availability of labile carbon fractions. Nitrogen addition may not have any long term stimulating effect on microbial activity (Berg and Matzner, 1997). However, our case study indicates that nitrogen effects in concert with climatic factors had stimulating effects on microbial activity that were sustained over the field period of the project (1996–1998). What are the possible implications of nitrogen deposition for the terrestrial carbon balance? To estimate an overall terrestrial carbon balance, decomposition has to be accounted for. However, this is difficult because the net carbon balance of soils is determined by litter input as well as by mineralization through the decomposer food web—both processes may not be in phase, resulting in organic matter accumulation or loss (Schulze, 2000). Nitrogen deposition may accelerate decomposition of light soil fractions (decadal turnover times), but it further stabilizes soil carbon compounds in heavier, mineral associated fractions (e.g., lignin and humus, with multi-decadal to century lifetimes) (Berg and Matzner, 1997; Neff et al., 2002). Hence, the relationship between soil carbon storage and nitrogen availability is complex and current models of terrestrial carbon cycling do not adequately represent soil processes (Dufresne et al., 2002; Neff et al., 2002). The effect of nitrogen deposition in concert with climatic change on the terrestrial carbon balance is currently unclear and an area of intense research (Ågren et al., 2001). The structure and functioning of the soil food web deserves prime consideration in this respect (Brussaard et al., 1997).

FOOD WEB STRUCTURE AND STABILITY

Community structure and the flow of energy, matter, and nutrients are closely connected to one another. Besides the effect of nitrogen deposition on food web structure and subsequent changes in function like altered mineralization rates, society has an interest in food web structure and biodiversity *per se* (Convention on Biological Diversity, 1992). In this respect, it is interesting to look at food web stability (de Ruiter et al., 1998). Food webs can be represented by community (Jacobian) matrices in which each element represents the effect of species on each other's dynamics near equilibrium (May, 1973). The effects of species or functional groups on each other's dynamics near equilibrium are called interaction strengths. Diagonal elements of the community matrix refer to intraspecific interactions. In this study, we look at stability as the minimum degree of intraspecific interaction needed for matrix stability; the smaller this intraspecific interaction, the more stable the matrix (Neutel et al., 2002). Long loops with relatively many weak links reduce the amount of intraspecific interaction needed for matrix stability. Food webs are more stable when they contain long loops with relatively many weak links (Neutel et al., 2002). It may be expected that increased nitrogen deposition lowers food web stability by simplifying food web structure and loss of food web complexity (Matson et al., 2002). However, food web complexity (i.e., the number of functional groups times connectance) was very similar at our case study sites. Merely the functional group of predaceous Nematoda was absent at FR and S-SE (Schröter et al., 2003). Any difference in food web stability will therefore be brought about by the differences in biomass structure at the different sites (see Table 2). Steep biomass pyramids can result in low minimum degrees of intraspecific interaction, or high stability—roughly a factor of 10 biomass decrease over trophic level can generate strong stabilizing patterns in interaction strength (Neutel et al., 2002). In the following, we therefore analyze the major bacterial and fungal trophic channels for steepness in their biomass pyramids.

Table 3 shows annual average biomass ratios between trophic levels for bacterial and fungal channels within the food webs at our case study sites. The smaller the biomass ratio of predator to prey, the steeper the biomass pyramid. When comparing these ratios we look for changes in order of magnitude that hint at differences in biomass pattern that may affect food web stability. Overall, the biomass ratios between trophic levels were similar between sites, however there were a few exceptions. In the first bacterial channel, the biomass ratio between panphytophagous (we refer to any organisms that feed on both detritus and microorganisms as *panphytophagous*) testate amoebae and bacteria was an order of magnitude higher at DE than at the other sites (see Table 3). In the channels from fungi to Enchytraeidae, as well as from fungi to panphytophagous collembola, the biomass ratios were an order of magnitude higher at FR and DE, indicating fewer fungal biomass available per fungivore, and therefore probably a larger per capita effect of

Table 3. Annual average biomass ratios between trophic levels for two bacterial and three fungal channels within the food webs. The ratios are biomass of the predator divided by its prey; the smaller the ratio, the steeper the biomass pyramid. Mean values of four sampling occasions are shown, standard deviations are given in parentheses (see Table 1 for site abbreviations). Predaceous nematodes were not present at S-SE and FR. Abbreviations: pan. = panphytophagous, t. = testate, pred. = predaceous, bact. = bacterivorous, fung. = fungivorous.

	N-SE		S-SE		FR		DE	
bacterial channels								
pan. t. amoebae/bacteria	0.3	(0.2)	0.7	(0.6)	0.4	(0.2)	1.9	(1.0)
pred. t. amoebae/pan. t. amoebae	0.2	(0.1)	0.4	(0.2)	0.2	(0.0)	0.3	(0.1)
bact. nematodes/bacteria	0.03	(0.01)	0.01	(0.00)	0.01	(0.00)	0.01	(0.02)
pred. nematodes/bact. nematodes	0.08	(0.00)					0.02	(0.00)
fungal channels								
fung. nematodes/fungi	0.0005	(0.0002)	0.0002	(0.0001)	0.0003	(0.0002)	0.0002	(0.0002)
pred. nematodes/fung. nematodes	0.30	(0.00)					0.05	(0.00)
Enchytraeidae/fungi	0.001	(0.002)	0.004	(0.003)	0.023	(0.006)	0.023	(0.006)
pan. collembola/fungi	0.005	(0.003)	0.004	(0.002)	0.017	(0.011)	0.019	(0.016)
pred. collembola/pan. collembola	0.01	(0.02)	0.01	(0.01)	0.09	(0.06)	0.02	(0.02)
pred. acari/pan. collembola	0.07	(0.03)	0.26	(0.18)	0.10	(0.10)	0.17	(0.18)

fungivores on fungi. Such strong effects may have destabilizing effects in a food web, which is expressed by high minimum degrees of intraspecific interaction (i.e., low stability) (Figure 2). Generally, high biomass ratios varied strongly, as indicated by high standard deviations. The biomass ratio of Enchytraeidae to fungi was an exception to this; here the biomass ratio fluctuated less strongly (see Table 3). The high biomass ratio of predaceous to fungivorous nematodes at N-SE may contribute to the lower stability of the food web at this site compared to S-SE (see Table 3, see Figure 2). The strong per capita effect of predaceous Acari on panphytophagous Collembola indicated by an increase of an order of magnitude in the biomass ratios at S-SE and DE relative to N-SE and FR is not directly related to the stability pattern depicted in Figure 2, probably because of the higher trophic position and the size of this effect in relation to all other effects within the food web.

Figure 2 indicates that the food web at S-SE was slightly more stable than at N-SE. The lowest stability is indicated for the food webs at FR and DE, the sites with high total mineralization rate, high amount of carbon mineralized per unit carbon biomass per year, and intermediate to high loads of nitrogen deposition. Variation in food web stability as indicated by standard deviation represents general and temporal variability over four sampling times and was very large at FR and DE. No consistent pattern indicated an effect of nitrogen deposition among the confounding factors. In some cases the relatively large biomass of microbivores seemed to result in decreased food web stability due to increased per capita effects of this trophic level. An in-depth analysis of loop weights

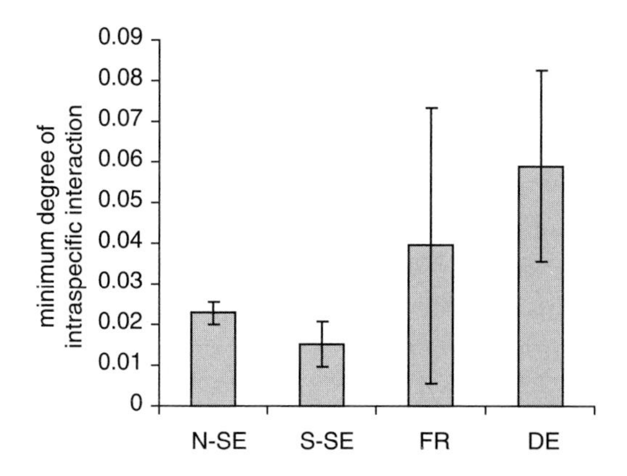

FIGURE 2 | Minimum degree of intraspecific interaction in the food webs of the different sites. The smaller is this intraspecific interaction, the more stable is the matrix. Mean values of four sampling occasions are shown, bars represent standard deviations (see Table 1 for site abbreviations).

and interaction strengths of the food webs may yield further insights, but is beyond the scope of this chapter. It further remains to be investigated to what extent the indicators used here represent food web stability as measured in the field.

DISCUSSION AND CONCLUSIONS

Nitrogen deposition can cause structural changes in food webs that alter ecosystem functioning. In soil, increased nitrogen input may result in shifts from fungal-based to more bacterial-based decomposer food webs. Increased nitrogen availability has direct effects on microbial activity and can enhance mineralization rates through stimulating effects on the bacterial and fungal subsystem at least in the short term (we provide evidence that this effect could be sustained for about two years). The overall long term effect of nitrogen deposition on the soil food web, decomposition and terrestrial carbon storage is not yet understood. Nitrogen deposition will impair food web stability when species or functional groups are lost. Nitrogen deposition may also impact stability through affecting the biomass pattern within a food web. First results show no consistent effects on biomass structure and food web stability that can be attributed to nitrogen deposition alone. However, when increased nitrogen availability feeds up to increased biomass of microbivores, overall stability may be weakened due to increased per capita effects of this trophic level. Nitrogen deposition is an important global change driver and its effects on ecosystems deserve further attention. Eutrophication due to nitrogen deposition is projected to continue through mid century (Alcamo, 2002). Differences in nitrogen availability contribute to a heterogeneous environment, and thereby to biodiversity. Therefore continuous nitrogen deposition could homogenise the environment—resulting in loss of unique habitats, species, food webs and ecosystems.

ACKNOWLEDGMENTS

Carlo C. Jaeger is recognized with sincere gratitude for company and guidance to DS in the world of matrices. The data on Collembola presented in this study were determined by Anne Pflug. Astrid Taylor produced the data on Acari. The field work was carried out as part of the CANIF project (EU funded, Project No. ENV4-CT95-0053). We dearly thank all project partners. The ATEAM project (EU funded, Project No. EVK2-2000-00075) provided travel support and data on nitrogen deposition.

5.4 | DIFFERENTIAL EFFECTS OF CONSUMERS ON C, N, AND P DYNAMICS: INSIGHTS FROM LONG-TERM RESEARCH

Wyatt F. Cross, Amy D. Rosemond, Jonathan P. Benstead, Sue L. Eggert, and J. Bruce Wallace

From the perspective of ecosystem energetics, animals are often regarded as relatively unimportant. Loss of energy with each trophic transfer results in minor contributions by higher trophic levels to the total energetic budget of a given system (Lindeman, 1942; Teal, 1962; Fisher and Likens, 1973). Increasingly, however, ecologists are recognizing that animals can play very prominent roles in the transformation, translocation, and cycling of essential elements such as carbon (C), nitrogen (N), and phosphorus (P) (Kitchell et al., 1979; Seastedt and Crossley, 1984; Jones et al., 1994; Andersen, 1997; Elser and Urabe, 1999; Strayer et al., 1999; Wallace and Hutchens, 2000; Crowl et al., 2001; Vanni, 2002; Hall et al., 2003). In many cases, the importance of these functional roles far exceeds that of animal biomass in ecosystem energy budgets, and can have major consequences for energy flow, nutrient cycling, and food web dynamics (Power et al., 1996b; Wootton and Downing, 2003). In extreme cases, animals may mediate the availability of essential nutrients (Elser et al., 1998) and ultimately determine the productivity and community composition of entire systems.

When assessing the role of animals in elemental dynamics, it is necessary to recognize that the importance of a given animal or assemblage may vary considerably depending on the particular element or elemental ratio of focus. Moreover, animals may strongly influence a single ecosystem process, but have relatively little influence on others. For example, some lentic fish, such as gizzard shad, are known to be extremely important in the translocation and recycling of P (Shaus and Vanni, 2000), but likely play a comparatively small direct role in lake C dynamics. Thus, in order to elucidate the dominant role of a given species or assemblage, one must consider their effects on multiple elemental currencies and a suite of ecosystem processes and fluxes (Duffy, 2002).

The recent development of ecological stoichiometry (Sterner and Elser, 2002, and references therein) has been a major catalyst for assessing the role of animals in food web and ecosystem dynamics. Although many of the conceptual ideas of ecological stoichiometry date back to Lotka (1925) and Redfield (1958), more recent theoretical and empirical work (Reiners, 1986; Sterner et al., 1998; Elser and Urabe, 1999) has provided a solid framework for hypothesis testing and application. Investigators are now measuring the elemental composition of organisms at all levels within food webs, from basal resources to top predators. This advance has led to considerable progress in estimating the role of animals in nutrient recycling (Elser and Urabe, 1999), understanding elemental constraints in food web and population dynamics (Andersen, 1997; Elser et al., 1998; Sterner et al., 1998; Stelzer and Lamberti, 2002), and comparing consumer-resource elemental imbalances among different ecosystem types (Elser and Hassett, 1994; Elser et al., 2000; Cross et al., 2003; Frost et al., 2003). In addition, by combining estimates of biomass or production of food web components with their elemental composition, assessments can be made about the relative contribution of animals to total particulate pools of C, N, and P (Hessen et al., 1992; Hassett et al., 1997), and the relative efficiency with which these elements are utilized or retained in the system.

Forested headwater streams are important sites of nutrient and organic matter input, storage, transformation, and export (Meyer and Wallace, 2001; Peterson et al., 2001). These first- and second-order streams are also the most abundant aquatic features in many landscapes, and often comprise >75% of the total stream length of larger forested basins (Wallace, 1988). Although animals (primarily invertebrates) inhabiting small streams make up a minor direct contribution to whole-stream carbon or energy budgets (Fisher and Likens, 1973; Webster et al., 1997), they are known to play a major role in the removal, translocation, or downstream export of carbon via leaf-eating (shredding) or particle-gathering activities (Wallace et al., 1982; Webster, 1983; Wallace et al., 1991; Pringle et al., 1999). Little is known, however, about the influence of stream invertebrates on the dynamics of other important elements, such as N and P (but see Grimm, 1988; Crowl et al., 2001), which may limit the productivity of small headwater streams and larger recipient downstream ecosystems.

In this chapter, we examine the role of stream invertebrate food webs in the storage, utilization, and translocation of C, N, and P in small Appalachian headwater streams. We combine a number of studies (both published and unpublished) on the production, storage, export, and elemental composition of stream food web components to gain a broader understanding of the role of invertebrates in elemental dynamics of headwater streams. Our basic approach was to combine long-term measurements of food web component biomass and secondary production with their C, N, and P content (Cross et al., 2003). These data were used to assess general patterns in elemental storage and utilization by stream invertebrates. In addition, unpublished data from an experimental invertebrate removal were used to quantify the effects of invertebrates on the downstream translocation of C, N, and P, and the elemental stoichiometry of bulk organic matter export.

METHODS

Study sites

Studies were conducted in three first-order streams draining catchments (C) 53, 54, and 55 at the Coweeta Hydrologic Laboratory, North Carolina. Coweeta is a large (2185 ha), heavily forested basin located in the Blue Ridge physiographic province of the southern Appalachian Mountains. Forest vegetation is dominated by mixed hardwoods (primarily oak, maple, and poplar), and a thick under-story of rhododendron that shades the streams throughout the year. First-order streams at Coweeta are extremely heterotrophic, and receive >90% of their energy base in the form of allochthonous leaf litter from the surrounding catchment (Webster et al., 1997; Hall et al., 2000). In-stream primary production (algae and bryophytes) is very low, and contributes <1% to ecosystem carbon budgets (Webster et al., 1997; Hall et al., 2000). Streams that drain C53, C54, and C55 have very similar physical and chemical characteristics (see Lugthart and Wallace, 1992), but differ in their history of experimental manipulation. This study includes 29 stream-years of data collected from all three catchments between 1984 and 2002 (C53: 11 years, C54: 6 years, C55: 12 years). Years of whole-stream experimental manipulations that were not considered 'reference' conditions included:

1986–1988 in C54 (treated with insecticide, see Cuffney et al., 1990; Wallace et al., 1991), 2000–2002 in C54 (enriched with nitrogen and phosphorus, see Cross et al., 2003; Gulis and Suberkropp, 2003), and 1993–2000 in C55 (exclusion of leaf litter and wood, see Wallace et al., 1999). All other stream-years (n = 20) were considered unmanipulated 'reference' conditions.

Background Data

During each year of study, invertebrates and benthic organic matter (primarily detritus) were quantitatively sampled on a monthly basis from the two dominant stream habitats: bedrock outcrops and mixed substrates (i.e., cobble, pebble, gravel, and sand) (see Wallace et al., 1999, for a complete explanation of methods). Annual averages were compiled for the biomass of basal food web components (i.e., total fine particulate organic matter [FPOM, <1mm], total coarse particulate organic matter [CPOM, > 1 mm], leaf litter, moss). Epilithon biomass was quantified every eight weeks during 1999–2002 in C53 and C54 with submerged ceramic tiles (J. Greenwood, University of Georgia, unpublished data). Secondary production of all invertebrate taxa was estimated each year using the size-frequency method (Hamilton, 1969) corrected for the cohort production interval (Benke, 1979) (except non-Tanypod Chironomidae in which the instantaneous growth method was applied, see Huryn, 1990; Benke, 1993).

Carbon, N, and P content of basal food resources and invertebrates were analyzed in C53 and C54 during the period of 1999–2002 (Cross et al., 2003).

Assessing the Role of Invertebrates in Elemental Storage, Utilization, and Export

The percent contribution of invertebrates to total particulate pools of C, N, and P was assessed by calculating the total nutrient content of invertebrates and basal food resources for each year of study. Mean percent C, N, and P of basal food resources and invertebrates in C53 and C54 from 1999–2002 (see Cross et al., 2003) was multiplied by basal resource and invertebrate biomass (dry mass/m^2) for each year. Storage of particulate C, N, and P was examined in two distinct habitats (mixed substrates, bedrock outcrops) because these habitats contain functionally distinct invertebrate communities, habitat structure, and fluvial geomorphology (Huryn and Wallace, 1987). For analysis of particulate C, N, and P storage we restricted our data set to the 20 stream-years of 'reference' data from C53, C54, and C55.

Relative utilization of C, N, and P by invertebrates was examined indirectly by regressing average annual leaf litter C, N, and P (resource availability) with annual production of invertebrate C, N and P (resource utilization) among years. A comparison of the linear slopes of regressions was used to make general conclusions regarding relative differences in elemental utilization.

Data from an experimental removal of invertebrate consumers (Cuffney et al., 1990) allowed us to estimate the effects of invertebrates on export of fine particulate C, N, and P, as well as elemental stoichiometry of bulk organic matter export. An insecticide (Methoxychlor) was applied seasonally to C54 for a period of three years (1985–1988, Cuffney et al., 1990; Wallace et al., 1991). Insecticide treatment resulted in massive mortality and downstream drift of invertebrates (Wallace et al., 1989), but had no effect on stream microbial activity (Cuffney et al., 1990; Suberkropp and Wallace, 1992). During the treatment, total invertebrate production and shredder production were reduced to 38% and 13% of pretreatment values, respectively (Lugthart and Wallace, 1992). Functional recovery of the invertebrate community was rapid and returned to pre-treatment levels during the second year of recovery (Whiles and Wallace, 1992; Whiles and Wallace, 1995). Cumulative export of fine particulate organic and inorganic material was quantified in the treatment stream (C54) and a reference stream (C55) biweekly during 1987–1990 with a Coshocton proportional sampler (4 mm pore size) connected to three large settling barrels (Cuffney et al., 1990; Wallace et al., 1991). Material at the base of settling barrels was analyzed biweekly for total N and P (APHA 1998); C was assumed to be a constant 50% of organic matter (ash-free dry mass) (Cross, unpublished data). Export data included one year during the insecticide treatment (December 1987–October 1988), and two years after treatment (November 1988–December 1990) during recovery of the invertebrate community. Time series of C, N, and P export in C54 and C55 were analyzed with randomized intervention analysis (RIA, Carpenter et al., 1989). RIA uses paired before-and-after time series data from a manipulated system and a reference system to detect changes caused by the manipulation. In this study, differences between streams during insecticide treatment were compared to differences after treatment.

Previous work demonstrated that invertebrate removal (and consequent declines in detritus consumption) had a large effect (240% increase) on the ratio of CPOM:FPOM exported from the treated stream relative to the reference stream (Wallace et al., 1995). We applied mean molar C:P, C:N, and N:P ratios of CPOM and FPOM (Cross et al., 2003) to export ratios of CPOM:FPOM to obtain weighted estimates of C:P, C:N, and N:P of bulk (CPOM + FPOM) organic matter export before and during insecticide treatment in both streams.

RESULTS

Elemental Storage

In the mixed substrate habitat, invertebrates contributed relatively small amounts (0.1%, 1.3%, and 2.2%) to total pools of C, N, and P (Figure 1). Most C, N, and P was present in large pools of FPOM and CPOM (primarily leaves and wood). Epilithon was a minor component of particulate C, N, and P in this habitat.

On bedrock outcrops, invertebrates were a very minor proportion of total C (Figure 1). Storage of C was dominated by FPOM, CPOM, and aquatic moss (primarily *Platylomella* sp.). Storage of N and P by invertebrates on bedrock habitat was considerably higher than that in mixed substrates, at 8.2% (N) and 11.2% (P) of total particulate nutrient pools (see Figure 1). However, the majority of N and P was stored in FPOM, CPOM, and moss. Bedrock epilithon was a minor contributor to total pools of particulate C, N, and P (see Figure 1).

Elemental Utilization

A significant positive relationship ($p < 0.001$) was found between mean annual standing crop of leaf litter C and annual invertebrate secondary production of C (Figure 2A). The slope of this relationship was much less than 1 (0.036), indicating that a relatively small proportion of available leaf litter C is converted to invertebrate C in these headwater streams. Leaf litter N and P were also positively related to invertebrate production of N and P ($p < 0.001$, Figure 2B, C), and the slopes of these relationships (N: 0.52, P: 0.90) were at least an order of magnitude higher than that of C, demonstrating relatively high sequestration of N and P by stream invertebrates.

Effects of Invertebrates on Particulate C, N, and P Export

Invertebrate removal via seasonal application of insecticide resulted in significantly lower export concentrations (g/L) of particulate C, N, and P from the treated stream relative to the reference stream (RIA, p values <0.01 for all 3 elements, Figure 3). During the insecticide treatment, concentrations of particulate C, N, and P were reduced to 33%, 42%, and

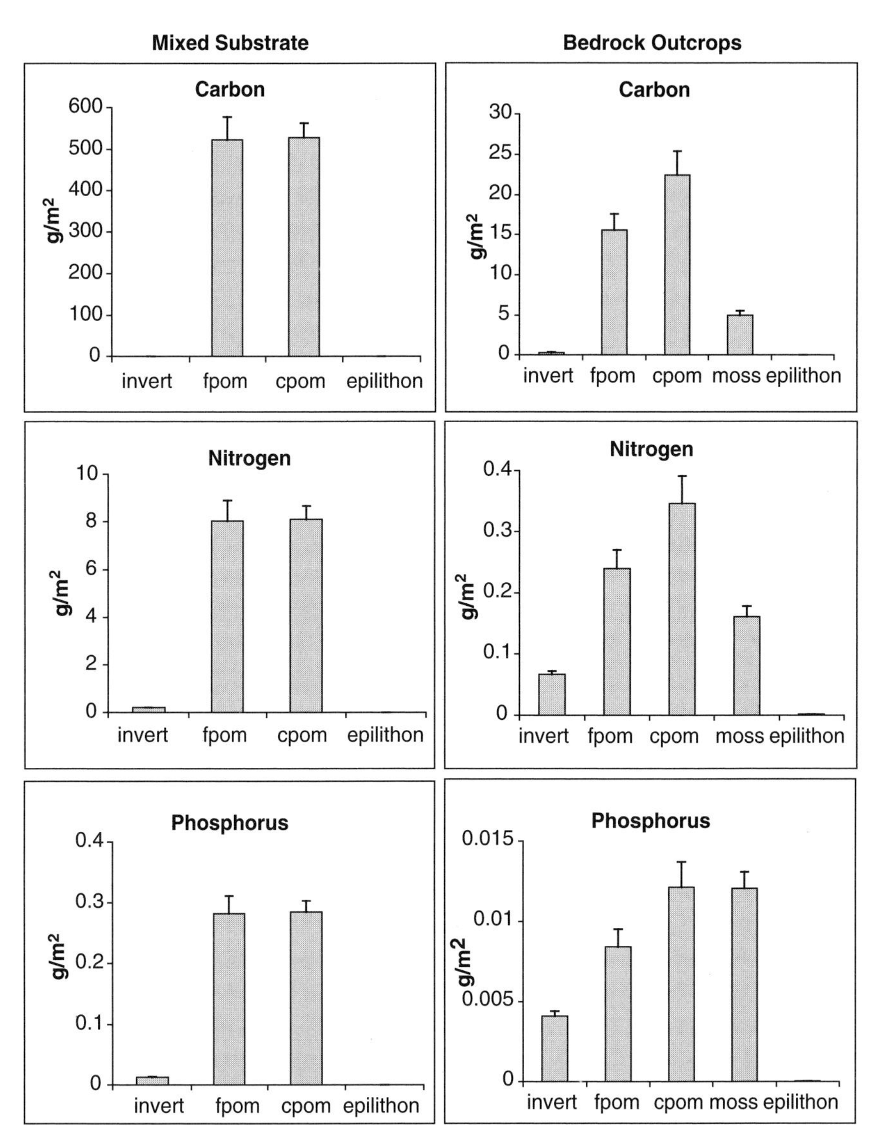

FIGURE 1 | Mean (+ISE) storage of carbon, nitrogen, and phosphorus among food web components in C53, C54, and C55 from the mixed substrate habitat and bedrock outcrops, invert-invertebrates, fpom-fine particulate organic material (<1 mm), cpom-coarse particulate organic material (>1 mm). N-20 years, except epilithon which is based on 3 years of data from C53 (J. Greenwood, University of Georgia, unpublished data). Note large differences in scale among graphs.

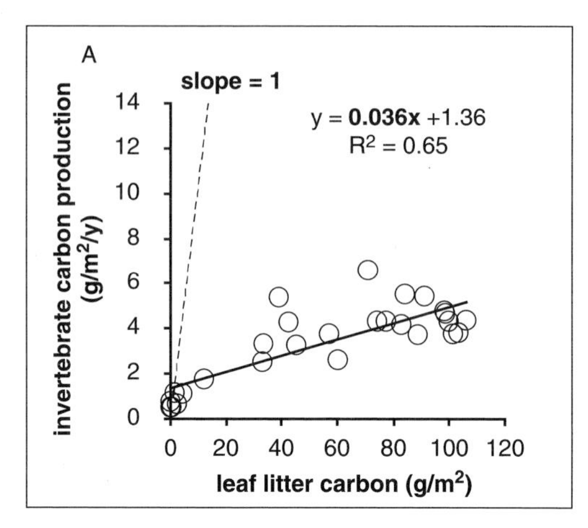

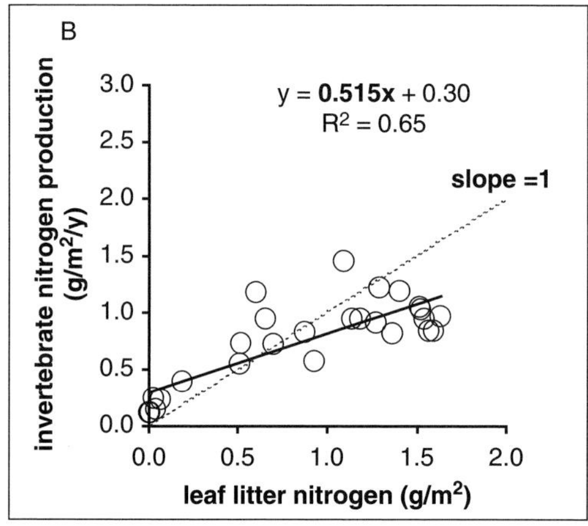

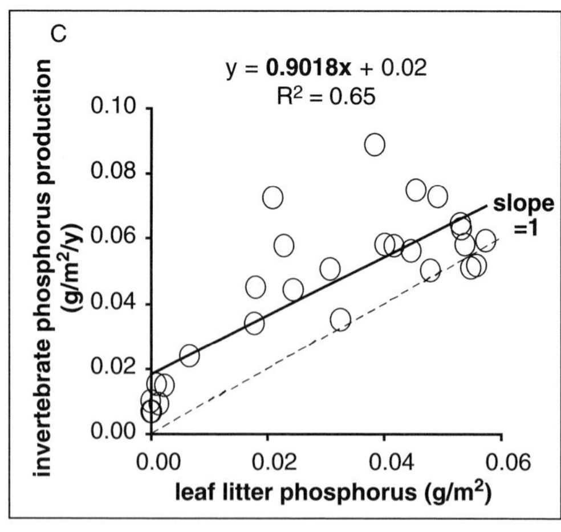

FIGURE 2 | Relationship between mean annual standing crop of leaf litter carbon (A), nitrogen (B), and phosphorus (C) and annual invertebrate secondary production of carbon, nitrogen, and phosphorus in C53, C54, and C55. N-29 years. All linear regressions are significant at $P < 0.001$. For comparative purposes, the dashed line has a slope of 1. Note large differences in scale among graphs.

39% of reference stream concentrations. Throughout the post-treatment period, export concentrations did not differ between the treated and reference streams, although concentrations tended to be higher in the treated stream during this period of rapid invertebrate community recovery (see Figure 3; Whiles and Wallace, 1992; Whiles and Wallace, 1995; Hutchens et al., 1998).

The average ratio of CPOM:FPOM exported from the treated stream increased from 0.053 before treatment to 0.18 during treatment, while little change occurred in reference stream ratios (0.028 before treatment, 0.048 during treatment) (Wallace et al., 1995). Altered export CPOM:FPOM ratios in the treated stream caused relatively large increases in C:P (40%, from 1219 to 1706), C:N (14%, from 36 to 41), and N:P (16%, from 30 to 35) ratios of total (CPOM + FPOM) exported particulate organic matter.

In order to gain a basin-wide perspective on the importance of invertebrates to elemental export, we calculated mean export of C, N, and P per square meter, and extrapolated these data to all headwater streams within the Coweeta basin. Mean annual export of fine particulate C, N, and P ($g/m^2/y$) was quantified from headwater streams during 13 unmanipulated stream-years in C53, C54, and C55 (Table 1). On a per-square meter basis, these first-order streams deliver a large subsidy of fine particulate C, N, and P to downstream river ecosystems (see Table 1). When all first-order streams within the Coweeta basin (length: 33.2 km; average wetted-channel width: 1.5 m) are considered, massive amounts of particulate C, N, and P are exported each year (see Table 1). Based on results from the insecticide treatment, it is possible to estimate the role of invertebrates in the total particulate C, N, and P export from the Coweeta basin. Applying percent differences in the treated versus the reference stream from above, it is evident that a major proportion of particulate C, N, and P export is due to invertebrate feeding and/or bioturbation activities (see Table 1).

DISCUSSION AND CONCLUSIONS

The implication that animal-derived effects on nutrient fluxes can dominate in ecosystems is receiving a considerable amount of support. In diverse ecosystems including lakes (Kitchell et al., 1979; Strayer et al., 1999; Vanni, 2002), streams (Grimm, 1988; Cuffney et al., 1990; Gende et al., 2002; Hall et al., 2003; this study), forests (Swank et al., 1981; Lovett et al., 2002), coral reefs (Meyer and Schultz, 1985), and estuaries (Dame

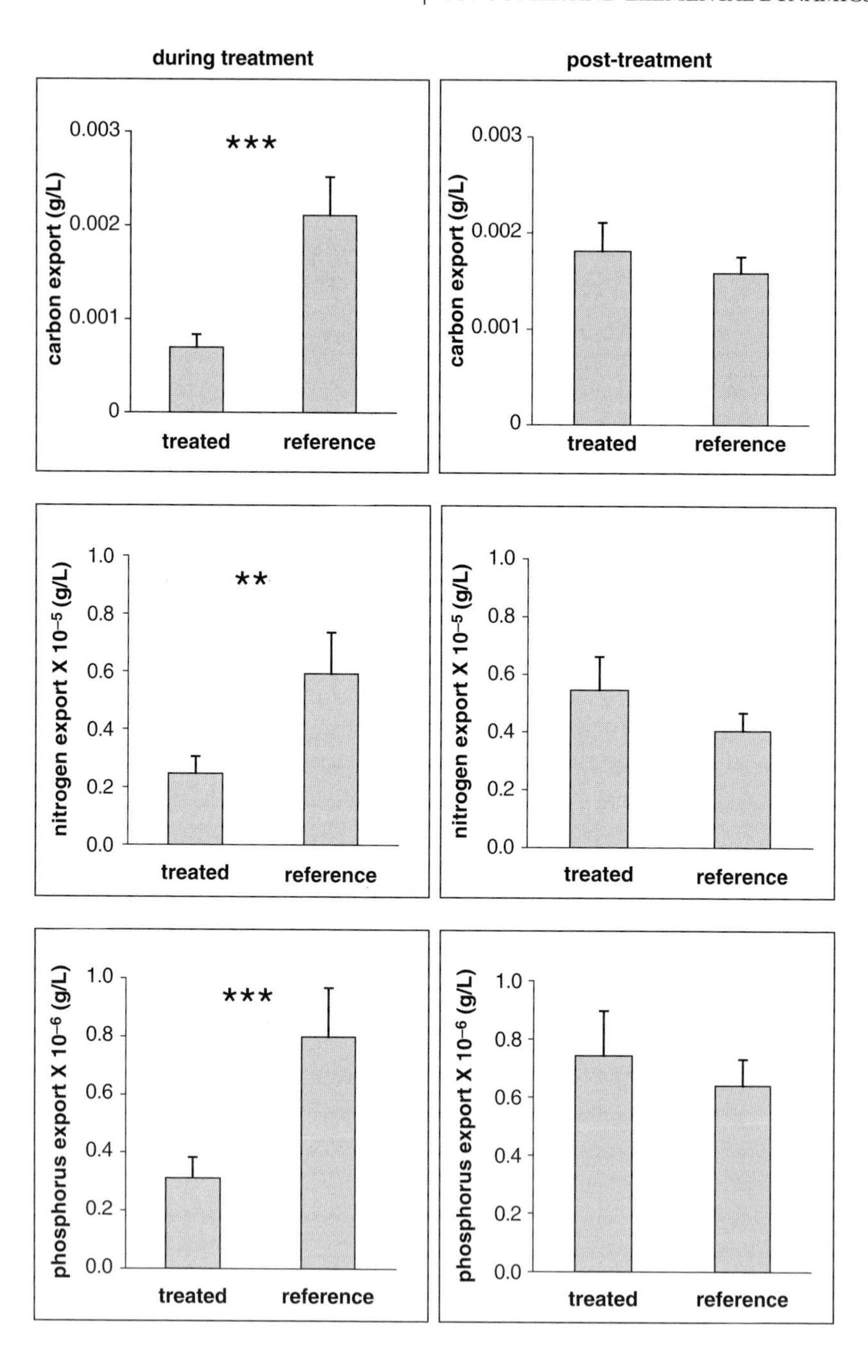

FIGURE 3 | Mean (+ISE) export concentration (g/L) of carbon, nitrogen, and phosphorus in the treated stream (C54) and a reference stream (C55) during invertebrate removal treatment (1987-1988) and after treatment (1988-1990). $^{**}P < 0.01$, $^{***}P < 0.001$, randomized intervention analysis. Note large differences in scale among graphs.

Table 1. Export of particulate carbon, nitrogen, and phosphorus (<4mm) from headwater streams at Coweeta Hydrologic Laboratory.*

	Mean export ($g/m^2/y$) ±SE	Total area of 1st order streams at Coweeta‡ (m^2)	Total basin wide export from 1st order streams (Kg/y) ±SE	Predicted basin-wide export from 1st order streams without invertebrates (Kg/y)
Carbon	173.8 ± 20.6	49800	8655.2 ± 1024.2	2856.2
Nitrogen	3.6 ± 0.7	49800	179.3 ± 32.4	75.3
Phosphorus	0.5 ± 0.09	49800	24.9 ± 4.7	9.7

*Based on 13 reference condition stream-years in C53, C54, and C55.
‡ Wallace 1988, assuming average wetted-width of 1st order streams is 1.5 m.

and Libes, 1993), animals contribute significantly to the dynamics of ecosystem-wide biogeochemical cycles, as well as many other ecosystem processes (Duffy, 2002). The magnitude of these contributions, however, can vary tremendously in space and time, and may be underappreciated if focus is limited to one elemental currency or ecosystem process.

One way animals can affect nutrient dynamics is by preferential utilization and storage of elements that are limited in supply. Essential elements, such as N or P, are often more efficiently or tightly cycled in ecosystems than other more abundant elements, such as C. In this study, detrital N and P were sequestered in invertebrate biomass much more readily than C. This is to be expected if N and P are in high demand relative to their availability in the system. Although stream microbes (i.e., bacteria, fungi, algae) can dominate the short-term uptake and cycling of N and P (Mulholland et al., 2000; Peterson et al., 2001; Gulis and Suberkropp, 2003), their turnover rates may be sufficiently high such that elements are quickly recycled and not retained for any significant period of time. In contrast, many invertebrates in headwater streams live for a longer period of time (approximately one to three years), and elemental storage by these invertebrates may be an important mode of nutrient retention in the system (Peterson et al., 1997) as well as a stable source of nutrients for in-stream predators.

We also found that invertebrates played a minor role in the storage of C in both dominant stream habitats, but contributed moderately to storage of N and P on bedrock outcrops (N = 8.2%, P = 11.2%). The contribution of invertebrates to particulate pools of N and P on bedrock outcrops is comparable to average values reported for zooplankton N and P in lake pelagic zones (N ~ 6%, P ~ 20%, Hessen et al., 1992; Andersen, 1997). These two habitats are somewhat similar in that nutrients are readily lost through physical erosive (to downstream) or sedimentary (to the lake benthos) processes. In these 'erosional' habitats, the retention of nutrients by consumers makes nutrients available to higher trophic levels and provides a source of recycled nutrients. In contrast, the mixed substrate stream habitat acts as a depositional zone for organic matter, and may be similar in function to lake benthic zones. In these 'depositional' sink habitats, the retention of nutrients by biota is potentially less critical for the maintenance of consumer productivity and nutrient recycling. More studies are needed which quantify the contribution of animals to total pools of particulate nutrients to assess patterns in nutrient storage among systems.

A number of recent studies have demonstrated that small headwater streams are extremely efficient sites for the retention of dissolved carbon and nutrients (Hall and Meyer, 1998; Peterson et al., 2001; Valett et al., 2002; Bernhardt et al., 2003; Webster et al., 2003). These streams are also the most prevalent aquatic features in many regions, and therefore play a strong role in the uptake and retention of these dissolved elements at a landscape scale (Meyer and Wallace, 2001). In this study we have shown that in addition to the importance of headwater streams in the retention of dissolved nutrients, these streams act as critical providers of particulate C, N, and P to downstream food webs (also see Wipfli and Gregovich, 2002). Animals (i.e., invertebrates) are prominent players in this process because they facilitate the transformation of large organic particles to small egested particles which are more readily exported downstream. Indeed, our results indicate that >50% of total fine particulate C, N, and P export from Coweeta headwater streams can be attributed to invertebrate consumptive processes (also see Cuffney et al., 1990). In contrast to the fate of dissolved nutrients, these particles are likely to be physically retained in larger streams and provide a dependable food resource for downstream consumers.

Our ability to quantify the large importance of invertebrates to downstream elemental export was made possible by a large-scale experimental invertebrate removal (Cuffney et al., 1990). Such 'deletion' experiments are critical for assessing interaction strengths among consumers and resources, and may be considerably more informative than other food web approaches (e.g., energetic flow food webs or topological webs) in determining the regulation of material flows (Paine, 1992; Polis, 1994). In forested headwater streams, the actual flow of energy (or carbon) to invertebrates is insignificant in terms of the whole-system energetic budget. In sharp contrast, their influence on the dynamics of detrital food resources and nutrients is highly significant.

Animals are also capable of affecting ecosystem nutrient dynamics by altering elemental ratios and stoichiometric balance of available nutrients (e.g., Elser et al., 1998, Vanni, 2002). In this study, invertebrates affected bulk elemental ratios of exported particulate nutrients by maintaining low CPOM:FPOM export ratios. In Coweeta streams, average C:P (1015 or 0.09% P) and C:N (34 or 2.9% N) of FPOM are much lower than CPOM (C:P–4858, or 0.02% P; C:N–73, or 1.37% N) as a result of increased surface area:volume ratios and a higher proportion of microbes on these small particles (Sinsabaugh and Linkens, 1990; Cross et al., 2003; 2005). Transformation of CPOM to FPOM by feeding activities of detritivorous invertebrates effectively facilitates the production and export of small nutrient-rich particles. Invertebrate removal caused a major reduction in this transformation, and a consequent increase in C:P, C:N, and N:P ratios of bulk organic matter export because nutrient-poor CPOM constituted a higher proportion of the total particulate organic matter being exported.

In summary, our analyses support the idea that animals play diverse roles in ecosystems and may have surprisingly large effects on processes (i.e., nutrient dynamics) that would not be predicted from their limited energetic contribution to many systems. Understanding linkages between animal species or assemblages and ecosystem processes will benefit greatly from studies that examine a suite of response variables, including multiple elemental currencies and multiple processes.

ACKNOWLEDGMENTS

We thank the European Science Foundation for funding support. We are indebted to the many previous investigators at Coweeta, whose published data were included in our analysis. We also thank two anonymous reviewers for constructive criticism. Research reported in this paper was largely funded by the National Science Foundation. Additional support was provided by NSF (DEB-9806610) during the preparation of this manuscript.

5.5 | MEASURING THE ABILITY OF FOOD TO FUEL WORK IN ECOSYSTEMS

Steven H. Cousins, Kathryn V. Bracewell, and Kevin Attree

This chapter reports early approaches to developing measurement methods that allow a fundamentally new way of understanding trophic interaction. Measurement has not always had a central place in food web studies; counting organisms and classifying species and their trophic relationships has been more important both recently and in the distant history of our subject. We could probably point to three phases in studying food webs. The initial phase of counting and description, which we might call an 'intelligent natural history' of food webs, led to the development of a number of core concepts including food webs themselves (Elton, 1927). The middle phase, where measurement was paramount, quantified the energy transfers between species using the bomb calorimeter to establish unit mass energy contents (Teal, 1962). The more recent (post-International Biological Programme) phase is one where we have returned to the art of classification and observation in food webs in order to identify pattern and universal properties (Cohen, 1978).

DeAngelis (1992) has classed these phases as contrasting the 'process-functional' school with the 'population-community' school. In recent years the latter approach has tended to dominate but with some innovative work also continuing in the process-functional tradition (DeAngelis, 1992; de Ruiter et al., 1993a). This tradition remains important since if feeding is about anything it is about gaining energy and nutrients for maintenance, movement, growth, and reproduction. Food webs are structures which only exist such that members of the web gain energy from others, or if they are predated, 'donate' energy and nutrients to their predators. We suggest that additional measurement techniques can clarify the energetics of trophic transfers. Although categories of stored energy can be identified in ecosystems, conventional calorimetry only provides a first law of thermodynamics description suitable for energy accounting in food webs in which energy is neither created nor destroyed. We therefore ask how can we make appropriate measurements for a second law approach to the operation and fuelling of food webs.

EARLY DESCRIPTIVE WEBS

There are some early observations (Elton, 1927), based on body size, which may help structure how these new methods can be applied to food webs. Elton made the very simple point that predators are normally larger than their prey. When this is combined with another observation that there is a limit to both the largest and smallest creatures in an ecosystem, then the number of possible steps in a trophic chain is constrained. The outcome of these feeding relationships is represented by the pyramid of numbers of animals; a distribution with larger numbers of smaller creatures which is fed upon by a progressively fewer number of larger ones. However this simple model does not apply to animals eating plants since this relationship is not directly size structured. Animals of all sizes eat plants although Elton observed that different plant parts had characteristic species of herbivore and that depending on the minimum size of the herbivore then the number of possible steps in a trophic chain was again constrained by size. Finally he noted that all life ultimately relied on energy derived by photosynthesis in the green plant.

This early view of food webs provides us with three distinct cases. First there is the measurement of the properties of *different parts* of the green plant as distinct entry points into the food chain, then the characteristics of the herbivore in linking the plant to carnivores and finally there is size-based measurement for predictive carnivory.

WORK IN ECOSYSTEMS

Work is measured in joules and when one gram of pure water at room temperature is raised by one degree then 4.185 joules are required to do so. For heat to flow there must be a gradient of temperature from hot to cold and the steepness of the gradient determines the rate at which the heat flow occurs. Importantly therefore, the external conditions affect the rate at which these energy transfers occur. In ecosystem energetics the standard measurement instrument is a bomb calorimeter (Phillipson, 1966; Petrusewitz, 1970). The heat of combustion of a small dried pellet of material burnt in an environment of oxygen is measured by the temperature increase in the mass of water in the water jacket and appropriately converted to Joules. Thus the calorimeter only measures the quantity of heat that is released while the rate of energy transfer is not considered.

What is perhaps surprising is that when materials as different as young leaves and old bark are dried and a unit mass burnt in a calorimeter, the heat output of each is remarkably similar (see Table 1). However, a unit mass of coal can also be burnt in the device and energy content measured but this does not predict the ability of that material to do work in an ecosystem. Similarly why should we believe that the measurement of energy content of bark, rather than coal, is meaningful in terms of fuelling work in the prevailing external conditions of the ecosystem?

SPECIFYING EXTERNAL CONDITIONS

External conditions affect the *rate* at which energy transfers take place by determining the source:sink gradients over which energy flows. One obvious condition is the temperature at which organisms function which is determined by the environment for poikilotherms and held generally at a somewhat higher and more constant value for

Table 1. Data for oak tree parts (*Quercus robur*) showing energy content, nitrogen dry matter, in relation to the rate of energy release $W.kg^{-1}$ by decomposition in soil averaged over the single period from 18–115 days after burial.

Oak tree part	Nitrogen (%dm)	Energy content (MJ/kg)	% mass loss per day	Watts/kg
acorn	1.333	18.6	0.0046	0.99
root	1.011	19.4	0.0009	0.20
bark	0.676	19.4	0.0005	0.11
twig	1.003	20	0.0011	0.25
brown leaf	1.651	20.3	0.0011	0.26
green leaf	2.853	20.2	0.0044	1.02

homeotherms. Another important condition is that the energy in food must transfer to any ingesting organism within the time that the food is present in the gut. Whereas energy transfer in the calorimeter is treated as instantaneous and measured in $J.kg^{-1}$ we are instead interested in $J.kg^{-1}.sec^{-1}$ flows under given external (biological) conditions. Since one joule per second is one watt then for ecosystem purposes we are interested in measuring the properties of materials in $kW.kg^{-1}$ rather than the $J.kg^{-1}$ that the calorimeter provides. The problem becomes how do you standardize and make relevant a power measurement when the size or design of the measurement instrument (the external conditions) will determine the power output that is achieved.

The measurement of power per unit mass of biological material entering a food chain (Figure 1) is dependent on the type of material (fuel) is supplied and what specific device (i.e., which organisms) release the energy. A fundamental question is whether there are species of creature evolved to digest bark at the same rate as other organisms might digest seeds? Or, has some irreversible change occurred when the simple sugars produced by photosynthesis are used to make complex long chain and cross-linked molecules such as lignin that affects all species feeding on it? Thus although these bonds can be broken at the temperatures experienced during combustion, what are the implications for biochemical reactions at the temperatures real organisms operate? An intermediate position is provided by plant defense compounds where evolution is known to play a role and an evolved suite of herbivore species can detoxify those compounds, while the bulk of herbivore species cannot do so.

EXPERIMENTS IN POWER MEASUREMENT

We investigated power density measurement in the three aspects of trophic interaction arising from Elton's analysis of food webs; the difference between plant parts, a herbivorous food chain step, and size-based carnivorous food chain step.

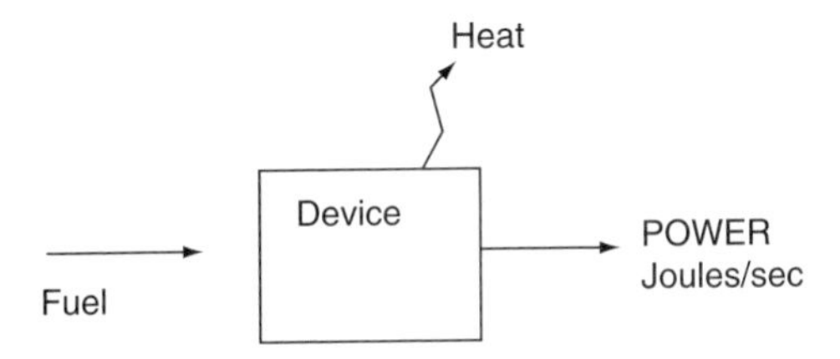

FIGURE 1 | Conceptual diagram showing how power output is dependent on the fuel and the conversion device used.

Comparing Plant Parts

To establish if there is a substantial difference in the properties of different plant parts as food we needed a single 'device' which we could apply to each of the plant parts of a single plant species. A bacterial digester looked appropriate since bacteria are associated with digestion of plants in soil during decomposition and bacteria are also associated with anaerobic digestion of plant material in ruminants. In this case we used soil itself as the physical medium of the digester. We took samples of different parts of the English oak (*Quercus robur* Linnaeus), dried, fragmented them and placed them in nylon bags. These were buried in soil in an ancient oak woodland where by definition the soil faunal community was dependent on that type of food supply (Bracewell, 2000) and could be said to be adapted or evolved to digest oak derived matter. Table 1 shows the energy content of the different materials as measured using a bomb calorimeter. Conventional plant quality measures of plant material composition (Cadisch, 1997) were also made and represented here as percentage mass of Nitrogen in the dry matter. The rate of weight loss was measured over a 115-day period and the loss dimensioned in $kW.kg^{-1}$ which was linear after the initial 18-day period when a more rapid loss occurred.

This experiment showed that whereas the energy content of the plant materials differed by only 10%, with bark being more energy dense than acorns for example, the rate at which this energy could be extracted by soil biota in the decomposition bags differed by some 900%. Of these materials the energy in bark was the least available and that in young leaves and acorns the most available. Thus the ability of the plant parts to power activity in the ecosystem were very different and not predictable from energy content analysis.

The Herbivorous Food Step

We investigated the silkworm (*Bombyx mori* Linnaeus) food chain as the reference herbivorous interaction and looked at the transformation of mulberry leaf into silkworm and silkworm faeces (frass). In a similar attempt to maintain evolutionary familiarity with the test substrates we extracted soil bacteria from soil taken from underneath our local mulberry tree (*Morus alba* Linnaeus) and grew them with the test substrates in digesters. The extraction process detached bacteria from soil by sonication in a mild detergent, then washed, filtered and centrifuged them to a pellet (Attree, 1998). The bacteria were re-suspended in a stock solution and added to flasks containing equal masses of powdered materials; leaf, silkworm, and frass plus bacteria-only controls. The respiration of the soil bacteria was measured over five days by passing carbon dioxide free air into the digester flasks and the CO_2 produced was measured by a precipitation of barium carbonate in the outlet flasks. This experiment has developed some basic techniques to measure the power output per unit mass of the materials in the food chain step and shows that while the energy in the system is conserved, the power density of the materials is not conserved. The power density of the silkworm is raised relative to the leaf and that of the frass lowered relative to both. These are, however, very early results only.

Muthukrishnan (1978) give energy accounting (first law) data for final instar silkworm larvae free-feeding on mulberry leaf and show: 1000 gcal of fresh leaf results in 462 gcal assimilated and 538 gcal frass and a gross conversion efficiency of 16% into silkworm mass. We approximate this to 1g leaf = 0.16g larva + 0.54g frass + 0.3g CO_2. The power densities of the materials produced and as measured above have been transformed from; leaf 12.6 W.kg-1 (estimated peak 16 W.kg-1) to larvae 13.0 W.kg-1 (estimated peak 22.0 W.kg-1) to frass 6 W.kg-1 (estimated peak 7.0 W.kg-1) (Attree, 1998). Thus a unit mass of leaf is transformed into a small quantity of material with a higher power density, the silkworm, and a large quantity of low power density material, the frass, while, by the first law, the energy contents are conserved.

The Carnivorous Food Step

Carnivory represents a very different problem compared to the last two cases. The proteins and fats that make up the non-structural components of animals are very similar in all species. If we dry and powder a unit mass of a mammal we would expect the same $kW.kg^{-1}$ output when placed in a bacterial digester regardless of its body size or position in a food chain. Unlike herbivory, where the herbivore has a different chemical make up to the plant, carnivores are similar in their composition to the things that they eat. However, whereas it may be easy for the herbivore to find the plant, the problem for the herbivore is extracting the energy from the plant during digestion. Conversely the problem for carnivores is to find and capture the prey, which is then relatively easily digested and assimilated. The factors which determine the rate at which energy can be supplied to a predator are dependent on foraging rather than digestion and the measure of rate of energy supply in $kW.kg^{-1}$ can be derived via foraging theory rather than microbial reactors.

If we adopt Elton's proposition that carnivorous food webs are size structured then we can take the pyramid of number and turn it through 90 degrees to provide a graph of organism abundance against body weight (Figure 2B). If this graph represents a real ecosystem and is viewed at successive time intervals, then energy will pass from smaller creatures to the larger ones by process of predation and animal growth. However to model this system, we recognize that any organism can be either prey if it is predated in the next time interval or a predator if it finds prey to eat. To distinguish these trophic roles *of the same organisms* we indicate when we are viewing them as prey, by showing a bar under the variable for the organism (such as weight, $\underline{w}$), or when viewing them as a predator we place the bar above the variable e.g. $\bar{w}$. Unit mass prey handling time, $\underline{\bar{h}}$, is a function of both by predator and prey states, so bars are shown above and below the variable concerned.

The process of predation by an individual can be decomposed into a search phase to encounter the prey and a handling phase where time is spent getting the prey into the mouth and swallowing it. For any given predator, the encounter rate p with prey is proportional to prey abundance N_1; the time between encounters given by $1/pN_1$. Search and handling rates are related since it is not viable to search for prey which cannot be handled and ingested sufficiently quickly per unit mass of prey obtained, represented by a lower limit α and upper limit β on prey size for a given size of predator. Similarly, the time used capturing and handling the prey is a time delay that cannot be spent searching for new prey. Allowing for handling time, encounters occur at $1/hwpN_1$ where h, the unit biomass handling time is multiplied by the mass w of the individual prey of N_1. The handling component is of primary importance in determining predation interactions and is fundamentally constrained by the predator morphology.

If instead of a single predator we take $\bar{n}$ predators at a particular weight, then over the whole distribution of all weights of all individuals n we get (Cousins, 1985)

$$\frac{\partial n}{\partial t} = -n \int_0^{\infty} \frac{\bar{p}\bar{n}\,d\bar{w}}{1+\int_{\underline{\alpha}}^{\beta} \underline{\bar{h}}\,\underline{w}\,\underline{p}\,\underline{n}\,d\underline{w}} \tag{1}$$

For a predator organism with an optimal prey size with least handling time h_{min} (see Figure 2A), then smaller prey will by definition, take more time to handle per unit biomass consumed; and similarly for larger individuals. As the width of the diet is expanded more prey individuals will be encountered but there will come a point where the time spent handling very small or very large prey will be so great that the total amount of flesh ingested will decrease. If the predator's demand for food cannot

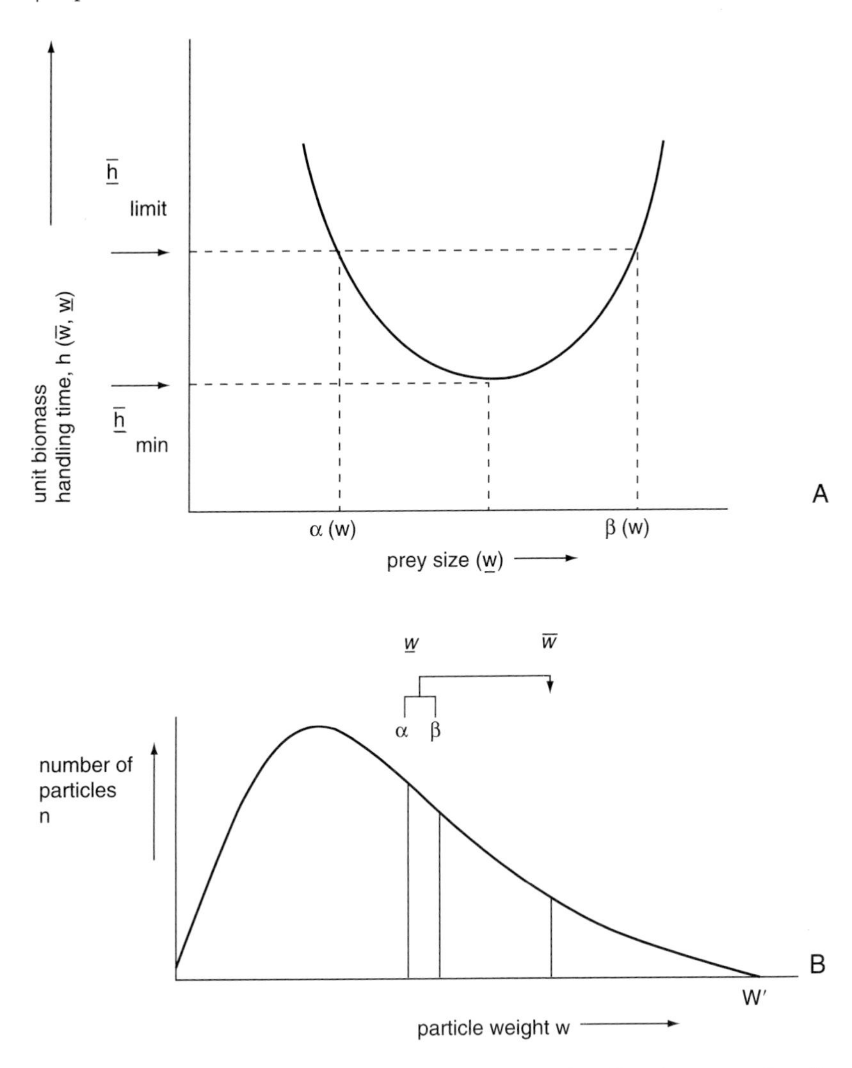

FIGURE 2 | A, Foraging window, α–β, for a predator $\overline{w}$ defined by the time to handle one unit of prey biomass $\bar{h}$ for prey $\underline{w}$ of different sizes (Parkin, 1981; Cousins, 1985). **B,** Number of all particles (organisms) for different weights w up to the maximum size of top predator W' after (Cousins, 1985).

be satisfied within this feeding limit (see Figure 2A) then it will eventually starve; if it can easily be met then it will select food from a narrower range of its possible diet. It is an axiom of this foraging approach (Pulliam, 1974; Charnov, 1976; Silvert, 1980; Cousins, 1985) that any individual of any species which is encountered by the predator and can supply 'flesh' within the current unit biomass handling rate will be eaten.

This provides a foraging based mechanism for interval food webs (Cohen, 1978) provided that other mechanisms for herbivory and detritivory are also included.

This type of approach to individual prey choice (see Figure 2A, B) (Parkin, 1981, Cousins, 1985) has the same basic form as (Williams, 2000) and with it the same implication of simple rules for diet choice where any species encountered is eaten if it falls within the window that the predator morphology (Warren, 1995; Cattin, 2004) determines.

Complex multi-parameter models of can be readily criticized (May, 2004) as un-testable although heuristically useful in some cases. In essence eqn (1) can be reduced to eqn (2) below.

$$E = apN_1N_2 \tag{2}$$

where $pN_l . N_2$ is the encounter rate of the N_l prey individuals in a weight interval of the distribution with N_2 larger predator individuals in a second weight interval. The term p is the predation rate. E can be standardized as the flow of assimilated energy for one kg of prey N_1 to one individual predator (i.e., when $N_2 = 1$). By knowing the assimilation constant for digesting flesh which typically has the value 0.8 (Heal, 1975) and knowing the energy content of the flesh then this predation rate can be dimensioned as energy per unit time per kg of prey biomass which has the required dimensional form, $kW.kg^{-1}$.

Any process which changes the value of p will therefore alter the power output of the 'device' in which the prey are located. So for example if predator and prey are located in water in a microcosm and the volume of water is increased then, even though the numbers of predators and prey stay the same, this will reduce the ability of the prey as fuel to sustain the predator's food web. Thus p will have been reduced and with it the value of $kW.kg^{-1}$ of the prey.

ECOSYSTEM AS OBJECT OR DEVICE

We began with the important point that to get an energy flow you need a source:sink gradient, and to measure a flow you need a device of some kind which you can standardize in order to get repeatable values and values which can be meaningfully compared between prey and between systems. Whilst laboratory equipment may provide the standardized measurement of plant parts and of herbivory (and by extension of detritus and detritivory), there is no immediate device that simulates carnivory other than treating the ecosystem itself as the device (i.e., the set of external conditions under which the trophic interactions occur). The dilution or cryptic location of organisms in an area and the relative abundance of organisms of different sizes found there (the pyramid of number) determines the power output of the system. The pyramid provides a gradient where there is a source of concentrated biomass in large particles versus a progressive sink of diffused material in smaller particles. The notion of the ecosystem as a 'device' is perhaps strengthened by considering the ecosystem as an observable spatial object derived from the food web (Cousins, 1990).

DISCUSSION AND CONCLUSIONS

This chapter has sought to strengthen the case for making direct measurement of the energetics of food webs. In making compatible measurements on plants, herbivores and carnivores we have attempted to move the study of food web energetics from a description of conserved energy balances (first law description) to a non-conserved set of power outputs (second law description). Different plant parts, animals, and by extension detritus, can be treated as materials, distributed in an ecosystem, each with a different potential to fuel organism metabolism and growth. These differences in unit power outputs $kW.kg^{-1}$ are seen to be large and therefore have the potential to explain many aspects of feeding interactions. Since energy has also been used as a driver for biodiversity studies then energy availability may also have potential for further exploration of diversity relationships.

ACKNOWLEDGMENTS

To our Cranfield University colleagues who have advised and helped on this project: Frank Taylor, Peter Allen, Mark Strathern, Stefan Dekker, Howard Parkin, and Dawn Fowler.

5.6 | TOWARDS A NEW GENERATION OF DYNAMICAL SOIL DECOMPOSER FOOD WEB MODELS

Stefan C. Dekker, Stefan Scheu, Dagmar Schröter, Heikki Setälä, Maciej Szanser, and Theo P. Traas

In the last decades, several soil research programs focused on the interaction between soil organisms, plants, and detritus in agricultural soils. Central in this research were the study of nutrient flows and cycling, as well as the availability of nutrients for crop productivity. Soil ecologists were especially interested in energy flow and the structure of food webs. With these foci in mind, food web models were constructed. One example is the soil Detrital Food Web (DFW) model for grass-prairie ecosystems (sensu Hunt et al., 1987; de Ruiter et al., 1993a). The model analyzes the relationships between community structure and ecosystem processes for different agricultural (de Ruiter et al., 1993b) and forest soils (Schröter et al., 2003). For some situations, model results were verified at the level of overall respiration and nitrogen mineralization rates (de Ruiter et al., 1994). However, verification on overall respiration and mineralization fluxes only gives confidence in part of the model behavior, due to the fact that most important fluxes take place among bacteria, fungi, and protozoa. Other consumer groups appear to have an insignificant direct effect on these fluxes. Nevertheless, the power of these models in predicting is surprisingly good, considering that these models have important shortcomings.

One of the reasons for the success of the DFW model is that it uses the equilibrium approach. With this approach, it is possible to start from observed equilibrium population abundances to calculate energy flows and interactions. The observed population abundances reflect, at least in part, all kind of processes in soil that impact populations and processes such as various indirect effects, as well as temporal and spatial heterogeneity. Although these processes are not explicitly modelled, they are implicitly taken into account due to the use of observed biomasses. Model input is based on yearly or seasonal averages meaning that model results may only be interpreted as yearly or seasonal means.

A shortcoming of the DFW model is that it does not address temporal variations within years, which is important, given the objective of the model to relate soil food web structure to nutrient availability in agroecosystems.

Hunt et al. (1987) developed the equilibrium DFW model based on Lotka-Volterra equations. The model calculates the feeding rates (i.e., the rate at which material is taken from an energy source), which is split into an excretion rate, a biomass growth rate and a mineralization rate. Annual feeding rates for every functional group are calculated assuming that the biomass production rate of a group balances the rate at which material is lost through natural death and predation:

$$F = (d\,B + P) / (a + p) \tag{1}$$

where F is the feeding rate (kg C ha^{-1} y^{-1}), d is the specific natural death rate (y^{-1}), B is the biomass (kg C ha^{-1}), P is death rate due to predation (kg C ha^{-1} y^{-1}), and a is the assimilated carbon per unit consumed carbon, p is the biomass-C production per unit assimilated carbon.

The consecutive step of adding dynamics to the model starts from observed biomass dynamics (O'Neill, 1969; de Ruiter and van Faassen, 1994). By linear interpolation between two measurement points of biomass, a dynamic feeding rate can be calculated. However, still such models start with observed biomasses.

A more general idea about dynamics in populations is to be preferred, for instance by describing the population dynamics in terms of differential equations of the Lotka-Volterra type (Moore et al., 1993):

$$dX_i / dt = -\,dX_i - \sum_{j=1}^{n} c_{ij} X_i X_j + a_i p_i \sum_{j=1}^{n} c_{ji} X_j X_i \tag{2}$$

in which c_{ij} is the coefficient of predation by group j on group i, and X_i, X_j are population sizes.

Starting with this full dynamic model, we do not use observed biomasses as input anymore. For this reason, important ecological processes that were taken into account implicitly in the DFW model via using observed biomasses, now have to be addressed explicitly. In this chapter we give suggestions about what kind of ecological processes need to be reconsidered. The new dynamical soil detritus food web model, (i) should model indirect effects explicitly, (ii) should incorporate spatial heterogeneity to predict nutrient availability, and (iii) should reproduce and estimate how soil food web structure evolves over time.

There are a number of other shortcomings (e.g., that the functional group approach assumes similar contributions within the group) independent on life status, while there is intra-group differentiation in feeding preferences, reaction to environmental stress factors, etc. Furthermore, the lack of loops such as cannibalism and intra guild predation, grazing-induced changes in the structure of microbial communities and the lack of consideration of mutualistic interactions, are not considered here because of limitations of space.

INDIRECT EFFECTS

Detritivorous and microbivorous decomposer fauna significantly enhance decomposition through so-called 'indirect effects' of their feeding activity (Anderson et al., 1981; Petersen and Luxton, 1982; Anderson, 1995). First of all, reducing the population sizes of microbes due to grazing fauna will enhance the growth rate of microbes. This indirect effect is density dependent and captured in the present DFW model. However, other important indirect effects are litter fragmentation or comminution, translocation and mixing of litter material, improvement of soil structure, re-concentration of limiting nutrients, and stimulation, transport, and inoculation of microbes, as well as stimulation of the humification processes (Anderson and Ineson, 1984; Lussenhop, 1992; Zechmeister-Boltenstern et al., 1998; Kusinska and Kajak, 2000; Scheu and Setälä, 2002). These indirect effects are important also at the ecosystem scale (Bengtsson et al., 1995). We believe that abiotic processes can affect soil food web structure and function and are common in other food webs as well. In this chapter, we will focus on the indirect effects controlled by biotic processes. Recently Pavao-Zuckerman and co-workers included specific indirect effects, such as microbial grazing, in a numerical food web modelling approach based on the HSB-C model (Fu et al., 2000). Including these indirect effects had considerable influences on the estimated carbon fluxes (Chapter 2.1); however, mechanisms were not included in the model. Moreover, microcosm studies prove the importance of Enchytraeidae in boreal systems (Huhta et al., 1998; Laakso, 1999), but food web simulations calculate only minor contributions of Enchytraeidae to carbon and nitrogen mineralisation in such systems (Schröter et al., 2003).

A systematic representation of indirect effects in DFW models requires a better understanding of the underlying mechanisms such as litter fragmentation and translocation, transport, inoculation, and growth limitations. The present DFW-model has six different model parameters that can be changed to take account of theses indirect effects: the input rate of substrate (I), the natural death rates (d), the feeding preferences (w), and the assimilation (a) and production efficiencies (p).

A particularly challenging task in future soil food web models is to incorporate non-trophic indirect interactions, which is probably indispensable for depicting and forecasting the dynamics of populations. In the following we make a few suggestions how an indirect effect could be modelled, and which environmental and population dynamical model parameters should change:

(i) The most important indirect interactions in soil food webs are those that are transferred in the food web through the abiotic forces, namely nutrients. By feeding on microbes, grazers produce nitrogen rich faeces. In a closed system this N can be taken up by prey species (microbes), which escaped from consumption. These "escaped" microbes compensate the loss of prey individuals by increased growth due to the N enrichment. Bacteria or plants that are not grazed can take up the same N that is liberated after the grazing process (Laakso and Setälä, 1999a). This indirect effect is typical in soil food webs. Trophic dynamic theory (Oksanen et al., 1981) assumes that in order to affect a prey, predator biomass has to change due to predation by top-predators. In soils no such assumption is needed. The DFW model, developed for agricultural and grassland soils, is based on energy (carbon) fluxes for an open N-system. Bacterial immobilization of N seems to be common in forest soils (Berg et al., 2001; Schröter et al., 2003), due to the larger C:N ratios of the detritus (humus). Because the resources of secondary consumers are relatively nitrogen rich, mineralization by soil fauna is more dependent on the availability of N than of C (Anderson et al., 1981; Setälä et al., 1990). This means that for closed N-systems, nitrogen limitation must be explicitly integrated into the model, for instance via functional responses on the a, p and growth rate parameters, which was already considered in previous model work (Smith, 1982; Knapp et al., 1983; Cochran et al., 1988).

(ii) Indirect effects are not only based on limiting N, but can also on other limiting resources like oxygen or phosphorus. Due to feeding and engineering of for instance earthworms, oxygen becomes more available, relaxing the limitation of growth rates by oxygen. A new model should not only capture C and N limitations but also other limiting resources.

(iii) Different field mesocosms experiments show that in the presence of epigeic macroarthopods the decomposition and humification processes both in litter and underlying soil are significantly promoted (Kajak et al., 2000; Kusinska and Kajak, 2000) (Szanser, 2000b). The amount of remaining litter was negatively correlated with mass of dead arthropods and faecal pellets (Szanser, 2000b, 2003). Additional laboratory and field experiments showed that input of insects' faeces promoted litter loss (Szanser, 2000a). Other studies support the opinion that input of liquid and structural faeces by phytophagous and predatory macroarthropods living aboveground into soil system might distinctly change soil microbial functioning and soil micro- and mesofaunal density and community structure (Dighton, 1978; Andrzejewska, 1979; Chmielewski, 1995; Szanser, 2000a, b). Field experiments indicate that the density and biomass of microarthropods and the input of dead animal mass into litter and soil system are regulated by epigeic macroarthropods' predatory activity (Kajak and Jakubczyk, 1977; Kajak, 1995; Wardle and Lavelle, 1997; Kajak et al., 2000; Szanser, 2000b, 2003). As a result, it may be concluded that epigeic fauna might have substantial indirect effects on soil organic matter cycling and soil food webs not only by predatory activity, but also by dead biomass and faeces. In the normal food web model a detritus feedback is modelled, meaning that the biomass and faeces of all functional groups are returned to the detritus pool. However, it seems to be more feasible to return these faeces to a second detritus pool that is more easily available and decomposable (e.g., a pool with al lower C:N ratio). Moreover, due to for instance litter fragmentation contact area is increased, meaning increased resource availability. In modelling terms these effects can be represented as an increase in substrate input or a change in substrate quality, meaning that a and p should be modelled as a function of the C:N ratio of the substrate. This requires that at least two, and possible more pools of litter with different C:N ratios are distinguished. However, this creates a new problem, because we then need to model the preferences of functional groups to the different pools.

Other indirect effects related to changes in the microbial community composition, changes in the distribution of litter resources in soil (bioturbation) and indirect mutualistic interactions may also need to be incorporated into future food web models to better understand and predict decomposer food webs (Bonkowski and Brandt, 2002; Scheu and Setälä, 2002). In addition, it is increasingly realized that the above- and belowground food web are much closer linked than previously assumed and therefore, interactions between these two subsystems of terrestrial ecosystems need to be incorporated into food web models for a more comprehensive understanding of terrestrial systems (Scheu, 2001; Wardle, 2002a; Wardle, 2002b).

SPATIAL HETEROGENEITY

Given the objective of the new model (i.e., to simulate temporal variation within years), spatial heterogeneity is important (Polis et al., 1996a). In soils, spatial variability can occur both vertically and horizontally. It seems to be logic that soil resource gradients will result in soil biota gradients. Input sources of soil food webs can be detritus, roots and algae, which can have large spatial variability. For instance, Wardle (1995) showed that an increase of recalcitrant detritus result in an increase of fungi. Furthermore, Berg et al. (2001) and Bengtsson and Berg (Chapter 5.1) have measured and modelled soil food webs in different organic horizons (LFH), from fresh litter (L) to humified organic matter (H) and found differences in species and food web structure. Spatial separation of resources can also decrease competitive pressure (e.g., Sulkava and Huhta, 1998).

The spatial variability of soil biota is not only dependent on the spatial patterning of detritus or roots, but of all resources. Bengtsson and Berg (Chapter 5.1) showed that the horizontal spatial variability of their soil food webs in forests is very low, while large gradients, for instance in soil water, temperature and litter input can be expected due to the gap structure of trees. A possible explanation of this low soil biota variability can be that all systems are under stress by an overwhelming factor, such as low pH values. Furthermore, spatial patterns of resources like water content or temperature are very dynamic in time, and often show greater fluctuations than the limited resources (Ettema and Wardle, 2002). This suggests that population dynamic processes and their forcing variables (e.g., temperature) can have large effects on the spatial pattern of the soil biota.

Due to the spatial heterogeneity of the environmental conditions and the dynamics in trophic interactions, food webs measured at shorter time intervals will be more realistic (Polis et al., 1996a). For instance temperature is directly related to the growth rates of soil biota, which can be indirectly modelled as a function of the growth rate parameters in the model, while temperature can be taken as forcing variables in the model (Schröter et al., 2003). Keeping the spatial heterogeneity of the resource inputs in mind, which have no large fluctuations in time due to the high stocks, food webs measured at small spatial scale will also be more realistic.

As a result, if we want to simulate ecosystem processes on a large spatial scale then we should use many point models in space. The output of these point models must be scaled up to the larger spatial scale. For every point model, forcing variables as resource input and environmental conditions are used as input. Given the objective to relate nutrient availability within years, these point models should have a dynamic character.

As also stated in the indirect effect section, a disadvantage of including more organic matter pools in the DFW model is the increase of model parameters. To solve parameter identification problems, we should have other experiments and observations. Cousins et al. (see Chapter 5.5) can measure besides energy content the energy availability (i.e., the rates of energy release by decomposition) for as well different pools of organic matter and functional groups. Measurements of energy availability of the different pools can be used to parameterize or verify the DFW model. The method of Cousins et al. could also imply a modification of the DFW model. Energy conversion is currently only dependent on the consumer, not on the food, while Cousins et al. implies that energy conversion efficiency is a function of the interaction between consumer and food properties. Food properties are for instance the energy availability of the food, how much energy a consumer needs to chase or capture a prey, etc. As a result, the efficiency parameters of the DFW model should be modelled as a function of consumer and prey. Moreover, excretion fluxes will also have different energy availabilities and will flow back in different pools of detritus. With a closer integration between experimentation and modelling we can develop a new generation of models, based on the combination of energy content and availability.

SUGGESTIONS FOR IMPROVEMENT IN THE CONSTRUCTION OF DFW MODELS

Building complex interactions in belowground systems into a food web model to predict fluxes of matter and dynamics of populations is a challenging task. The basis of each food web model is that trophic species are linked by interactions. Due to the exceptionally high number of species in soil, biological species have to be aggregated to trophic species sharing prey and predators. Until today this was uniformly done by using higher taxonomic units as trophic species assuming that taxonomically similar species feed on a similar spectrum of prey and share a similar range of predators. Recent studies using stable isotope analysis suggest that this approach is of very limited use to construct soil food webs (Scheu and Falca, 2000; Scheu and Setälä, 2002). Higher taxonomic units in fact may only reflect very general feeding groups, such as predators and decomposers (detritivores and microbivores). Species of each of the major higher taxonomic groups of decomposers, including Collembola, Oribatida, Enchytraeidae, Lumbricidae, Isopoda, and Diplopoda, appear to include those that predominantly feed on dead organic matter (litter and detritus; primary decomposers) and those that feed on micro-organisms (fungi and bacteria; secondary decomposers). Similarly, in each of the major higher taxonomic groups of predators, including Araneida, Gamasida, Staphylinidae and Chilopoda, Formicidae there appear to be species of the first and second predator level. In both decomposers and predators, generalist feeders predominate and at least in species rich natural systems, such as forests, it appears premature to draw links between certain species; knowledge on food relationships simply is too incomplete. Considering these limitations, we call for new food web models, which incorporate these uncertainties, possibly a statistical rather than an explicit representation of trophic species and links has to be employed.

DISCUSSION AND CONCLUSIONS

In conclusion, at present there are enough opportunities to build dynamic soil food web models, which are able to depict and predict the dynamics and control of invertebrate populations. However, we call for a closer integration of experimentation and modelling to allow for model verification. The new model concept can be extended with mechanisms that appear important from the conducted experiments. Until now, only a few studies have verified the equilibrium DFW model against overall ecosystem fluxes (de Ruiter et al., 1994).

In contrast, the dynamical model does not start with observed biomasses and therefore verification is even more important. This should not only be based on overall ecosystem fluxes but on the observed dynamics of individual populations. Other independent estimates of model parameters will depend on what kinds of experiments or observations can be carried out. For example, Cousins et al. (see Chapter 5.5) shows that measurements of the rate of energy release by decomposition in different pools of organic matter give more information than the energy content of these pools. These measurements can be used to parameterize or verify the feeding rates of the DFW model. Consideration of experimental results in food web models will allow a better integration of interactions in belowground communities into food web models and improve their acceptance by the community of soil ecologists, thereby achieving the ability to build more general soil food web models in future.

6.0 | FOOD WEBS, BIODIVERSITY, AND ECOSYSTEM FUNCTIONING

Peter J. Morin

The diverse chapters in this section of the book share the theme of exploring various aspects of ecosystem functioning in food webs of different complexity. The key common features of these contributions are the recognition that changes in diversity across ecosystems will almost certainly be distributed across multiple trophic levels, and that the interactions among species on those different trophic levels greatly complicate the prediction of possible consequences for ecosystem functioning. This focus contrasts with the prevailing tendency for recent models and empirical studies to consider diversity and functioning patterns within single trophic levels, usually primary producers. There is of course a long history of interest in how the properties of food webs can influence certain aspects of community and ecosystem functioning. Some of this traces back to the venerable stability-complexity debate (MacArthur 1955; Elton 1958; May 1972; Pimm 1991), a topic that is revisited and extended by Dell et al. (see Chapter 8.1). Dell et al. show how nonrandom patterns in food webs may contribute to system-wide stability. Other contributions use models to explore how relatively small (see Fox, Chapter 6.2) or larger (see Loreau and Thebault, Chapter 6.1) changes in the diversity of coupled consumer-victim systems might depart from the predictions made by models that focus on diversity change in systems consisting of a single trophic level. In these cases, the function of convenience and interest is biomass production, although effects on other processes can certainly be imagined. Other studies experimentally explore the consequences of variation in diversity in aquatic systems while focusing on either invasibility (see Beisner and Romanuk, Chapter 6.5) or total system productivity (see Downing and Wootton, Chapter 6.3), while exploring the possibility of using position in the food web as a predictor of ecosystem responses to species loss or gain. Setälä (see Chapter 6.4) also provides similar insights about functioning in the context of complex soil food webs. Petchey et al. (see Chapter 6.6) provide a preliminary exploration of the ability to predict ecosystem consequences from aggregate properties of food webs, in this case using a measure of functional diversity (as in Petchey and Gaston, 2002) based on food web linkages.

It is unclear why studies of food webs (Polis and Winemiller, 1996) and biodiversity and ecosystem functioning (Loreau et al., 2002) have proceeded along largely separate lines, but there are probably some practical reasons for the oversight. Many prominent studies of biodiversity and ecosystem functioning focus on plants (Tilman et al., 1996; Hector et al., 1999; Mulder et al., 2001), and occur outside of an explicit food web context. This is due partly to the relative ease of conducting experimental diversity studies with plants, and also reflects the relative simplicity of explaining these results with mathematical models. Experiments and models involving multiple trophic levels, let alone entire food webs, become dauntingly complex. Nonetheless, we can hazard a few guesses about how consumers will influence biodiversity and functioning across entire food webs.

Experiments in a number of systems show that predators can either increase or decrease diversity across the entire community (Paine, 1966; Lubchenco, 1978; Tansley and Adamson, 1925). It is tempting to speculate that such predator-mediated diversity patterns will have consequences for diversity-dependent functioning. However, Paine (2002) has recently pointed out that generalizations about plant diversity and functioning (specifically productivity) made in terrestrial systems may not readily apply to marine systems, where the most productive communities consist essentially of macro-algal monocultures. In these systems, consumers increase producer diversity but decrease productivity. Different studies of grazers in terrestrial systems show that consumers can augment plant diversity and productivity in grasslands (McNaughton 1977, 1979). It seems clear that interactions within food webs can influence productivity and other aspects of ecosystem functioning, though the direction of those effects may be strongly system-specific. Effects of other sorts of interactions on functioning within complex webs of interacting species are far from clear, although the mutualists that are frequently ignored in depictions of food webs also appear to have important effects (van der Heiden et al., 1998).

Consequences of increased diversity within higher trophic levels are more difficult to predict, partly because of the complex ways that consumers can interact. Remarkably few empirical studies have compared the separate and aggregate impacts of different consumers on prey diversity (Morin, 1995), and even fewer studies have explored the consequences of multiple consumers on functioning. A number of recent studies have manipulated diversity across multiple trophic levels, but these studies cannot separate effects of consumers and primary producers because of design constraints. Nonetheless, these studies often show diversity-functioning effects that differ from those observed in studies of terrestrial plants (Naeem and Li, 1997; McGrady-Steed et al., 1997; Hulot et al., 2000; Downing and Leibold, 2002). These findings mirror the results of theoretical work (see Fox, Chapter 6.2; Loreau and Thebault, Chapter 6.1), which shows that effects of increasing consumer diversity depend critically on assumptions about the extent of consumer specialization.

Although many empirical studies focus on the impacts of single consumers, most natural systems are replete with many consumer species that can interact in a variety of ways. Part of the challenge in developing predictions about how consumer diversity will influence ecosystem functioning lies in predicting how multiple consumers will affect prey assemblages (Sih et al., 1998). Both interference among consumers and the conflicting demands of mounting defenses against different kinds of consumers can lead to emergent effects that are not readily predictable from the traits of individual consumer species. Despite these challenges, the chapters in this section show that links between food web dynamics and ecosystem functioning are a promising and necessary area for much new research.

6.1 | FOOD WEBS AND THE RELATIONSHIP BETWEEN BIODIVERSITY AND ECOSYSTEM FUNCTIONING

Michel Loreau and Elisa Thébault

BIODIVERSITY AND ECOSYSTEM FUNCTIONING: A NEW PARADIGM

The relationship between biodiversity and ecosystem functioning has emerged as a new research area at the interface between community ecology and ecosystem ecology which has expanded dramatically during the last few years (see syntheses in Loreau et al., 2001; Kinzig et al., 2002; Loreau et al., 2002). This new area finds its origin in a questioning that started only about a dozen years ago on the potential consequences of biodiversity loss which results from the increasing human domination of natural ecosystems, a domination that is likely to further develop considerably during the twenty-first century.

Three types of reasons have been put forward to justify current concerns about threats to biodiversity. First, biodiversity provides us with a number of natural resources that lead to the production of use values, whether as food, new pharmaceuticals, genes that improve crops, or organisms that perform biological control. Second, it is intricately linked to human well-being for aesthetic, ethical, cultural and scientific reasons. Third, it contributes to the provision of ecosystem services that are generally not accounted for in economic terms, such as primary and secondary production, plant pollination, climate regulation, carbon sequestration, the maintenance of water quality, and the maintenance of soil fertility. It is this third category of potential impacts of biodiversity which gave rise to the emergence of the biodiversity and ecosystem functioning area: could biodiversity loss alter the functioning of ecosystems, and thereby the ecological services they provide to humans?

When this question was posed in the early 1990s, scientific ecology had a number of theories and empirical data that clearly showed the importance of "vertical" diversity (i.e., functional diversity across trophic levels along the food chain) in ecosystems. An eloquent example of the dramatic impacts that changes in vertical diversity can have is provided by the kelp—sea urchin—sea otter food chain in the Pacific. Removal of sea otters by Russian fur traders allowed a population explosion of sea urchins that overgrazed kelp (Estes and Palmisano, 1974). Reduction in kelp cover in turn leads to extinction of other species living in kelp, as well as increased wave action, coastal erosion, and storm damage (Mork, 1996). More intense herbivory in the absence of sea otters has also been shown to trigger evolution of chemical defences in kelp (Steinberg et al., 1995). Thus, removal of a single top predator generates a cascade of population dynamical, physical, and even evolutionary effects within ecosystems.

In contrast, little was known on the ecological significance of "horizontal" diversity (i.e., taxonomic and functional diversity within trophic levels). Different theories of coexistence among competing species have vastly different implications for the relationship between species diversity and ecosystem processes. To take two extreme examples, neutral theory assumes that all species in a community are equivalent (Hubbell, 2001). This implies functional redundancy among species, and hence an absence of any effect of changes in diversity on aggregate community or ecosystem properties. At the other extreme, niche theory postulates that all species differ to some extent in the resources they use. This implies functional complementarity among species, and hence increased productivity and other ecosystem processes with diversity (Tilman et al., 1997; Loreau, 1998a).

To investigate the effects of "horizontal" diversity on ecosystem processes, a new wave of experimental studies was developed using synthesised model ecosystems. Many of these studies were focused on effects of plant taxonomic and functional-group diversity on primary production and nutrient retention in grassland ecosystems. Because plants, as primary producers, represent the basal component of most ecosystems, they represented the logical place to begin detailed studies. Several, though not all, experiments using randomly assembled communities found that plant species and functional-group richness has a positive effect on primary production and nutrient retention (Tilman et al., 1996; Hector et al., 1999). Although the interpretation of these experiments was much debated (Huston, 1997; Hector et al., 2000; Huston et al., 2000), this controversy has been largely resolved by a combination of a consensus agreement on a common conceptual framework (Loreau et al., 2001), the development of a new methodology to partition selection and complementarity effects (Loreau and Hector, 2001), and new experimental data (Tilman et al., 2001; Van Ruijven and Berendse, 2003). These new studies all showed that plant diversity influences primary production through a complementarity effect generated by niche differentiation and facilitation. Thus, there is little doubt that species diversity does affect at least some ecosystem processes, even at the small spatial and temporal scales considered in recent experiments.

Even when high diversity is not critical for maintaining ecosystem processes under constant or benign environmental conditions, however, it might nevertheless be important for maintaining them under changing conditions. The insurance hypothesis proposes that biodiversity provides an "insurance," or a buffer, against environmental fluctuations because different species respond differently to these fluctuations, leading to functional compensations between species and hence more predictable aggregate community or ecosystem properties (McNaughton, 1977; Yachi and Loreau, 1999). A number of studies have recently provided theoretical foundations for this hypothesis (Doak et al., 1998; Yachi and Loreau, 1999; Lehman and Tilman, 2000). Several empirical studies have found decreased variability of ecosystem processes as diversity increases, despite sometimes increased variability of individual populations, in agreement with the insurance hypothesis (Tilman, 1996; McGrady-Steed et al., 1997). The interpretation of these patterns, however, is complicated by the correlation of additional factors with species richness in these experiments, which does not fully preclude alternative interpretations (Huston, 1997).

Virtually all of these recent theoretical and experimental studies on the effects on biodiversity on ecosystem functioning and stability have concerned single trophic levels, primary producers for the most part. Although they have contributed to merging community and ecosystem ecology, they have unintentionally disconnected "vertical" and "horizontal" diversity and processes. Yet trophic interactions can have important effects on the biomass and productivity of the various trophic levels (Abrams, 1993) as well as on ecosystem stability (MacArthur, 1955; May, 1973). An important current challenge is to understand how trophic interactions affect the relationship between biodiversity and ecosystem functioning. A few recent experiments have started to investigate biodiversity and ecosystem functioning in multitrophic systems (Naeem et al., 2000a; Downing and Leibold, 2002; Duffy et al., 2003). Theory, however, is sorely missing on these issues. The new biodiversity–ecosystem functioning area has developed independently of classical food web theory, while food web theory has largely ignored recent developments on biodiversity and ecosystem functioning. We suggest that it is high time to lay a bridge between these two approaches to foster cross-fertilization and build a broader theoretical framework that has greater relevance to natural ecosystems.

FOOD WEB CONSTRAINTS ON THE RELATIONSHIP BETWEEN BIODIVERSITY AND ECOSYSTEM FUNCTIONING

To meet this challenge, we have developed an ecosystem model which allows us to analyze the impacts of food web structure on both the relationship between species diversity and total biomass at the various trophic levels and the relationship between species diversity and the temporal variability of these biomasses (Thébault and Loreau, 2003). This model is an extension of the model proposed by Loreau (1996) for a nutrient-limited ecosystem containing an arbitrary number of plants and specialised herbivores in a heterogeneous environment. In this model, plant nutrient uptake is assumed to decrease the soil concentration of a limiting nutrient in the immediate vicinity of the rooting system, thus creating a local resource depletion zone around each plant and allowing plant coexistence under some conditions. Here we generalize it by allowing herbivores to be generalists, but we do not consider food web configurations that include carnivores for simplicity. We present results for total plant and herbivore biomasses as ecosystem properties for comparison with experimental studies. Note that total plant biomass may be a poor approximation of primary production when consumed by higher trophic levels, but total biomasses at producer and consumer trophic levels are often used for convenience as ecosystem properties in experiments on the relationship between species diversity and the functioning of both single-trophic-level and multitrophic ecosystems (Duffy et al., 2003; Finke and Denno, 2004). In the scenarios analyzed later (in which all plants have equal nutrient loss rates), it can further be shown that primary production is simply proportional to total ecosystem biomass (i.e., the sum of total plant and herbivore biomasses). Total ecosystem biomass can thus be used as a surrogate for primary production here.

We first examine how changes in species richness influence ecosystem properties at equilibrium for different food web structures when all plant and herbivore species can coexist (Thébault and Loreau, 2003). Different scenarios of biodiversity changes are considered: either plant species richness and herbivore species richness vary in parallel, or herbivore species richness varies alone. Changing plant richness only leads to unfeasible food web configurations in our model because there cannot be more herbivore species than plant species at equilibrium. To analyze expected ecosystem responses to changes in species richness, we calculate, at each diversity level, the mean of plant and herbivore biomass across random compositional assemblages, as is often done in experiments (Tilman, 1996; Hector et al., 1999; Tilman et al., 2001).

Effect of Top-Down Control

When each plant is consumed by a specialist herbivore, mean total plant biomass increases linearly with species richness (Figure 1). Each plant species is then controlled by its own herbivore and is unaffected by the addition of other species, which causes this linear increase in total plant biomass. The corresponding complementarity effect (as defined by Loreau and Hector, 2001) is positive while there is no selection effect. However, this complementarity is generated by a very different mechanism than in simple competitive systems: it does not arise here from resource partitioning or facilitation, but from top-down control by different species of the upper trophic level.

In contrast, when the same food web comprises an inedible plant, mean total plant biomass does not increase linearly and can even decrease at high diversity (see Figure 1). In the latter case, the biomass of the inedible plant is controlled by resource availability which decreases as plant richness increases. This also leads to a negative selection effect because the inedible plant is dominant but is most affected by an increase in diversity.

In both cases, total herbivore biomass can show complex relationships with diversity. In the scenarios examined in Figure 1, it decreases at high diversity when all plants are edible, and it also decreases at intermediate species richness when the food web comprises an inedible plant. Total ecosystem biomass, however, and hence also primary production, increases monotonically with diversity until saturation, just as in systems with a single trophic level (Loreau, 1998a).

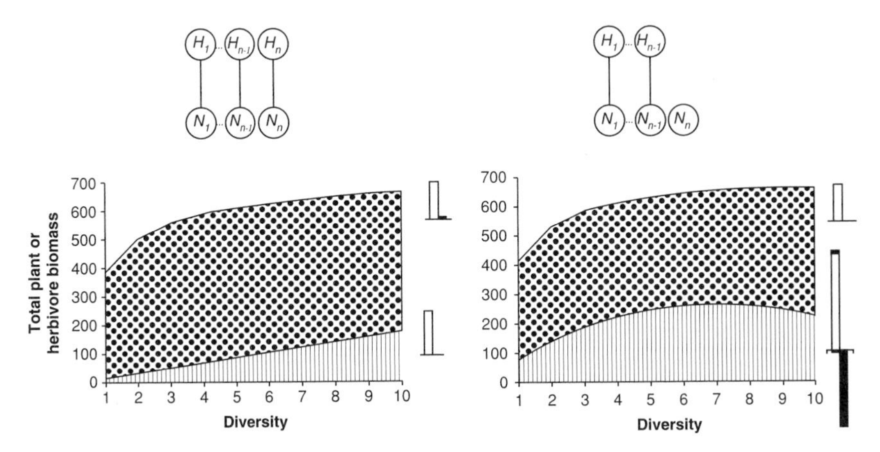

FIGURE 1 | Expected total plant biomass (hatched area) and total herbivore biomass (dotted area) as functions of plant species richness for a food web with specialist herbivores and no inedible plant (left panel) or one inedible plant (right panel). Herbivore species richness varies parallel to plant species richness to keep the same food web configuration along the diversity gradient. Top panels show the food web configurations analyzed in the corresponding column. N_i and H_i denote plant and herbivore species *i*, respectively. Small histograms on the right of the panels show the strengths of the complementarity effect (in white) and the selection effect (in black) for the highest diversity treatment (10 plant species). These effects are measured on the same scale as the *y*-axis, except for plant biomass in case of a food web with one inedible plant, where they are reduced by a factor of 2.

Thus, the nature of population controls (top-down vs. bottom-up) in an ecosystem can profoundly affect ecosystem responses to changes in species richness. Heterogeneity within trophic levels and the presence of inedible species are important to consider as they modify top-down control and trophic cascades in food webs (Leibold, 1989; Abrams, 1993).

Effect of Food Web Connectivity

The degree of generalization or specialization of herbivores, which is a measure of food web connectivity, also has a strong impact on the relationship between diversity and ecosystem processes (Figure 2, top). When herbivores are generalists, mean total plant biomass does not increase linearly anymore and can even decrease at high diversity levels. In this case, the biomass of each plant species is still controlled by herbivores but it decreases with the addition of other herbivore species because plant consumption increases. This in turn can result in a decrease in total plant biomass. Mean total herbivore biomass is generally higher when herbivores are generalists than when they are specialists but it also increases less with diversity and can decrease at high diversity.

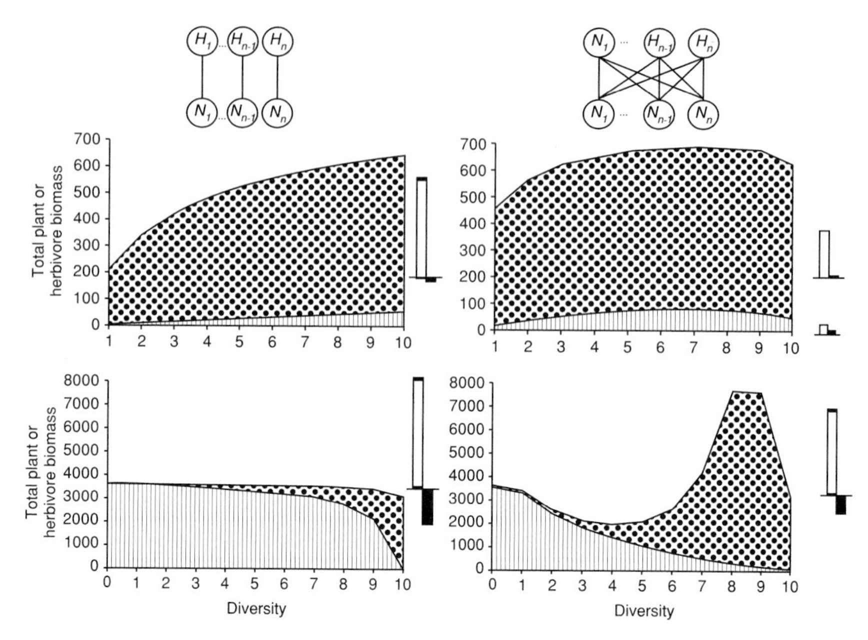

FIGURE 2 | Expected total plant biomass (hatched area) and total herbivore biomass (dotted area) as functions of species richness for a food web with specialist herbivores (left panels) and a food web with generalist herbivores (right panels). Species richness varies either at the two trophic levels simultaneously (top panels) or at the herbivore trophic level only while plant species richness is held constant at 10 species (bottom panels). Small histograms on the right of panels show the strengths of the complementarity effect (in white) and the selection effect (in black), measured on the same scale as the *y*-axis, for the highest diversity treatment (10 herbivore species).

Competition between generalist herbivores is important and resource-use complementarity is lower, as indicated by the smaller complementarity effect. Total ecosystem biomass, and hence primary production, can also decrease at high diversity when herbivores are generalists.

Effect of Trophic Position of Species Loss

Changes in diversity at the consumer trophic level alone (see Figure 2, bottom) have very different effects than do simultaneous changes at both the plant and herbivore trophic levels (see Figure 1, Figure 2, top). In this scenario, mean total plant biomass always decreases upon herbivore addition, whether herbivores are specialists or generalists. These results are consistent with recent experiments on the effects of consumer diversity (Duffy et al., 2003). However, it decreases faster at low diversity when herbivores are generalists because consumption of each plant is higher. Mean total herbivore biomass always increases with diversity when herbivores are specialists, but it can strongly decrease at high diversity when herbivores are generalists (see Figure 2, bottom). Again, when herbivores are generalists, resource-use complementarity is smaller, which can result in a decrease in total herbivore biomass at high diversity and a smaller complementarity effect. In both cases, total ecosystem biomass and primary production can decrease at high diversity and show complex relationships with diversity.

Thus, biomass variations depend strongly on the trophic position of species loss and on herbivore generalism. The strong increase in total herbivore biomass at intermediate diversity levels when herbivores are generalists may be explained by a strong increase in herbivore consumption together with more favorable conditions for herbivore-mediated plant coexistence. The mechanism that maintains coexistence is an important determinant of the relationship between diversity and ecosystem functioning, as suggested by a few earlier studies (Mouquet et al., 2002; Fox, 2003).

FOOD WEB CONSTRAINTS ON THE RELATIONSHIP BETWEEN BIODIVERSITY AND ECOSYSTEM STABILITY

So far we have focused on the equilibrium level of aggregate ecosystem properties in a constant environment. We now shift to the stability—as assessed by the ability to reduce temporal variability—of these properties in a fluctuating environment. In our model, we include environmental fluctuations in the form of sinusoidal variations in temperature or other abiotic variables on which plant growth rates and herbivore consumption rates depend. We assume that these rates have a Gaussian dependence on temperature with identical standard deviations. Temporal niche differentiation among species is then determined by their overlap in these responses to temperature variations. Three scenarios of niche differentiation are considered: (1) no niche differentiation, complete niche overlap: the niches of all species are centered on the average temperature experienced by the system; (2) intermediate niche differentiation and overlap: the means of the Gaussian responses are regularly distributed over half the temperature gradient; (3) high niche differentiation, low niche overlap: the means of the Gaussian responses are regularly distributed over the whole temperature gradient. We also study three food web configurations: (1) herbivores are strict specialists; (2) herbivores are generalists and compensate plant species loss by increasing their consumption on other species; (3) herbivores are generalists and their consumption rates on the various plant species do not depend on plant diversity, such that there is no trade-off among their abilities to consume different plants. For each scenario and configuration, we report the results from numerical simulations of the model for the coefficient of temporal variation (CV)—a common standardized measure of variability (Doak et al., 1998; Lehman and Tilman, 2000)—of both the biomasses of individual species and the total biomass at each trophic level. Analytical results will be presented elsewhere (Thébault and Loreau, 2005).

Effect of a Trade-off Among Herbivore Consumption Rates

There are significant differences in the relationship between species diversity and temporal variability between the case where herbivores are generalists without trade-off and the other configurations (Figure 3). In the former configuration, the CVs of total and species-specific biomasses at both the plant and herbivore trophic levels increase strongly with species richness (see Figure 3). When herbivores are either specialists or generalists with a trade-off, CVs vary far less with diversity. In both configurations, the relationships between diversity and stability have similar forms. In plants, the CV of total biomass decreases with diversity whereas the CV of individual species biomasses increases (see Figure 3). In herbivores, the CV of total biomass also decreases with species richness but the CV of individual species biomasses depends on niche differentiation and can either increase or decrease with diversity (see Figure 3).

Thus, the classical result that the stability of individual populations decreases as diversity increases (May, 1973; Tilman, 1996) holds generally. This destabilizing effect of diversity is counteracted for aggregate

→

FIGURE 3 | Coefficient of variation of species-specific (in grey) and total (in black) plant and herbivore biomass as functions of plant species richness for different food web configurations. Herbivore species richness varies parallel to plant species richness to keep the same food web configuration along the diversity gradient. Plant and herbivore species can have different degrees of temporal niche differentiation: no niche differentiation (solid lines), intermediate niche differentiation (dotted lines with full triangles or diamonds), and high niche differentiation (dotted lines with open triangles and diamonds). Left panels: food web with specialist herbivores; middle panels: food web with generalist herbivores with a trade-off among consumption rates; right panels: food web configuration with generalist herbivores without this trade-off.

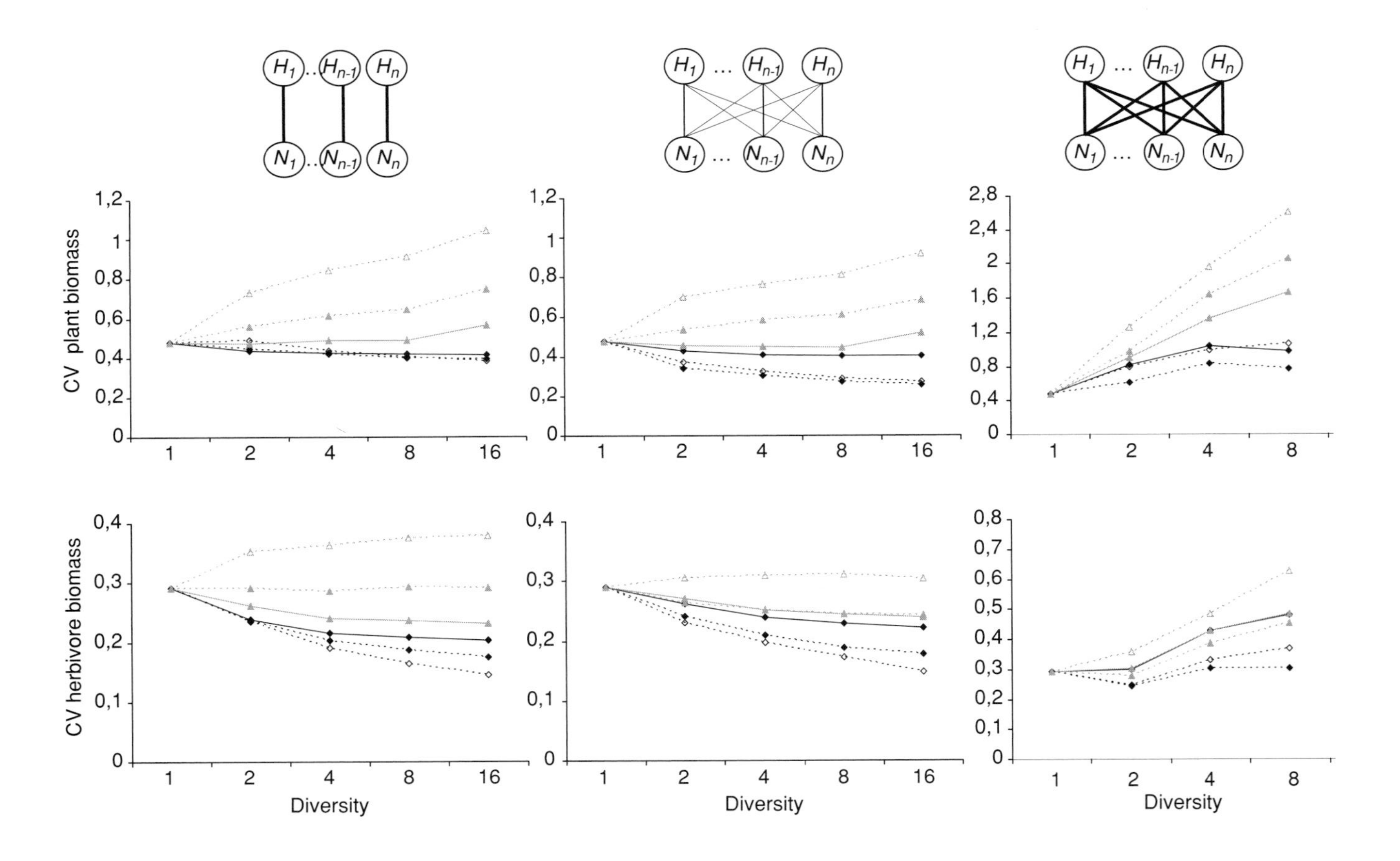
CV plant biomass
CV herbivore biomass
Diversity
Diversity
Diversity

ecosystem properties in our model when mean herbivore consumption rates do not increase with diversity. If herbivore generalism has no cost, however (a questionable assumption), the stability of total biomass also decreases with diversity because the mean interaction strength increases. Our model thus confirms the classical result that stability requires low interaction strength (Ives et al., 2000; Kokkoris et al., 2002).

Effect of Herbivore Generalism when There Is a Trade-off among Herbivore Consumption Rates

Despite strong similarities between the effects of species diversity on temporal variability in the configurations where herbivores are either specialists or generalists with a trade-off, food web connectivity has an influence on temporal variability and the strength of diversity effects. Herbivore generalism has a stabilizing effect on plant and herbivore individual species when there is niche differentiation: the CV of species-specific biomasses increases less with diversity when herbivores are generalists (see Figure 3). Individual species are then stabilized by a lower synchrony in prey availability (Petchey, 2000) and predator consumption. These results support the old idea that communities with many interacting species can be less prone to large fluctuations than communities with fewer species (MacArthur, 1955). This stabilizing effect of generalism results in a strong decrease in the variability of total plant biomass with diversity, but a somewhat weaker decrease in the variability of total herbivore biomass. Food web connectivity increases the variability of individual herbivore species when these have identical niches. This destabilizing effect might be due to increased competition among herbivores.

Effect of Niche Differentiation

Niche differentiation has similar effects in all food web configurations: the CVs of total plant and herbivore biomasses decreases more, or increases less, with diversity when plants and herbivores respond differently to environmental fluctuations (see Figure 3). At the individual species level, however, CVs generally increase more when niche differentiation is greater (see Figure 3). Thus, the insurance effect of biodiversity at the ecosystem level is partly generated by temporal complementarity among species in our system. However, even when species respond identically to environmental changes, the temporal variability of total plant and herbivore biomasses decreases with diversity if herbivores are specialists or generalists with a trade-off. This pattern differs from predictions for purely competitive systems, in which in theory a decrease in variability can occur only when species responses to environmental fluctuations differ (Ives and Hughes, 2002).

DISCUSSION AND CONCLUSIONS

Recent theoretical and experimental work provides clear evidence that biodiversity loss can have profound impacts on functioning of natural and managed ecosystems and their ability to deliver ecological services to human societies. Work on simplified ecosystems in which the diversity of a single trophic level—mostly plants—is manipulated shows that taxonomic and functional diversity can enhance ecosystem processes such as primary productivity and nutrient retention. Theory also strongly suggests that biodiversity can act as biological insurance against potential disruptions caused by environmental changes.

One of the major current challenges, however, is to extend this new knowledge to multitrophic systems that more closely mimic complex natural ecosystems. This requires merging biodiversity–ecosystem functioning and food web theories. We make a first step in that direction in this chapter. Our new theoretical work shows clearly that trophic interactions have a strong impact on the relationships between diversity and ecosystem functioning, whether the ecosystem property considered be total biomass or temporal variability of biomass at the various trophic levels. In both cases, food web structure and trade-offs that affect interaction strength have major effects on these relationships.

Multitrophic interactions are expected to make biodiversity–ecosystem functioning relationships more complex and nonlinear, in contrast to the monotonic changes predicted for simplified systems with a single trophic level. Furthermore, these relationships depend strongly on the ecosystem properties considered; total plant biomass, total herbivore biomass, and primary production can show very different patterns as diversity varies. These relationships, however, are predictable provided environmental conditions and food web structure are known. It is also remarkable that, despite the complexity generated by trophic interactions, biodiversity should still act as biological insurance for ecosystem processes against environmental fluctuations in multitrophic systems, except when consumers are generalists and generalism has no cost. The relationships between diversity and ecosystem variability also seem to be less dependent on the ecosystem properties considered since the variability of primary and secondary production shows the same trends as the variability of total plant and herbivore biomass (results not shown) as diversity varies.

Our model makes several predictions that would deserve to be tested experimentally to gain better knowledge of the impacts of biodiversity changes on ecosystem functioning. Merging food web and biodiversity–ecosystem functioning approaches is an exciting challenge that offers promising perspectives in both areas.

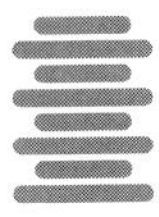

6.2 | BIODIVERSITY, FOOD WEB STRUCTURE, AND THE PARTITIONING OF BIOMASS WITHIN AND AMONG TROPHIC LEVELS

Jeremy W. Fox

Two of the most fundamental properties of an ecological community are diversity and total biomass. Classical community ecology treats questions about diversity (defined here as species richness) as questions about the rules that govern how the total biomass of a community or assemblage is partitioned among species (Tilman, 1982; Holt et al., 1994). Diversity and biomass are typically linked through the assumption that both total biomass, and the rules that govern the partitioning of that biomass, reflect the availability of limiting resources (Holt et al., 1994). In contrast, much recent work treats diversity as an independent rather than a dependent variable (Loreau et al., 2001). On this view, resource availability may determine the potential biomass a system can support, but species diversity and composition determine realized biomass. This view also emphasizes the fact that total biomass within one or more trophic levels is often correlated with the rate or level of ecosystem functions like primary and secondary productivity, CO_2 flux, and nutrient retention (Naeem et al., 1994; McGrady-Steed et al., 1997). Understanding the relationship between diversity and total biomass therefore has practical as well as fundamental implications.

Experimentally reducing plant diversity often reduces total plant biomass, productivity, and nutrient retention (reviewed in Loreau et al., 2001). Often, but not always: results may depend on the composition of the heterotrophic community (Mulder et al., 1999). Further, changes in plant biomass due to the removal or altered composition of higher trophic levels often exceed the changes in plant biomass produced by experimental manipulations of plant diversity (Duffy, 2003). Since animals are often at greater risk of extinction than plants due to human activities (Jackson et al., 2001), loss of animal diversity may be the greatest risk to ecosystem function in many systems.

A simple food web model is used here to predict the joint effects of loss of diversity from each of two adjacent trophic levels ("prey" and "predators") on partitioning of biomass among trophic levels. Does loss of species from one trophic level necessarily reduce total biomass within that level, or increase total biomass on the next-lower level? What are the consequences of correlated losses of diversity across multiple trophic levels? The model I use to answer these questions is obviously too simple to be a quantitatively realistic description of any particular system, but simple, analytically tractable models can sharpen our intuitions, provide general insights, and help us understand the behavior of more elaborate models. Further, experience shows that the ability of simple models to correctly predict qualitative features of the behavior of natural systems should not be discounted, a point to which I return later. My approach is complementary to models that allow examination of a greater range of diversity levels, at the cost of examining a more limited range of food web structures, parameter values, and extinction scenarios (Thébault and Loreau, 2003; see Loreau and Thébault, Chapter 5.1).

THE MODEL

The dynamics of three prey species are considered, N_i (i=1,2,3), competing for a shared limiting inorganic resource R and consumed by two predators, P_1 and P_2 (Figure 1). Predator 1 is a specialist on prey species 1, while predator 2 is a generalist consuming prey species 1 and 2 when they are available. Previous work (Fox, 2003, 2004b) considered webs with only prey species 1 and 2; here I build on this work by considering a web with a third, inedible prey species. The model is defined by

$$\frac{dR}{dt} = D(S-R) - R(u_1 N_1 + u_2 N_2 + u_3 N_3) \tag{1a}$$

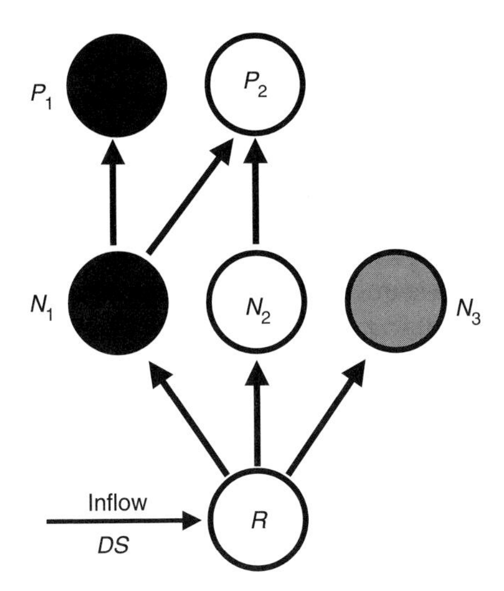

FIGURE 1 | Food web of equation (1a-f). Labeling, and shading of circles, identify species within each of the top two trophic levels. Resources flow into the system at rate *DS*. Flow out of all compartments (circles) occurs at per-unit rate *D* (not shown).

$$\frac{dN_1}{dt} = N_1(u_1c_1R - D - f_{11}P_1 - f_{12}P_2) \tag{1b}$$

$$\frac{dN_2}{dt} = N_2(u_2c_2R - D - f_{22}e_{22}P) \tag{1c}$$

$$\frac{dN_3}{dt} = N_3(u_3c_3R - D) \tag{1d}$$

$$\frac{dP_1}{dt} = P_1(f_{11}e_{11}N_1 - D) \tag{1e}$$

$$\frac{dP_2}{dt} = P_2(f_{12}e_{12}N_1 + f_{22}e_{22}N_2 - D) \tag{1f}$$

where u_i is the per-unit resource uptake rate of prey species i, c_i is the conversion efficiency of prey species i, f_{ij} is the per-unit feeding rate of predator j on prey i, and e_{ij} is the conversion efficiency of predator j on prey i. The resource is supplied from an external source in chemostat fashion at incoming concentration S and dilution rate D, and both resource and organisms leave the system at the same rate. A chemostat model with linear functional responses is used for the sake of simplicity; the likely effects of allowing each species its own density-independent loss rate are discussed later.

Attention is restricted to that portion of parameter space that allows all species to coexist stably at equilibrium, if they are present initially. It is therefore assumed that both prey and predators exhibit appropriate niche differences. Stable coexistence of prey reflects a trade-off between competitive ability and predation resistance or tolerance: prey species 1 would outcompete the others in the absence of predators, and prey species 2 would outcompete 3 (i.e., it is assumed $u_1c_1>u_2c_2>u_3c_3$, $u_2 f_{12}e_{12}>u_1 f_{22}e_{22}$, and $f_{22}(u_1c_1-1)>f_{12}(u_2c_2-1)$). Coexistence of predators reflects differential use of prey: the specialist predator 1 would outcompete the generalist predator 2 if only prey species 1 were present ($f_{11}e_{11}>f_{12}e_{12}$). Many examples of such trade-offs and differences in resource use are known (Chase et al., 2000a; Norberg, 2000). Coexistence of all species also requires an intermediate enrichment (S) level. Fox (2004b) describes constraints on enrichment levels for webs lacking prey species 3; incorporating species 3 implies three additional constraints:

$$S > \frac{D}{u_3c_3}\left[1+\frac{u_2}{f_{22}e_{22}}\right] \tag{2a}$$

$$S > \frac{D}{u_3c_3}\left[1+\frac{u_1}{f_{1j}e_{1j}}\right] \tag{2b}$$

$$S > \frac{D}{u_3c_3}\left[\frac{f_{11}e_{11}(f_{22}e_{22}+1)+u_1f_{22}e_{22}-u_2f_{12}e_{12}}{f_{11}e_{11}f_{22}e_{22}}\right] \tag{2c}$$

where j=1,2.

By restricting attention to the parameter values that allow coexistence, we can to use analytical model solutions to predict the results of a hypothetical experiment in which the experimenter measures long-term (equilibrial) total predator biomass ($=\sum_j P_j^*$, where asterisks denote equilibrial values of state variables), total prey biomass ($=\sum_i N_i^*$), and resource level ($=R^*$) in the most diverse web, and in all possible subsets of that web. This experiment reveals the long-term effects of every possible loss of one or more species from the original, maximally diverse food web. Note that some possible subsets are always infeasible and collapse to simpler webs. For instance, a web with both predators but only prey species 1 will lose predator 2 to competitive exclusion (Fox, 2004b).

The effects of changes in species composition on ecosystem-level properties like total biomass fall into two categories. Substitution effects arise when one species is replaced by another that has the same prey and predators, so that there is no change in the food web topology. This results only in a quantitative change in the model parameters, with a corresponding change in the equilibrial biomasses of one or more species. The substitution effect is essentially synonymous with the more familiar "selection effect" (Loreau and Hector, 2001), and I will use the two terms interchangeably. Structural effects arise when adding, removing, and/or replacing species changes the topological structure of the food web. Holt and Loreau (2002) analyze selection/substitution effects in a simple food chain model. Here I examine the relative importance of selection and structural effects.

Note that structural effects are not synonymous with "niche complementarity" (sensu Loreau and Hector, 2001). My model does not consider many of the niche differences that can allow species to coexist, and therefore have complementary effects on ecosystem properties. More importantly, "niche complementarity" typically refers to niche differences among species within the same trophic level. Many structural changes have complex effects on total biomass that cannot be attributed solely to "niche complementarity" among species within any one trophic level.

MODEL PREDICTIONS

Because I consider a chemostat model, the total amount of resource in the system at equilibrium is fixed at S (Grover, 1997). Changes in species composition only affect how the total amount of resource is partitioned among species and the free resource pool R, and the fraction of species' biomass comprised of the limiting resource. Generating predictions about how changes in species diversity and composition affect total biomass within each trophic level amounts to determining how the resources bound within one or more species at equilibrium are redistributed when those species are removed.

I first consider effects of diversity and species composition on equilibrial resource level R^* (Figure 2). Structural effects are more important than selection effects in determining equilibrial resource level. For any set of parameter values within the region of parameter space considered here, food webs with an uneaten prey species maintain lower resource levels than webs without uneaten prey (a structural difference; see Figure 2). Considering only the webs with uneaten prey, a selection effect determines the rank-order of resource level among webs: the more competitive the uneaten species, the lower the resource level (see Figure 2). Considering the webs with no uneaten species, reticulate food webs have resource levels intermediate between different linear food chains (see Figure 2). Therefore, both structural and selection effects drive variation in resource levels among webs lacking uneaten prey.

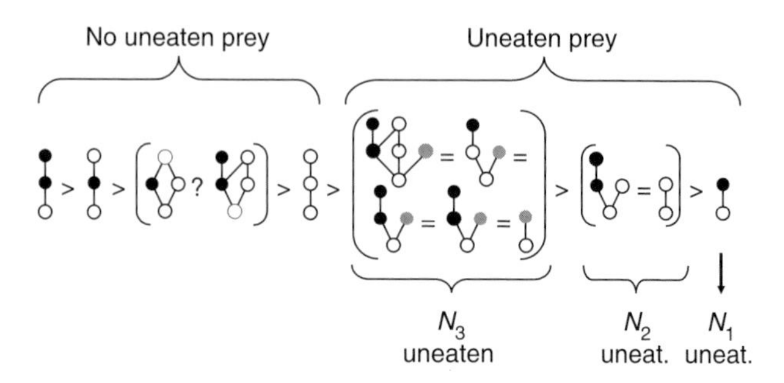

FIGURE 2 | Food web of equation (1a-f), and all stable, feasible subsets thereof, in rank order of equilibrial resource level, within the region of parameter space allowing all species to coexist. Braces above and below the webs respectively highlight structural and substitution effects on equilibrial resource level. A question mark indicates two webs whose rank order depends on parameter values. Within each trophic level, differently shaded circles represent different species, as in Figure 1.

Next I consider effects of diversity and species composition on equilibrial total prey biomass. To ease interpretation, I consider total prey biomass measured in resource units. This effectively sets all prey conversion efficiencies equal to 1, so that total prey biomass cannot vary due to variation in conversion efficiency among prey (see Fox, 2004b, for analyses relaxing this assumption). As with resource levels, structural effects are more important than selection effects in determining total prey biomass (Figure 3). Food chains lacking predators (i.e., a single prey species on its own) produce the highest total prey biomass, followed by linear food chains augmented with an uneaten prey species, followed by linear food chains and reticulate food webs. Within each of these four structural categories, selection effects operate. For instance, in the absence of predators, more competitive prey species produce higher total prey biomass (see Figure 3). Food chains augmented with an uneaten prey species produce higher total prey biomass, the better the average competitive ability of the two prey (see Figure 3). Counterintuitively, of the two augmented food chains with the same eaten and uneaten prey species (prey species 1 and 3, respectively), the chain with the more effective predator species 1 produces higher total prey biomass (see Figure 3). This result occurs because predator 1 is more effective at releasing prey species 3 from competition with prey species 1. Contrary to the predictions of models that only consider two trophic levels (Tilman et al., 1997), the rank order of webs in terms of total prey biomass is not simply the reverse of the ranking in terms of resource level (see Figure 2, 3). High biomass of prey is not necessarily

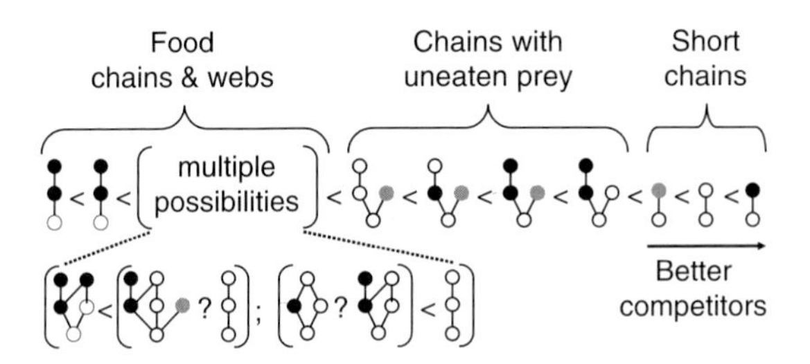

FIGURE 3 | Food web of equation (1a-f), and all stable, feasible subsets thereof, in rank order of total prey biomass measured in resource units, within the region of parameter space allowing all species to coexist. Text above and below the webs respectively highlights structural and substitution effects on equilibrial resource level. Symbols as in Figure 2. Webs are displayed in two groups to enable clear display of all the information on rank ordering that can be obtained analytically.

associated with low resource levels, because predator biomass also contains resources.

Analysis of the effects of diversity and food web structure on total predator biomass is algebraically challenging, even when predator biomass is measured in resource units. However, one common effect of structural change within the prey trophic level is to reduce total predator biomass (whatever the units of measurement). For example, augmenting a linear food chain with a less-edible or inedible prey species always reduces predator biomass (Grover, 1994; Holt et al., 1994; Leibold, 1996; Fox, 2003).

Three general predictions emerge from the theoretical analyses. First, and most importantly, the effects of changes in diversity (species richness) within a single trophic level depend on the species composition of other trophic levels. Removing a species from a trophic level does not always decrease total biomass within that level, or increase total biomass on the next-lower level, as suggested by models that consider only two trophic levels (Tilman et al., 1997). Indeed, it may be best not to express model predictions in terms of species richness at all, but in terms of structural and selection effects (see Figure 2, 3).

Second, the most diverse food webs (prey + predator richness) usually exhibit intermediate values of resource level, total prey biomass, and total predator biomass, whatever the units of measurement. In general, producing a stable, feasible, diverse food web requires satisfaction of numerous constraints, entailing a fine balancing of parameter values (Fox, 2004b). The extreme parameter values required to produce very high biomass for one species or trophic level are unlikely to be compatible with positive biomass for all other species. Experiments manipulating diversity in a correlated fashion across all trophic levels therefore are likely to find that the most diverse webs are intermediate in function to less-diverse webs. The utility of interpreting model behavior in terms of species richness is questionable.

Third, the constraints on parameter values (i.e., the trade-offs) that ensure coexistence strongly constrain the possible rank-orderings of webs in terms of equilibrial resource level or total prey biomass. The 14 feasible food webs considered here (see Figure 2, 3) could be ranked in 14! ($>8.7\times10^{10}$) different orders with respect to any given ecosystem-level property or function, barring ties. But $<<0.01\%$ of those rank-orders can occur within the region of parameter space that allows all species to coexist (see Figure 2, 3). This is not the case in single-trophic level models such as the Lotka-Volterra competition model, where assuming that species coexist only weakly constrains the relationship between species composition and ecosystem function.

How robust are these general predictions to changes in model assumptions? The most crucial assumptions are (i) the system contains at most five species, (ii) coexistence depends entirely on a fine balancing of parameters governing trophic interactions, (iii) all species, and the basal resource, experience the same density-independent loss rate, and (iv) resource inputs occur only in inorganic form (i.e., no immigration). All these assumptions are unrealistic: real communities contain many species, species coexist via non-trophic mechanisms (e.g., spatial refugia), ecosystems gain and lose resources through processes characterized by very different rates (e.g., atmospheric deposition, fixation of nutrients, leaching, water flow, animal movement), and ecosystems are open to immigration. Robustness has theoretical and empirical aspects. We can ask whether more realistic models make the same predictions as equation (1a-f). We can also ask whether real ecosystems behave in the way equation (1a-f) predicts. Next I discuss theoretical robustness; I discuss empirical robustness in the following section.

The general prediction that the ecosystem-level effects of species loss from one trophic level depend on the diversity and composition of other trophic levels should be highly robust. The species diversity and composition of one trophic level likely will affect the diversity, species composition, and biomass on other trophic levels, as long as species' functional and numerical responses take biologically reasonable forms. However, current theory focuses on the determinants of diversity within a single trophic level, or on the consequences of removing entire trophic levels (but see Grover, 1994; Steiner and Leibold, 2004). Understanding the connections between the diversity and composition of multiple trophic levels, and the consequences of these connections for ecosystem function, is an important area for future theoretical work.

The general prediction that diverse food webs function at an intermediate level is sensitive to assumptions (iii-iv). This prediction arises from the fact that the total amount of limiting resource in a chemostat at equilibrium is fixed, so that changes in food web structure merely redistribute the resource within the web. Relaxing assumption (iii) allows changes in species composition to change the total amount of resource in the system at equilibrium, as well as the partitioning of the resource within the system (de Mazancourt et al., 1998; de Mazancourt and Loreau, 2000). Relaxing assumption (iv) can allow implausible outcomes such as infinite biomass (Loreau and Holt, 2004), suggesting that we still have much to learn about how within-system demographic processes interact with between-system flows of organisms and materials. The sensitivity of this prediction to assumptions (i-ii) is more difficult to evaluate (but compare Figures 2 and 3 to Thébault and Loreau, 2003; see Loreau and Thébault, Chapter 5.1).

The general prediction that constraints on species coexistence strongly constrain the ecosystem-level effects of species composition obviously is sensitive to assumption (ii), but may be less sensitive to the other assumptions. For instance, allowing resource input and output (loss) rates to vary (assumptions iii-iv) would increase the number of free parameters, thereby increasing the number of possible orders in which webs could be ranked with respect to any given ecosystem property. But as long as species coexistence depends at least in part on trade-offs in resource use, and between resource use and predator resistance, the number of possible rank-orders should be relatively limited.

EXPERIMENTAL TESTS

The results of several published experiments are consistent with the model predictions, suggesting that equation (1a-f) is at least somewhat robust in an empirical sense. Mikola and Setälä (1998a) manipulated prey diversity (2 or 6 prey species, all edible) in a web with a generalist top predator. Their results strongly support the model predictions: total prey and predator biomasses in diverse webs were intermediate to the levels exhibited by less-diverse webs. Three tests for effects of prey diversity on predator biomass which included inedible prey found that increasing prey diversity decreases predator biomass because of competition between edible and inedible prey, as predicted by the model (Bohannan and Lenski, 2000; Steiner, 2001; Fox, 2004a). Reductions in predator biomass due to increasing prey diversity probably occur in many other systems. The main prerequisite for predator biomass to decline with increasing prey diversity is a trade-off among prey between competitive ability and resistance to consumers, with some prey being nearly or completely inedible. This trade-off is common (Proulx and Mazumder, 1998; Bohannan and Lenski, 2000; Chase et al., 2000a; Steiner, 2001; Fox, 2004a) and is well-known to promote prey coexistence, but its effects are far more wide-ranging (Leibold, 1996). By clearly demonstrating general, previously-unsuspected consequences of the competitive ability-predation resistance trade-off, simple models like equation (1a-f) provide a powerful motivation for empirical work (Leibold, 1989; Leibold, 1996; Bohannan and Lenski, 2000; Steiner, 2001; Fox, 2004a).

Tests of the effect of predator diversity on total prey biomass or abundance yield mixed results—but mixed in a manner largely consistent with equation (1a-f) (Cochran-Stafira and von Ende, 1998; Naeem and Li, 1998; Norberg, 2000; Duffy et al., 2001; Sommer et al., 2001; Fox and McGrady-Steed, 2002; Duffy et al., 2003; Fox, 2004a). Experimental studies fail to find significant effects of predator diversity on total prey biomass due to compensatory dynamics among competing prey (Norberg, 2000), and predators that differ only moderately (Norberg, 2000; Duffy et al., 2001), or very little (Fox, 2004a) in diet. Equation (1a-f) predicts that increasing predator diversity will fail to reduce total prey biomass in just these circumstances (Fox, 2004b). Significant effects of predator diversity on total prey biomass can arise from selection effects for a single, highly-effective predator (Fox and McGrady-Steed, 2002), and from predators with very different, complementary diets (Sommer et al., 2001)—again, the same circumstances in which equation (1a-f) predicts significant effects of predator diversity. Significant effects of predator diversity can also arise from the complex structural effects that can occur when comparing food webs more reticulate than the webs considered here (Naeem and Li, 1998; Duffy et al., 2003).

Studies manipulating diversity in a correlated fashion across multiple trophic levels are necessarily more difficult to interpret, particularly since such experiments often consider highly reticulate food webs. McGrady-Steed et al. (1997) performed such an experiment in aquatic microcosm communities, and found that more diverse communities exhibited intermediate levels of several key ecosystem properties and functions: bacterial (decomposer) abundance, algal (producer) abundance, herbivore abundance, and CO_2 flux. These results of McGrady-Steed et al. (1997) are broadly consistent with the model prediction that diverse communities will function in an intermediate fashion. However, other ecosystem properties and functions (e.g., detrital decomposition rate, total predator abundance) increased monotonically with increasing diversity. Downing and Leibold (2002) manipulated diversity in a correlated fashion across trophic groups in aquatic mesocosms and found little systematic effect of diversity on total biomasses of groups, primary productivity, decomposition rate, or O_2 flux. Naeem et al. (1994) performed a similar manipulation in terrestrial mesocosms and found that increasing diversity increased CO_2 uptake and primary productivity. However, Naeem et al. (1994) attributed their results largely to changes in plant diversity, with other trophic levels apparently playing a minor role.

DISCUSSION AND CONCLUSIONS

The main conclusion of the analysis the ecosystem-level effects of species loss from one trophic level depend on the diversity and composition of other trophic levels. It would perhaps be more surprising if this were not true. Equally important, and less intuitive, are the intimate connections between the constraints on parameter values (=trade-offs) required to maintain species coexistence, and the ecosystem-level function of those species. The mere fact that a set of species coexists within a food web strongly constrains the relationship between species composition and ecosystem function.

Developing closer links between experiments and theory will be a crucial next step. Experimental work on biodiversity and ecosystem function has outpaced theoretical work, particularly when considering biodiversity on multiple trophic levels. The pioneering experiments of Naeem et al. (1994) and McGrady-Steed et al. (1997) were originally interpreted in light of heuristic verbal models rather than mathematical theory. We need more models exploring the links between species diversity and composition on multiple trophic levels. What happens when one or a few species (as opposed to whole trophic levels) are removed from a reticulate food web? We also need experiments that explicitly consider structural vs. selection effects on ecosystem function, rather than the effects of species richness. Such experiments will need to carefully consider how to identify the empirical counterparts to the state variables in theoretical models. How to conduct this identification is not obvious when the species richness of the real food web exceeds the number of state variables in the theoretical model. One approach, successful in aquatic food web ecology but less often applied elsewhere, is to aggregate species into trophic groups for purposes of data analysis, and identify the aggregations (e.g., "small algae", "large algae", "*Daphnia* spp.") with the "species" (state variables) in simple mathematical models

(Leibold 1989). Alternative models (alternative trophic aggregations) can be treated as alternative, testable hypotheses (Hulot et al., 2000). Bringing experimental data to bear on the growing body of mathematical theory on biodiversity, ecosystem function, and food web structure (Holt and Loreau, 2002; Fox, 2003; Thébault and Loreau, 2003; Fox, 2004b; see Loreau and Thébault, Chapter 5.1) will be an exciting challenge for future work.

6.3 | TROPHIC POSITION, BIOTIC CONTEXT, AND ABIOTIC FACTORS DETERMINE SPECIES CONTRIBUTIONS TO ECOSYSTEM FUNCTIONING

Amy Downing and J. Timothy Wootton

Ecologists have long recognized the potential importance of species for ecosystem dynamics (Lotka, 1925; Tansley, 1935; MacArthur, 1955), and have recently devoted considerable effort towards understanding species effects on ecosystem processes (Schulze and Mooney, 1993; Jones and Lawton, 1995; Kinzig et al., 2002; Loreau et al., 2002). These efforts are driven largely by the desire to predict consequences of biodiversity loss for ecosystems, and to help target conservation efforts towards species whose losses will have large consequences for ecosystems.

Research has begun to lay the groundwork for understanding consequences of biodiversity loss by first exploring the relationship between biodiversity and ecosystem functioning (Schwartz et al., 2000). Results reveal that both species richness (Tilman et al., 1996; McGrady-Steed et al., 1997; Naeem and Li, 1997; van der Heijden, 1998; Hector et al.,

1999; Naeem et al., 2000a; Engelhardt, 2001; Cardinale et al., 2002) and species composition contribute to biodiversity–ecosystem functioning relationships (McGrady-Steed et al., 1997; Hooper and Vitousek, 1998; Mikola, 1998; Mikola and Setälä, 1998a; Symstad et al., 1998; Sankaran and McNaughton, 1999; Norberg, 2000; Emmerson et al., 2001; Downing and Leibold, 2002). The experimental results have fueled a vigorous debate as to whether 'richness' and 'composition' effects are due to (1) the richness *per se* of the assemblage, (2) the unique combination of species that make up the assemblage, or (3) the presence of one or two species with strong effects on ecosystem processes. Determining the relative importance of these three factors can be quite difficult, and is at the heart of the sampling vs complementarity debate (Aarssen, 1997; Huston, 1997; Tilman, 1997a; Tilman et al., 1997; Loreau and Hector, 2001). The sampling hypothesis (sometimes referred to as selection effect) states that diverse communities are statistically more likely to contain species with strong effects on a given ecosystem function, therefore diversity influences ecosystem functioning primarily through the effects of a particular species (Tilman, 1997a; Hector, 1998; Loreau, 1998b; Wardle, 1999b; Loreau and Hector, 2001). Alternatively, the resource complementarity hypothesis states that diverse communities display greater niche diversification and/or facilitative interactions between species, therefore diversity influences ecosystem functioning through interactions among the assemblage of species in the ecosystem (Aarssen, 1997; Tilman et al., 1997; Hooper, 1998).

In some cases, the presence of one or a few species can clearly determine ecosystem processes. For example ecosystem engineers (Lawton, 1994; Lawton and Jones, 1995) and keystone species (Power et al., 1996b; Fauth, 1999), have obvious and dramatic effects on community structure and ecosystem processes and the loss of these species from communities often have known consequences. However, species other than keystones and engineers can determine ecosystem dynamics. In order to predict consequences of biodiversity loss, we still need to address key questions about individual species effects on ecosystem processes. For example, are individual species' effects consistent in different biotic and abiotic contexts, and can we predict species with strong effects based on their association with a particular functional group?

The goal of this chapter is to explore individual species' effects on ecosystem processes in pond food webs, and to determine if these effects are predictable and consistent with respect to trophic position, biotic context, and abiotic factors. We present data from two aquatic mesocosm experiments designed to disentangle the effects of species richness and species composition on ecosystem dynamics. The main results from these experiments show that while species richness can be important, unique *combinations of species* have distinct effects on ecosystem functioning (Downing and Leibold, 2002; Downing, 2005). Here we evaluate the average effect of *individual species* on ecosystem processes in a background of randomly assembled communities that differ in average productivity.

METHODS

The effects of individual species of macrophytes, mesograzers, and invertebrate predators on ecosystem functioning were evaluated from data collected in two separate aquatic mesocosm experiments conducted at Kellogg Biological Station in Michigan (Downing and Leibold, 2002; Downing, 2005). In both experiments, species composition was nested in species richness, and cross-factored with disturbance (control vs. acidification). Species richness and composition were simultaneously manipulated within all three functional groups; rooted macrophytes, mesograzers feeding primarily on periphyton, and invertebrate predators. The species pool used to establish the diversity and composition treatments in the two experiments consisted of 21 species and 24 species respectively, with an equal number of species in the trophic groups macrophytes, mesograzers, and invertebrate predators (Table 1). Species were assembled in communities that differed in terms of species richness and species composition. In the low productivity experiment, communities consisted of 1, 3, or 5 species per functional group, creating a richness gradient of 3, 9, or 15 species. In the high productivity experiment, communities consisted of 2 or 6 species per functional group, creating two richness treatments of 6 and 18 species. In addition to the richness treatments, seven unique species compositions were nested within each richness level and replicated four times. Each composition was chosen randomly from the larger species pool, therefore species associations in treatments can be treated as a random effect. Half of each experiment received a pulse acidification event part way through the experiment, lowering and raising the pH from 9 to 3.5 over 48 hours. The disturbance was part of a separate experiment designed to explore the stability response of ecosystem processes in ecosystems that vary in species diversity and composition (Downing, 2001). For the purposes of the analyses describe here, the disturbance provides another source of ecosystem variability such as might be experienced in ponds with different buffering capacity, or in ponds that differ in their exposure to short and severe disturbances.

Table 1. Species pools in low and high productivity experiments

Low productivity		
Macrophytes	Grazers	Predators
Potamogeton compréssus	*Crangonyx richmondensis*	*Neoplea striola*
Utricularia vulgaris	*Rana catesbeiana*	*Ambrysus*
Myriophyllum verticullatum	*Rana clamitans*	*Gyrinus*
Elodea occidentale	*Physa gyrina*	*Ascilius*
Vallisneria americana	*Hyallela azteca*	*Belostoma flumireum*
Ceratophyllum demersum	*Helisoma trivolis*	*Notonecta undulata*
Potamogeton crispus	*Trichocorixa*	*Notonecta* sp.
		*Coenagrionidae**
		*Libellulidae**
High productivity		
Macrophytes	Grazers	Predators
Sagittarria rigida	*Hesperocorixa*	*Neoplea striola*
Utricularia vulgaris	*Rana catesbeiana*	*Ambrysus*
Myriophyllum verticullatum	*Rana clamitans*	*Gyrinus*
Potamogeton natans	*Physa gyrina*	*Coptotomus*
Vallisneria americana	*Hyallela azteca*	*Dineutus*
Ceratophyllum demersum	*Helisoma trivolis*	*Notonecta undulata*
Potamogeton crispus	*Trichocorixa*	*Buenoa*
Elodea occidentale	*Crangonyx richmondensis*	*Belostoma flu*
mireum		

*Indicates species not originally stocked in the experiment. Underlined letters indicate abbreviations used in Figures.

The mesocosms consisted of 300 liter polypropylene cattle watering tanks, and are described in detail elsewhere (Downing and Leibold, 2002). All communities were inoculated with diverse mixtures of phytoplankton, zooplankton, periphyton, and microbes collected from seven local ponds, and all communities began with representation in the macrophyte, mesograzer, and predator communities. The species were added from early May through early June, as they became naturally abundant during the spring season. After the communities were established, six ecosystem response variables were measured repeatedly through the rest of the active growing season (July–September) for each experiment. The response variables included productivity, respiration, decomposition, phytoplankton, zooplankton, and periphyton. By late September, reduced temperatures and short days slow activity levels of organisms in the mesocosms as well as natural ponds, and serves as an appropriate ending point for the experiments. At the termination of the experiments, each mesocosm was sieved and all organisms remaining were recorded. On average, mesocosms retained 75 % of species initially stocked in the treatments, although 65 % of all mesocosms experienced at least one invasion of an unintended species not originally stocked in the treatment. The majority of invasions were by pleid beetles (*Neoplea striola)* and damselflies (*Coenagrionidae* sp.).

The two experiments differ in two potentially important ways. First, average productivity levels varied between experiments. In the low productivity experiment, nutrients were added at the beginning of the experiment to approach average natural concentrations of nitrogen and phosphorus (1500 μg N/liter and 150 μg P/liter) in local ponds. These mesocosms were nutrient-limited because the original nutrients were never replenished. In the second experiment, nutrients were added at the same concentration at the beginning of the experiment, but were also added biweekly at concentrations of 500 μg N/l and 50 μg P/l. The additions were to replace available nutrients removed from the water column due to processes such as denitrification, sedimentation, and conversion into biomass throughout the season. Freshwater productivity is strongly influenced by nutrient loading (Carpenter and Kitchell, 1993b), therefore the difference in the frequency of nutrient additions has a large effect on productivity in these experimental ponds. The second major difference between experiments is that the species pool and richness levels were unique to each experiment (see Table 1). Therefore, not all species can be compared across years as some species were present in only one of the two experiments.

In order to detect effects of individual species on ecosystem functioning, we first reduced the six ecosystem response variables to two principle components for each year (Figure 1a, c). Variables used in the principle components analysis consisted of time-averaged means of each response variable over the last four weeks of the experiments. The principle components analysis produced three principal component (PC) scores in the low productivity experiment, and two PC scores in the high productivity experiment with eigen values greater than one. At low productivity, PC1 and PC2 explain 39% and 18% of the variation respectively. At high productivity, PC1 and PC2 explain 46% and 18% of the variation respectively.

To test for significant effects of each species on the first two principal component estimates of ecosystem functioning, we performed t-tests

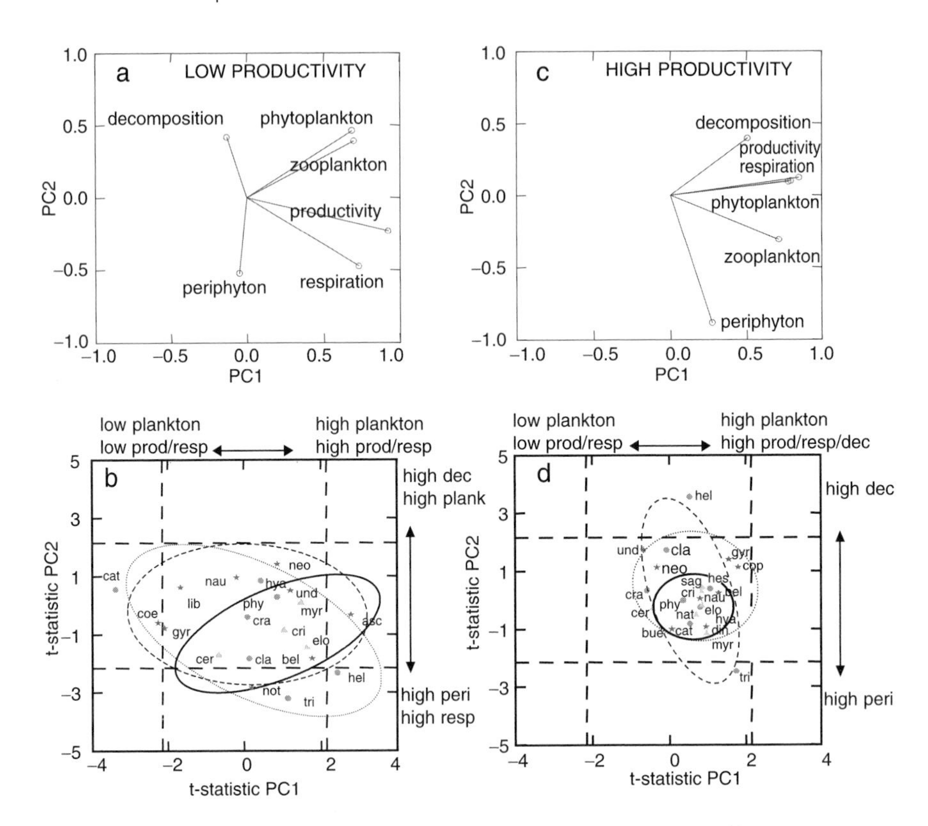

FIGURE 1 | Principal components analysis and individual species effects on ecosystem functioning. **A,** Factor loadings at low productivity. **B,** Strength of species effects on PC1 and PC2 at low productivity. See Table 1 for species abbreviations. Values plotted are the t-test statistic for the difference in PC scores in the presence and absence of each species. The values plotted indicate the direction of the PC response in the presence of each species, e.g. communities with bullfrogs (*Rana catesbeiana*) at low productivity had on average lower productivity, respiration, and plankton biomass than did communities with bullfrogs. Those species falling on or outside the dotted lines have significant effects on PC scores at the $p < 0.05$ level. Solid circle = confidence ellipse for macrophytes set at probability value of 0.68 based on sample means (equivalent to standard deviations). Short dash circle = confidence ellipse for grazers. Long dash circle = confidence ellipse for predators. **C,** Factor loadings at high productivity. **D,** Strength of species effects on PC1 and PC2 at high productivity.

on the mean PC scores for tanks with and without each species for each experiment. Only species that were still present at the end of the experiment were used in the analysis, including the few species that successfully invaded the mesocosms. A species is identified as being 'potentially important' for ecosystem functioning if t-tests are significant at the $p < 0.05$ level.

Finally, we conducted a hierarchical cluster analysis to determine if our *a priori* classification of species into macrophyte, mesograzer, and predator functional groups was a good predictor of species effects on ecosystem functioning. Hierarchical cluster analysis was conducted separately for species in the low and high productivity experiments. The data used in the cluster analysis were the time-averaged means of the six original ecosystem variables in the presence of each species. The data were standardized and the analysis was conducted using complete linkage based on euclidean distances. All statistical analyses were conducted in SYSTAT.

RESULTS

The high productivity experiment had higher productivity, respiration, and standing crops of plankton and periphyton biomass, and a small but significant decrease in decomposition rates (Table 2). In both experiments, the first principal component (PC1) distinguishes primarily between high versus low ecosystem rates and plankton biomass, and the second principal component (PC2) distinguishes primarily between high decomposition rates and high periphyton biomass. Periphyton and decomposition rates have slightly different loadings with respect to plankton biomass, and productivity and respiration rates in both experiments, but the remaining overall structuring of ecosystem responses is similar (see Figure 1a, c). The first two principal components combined explain about 58% and 64% of the original variation in ecosystem response in low and high productivity respectively.

A larger proportion of species had detectable effects on ecosystem functioning at low productivity (see Figure 1b, d). The macrophyte species had no significant effects on ecosystem functioning. In contrast, three out of seven grazer species, and three out of nine predator species showed significant effects on ecosystem functioning. At high productivity, species' effects were generally not as strong; zero macrophyte species, two out of eight species of grazers, and zero predators revealed detectable effects on ecosystem functioning. The same trends are indicated by the 95% confidence ellipses around each trophic group. The predator and grazer ellipses are much larger and fall outside the significant lines for each experiment (see Figure 1 b, d). These results indicate that despite substantial variation between replicates in terms of disturbance treatments, richness treatments, and compositional treatments, some species show consistent and detectable effects on ecosystem properties within experiments.

Table 2. Ecosystem rates and trophic variables between experiments.

	Low productivity	High productivity		
Ecosystem variables	mean ± st dev	mean ± st dev	F*	p
productivity rates (mg O_2/hr)	0.199 ± 0.039	0.223 ± 0.040	19.5	<0.001
respiration rates (mg O_2/hr)	0.149 ± 0.023	0.339 ± 0.063	643.4	<0.001
decomposition rates (mg dry leaf wgt/day)	0.044 ± 0.007	0.041 ± 0.005	5.2	<0.001
log phyto-plankton biomass (μg chla/l)	0.811 ± 0.548	1.606 ± 0.414	85.2	<0.001
log periphyton biomass (μg chla/m^2)	2.565 ± 0.315	3.302 ± 0.057	139.5	<0.001
log zooplankton biomass (μg dry weight/l)	3.028 ± 0.415	3.530 ± 0.443	46.5	<0.001

*Univariate ANOVA results on year effects with df = 1, 138.

In contrast, individual species effects were not consistent between the low and high productivity ecosystems. To illustrate this point, we focus on two species, *H. trivolis* (snails), and *R. catesbeiana* (bullfrog tadpoles). First, *H. trivolis* had significant effects on ecosystem functioning in both years, although the effects on ecosystem processes differed between years. In low productivity systems, *H. trivolis* is associated with high plankton biomass, and high productivity and respiration. At high productivity, *H. trivolis* is associated with high decomposition rates. *R. catesbeiana* had detectable effects on ecosystem functioning in low productivity ecosystems, but not at high productivity. At low productivity, *R. catesbeiana* is associated with low plankton biomass, low productivity, and low respiration.

Despite inconsistent effects of individual species from year to year, species within trophic groups showed similar patterns between experiments. Macrophytes had consistently small and undetectable effects on ecosystem functioning (smaller confidence ellipse), and the response is centered around slightly positive values on the PC1 axis (see Figure 1 b, d).

In contrast, grazers and predators had more variable and less consistent effects (larger confidence ellipses) than macrophytes in both years. In addition to having more variable effects, grazer and predator functional groups also contained more species with potentially important effects on ecosystem functioning. The cluster analysis confirmed the results observed in Figure 1. Post hoc clustering of species based on their effects on ecosystem functioning appears largely random with respect to the functional groups macrophytes, grazers, and predators.

DISCUSSION AND CONCLUSIONS

Several notable results emerge from the analyses of individual species effects on ecosystem functioning. First, the effects of individual species varied from experiment to experiment and were stronger in experimental ponds with lower total productivity. Second, species at higher trophic levels had more significant and variable or less predictable effects on ecosystem functioning. Third, species within functional groups have very different effects on ecosystem functioning, indicating a lack of redundancy within functional groups.

The fact that individual species effects were unique between experiments might be explained by differences in the experimental conditions. The two experiments differed in terms of average total productivity, the species pool from which the communities were randomly assembled, and the particular weather patterns present in each year. All other aspects of the experiments were similar. The same functional groupings were used in both years and all replicates were assembled and sampled in a similar manner. In addition, in both years half of the replicates received a pH manipulation which, for the purposes of this analysis, creates additional variability between replicates within experiments in the environmental conditions each species experiences.

Perhaps the most substantial difference between experiments that might explain why the effects of some species were inconsistent across experiments is the average difference in productivity between years. In high productivity aquatic ecosystems, ecosystem variables such as plankton biomass, productivity and respiration may be determined largely by the overall rate of nutrient input rather than the presence of certain species, analogous to the 'bottom-up' perspective of communities (McQueen et al., 1989). In contrast, for nutrient-limited ecosystems, an organism's role as a nutrient recycler is likely to have a proportionally larger effect on nutrient availability, therefore individual species may be more likely to have larger effects on ecosystem processes. In addition, the role of organisms as consumers, predators, and decomposers will also have a proportionally larger effect if nutrients are less important in driving an ecosystem process such as productivity. Other aquatic studies have found that the importance of keystones decreases with increasing productivity (Carpenter, 1992; Lathrop and Carpenter, 1992), suggesting that potential productivity may be an important factor determining the context-dependency of species-specific effects in ecosystems.

The different species backgrounds of the two experiments may also play some role for the inconsistency of species effects between experiments. About 62% of species, (16 out of 26), were present in both experiments (see Table 1). Species effects in ecosystems will ultimately depend on the identities of other species in the food web because species interactions determine how species participate in material and energy flow (Chapin III et al., 2000; Daugherty and Juliano, 2002; Poff et al., 2003). For example, a species may be present in both experiments, but in one experiment the species may be frequently paired with a strong competitor or a strong predator, both of which will affect the net impact of the species on ecosystem functioning. The prevalence of such context-dependent species interactions is often due to a combination of direct and indirect effects, and has been well documented in food webs (Yodzis, 1988; Leibold and Wilbur, 1992; Menge et al., 1994; Wootton, 1994b; Power et al., 1996b; Chalcraft and Resetarits, 2003).

In contrast to the inconsistent effects of individual species between experiments, higher trophic levels (mesograzers and predators) consistently had stronger and more variable effects within years, even though the ecosystem response variables we measured were largely influenced by primary producers. These strong and variable species effects are exemplified by the grazers *Helisoma* (snail) and *R. catesbeiana* (bullfrog tadpole), and the predators *Ascilius* (ditiscid) and *Coenagrionidae* (damselfly) (see Figure 1). The strong effects of higher trophic levels is not particularly surprising. However, the large variability in species effects within functional groups is striking. Despite obvious differences in the behavior and ecology of grazers such as tadpoles and snails, one might expect two grazers to have more similar effects on ecosystem functioning than a grazer and a predator species. The hierarchical cluster analysis provides additional support for the variable effects of grazers and predators on ecosystem functioning. Cluster analysis reveals that species do not cluster into the *a priori* defined functional groups based their effects on ecosystem functioning. In contrast, species grouped together through cluster analysis are often composed of two or three functional groups (Figure 2), indicating that species within functional groups do not influence ecosystem functioning similarly. These results must be interpreted

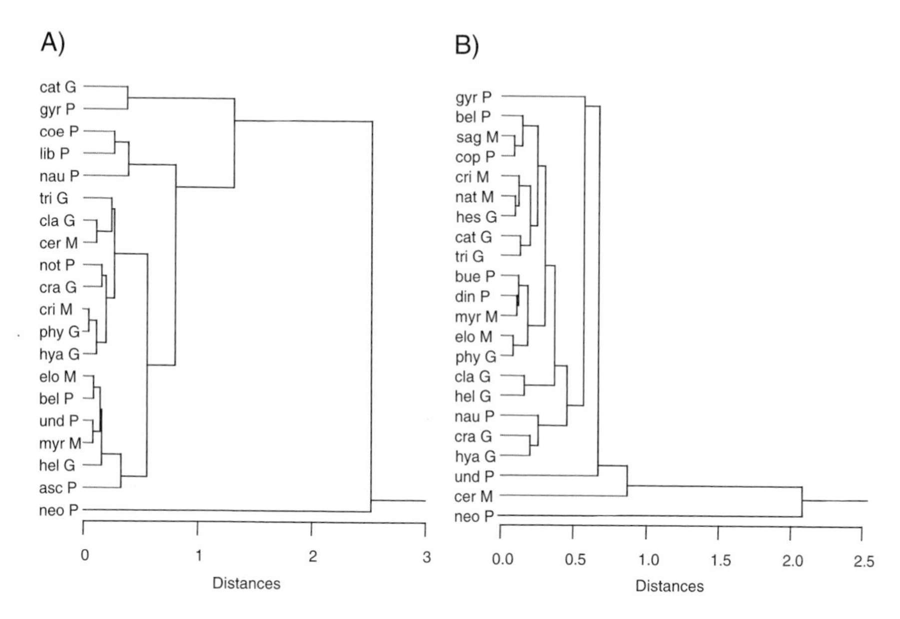

FIGURE 2 | Hierarchical cluster analysis of species using the six original standardized ecosystem variables, using Euclidean distances and Complete linkage. **A,** Species in 1998. **B,** Species in 1999.

with caution given the limited number of species tested (26 species) under a relatively narrow set of conditions (i.e., low vs high productivity, presence vs absence of pH disturbance, and relatively low variation in richness from 3–18 species) used between these two experiments. However, our results are also supported by theory that indicates species effects will depend strongly on their position in the food web and the structure of the food web (Thébault and Loreau, 2003).

The variability in direction and magnitude of species effects within functional groups other than primary producers suggests that at least the species used in these experiments are not redundant with respect to their net effects on ecosystem functioning (Vitousek and Hooper, 1993). The lack of redundancy in species' effects within functional groups is at first surprising given that many of these species should have similar and potentially redundant direct effects on ecosystems. Importantly, redundancy in species' *direct* effects does not guarantee that those same species' will also have redundant *net* effects on ecosystems. In these experiments, the impacts of grazers and predators on ecosystem functioning are probably being mediated primarily through direct and indirect effects on the primary producer communities as the primary producers are either measured directly, or largely control the ecosystem variables monitored (Downing and Leibold, 2002). Species may affect primary producers through direct mechanisms such as the consumption of periphyton or phytoplankton by grazers, or through indirect mechanisms such as nutrient regeneration by grazers. Additionally, predators may enhance productivity by reducing abundances or foraging rates of grazers (Chalcraft and Resetarits, 2003). Ultimately, the number and relative strength of indirect effects will depend on the suite of species present. The greater variability of effects of species at higher trophic levels may be in part due to the greater variety and potential for indirect effects.

A first step towards determining potentially important species for ecosystem functioning may be to identify species with strong effects on the primary producer community. Many of the ecosystem processes that are of concern to humans such as productivity, CO_2 regulation, crop production, and nutrient retention will depend heavily on the primary producer community (Costanza et al., 1997; Daily, 1997). In systems where factors other than nutrient input (e.g., trophic interactions) affect primary production, grazers and predators may be very important in influencing ecosystem functioning through their direct and indirect effects on primary producers. Our results suggest that we should perhaps be concerned over the loss of species at trophic levels other than primary producers because they may show more variable and perhaps stronger effects on ecosystem functioning. In addition, our work suggests that a trophic level approach to predicting a species effect on an ecosystem may not be appropriate as species within trophic levels, defined here by functional groups, did not function similarly in these ecosystems. Clearly, additional work should be done in a variety of ecosystems using a larger set of species than we used in these experiments to determine if our findings apply more broadly. Most prior experimental investigations of the relationship between biodiversity and ecosystem function have explored only single trophic level (plant) systems (Schmid et al., 2002). We need to expand our focus of species effects on ecosystems to include species other than primary producers to determine if the patterns we found are consistent across ecosystems (Duffy, 2002; Schmid et al., 2002; Duffy et al., 2003; Thébault and Loreau, 2003). Importantly, species other than plants are probably most easily lost from ecosystems due to their lower numbers and higher susceptibility to extinction (Holt et al., 1999; Petchey et al., 1999).

Our method for exploring individual species effects on ecosystem dynamics in randomly assembled communities may provide a powerful way to screen large amounts of data for species-specific effects on ecosystems. Previous analyses have shown strong effects of replicated combinations of species for ecosystem functioning (Downing and Leibold, 2002). Here we explore effects of individual species in a background of randomized communities varying in composition and richness. Therefore, unambiguously attributing ecosystem responses to specific species is difficult because all other species and factors are not held constant. Similar approaches to the one described here may be most useful as a way to identify potentially important species. Once these species have been identified, more specific and mechanistic studies can be utilized to further explore and confirm a species' role in the ecosystem.

Exhaustively exploring individual species effects is obviously not practical or even possible for all species in all ecosystems, and species impacts in ecosystems may reveal very little generality or predictability beyond specific and well-studied ecosystems (Brown, 1995b). However, experimental and theoretical explorations of species' effects in ecosystems may reveal that certain species traits are associated with particular impacts in ecosystems and may help target studies towards particular suites of species. For example, in this study we show that plants have weaker and more consistent effects than grazers and predators, although our conclusions are based on a limited number of species and should be validated with additional studies. Furthermore, identifying important biotic and abiotic factors that are responsible for determining context-dependent species interactions may help to predict when and if a species' invasion or extinction will matter. Understanding species effects on ecosystem functioning will require that we expand our focus beyond obvious keystone species or ecosystem engineers, and beyond simplified descriptors of communities such as species richness. A promising place to begin may be to focus on individual species of grazers and predators with strong effects on primary producers. These approaches may help to predict the consequences of species invasions or extinctions for ecosystem functioning (Srivastava, 2002; Wootton and Downing, 2003), and may help guide our efforts towards preserving critical ecosystem functions and services (Balvanera et al., 2001).

6.4 | DOES BIOLOGICAL COMPLEXITY RELATE TO FUNCTIONAL ATTRIBUTES OF SOIL FOOD WEBS?

Heikki Setälä

Due to their extremely high number of organisms and the relatively high species richness soils have been described as "poor man's tropical rainforests" (Usher et al., 1979). A fistful of organic soil can contain tens or hundreds of species of soil fauna (Giller, 1996; Behan-Pelletier, 1999) and a gram of soil may harbor thousands of species of microorganisms (Torsvik et al., 1990). Trophic interactions between this numerically abundant and species diverse soil biota are responsible for carrying out important ecosystem processes, such as nutrient cycling, of which all terrestrial life depends on. A crucial question therefore is whether the high taxonomic richness is a prerequisite for a system to function properly, or would just a subset of biota be enough for nutrient cycling to proceed efficiently. According to ecological theory each of the diverse taxa should have not only their own ecological (Hutchinson, 1957) but also their functional niche (Rosenfeld, 2002). Whereas the Hutchinsonian niche defines where and under which circumstances a particular species can exist, the functional niche defines the ecological effects that a species/taxon has within that habitat. From a food web perspective the concept of biological diversity relates to the structural complexity (e.g., food chain length, heterogeneity within trophic levels, etc.) of food webs. For example, how precise—taxonomic wise—should soil food web ecologists be to accurately predict the functional consequences that may result from losses of particular taxonomic or functional units within a soil food web? Do soil food webs behave like any other food webs, or does the exceptionally high species diversity and individual density (Giller, 1996), in concert with some peculiarities of soils as a habitat (Setälä et al., 2005), make soil food webs special as regards to the biological complexity-ecosystem functioning interaction?

The aims of this chapter are to (i) give a review of the physical characteristics that can influence community structure and system processes in soils (donor control, energy flow controlled by microbes, links between omnivory, and soil heterogeneity), (ii) explore the effects of community complexity (i.e., the number of trophic levels, feeding guilds, and species in soil communities) on N-mineralization and NPP, and (iii) point out the strong context dependency of biodiversity-ecosystem functioning studies in soils arising from the conditional mutualistic-antagonistic effects of mycorrhizae.

CHARACTERISTIC FEATURES OF SOIL FOOD WEBS

Consumers Are of Little Importance in Terms of Biomass, Energy Flow, or Effect Strength on Detritivores

Soil food webs are traditionally considered as donor controlled systems with bottom-up forces playing a decisive role in structuring communities (Pimm, 1982): as detritus is dead it may be impossible for organisms that depend on this very basal resource to directly control its renewal rate (Begon et al., 1990; Scheu and Setälä, 2002). At first sight it would thus appear that only the nutritional quality, as well as the diversity of litter types with varying nutritional characteristics, would influence the structural and functional attributes of soil food webs. However, as previously pointed out, soil communities are species diverse and the food webs may appear seemingly complex in structure; not all organisms feed directly upon detritus but are, at least partly, biotrophs including microbial feeders, carnivores, root feeders, etc. (Verhoef and Brussaard, 1990). As a matter of fact, food webs composing of up to five potential trophic levels have been commonly described in soils (Coleman, 1996). Despite this complexity, soil food webs—when screened through an energetic perspective—appear boring: often about 95% of the energy passing through the soil food web flows from detritus to soil saprophytic microbes with an insignificant proportion utilized by consumers at the higher positions of the food web (Moore and De Ruiter, 1991; de Ruiter et al., 1995). In addition, as the biomass of saprophytic microbes can comprise up to 99% of the total biomass of the soil food web (Coleman and Crossley, 1996), the role of top-down regulation at higher trophic positions as affecting the structure and functioning of soil communities appears marginal. Furthermore, growing body of literature suggests that primary (soil microbes) and secondary (soil fauna) consumers of soil food webs are feeding generalists, even omnivores (Ponsard and Arditi, 2000; Scheu and Falca, 2000; Maraun et al., 2003; Setälä et al., unpublished manuscript). If omnivory is a prevailing feature in soil food webs, interpretation as to how the rate and magnitude of decomposer-driven processes are affected by the biological complexity of belowground food webs is unlikely to be straight forward.

We have good reasons to expect that soils—as a habitat—share features that promote omnivorous feeding habits among soil biota. First, soils are three-dimensional, opaque habitats being immensely heterogenous both horizontally and vertically (Anderson, 1978). In such an environment food resources are not only difficult to locate but are also unpredictably available (Lee and Pankhurst, 1992). Under such conditions organisms cannot be overly choosy as regards to their dietary preferences, which inevitably promotes feeding generalism among soil biota. Second, since detritus is dead, consumers and their detrital food resource cannot co-evolve, which can explain the lack of feeding specialists among detritivores (Scheu and Setälä, 2002). The three-dimensional structures of soils also modify the body structure of soil biota: due to the porous structure of soils virtually all soil inhabiting organisms are either microbes or small-sized invertebrates. The lack of vertebrates and endotherms in most soil food webs obviously relates to the physical constraints of soils.

Due to the various features that characterize (i) soil biota (at the individual and community level) and (ii) soils as a habitat, we may hypothesize that decomposition processes, including nutrient mobilization, are relatively insensitive to architectural changes of soil food webs. I will describe this hypothesis by reviewing studies in which the functional importance of interaction taking place between and within trophic groups has been empirically examined. In most cases net primary production represents the function against which the diversity/complexity measures are evaluated, simply because net primary production (NPP) is of ultimate importance for all heterotrophs in every ecosystem and is largely controlled by the activity of the detrital food web (Wardle, 2002a).

ECOSYSTEM FUNCTION IN RELATION TO BIOLOGICAL COMPLEXITY

Species diversity of soil biota appears far less influential in affecting soil processes and NPP as the trophic structure of the decomposer food webs, and predation (as affecting the numeric abundance of organism) may be not as important as excretion of N-rich materials by the consumers.

Interactions Between Trophic Levels and Ecosystem Functioning

Over four decades ago evidence started to accumulate showing the importance of litter-feeding fauna in stimulating the decomposition rate of organic matter and enhancing nutrient mobilization in that material (see the reviews by Verhoef and Brussaard, 1990; Coleman, 1996). Later on it became evident that not only detritivores but also other trophic groups of soil fauna, albeit having a marginal biomass, can control the rate of degradation processes. Importantly, micro- and mesocosm studies have repeatedly shown that the presence of secondary consumers can have a remarkable stimulative influence on not only soil processes but also on plant growth (Ingham et al., 1985; Setälä and Huhta, 1991; Alphei et al., 1996). In other words, by feeding upon detritus, soil microbes, or other animals, soil fauna speed up nutrient mobilization and thereby indirectly affect NPP by controlling the availability of nutrients for the uptake of the plants (Figure 1A). It is especially indirect interactions—mediated through nutrient availability—that characterize trophic interactions in soil food webs (Bengtsson et al., 1996). This enables the impacts of a consumer to propagate through the food web without affecting the biomass of adjacent trophic levels, which is in contrast to typical aboveground feeding interactions where the consumer directly affects the biomass of its victim (Wardle, 2002a; Maraun et al., 2003).

In all, even a minor change in the concentration of plant-available nutrients in the plant rhizosphere has a potential to bring about changes that are functionally relevant in the ecosystem. This is because primary producers compete with saprophytic microbes for nutrients, mostly N and P, and because microbes are often superior competitors to plants in nutrient acquisition (Kaye and Hart, 1997; Schmidt et al., 1997). Consequently, nutrients will be immobilized into microbial biomass and will be out of the reach of plant roots unless a third party, such

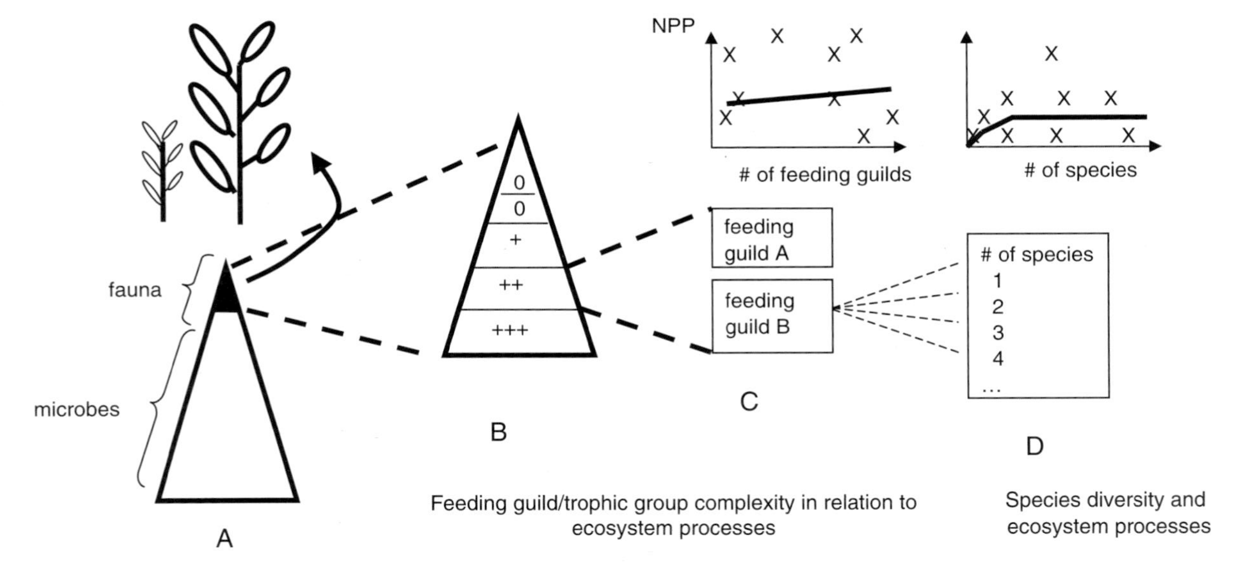

FIGURE 1 | The influence of biological complexity within the various levels (A-D) of the soil detrital food web on ecosystem functioning. **A,** Addition of secondary decomposers (black tip of the biomass pyramid; soil fauna) to systems with primary decomposers (microbes) can result a significant release of nutrients and a concomitant increase in plant growth. **B,** Trophic groups within secondary consumers close to the base of the food web tend to have a larger (+ + +) influence on ecosystem processes than trophic groups topping (0) the web. **C,** Heterogeneity within trophic groups can, e.g. via complementary resource use, influence the rate of ecosystem functioning. **D,** Diversity of species within functional groups is only weakly related to e.g. the rate of plant growth. Species-specific properties of taxa can, however, affect system-level processes in an idiosyncratic manner.

as microbial consumers, returns these nutrients back to the soil (Moore et al., 2003). As practically all trophic interactions are followed by excretion of growth-limiting nutritional compounds by the consumer, feeding interactions with insignificant energetic loadings can manifest themselves as trophic relationships with functional significance. This is the basic mechanism as to how trophic groups with low biomass and energetic activity (depicted as a solid small triangle topping the larger biomass pyramid in Figure 1A) has a potential to play a decisive role in controlling NPP.

The number of trophic levels/groups within secondary consumers can also affect decomposition processes (see Figure 1B). It is well established that removal of detritivores (Haimi et al., 1992; Scheu et al., 1999; Liiri et al., 2002b) and/or microbivores (Clarholm, 1985; Ingham et al., 1985; Bonkowski et al., 2001) from soil food web can exert significant negative influences on plant growth. However, the functional importance of predators topping the detrital food web appears to be smaller than that of the trophic groups harboring the lower positions in the food web. Removal of predatory mites or centipedes from micro- and mesocosms resulted in no significant change in plant growth despite of their relatively large impact on the biomass of their prey populations (Setälä et al., 1996; Laakso and Setälä, 1999a; Laakso et al., 2000). It is possible that the prey organisms, microbivores, and microbi-detritivores (shown to markedly contribute to nutrient dynamics in soils, see Laakso and Setälä [1999a]) can compensate for the biomass loss by increasing their biomass turn over rates. It is equally possible that the direct contribution of top predators to nutrient dynamics, through their N-rich excretory products, is large enough to make a substantial addition to the plant-available nutrient pool in the systems (Setälä, 2002). Whether the importance of top-down regulation and its functional importance gets diluted with increasing number of trophic steps in natural (field) communities remains to be tested (but see Santos et al., 1981; Kajak et al., 1993).

Feeding Guild Complexity in Relation to Ecosystem Processes

Although our knowledge of the influence of interactions taking place within trophic levels/groups on system level processes is limited, we have reasons to expect that trophic groups do not always act as uniformly behaving entities. For example, heterogeneity within primary decomposers (i.e., saprophytic microbes) is likely to relate to the functioning of the soil system. In this trophic group bacteria and fungi have functionally divergent properties giving rise to two different energy-nutrient channels (Coleman et al., 1983; Moore and Hunt, 1988). Bacteria (and the bacterial-based energy channel) are adapted to use easily degradable substrates and dominate in the plant rhizosphere, whereas the fungi, being more efficient in degrading recalcitrant plant material, dominate in the soil bulk (Ingham et al., 1989; Whitford, 1989). As the two microbial groups are also known to differ in their ability to resist and recover from disturbances (Moore and de Ruiter, 1997) it is obvious that biological complexity within trophic groups existing close to the base of the detrital food web should relate to the functioning of the system (Ingham et al., 1986; Beare et al., 1992).

The functional importance of biological complexity within the trophic groups of secondary consumers has received surprisingly little attention among soil ecologists. Scattered evidence suggests that at least some trophic groups of soil fauna can functionally be split further (see Figure 1C). For example, addition of fungal feeding nematodes to microcosms with bacterial-feeding nematodes resulted in higher mobilization of NH_4-N as compared to systems with bacterial-feeders only (Laakso and Setälä, 1999b). Alphei et al. (1996) showed in a microcosm experiment that when bacterial-feeding protozoans and nematodes existed together the growth of grass was greater as compared to situations when the two faunal groups existed alone. Laakso and Setälä (1999a) manipulated the complexity of the microbi-detritivore feeding guild in microcosms. They noted that variation in the composition within the microbi-detritivorous fauna resulted in slight but significant variation in the growth of birch seedlings. Further, microcosms with a complex set of soil fauna (including various representatives of bacterivores, fungivores, and microbi-deteritivores) were more efficient in stimulating birch growth than were systems in which each of the feeding groups was represented by one to five species of fauna (Laakso and Setälä, 1999a). The studies by Alphei et al. (1996) and Laakso and Setälä (1999a, Laakso and Setälä, 1999b) suggest that at least some representatives of fauna within the feeding guilds may not contribute equally to nutrient dynamics and therefore cannot be considered as functionally redundant entities.

It is worth noting, however, that categorization of soil fauna into well-defined trophic groups or feeding guilds is difficult. This stems partly from the fact that experimental evidence of the dietary requirements of most fauna is limited (Hunt et al., 1987; Lee and Pankhurst, 1992; Siepel and de Ruiter-Dijkman, 1993). The high prevalence of feeding generalists, even omnivores, and the almost complete lack of feeding specialists in most detrital food webs (Maraun et al., 2003; Setälä et al., unpublished manuscript) is likely to diminish the functional importance of biological complexity at all levels of the ecological hierarchy. This is also a possible explanation as to why trophic cascades—typically occurring in systems with well-defined trophic levels—are seldomly reported in belowground food webs (Mikola and Setälä, 1998b; Laakso and Setälä, 1999a; Wardle, 1999a). In summary, although nutrient cycles and plant growth can be sensitive to changes taking place within trophic groups of secondary consumers, these effects have proven to be weak as compared to effects resulting from interactions between true trophic levels and interactions taking place close to the base of the detrital food web.

Species Diversity and Ecosystem Processes

Based on the obvious non-specialized feeding habits of most soil organisms of the detrital food web it may be predicted that the relationship between species diversity and ecosystem functioning is weak. The few studies that exist of this topic indeed refer to high functional similarity between species across various trophic groups. Liiri et al. (2002a) manipulated the species number (1 to 51 species) of soil microarthropods (Acari and Collembola) in microcosms containing seedlings of silver birch. The tree seedlings grew significantly larger in systems with a couple of species of microarthropods as compared to microarthropod free controls. However, tree growth in systems with five or more species did not differ from those containing the full set of microarthropod community, indicating a high degree of functional redundancy among these fauna. Similarly, in their microcosm experiment Setälä and McLean (2003) manipulated the taxonomic richness of soil saprophytic fungi and followed decomposition activity (measured as CO_2 production) in the systems. They showed that decomposition activity was only weakly related to the diversity of soil fungi in the microcosms. As with the study with microarthropods (Liiri et al., 2002a) the only significant differences in decomposing activity between the diversity treatments took place at the species poor (1 to 5 species) end of the gradient (Setälä and MacLean, 2003) (see Figure 1D). The limiting data available thus suggests that the idea of complementary resource use (Tilman et al., 1997) (i.e., that each species possesses certain traits that allow them to utilize resources differently, holds poorly in soils).

ECOSYSTEM FUNCTION IN RELATION TO STABILITY AND CONTEXT DEPENDENCY OF DETRITAL FOOD WEBS

The stability of soils relate to the strength and diversity of trophic interactions rather than to the complexity of taxa within trophic groups. Furthermore, the role of mycorrhizal fungi in belowground food webs can change the nature of biodiversity-ecosystem functioning links across N-gradients.

Biological Complexity and Stability of Soil Food Webs

Food web theory predicts that food web control is intimately associated with stability of the webs (Rosenzweig, 1971; McCann et al., 1998), and that dynamic stability of food webs is, in turn, defined by the patterning of trophic interactions within the community (de Ruiter et al., 1995). Moore et al. (2003) specified the dynamic stability concept by including nutrient transfer rates as an important parameter controlling the structure and stability of not only soil food webs but also those of the above-ground systems. It is to be noted here that food web stability in its traditional form refers to the resistance and resilience of populations and communities, whereas in the current context the stability concept has strong annotations also to the functional attributes of food webs.

Irrespective of the perspective, food web theory and recent soil food web models (Moore and Hunt, 1988; Moore and de Ruiter, 1991; de Ruiter et al., 1995; Neutel et al., 2002) refer to a close connectance between soil food web architecture, ecosystem function and system stability. Does empirical data support the theory?

Despite its general interest only few well-designed experiments studying the complexity–stability relationships have hitherto been published. In their microcosm experiments Liiri et al. (2002a) and Setälä and McLean (Setälä and MacLean, 2003) attempted to shed light on the "biodiversity begets stability" statement by testing the so called "insurance hypothesis (Naeem and Li, 1997). According to the hypothesis the ability of a system to resist and recover from a disturbance is related to the biological diversity of the system. However, neither of the two studies (Liiri et al., 2002a; Setälä and MacLean, 2003) in which drought was applied as a disturbance found consistent evidence to support the hypothesis. Instead, it was the species composition rather than the numerical complexity of the community that explained the resistance and resilience of the systems (Liiri et al., 2002a; Setälä and MacLean, 2003). In contrast, Griffiths et al. (2000) showed in a microcosm experiment that soils with high complexity (i.e., including bacteria, fungi, protozoa, and nematodes) were more resistant to a perturbance (produced either by heat or a chemical) than soils with impaired (mostly secondary consumers removed) complexity. Scattered experimental evidence thus suggests that functional attributes, including stability, of soils relate to the strength and diversity of trophic interactions rather than to the complexity of taxa within trophic groups.

Conditionality of Food Web Complexity-Ecosystem Functioning Interaction

As the patterning of food webs is influenced by various environmental factors, such as the productivity of the system (Moore and de Ruiter, 2000; Wardle, 2002a), the ways as to how ecosystem functions are controlled by soil food webs are also likely to be context dependent.

This warrants us to take for granted that the weak diversity-ecosystem functioning relationship in soils (Setälä et al., unpublished manuscript), including the stability aspects, should hold beyond the more or less artificial microcosm conditions. However, as it is notoriously difficult to establish field experiments with properly controlled belowground communities (Lawton, 1996; Scheu and Setälä, 2002), soil ecologists have been obliged to test hypotheses and theory using a simplifying micro- and mesoscosm approach. Although the example given later derives from a microcosm experiment, it nevertheless emphasizes the context-dependent nature of interactions between food web structure and ecosystem functioning.

Thus far a belowground trophic interaction has been ignored that can—both from an energetic and nutrient dynamic point of view—be expected to exert the most influential control over the structure and functioning of entire ecosystems: the interaction between plants and mycorrhizal fungi. It is well established that without their fungal mutualists plants growing in nutrient deficient soils do worse than plants having a mycorrhizal association (Allen, 1991). From a food web perspective, the plant-fungi interaction provides another interesting and functionally important tri-trophic level interaction—that between plants, mycorrhizal fungi, and their grazers. As mycorrhizal fungi provide a quantitatively and qualitatively rich food resource for various soil animals (Moore et al., 1985; Schultz, 1991; Klironomos et al., 1999), it may be predicted that grazing by soil fauna on these fungi can negatively affect NPP by distorting the energy and nutrient pipe line between the fungi and their host plant. This trophic interaction would have a direct impact on plant performance and thereby a potential to control the successional development of plant communities (this effect parallelling to the effects of plant-root herbivore interactions described in Brown and Gange, 2002). An important lesson here is that the complexity of the (i) mycorrhizal and (ii) microbial feeding community, together with abiotic conditions of the soils, appears to determine whether grazing by fauna upon mycorrhizal fungi is positive or negative for the plant. Feeding by fauna upon ectomycorrhizal (EM) and vesicular-arbuscular (VA) fungi appears to be negative for the plant in biotically simple communities with one species of fauna and one species of mycorrhizal fungi (Finlay, 1985; Ek et al., 1994). However, in the presence of a diverse mycorrhizal community and a rich fungal-feeder community (composing of fauna capable of browsing, grazing, and piercing the fungi) the performance of plants can be better as compared to plants grown in the absence of fungal grazers (Setälä, 1995, 2000). This can be the case despite of the dramatically reduced mycorrhizal biomass in the grazed systems (Setälä, 1995). In other words, the outcomes of direct trophic interactions between fungal grazers and mycorrhizal fungi, and the indirect interactions between fungal consumers and plants, can be conditional to the biological complexity of the below-ground food web.

Interestingly, the outcomes of such multitrophic level interactions are likely to be more complex and context dependent than we may expect. Besides biological complexity (evidently related to grazing intensity) the importance of abiotic resources, nutrients in particular, can be responsible for this complexity. Fertility of the soil can determine whether mycorrhizal grazers are beneficial or detrimental for the plant. This is because depending on the availability of nutrients in the soil, the relationship between plants and mycorrhizal fungi can vary from mutualism via neutralism to parasitism (i.e., the outcome of the interaction is conditional to abiotic conditions). When nutrients are in short supply abundant fungal-feeding soil fauna can overconsume the beneficial mycorrhizal fungi and thus render the symbiosis ineffective (Setälä et al., 1997). In contrast, under nutrient rich conditions when mycorrhizal fungi may start behaving as a parasite, grazing by soil fauna on mycorrhizal fungi can be beneficial to the plant by constraining the growth of its parasite (Setälä et al., 1997; Scheu and Setälä, 2002). Besides direct effects on plant productivity, fertility of a site can thus indirectly—via regulating the multitrophic level interactions between plants, mycorrhizal fungi, and soil fauna—control the performance of plants and their fungal symbionts. This adds a new dimension as to how site fertility can control primary production and the consumers, both above- and below-ground (compare to Hairston et al., 1960; Oksanen et al., 1981; Moen and Oksanen, 1999; Wardle, 2002a).

DISCUSSION AND CONCLUSIONS

Soils and soil food webs appear to share features that differentiate them from aquatic and aboveground habitats and their food webs. The heterogenous resource supply and the three-dimensional structure of soils are likely to contribute to the high degree of omnivory within soil biota. This, in turn, impacts on how predictably soil food web structure relates to functional attributes, including stability, of the food web. Based on data of an extensive literature analysis, Neutel et al. (2002) showed mathematically that food webs are organized in trophic loops in a way that enhances food web stability. Indeed, omnivory represents a typical trophic loop in that the proportion of weak links—having a small contribution to the energetics of the system and shown to increase stability (Neutel et al., 2002)—necessarily increases with increasing degree of omnivory in a food web. If trophic loops enhance stability of community food webs, ecosystem processes can also be predicted to be resistant to perturbations in food webs where omnivores are common. Soil food webs serve as good examples of this.

Theoretical predictions, as previously presented, are, at least partly, supported by empirical data. Interactions between well-defined trophic levels existing close to the base of detrital food webs, such as those between microbial consumers and microbes, commonly manifest themselves as unambiguous alterations in system-level processes. However, removal or addition of trophic groups at the intermediate or high trophic positions is less likely to translate to substantial alterations in decomposition processes and NPP. The same applies with interactions within feeding guilds: although it is fairly well shown that the rate of decomposition processes can be sensitive to a specific composition of species (Laakso and Setälä, 1999a; Mikola et al., 2002), the influence of diversity *per se* within a functional group on soil processes is marginal. It is suggested that the omnivorous feeding habits and resource use of soil biota creates functional redundancy and enhances stability of detrital food webs. However, the context dependent nature of the outcomes of some important trophic relationships should not be ignored when evaluating the strength of the relationship between biological complexity and ecosystem functioning in soil food webs.

Many unresolved questions emerged from the various subsections of this chapter. For example, what limits the biomass of top consumers in a seemingly donor controlled system? Carefully planned experiments combined with a stoichiometrical approach could provide fruitful insights here. Further, does consumer biomass or consumer diversity more strongly control N-availability and NPP? The answer to this question would bear directly on biodiversity-ecosystem functioning studies as well as debates about whether the concept of trophic levels is indeed valid in soils. Finally, would external loading of nutrients determine the net effects of saprophytic microbes and mycorrhizae on NPP? Although all these questions can effortlessly be subjected to testable hypotheses under controlled experimental conditions, a comprehensive view of the role of the biological complexity within soil food webs is difficult to reach without larger-scale field experiments.

6.5 | DIVERSITY, PRODUCTIVITY, AND INVASIBILITY RELATIONSHIPS IN ROCK POOL FOOD WEBS

Beatrix E. Beisner and Tamara N. Romanuk

Unraveling the ecology of species invasions has as its ultimate pragmatic goal the ability to halt the spread of exotic species by predicting where invasions are likely to occur and which species are of greatest threat. Observations and theory suggest that aspects of food web structure such as species richness may be more important predictors of the extent and success of species invasions than the traits of invading species (Elton, 1958; Case, 1990; Jenkins and Buikema, 1998; Levine, 2000; Naeem et al., 2000b; Davis and Pelsor, 2001; Shea and Chesson, 2002; Stachowicz et al., 2002). However, species richness *per se* is unlikely to inhibit invasion. Instead, species-rich communities may have fewer available resources than species-poor communities, owing to greater niche overlap or to large effects of dominant species, thus limiting the success of invaders. Whether this same pattern should hold in a multi-trophic level system of wild organisms remains to be generalized. To determine whether food web properties such as species richness or resource limitation affect invasibility, it is necessary to manipulate both diversity and resources simultaneously. The search for general principles that govern invasibility however, is complicated by a lack of consistent measures for both invasion success and for the consequences of invasions at the community level. Furthermore, to determine whether species traits that shape food webs, such as trophic habits of invaders (Moyle and Light, 1996), are important components of invasion success, it is necessary to perform comparative studies in similar communities with different invaders.

Recent invasions of the fish hook waterflea, *Cercopagis pengoi* and the spiny waterflea, *Bythotrephes longimanus* into new habitats have generated interest in determining the mechanisms underlying invasibility in aquatic food webs. Propagule pressure is thought to be a key mechanism for zooplankton communities (Shurin, 2000). However, community resistance to, and consequences of invasion in aquatic food webs has been investigated only to a limited degree. Are the factors that confer resistance in other ecosystems, like limited resource availability or stress (Huenneke et al., 1990; Davis et al., 2000) and native species richness (Elton, 1958; Vermeij, 1991; Kennedy et al., 2002) also important in aquatic invertebrate communities? In two different experiments we tested whether (i) resource limitation at the base of the food web measured as nutrient availability and (ii) food web structure as determined by species richness across a multi-trophic level system influenced invasibility of tropical freshwater rock pool food webs. We also explore the consequences of invasion for these communities. In these experiments, reticulate invertebrate food webs consisting of detritivores, herbivores (both filter feeding and scraping types), and facultative predators were used to determine whether nutrient limitation or diversity plays a greater role in both the resistance to invasion and in the response of the community to invasion when diversity is spread over a number of trophic levels.

We made the following predictions: (1) augmented nutrient levels should increase the likelihood of successful invasion because, as for purely competitive systems, resource limitation may propagate through the food web; (2) higher levels of biodiversity (species richness or diversity) should decrease invasibility for the invader because of a greater presence of competitors and/or predators in rich food webs.

In this chapter, we examine a number of statistics to assess community-level resistance and response to invasion events. Until now, there has been no agreed upon measure for invasion success and consequences. Therefore we use several statistics to provide a comparison and contrast of measures on the same experimental data. We also compare and contrast data from two studies conducted in the same rock pool ecosystem but under different experimental invasion conditions to see if measured responses are robust to differences like the degree of control over community structure and invader identity. Differences in the trophic position and life history traits of invaders in each experiment will be considered as we compare invasion into similar communities by an obligately aquatic omnivore/detritivore with a rapid generation time (an ostracod species) to that by a longer-lived herbivore with a terrestrial adult stage (a midge species).

METHODS

Using the same set of tropical freshwater rock pool meiofaunal communities (Romanuk and Kolasa, 2001, 2002a, b, 2004, for more detail) we conducted two studies that differed in experimental design, community membership, and trophic habits of the invader (Table 1). Rock pool food webs are predominately detritus-based. In 15 years of sampling 49 natural rock pools, approximately 75 species have been identified with the meiofaunal component consisting primarily of copepods, cladocerans, ostracods, worms, and insect larvae. For both experiments we assembled microcosms using species from freshwater rock pools. In the controlled invasion experiment the species assemblage consisted of seven species: two ostracods (*Candona* sp., *Cypridopsis* sp.), one copepod (*Nitocra spinipes*), and three cladocerans (*Alona davidii, Leydigia leydigia, Ceriodaphnia* sp.) with a competitive ostracod invader (*Potamocypris* sp.). For the natural invasion experiment the species assemblage consisted of 11 species: three ostracods (*Candona* sp., *Cypridopsis* sp., *Potamocypris* sp.), two cladocerans (*Alona davidii, Ceriodaphnia* sp.), one copepod (*Orthocyclops modestus*), a larval decapod (*Armases miersii*), two worms (*Nematode* sp., *Oligochaete* sp.), and a dipteran larva (*Culex* sp.) with an insect invader (*Dasyhelea* sp). Trophic habits of the species varied from detritivores (worms, *N. spinipes*) filtering herbivores (cladocerans), detritivore-omnivores (ostracods), and predators (*O. modestus*). Figure 1 shows the probable food webs used in the experiments with trophic habitats based on preliminary stable isotope data (Beisner and Romanuk, unpublished data). Generation times for these taxa are rapid (approximately a few days to two weeks) owing to high water temperatures.

The experiments were conducted at the Discovery Bay Marine Laboratory in Jamaica and the experimental designs are contrasted in Table 1. The controlled invasion experiment was conducted in January 2003 and the natural invasion experiment was conducted in September–October 2001. Experimental communities were maintained outdoors in 500 mL plastic containers, and the microcosms were randomly redistributed every three days in order to minimize site-specific effects.

In the controlled invasion experiment the composition of the source community and the invader was controlled and constant across replicates.

Table 1. Design of invasion experiments. For the controlled invasion experiment, low richness communities were nested subsets of the more diverse ones. In the *Community composition* row for controlled invasion, the first number in the brackets indicates the number of individuals added and the second number (BT = #) is the biodiversity level at which that species was first incorporated.

	Experiments	
Experimental Design	Controlled Invasion	Natural Invasion
Factor Levels	4 X 3	5 X 3
Levels of biodiversity (BT)	4 levels: 2, 4, 6, 7 species	5 levels: 0, 25%, 50%, 75%, 100% density
Levels of nutrients (NT)	3 levels: Ambient, 10X, 100X (total)	3 levels: Ambient, 6X, 13X (soluble)
Replicates	10	3
Nature of invasion	Controlled	Natural
Frequency of sampling	One final sample (day 42)	Weekly, 4 samples
Microcosm volume	500 ml	500 ml
Sample volume	500 ml	30 ml
Container	Plastic with lid	Plastic without lid
Community Design		
Community assembly	Fixed number and composition	Random dilutions of natural assemblages
Invader species	*Potamocypris* sp. (3 individuals in each replicate)	*Dashelya* sp. (naturally oviposited in microcosms)
Community composition	*Cypridopsis* cf. *mariae* (5, BT = 2)	*Candona* sp.,
	Ceriodaphnia rigaudi (5, BT = 2)	*Cypridopsis* cf. *mariae*
	Leydigia leydigia (3, BT = 4)	*Potamocypris* sp.
	Candona sp. (3, BT = 4)	*Alona* sp. *Ceriodaphnia rigaudi.*
	Nitocra spinipes (3, BT = 6)	*Orthocyclops modestus* *Armases miersii*
	Alona davidii (2, BT = 6)	Nematode species
	Orthocyclops modestus (6, BT = 7)	Oligochaete species *Culex* sp.

At the beginning of the experiment community members were added to rock pool water filtered through 63 μm silk mesh at densities indicated in Table 1. Following 16 days of acclimation and growth, three individuals of the invader species (Ostracoda: *Potamocypris* sp.) were added to each container. Nutrients were manipulated by adding N and P (as solutions of

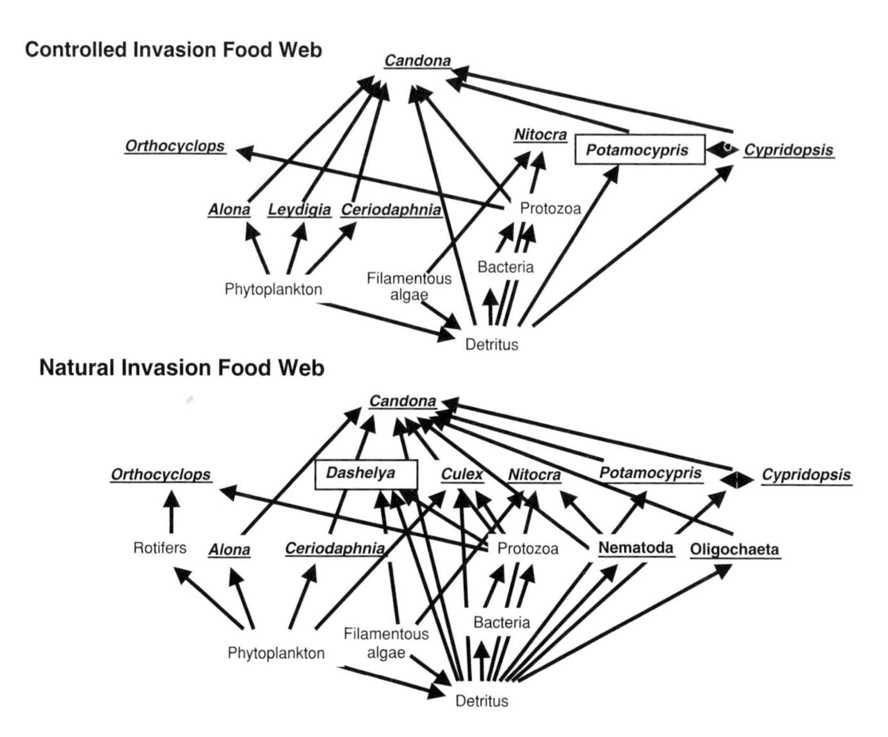

FIGURE 1 | A rough description of the rock pool food webs for each experiment based on preliminary stable isotope date (not presented). A full analysis of the rock pool food web structure has not yet been done. Manipulated species are underlined and the invasive species in each case is indicated with a box.

KH_2PO_4 and $NaNO_3$) at 10x and 100x ambient levels estimated from average values for 49 natural rock pools. Forty-two days after the start of the experiment, the entire container contents were filtered through a 63-μm mesh.

In the natural invasion experiment, communities were assembled using a dilution series (Romanuk and Kolasa, 2004). Rock pool water was first filtered (63 μm) to remove all target species. This filtered rock pool water was then added in varying degrees to unfiltered rock pool water to assemble five diversity levels (0%, 25%, 50%, 75%, and 100%). The percent represents the amount of unfiltered water with organisms included. The natural rock pool communities used in this experiment contained 11 species of invertebrates (see Table 1). Microcosms were left to be naturally invaded by a biting midge of the genus *Dasyhelea* (the invader), which has an aquatic larval but a terrestrial adult stage. *Dasyhelea* is a common native species in the area that uses rock pools, container habitats, rain pools, phytotelmata, and other small bodies of water to oviposit. Nutrients were manipulated by adding N and P in a 20:1 N:P ratio as dissolved NH_4NO_3 and H_2PO_4. Nutrient level I (NT = 1) was obtained adding nutrients once at the start of the experiment (n = 15). Nutrient level II (NT = 2) was obtained by adding nutrients at the start of the experiment and again after the first week (n = 15). The other 15 microcosms had no nutrients added and served as controls for the nutrient manipulations (NT = 0). The experiment was run for one month with weekly sampling. All statistics were calculated as averages over the last three weeks of sampling, except for extent of invasion, which only used the data from the final sample (see Table 1).

We used a variety of measures to assess resistance to and consequences of invasion (Table 2). The manipulation of species richness in both experiments will be collectively referred to as the biodiversity treatments (BT). Successful invasions are those where the invader population increased from initial densities (from inoculation levels or from 0). All analyses were repeated separately for each experiment.

RESULTS

Invader Response Measures

We compared the abundance of the invader relative to the total abundance of the entire community at the end of each experiment. Significant increases in invader abundance relative to other species' contribution imply a more successful invasion. Relative abundances were compared across treatments in separate 2-way ANOVA's with biodiversity and nutrient levels as treatments (see Table 2). Similar responses were observed in both experiments with regard to biodiversity effects (see Figure 1): a decline in invader relative abundance with increases in biodiversity. An increase in invader relative abundance in response to augmented resource levels was only observed in the natural experiment although trends to the same effect were observed under the controlled conditions.

Dominance of the Invader

A successful invasion can also be defined as a situation where the invader becomes the dominant species in the community. This idea is most related to classic Lotka-Volterra competition theory (extended to community assembly rules) where, in the absence of perturbation, one species eventually excludes the others. We determined the number of replicates

Table 2. Effect of nutrient treatment (NT) and biodiversity treatment (BT) on measured community properties in each experiment.

Factor	Measure	Controlled Invasion	Natural Invasion
A. Invader Response			
(i) Invader relative abundance (Figure 1)	Relative abundance of the invader	• negative effect of BT ($P < 0.0001$) • no NT effect	• negative effect of BT ($P < 0.0001$)* • positive effect of NT ($P = 0.032$)
(ii) Dominance of the invader	Number of replicates where invader becomes the highest ranked species	• no effect of BT or NT	• significant effects of NT and BT* ($\chi^2 = 33.409$, df = 15, $P = 0.004$) • negative effect of BT • positive effect of NT
(iii) Extent of the invasion (Figure 2)	Proportion of patches invaded	• not significant *Trends:* • negative effect of BT • positive effect of NT	• not significant *Trends:* • positive effect of NT
B. Community Response			
(i) Relative abundance	Evenness	• no change in E' with invasion	• no change in E' with invasion
(ii) Species composition	Observed population declines	• no significant differences in proportion of populations declining with invasion success ($\chi^2 = 2.773$, df = 1, $P = 0.0959$)	• significant difference in number of populations declining with invasion success ($\chi^2 = 22.303$, df = 10, $P = 0.0136$)* • significant decreases in *Alona*, *Orthocyclops*, and Nematode with invasion
(iii) Community displacement	Correspondence analysis	• displacement decreases with invasion success ($P < 0.0001$)	• no change in displacement with invasion success* • *Trend*: greater displacement with invasion success

* In the natural invasion experiment, we used mean values calculated over three weeks of sampling dates.

where the invasive became dominant and compared these across combined treatments of BT and NT using Chi-square analysis of the associated contingency table (Zar, 1984). Results differed between experiments for this measure with responses only appearing under natural invasion conditions, where in general, there were trends of negative effects of biodiversity and positive effects of nutrients on invasion success (see Table 2). Under natural invasion conditions, the invader numerically dominated 6 of the total 45 microcosms. Five of these were at the lowest biodiversity treatment (0%) and one at the 25% treatment. Nutrient enriched microcosms were dominated more often than low nutrient microcosms. In NT = 1 and NT = 2 the invader became dominant in 3 of 15 replicates, whereas the invader never became dominant in NT = 0. Under controlled invasion conditions, the invasive species only dominated the communities in two of the total 120 replicates, both at lowest biodiversity (BT=2), one under intermediate and one high nutrient conditions.

Extent of Invasion

The advantage of an experimental study of invasion is that it provides an opportunity to examine the repeatability of an event. We calculated the proportion of replicates in each treatment that were successfully invaded and compared these using Chi-square analysis of the associated contingency table of BT and NT vs. proportion invaded (Zar, 1984). Although in both experiments there was no significant effect, there was a trend in both cases for a positive effect of nutrient levels, with more replicates invaded under high nutrient conditions. In the controlled experiment, there was a further trend toward a negative effect of species richness (Figure 3).

Community Response Measures

In addition to gaining insight into how ecosystem properties prior to invasion influence the invasibility of communities (or the success of the invader), we can ask how successfully invaded communities were changed by the invasion itself. To do so, we compare several community-level responses between successfully vs. unsuccessfully invaded communities.

Community Relative Abundances

Responses by communities related to changes in relative abundance as those just discussed might most easily be observed using an estimate of community evenness like the Camargo index:

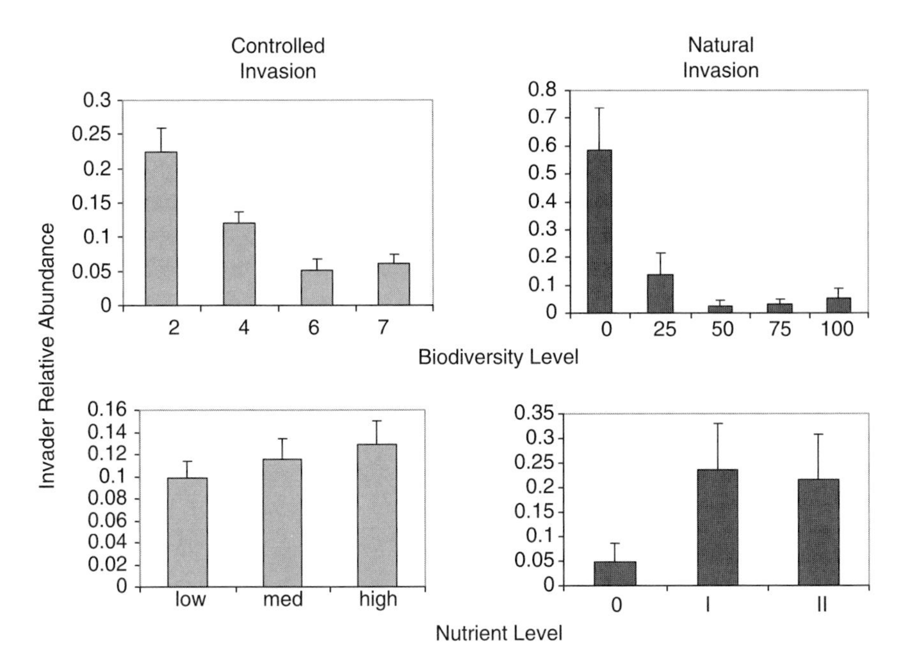

FIGURE 2 | The mean (± standard error) relative abundance of the invader species in each experiment showing the responses to biodiversity and nutrient addition treatments.

$$E' = 1.0\left[\sum_{i=1}^{s}\sum_{j=i+1}^{s}\left[\frac{|p_i - p_j|}{s}\right]\right] \quad (1)$$

where E' is the index of evenness, p_i is the proportion of species i in the total sample, p_j is the proportion of species j in the total sample and s is the total number of species in the sample. This index is useful for our purposes because it is unaffected by levels of species richness and is relatively unaffected by rare species (Krebs, 1999). The index ranges from 1 for completely even communities to 0 for a lack of evenness. Values of E' were calculated for each replicate community (final community structure including invader for the controlled invasion and average structure over three final dates for the natural invasion experiment). This community characteristic was unaffected (t-test) by the success of the invader in both the controlled invasion ($E'_{successful} = 0.55 \pm 0.03$ vs. $E'_{unsuccessful} = 0.49 \pm 0.03$) and the natural invasion ($E'_{successful} = 0.41 \pm 0.04$ vs. $E'_{unsuccessful} = 0.46 \pm 0.04$) experiments.

Species composition

Loss of species from the community is a potential response of communities to invasion. Documenting which species decline where the invader is successful provides useful biological information about susceptible species. Changes in species composition were assessed by determining which species were lost from communities that were invaded. Resident species only showed significant declines when *Dashelya* was the invader (see Table 2).

Community displacement

To estimate the effect of invasion on the entire community we used a multivariate statistic (Correspondence Analysis [CA]),to determine the consequence of invasion on the dissimilarity of invaded and uninvaded communities. We ran a CA to identify the locations of all replicate communities in multivariate space relative to the starting condition on abundance data that was relativized across species and downweighted for rare species. The length of the vector defining the distance of each replicate from starting conditions was calculated and the mean lengths were compared in a t-test for invasion effects. We found contrasting results in the two experiments. In the controlled invasion experiment the communities moved less between initial and final conditions when invasions were successful ($distance_{successful} = 67.73 \pm 6.5$, $distance_{unsuccessful} = 107.03 \pm 6.6$). For the natural invasion experiment successful invasion by *Dashelya* led to a greater change in the community in multivariate space than for communities that were not invaded successfully ($distance_{successful} = 149.3 \pm 39.8$, $distance_{unsuccessful} = 106.651 \pm 40.7$) but this was only a trend and was not a significant difference.

DISCUSSION AND CONCLUSIONS

Together, our studies demonstrate that for rock pool invertebrate food webs, species richness across various trophic levels inhibits the success of species invasions, and that this effect does not depend on differences in trophic habits of the invasive species. The effect was stronger in the natural community experiment (with more response variables showing a significant response) than in the controlled experiment—a difference that will be discussed later. Thus, we find a similar effect of richness in our multi-trophic level systems as has been demonstrated in purely competitive systems where species-rich communities have higher resistance to invasion (Tilman, 1997b; Kennedy et al., 2002; Stachowicz et al., 2002). However, these previous studies in competitive systems have shown that species diversity enhances resistance to invasion by crowding and more complete use of space which is not the case for our system where the limiting resource is nutrient availability, particularly phosphorus (Romanuk and Kolasa, 2004).

In the rock pools where competition for nutrient resources occurs, nutrient availability may only inhibit invasions for species with specific trophic habits. For example, nutrient limitation could inhibit invasions of herbivores or possibility predators but would be unlikely to inhibit invasion of detritivores. We found that in the natural invasion experiment relative abundance of the invader was affected by a strong interaction between species richness and nutrient conditions (see Romanuk and Kolasa, 2004, for more detail), while in the controlled invasion experiment there was no effect of nutrient conditions on any invasibility response measures. Thus, while trophic traits of the invasive species appear to be unrelated to whether species richness is an important determinant of invasibility, they do affect whether nutrient limitation plays a role. When the invader is competing for algal resources (natural experiment), nutrient levels were important, but they were not important when the invader competed primarily for detrital resources (controlled experiment) — at least in the short term. In the natural invasion experiment the invader was a herbivorous midge that oviposited in the microcosms, likely resulting in competition between the midge larvae and cladocerans for algal resources. In the controlled invasion experiment the invader was an obligately aquatic ostracod that competed with other detritivores-omnivorous ostracods in the resident community.

In contrast to our results for invader success and invasibility, our results for the response of each community to invasion were highly variable between the two experiments. Again, this was likely a result of the different trophic habits and life history strategies of the invaders used in the two experiments. In the controlled invasion experiment, we introduced an ostracod into a community that already contained other ostracods. While there is likely competition (and potentially predation) between the three ostracod species, these species are often found coexisting in natural pools suggesting that they can partition resources. This may explain why there was a smaller community shift and no significant declines in other populations when this “resident” invader species was introduced into the controlled microcosms. In the natural invasion experiment the invader was a temporary rock pool resident, only interacting with the rock pool food web directly during its larval phase. In the adult stage, midges are able to choose between many rock

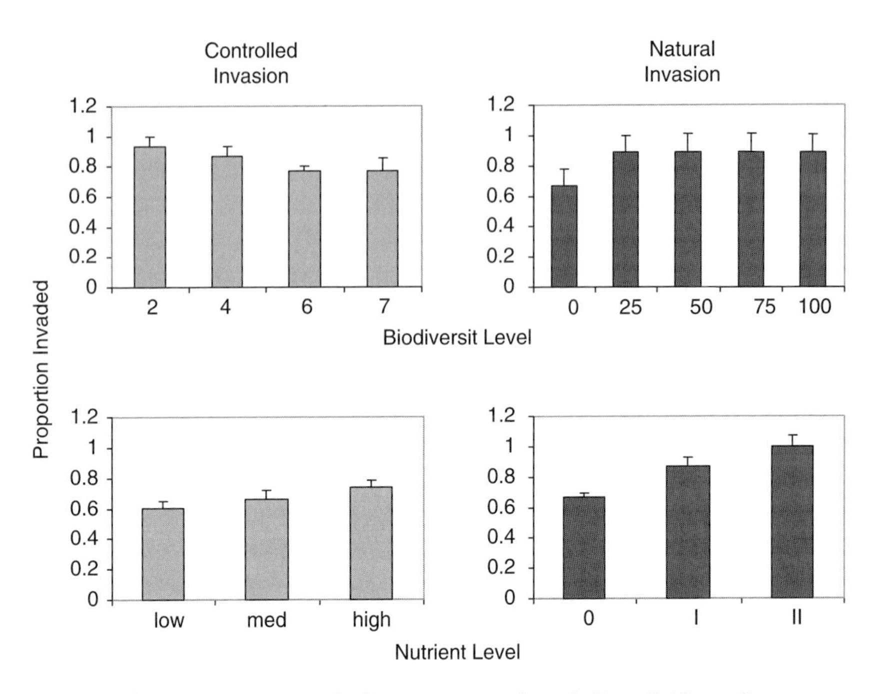

FIGURE 3 | The mean (± standard error) proportion of all available replicates successfully colonized by the invader species in each experiment showing the responses to biodiversity and nutrient addition treatments.

pools with different resource levels for ovipositing. In their larval phase, midges can induce strong negative interactions with other meiofaunal species in one particular rock pool without suffering individual mortality or fitness losses since, over their lifespan, they are operationally dealing with many different potential habitats and their populations are additionally subsidized by the terrestrial environment. In the natural invasion experiment, there was a trend towards greater losses of resident species and larger community displacement in multivariate space with invasion as compared to the ostracod controlled invasion experiment. Thus, our work shows that both susceptibility of a food web and their responses to invasion can be strongly dependent on the traits of an invasive species.

We can draw several conclusions based on a comparison of these studies in rock pools. Invasive species traits such as trophic habits in relation to actual resource availability appear to constrain invasibility. Similarly, the scale at which an invader operates in terms of reproduction and mobility relative to the characteristic scale of the native food web also appears to affect the response of the community to invasion.

Species richness appears to inhibit invasion regardless of the trophic status of the invader, whereas nutrient limitation is an important determinant of invasion success only for invaders with specific trophic traits (i.e., herbivores). We also found that the various measures of invasibility and community responses to invasion produced conflicting results, suggesting that a wide range of measures of invasibility should be used in diversity-invasion studies to allow an assessment of overall effects. Finally, by comparing invasion into more controlled and more natural food web assemblages, we find that overall, there are stronger responses observed for the main treatment effects in the natural food web experiment. We hypothesize that this may be a result of a greater maintenance of past community structure in the natural experiment, but the actual mechanisms for the effect remain to be examined in future work.

Invasion into food webs has been little studied outside of the assembly rules domain in ecology (Drake, 1990). Much work remains to be done to determine whether there are general rules about the relationships between invasibility and food web structure, as well as the influence of invader life histories on food webs. The understanding gained from such work will better enable ecologists to contribute to the pressing problem of a global homogenization of species currently underway (Rahel, 2002).

ACKNOWLEDGMENTS

We thank J. Kolasa and A. Hayward for useful discussion and help with field work and J. Hovius for sample processing. Thanks to two anonymous reviewers for comments. Financial support from NSERC (Canada) is gratefully acknowledged.

6.6 | MEASURING THE FUNCTIONAL DIVERSITY OF FOOD WEBS

Owen L. Petchey, Jill McGrady-Steed, and Peter J. Morin

What will be the functional consequences of extinctions? What is the role of biodiversity in determining ecosystem functioning? One way of approaching these questions is to understand relationships between biodiversity and ecosystem level processes (Schulze and Mooney, 1993; Jones and Lawton, 1995; Kinzig et al., 2002; Loreau et al., 2002), and this has been the subject of a decade of intense experimental and theoretical research. Many experiments have manipulated the number of species in a local community and observed the effect on ecosystem level variables, such as primary production, gas flux, decomposition rates, invasion resistance, and nutrient depletion (Wardle et al., 1997a; van der Heijden, 1998; Levine, 2000; Fridley, 2002). Some reviews of this work suggest a general positive relationship between species richness and levels of ecosystem processes, though the strength and even direction of the relationship can depend on factors such as spatial scale (Wardle et al., 1997b; Loreau, 2000). Where positive relationships between biodiversity and ecosystem functioning are clear, discussion turns to the mechanisms responsible: typically whether complementary use of resources or more probabilistic mechanisms cause the relationships (Aarssen, 1997; Huston, 1997; Hector et al., 2002). Both the observation of weak and variable effects of species richness on ecosystem level properties and theoretical recognition of the role of complementarity in creating these relationships (Loreau, 1998a) has led to calls for measures of biodiversity that better explain ecosystem functioning. Measures based explicitly on the extent of resource use complementarity among the species in a community provide one possible approach (Díaz and Cabido, 2001; Tilman, 2001; Hooper et al., 2002; Schmid et al., 2002).

One component of biodiversity that measures the extent of resource use differences among species is functional diversity (Tilman, 2001). Hence, it may also explain and predict ecosystem functioning more accurately than does species richness. Experiments, mostly in plant communities, show that variation in functional diversity can have substantial effects on ecosystem functioning, though again effects appear contingent upon factors such as spatial scale and environmental conditions (Loreau and Hector, 2001; Tilman et al., 2001; Fridley, 2002; Hector et al., 2002; Hooper et al., 2002; Petchey, 2004; Petchey et al., 2004a; Reich et al., 2004). Nevertheless, the concept of functional diversity is relatively well developed for single trophic level communities and is a way of relating the structure and functioning of these simple communities.

At present there appears to be no equivalent measure of the diversity of food webs and there is little sign that a structural description of food webs exists that is designed to predict ecosystem level properties. Here, we briefly review the concept of functional diversity as it applies to single trophic levels and then extend the concept to food webs. We draw links with existing food web concepts and suggest a continuous measure of the functional diversity of food webs. A re-analysis of an experiment that manipulated biodiversity in model aquatic communities tests whether this measure explains variation in ecosystem functioning. We then discuss lines of evidence that suggest functional diversity could be a useful predictive variable in food webs and evidence opposing this hypothesis.

Measuring Functional Diversity

Measures of functional diversity aim to estimate the extent of resource use differences among a group of species. An example is functional group richness, which counts the number of functional groups represented by the species in a community. These functional groups (guilds are very similar) are classifications of species into groups that share similar resource use requirements and play similar functional roles in communities (Adams, 1985; Simberloff and Dayan, 1991; Chapin et al., 1996; Díaz and Cabido, 1997; Root, 2001;, but see Blondel, 2003). The converse is that species in different functional groups play different functional roles in communities. An example is division of plant species into grass, herb, and legume functional groups, where evidence exists that this division provides information about the effects of functional diversity on primary productivity (Hector et al., 1999; Hooper et al., 2002; Petchey, 2004; Reich et al., 2004). One limitation of functional group richness, however, arises when the actual functional differences are distributed continuously among species, such that categorizing species provides a distorted representation of the variation (Petchey and Gaston, 2002b). This limitation makes functional group richness useless for answering some questions about functional diversity, for example the effects of extinctions on functional diversity (Petchey and Gaston, 2002a).

Continuous measures of functional diversity that are derived from information about species' resource use patterns avoid this and other limitations (Bengtsson, 1998; Walker et al., 1999; Díaz and Cabido, 2001; Hooper et al., 2002; Petchey and Gaston, 2002b; Mason et al., 2003). Essentially, these continuous measures estimate the dispersion of species in multivariate space, where the axes of that space are the resources that the species consume, or variables that are correlated with the resources that species consume. These measures can be generalized as functional diversity = f(***X***), where ***X*** is a matrix that describes the use patterns of s species on t resources such that

$$X = \begin{pmatrix} x_{1,1} & x_{1,2} & x_{1,3} & \cdots & x_{1,t} \\ x_{2,1} & x_{2,2} & x_{2,3} & \cdots & x_{2,t} \\ x_{3,1} & x_{3,2} & x_{3,3} & \cdots & x_{3,t} \\ \cdots & \cdots & \cdots & \cdots & \cdots \\ x_{s,1} & x_{s,2} & x_{s,3} & \cdots & x_{s,t} \end{pmatrix}$$

Here, $x_{i,j}$ represents where species i lies on resource axis j. Alternately, one can say that $x_{i,j}$ represents whether species i consumes resource j, or that $x_{i,j}$ is the trait value of species i for trait j, where the trait of a species determines or is correlated with its resource use patterns. The measures (Petchey and Gaston, 2002b) are calculated using explicitly quantitative methods, rather than the less quantitative approaches that are sometimes used to divide species among functional groups (Root, 1967; Simberloff and Dayan, 1991; though see Díaz and Cabido, 1997). Evidence suggests that these continuous trait based measures are better at explaining local variation in primary production than categorical measures of diversity, such as species richness or functional group richness (Petchey et al., 2004a). However, all of the continuous measures so far proposed have been limited to single trophic levels (Walker et al., 1999), or at least groups of species that share similar resource use patterns (Petchey and Gaston, 2002b).

The Functional Diversity of Food Webs

Many descriptions of food web structure fit into the general formula, structure = $f(X)$; for example connectance, species richness, trophic height, number of trophic groups, and number of trophic species (Cohen and Luczak, 1992). Here, X is the matrix description of the food web where element $x_{i,j}$ is one if species j consumes species i (or zero if not) (for consistency with the food web literature X here is the transpose of X above). However, no structural measure is designed specifically to predict ecosystem level properties of food webs, because none provide a measure of the diversity of resource use patterns exhibited by the species in a food web. One way to measure the functional diversity of food webs is to count the number of trophic groups present. This is somewhat analogous to counting the guilds or functional groups (Simberloff and Dayan, 1991; Tilman, 1997a; Hector et al., 1999). It forces the assumption that species occupy discrete functional groups, regardless of whether they do or not (Cousins, 1987). Another way is to count the number of trophic species present, where a trophic species is a group of species that share the same or similar predators and prey (Martinez, 1991; Yodzis, 1993). This approach requires a largely arbitrary decision about the level of similarity that groups species into trophic species. The result is a number of distinct groups of species which is, again, a categorization of the differences among species.

Another option that has not previously been explored is to use one of the continuous measures of functional diversity that have recently appeared. Here, we focus on one measure of functional diversity (FD) (Petchey and Gaston, 2002b), because of advantages over some other measures (Walker et al., 1999; Mason et al., 2003). The measure of functional diversity is the total branch length of the dendrogram that results from clustering species in trait (i.e., resource) space (Petchey and Gaston, 2002b). The clustering is accomplished by measuring distances between species (columns of matrix X) in resource space (rows of matrix X) followed by hierarchical agglomerative clustering of the species according to the distances. Some examples of the behavior of this measure are given in Figure 1. The measure is related to the number of trophic species, but does not apply any level of similarity to decide which species fall into the same or different trophic species. The measure does not consider who is eaten by what, only who eats what, in keeping with the notion that functional diversity measures the diversity of resource use strategies of species.

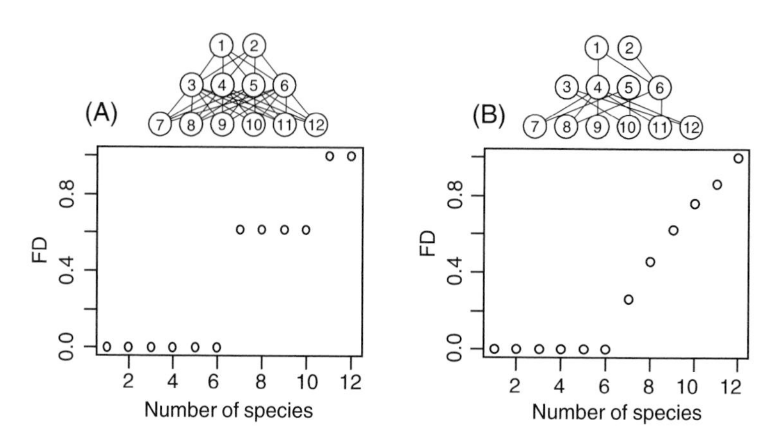

FIGURE 1 | Effects of food web structure on the decrease in functional diversity caused by losing species in numeric order. **A,** All species at a trophic level feed on all the species in the next lower trophic level; hence, all the species at a level are identical in their resource use patterns. Here, loss of a species has no effect unless it is the last in the trophic level. This illustrates that loss (or addition) of a new trophic level will cause a large increase in FD, whereas loss of species from within a trophic has little effect. **B,** When species within a trophic level have different resource use patterns, loss of a species always causes a loss of functional diversity.

MATERIALS AND METHODS

We re-analyzed data from an experimental manipulation of species richness and food web structure to investigate whether functional diversity explains a significant amount of variation in ecosystem processes. McGrady-Steed et al. (1997) varied the initial number of eukaryotic species (from 0 to 31) and the presence of some trophic groups among microbial microcosm communities that were housed in a laboratory incubator. These communities contained a variety of protist and rotifer species including algae, bacterivores, herbivores, and predators (Figure 2A). Respiration rate (CO_2 production over an 18-hour dark period) and decomposition rate (weight lost from a wheat seed) were measured over six weeks. The primary analysis (McGrady-Steed et al., 1997) found a significant relationship between species richness and both respiration and decomposition rate during the sixth week of the experiment, with an explanatory power of about 20% and 40% respectively. All other details of the experiments are contained in earlier publications (McGrady-Steed et al., 1997; McGrady-Steed and Morin, 2000).

The incidence matrix of the food web described in Fox et al. (2002) was used to calculate the functional diversity of each of the communi-

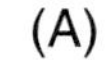

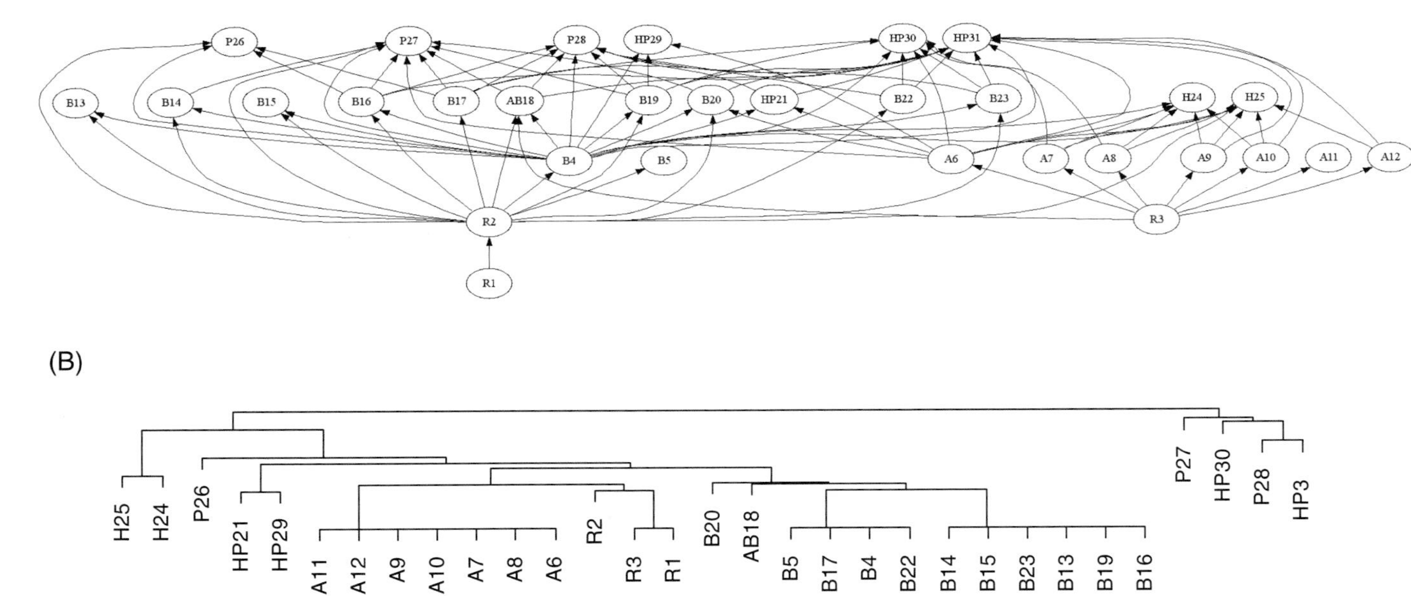

FIGURE 2 | A, The food web of McGrady-Steed et al. (1997). Letters refer to trophic groups: R = resource; B = bacterivores; A = autotroph; H = herbivore; and P = predators. Numbers refer to resource type or species: 1 = detritus; 2 = bacteria; 3 = light; 4 = microflagellates; 5 = gastrotrich; 6 = *Chlamydomonas*; 7 = diatom; 8 = *Ankistrodesmus*; 9 = *Scenedesmus*; 10 = *Staurastrum*; 11 = *Netrium*; 12 = *Euglena*; 13 = *Rotaria*; 14 = *Monostyla*; 15 = *Amoeba* small; 16 = *Aspidisca*; 17 = *Colpoda* small; 18 = *Paramecium bursaria*; 19 = *Halteria*; 20 = *Paramecium*; 21 = *Hypostome*; 22 = *Colpidium*; 23 = *Colpoda* large; 24 = *Brachionus*; 25 = *Frontonia*; 26 = *Amoeba* large; 27 = *Spirostomum*; 28 = *Heliozoan*; 29 = *Stentor* I; 30 = *Stylonychia*; 31 = *Stentor* II. **B,** The dendrogram that describes resource use relationship between the species in the food web.

ties. The measure of functional diversity used is the total branch length of the dendrogram that results from clustering species in trait (i.e., resource) space (Petchey and Gaston, 2002b). The clustering is accomplished by measuring distances between species (columns of matrix X) in resource space (rows of matrix X) followed by hierarchical agglomerative clustering. We selected the combination of distance measure (from Euclidean, Manhattan, and inverse of Pianka's niche overlap) and cluster algorithm (from average linkage, minimum variance, and single linkage) that maximized the correlation between the distance matrix and the phenetic distances across the dendrogram (Sneath and Sokal, 1973). Euclidean distance and average linkage clustering (UPGMA) resulted in the highest correlation ($r^2 = 0.96$) and the resulting dendrogram was used to calculate FD (see Figure 2B). Functional diversity was calculated from the composition of each community during each week of the experiment.

Various regression models tested the explanatory power of S and FD when respiration rate or decomposition rate was the response variable; data from each week were analyzed separately. The r^2 of the bivariate relationships between diversity and functioning indicate the relative explanatory power of S and FD. The two explanatory variables were correlated, making interpretation of the multiple regression difficult. Multiple regression gave the total explanatory power of both explanatory variables and the loss of explanatory power caused by removing each from a model containing both. There were insufficient numbers of communities to allow separate analyses within species richness levels (Petchey et al., 2004a).

RESULTS

Species richness and FD were strongly correlated across the different communities during the sixth week of the experiment ($r^2 = 0.94$; Figure 3A). This is caused by many resource axes (traits/columns) along which the species can separate and a lack of strong correlation between the resources that species use (Petchey and Gaston, 2002b). There was a relatively weak correlation between trophic group richness (species were classified as autotrophs, bacterivores, herbivores, or predators) and the functional diversity of the food web ($r^2 = 0.003$; see Figure 3B). This indicates that the trophic groups poorly represent the diversity of trophic strategies of the species in the food web. In particular, there is variation in feeding patterns within the herbivores and within the predator trophic groups.

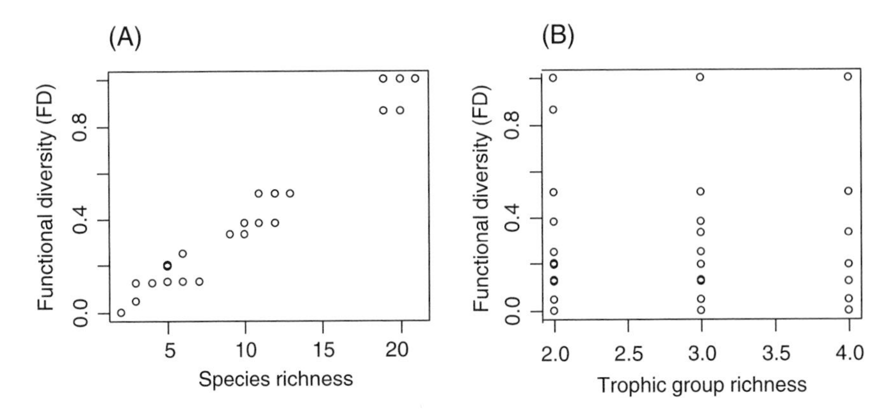

FIGURE 3 | The relationship between functional diversity (diversity of trophic strategies) and (**A**) species richness and (**B**) trophic group richness for the 50 communities in the sixth week of the experiment.

Modelling respiration rate as a function of only FD resulted in a positive relationships that were significant at $\alpha = 0.05$ during weeks three to six (Figures 4A and 5A). Using both species richness and FD resulted in a total explanatory power that varied from about 5% to 30% across the different weeks of the experiment (see Figure 5B). The total explanatory power and loss of explanatory power caused by removing either of species richness or FD from the model was significant at $\alpha = 0.01$ during the first and sixth week (see Figure 5B).

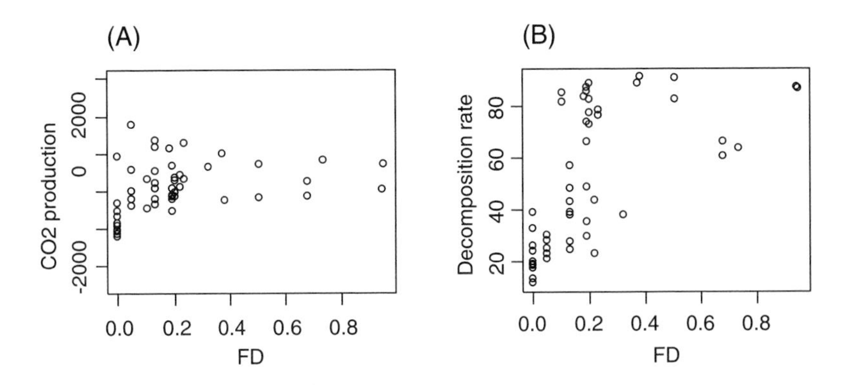

FIGURE 4 | Relationships between biodiversity and ecosystem functioning in an experimental manipulation of the biodiversity of microbes in aquatic microcosms (McGrady-Steed et al., 1997). **A, B,** Cumulative CO_2 flux (µl per 18hr). **C, D,** Decomposition rate as percent of initial dry mass of wheat seeds lost.

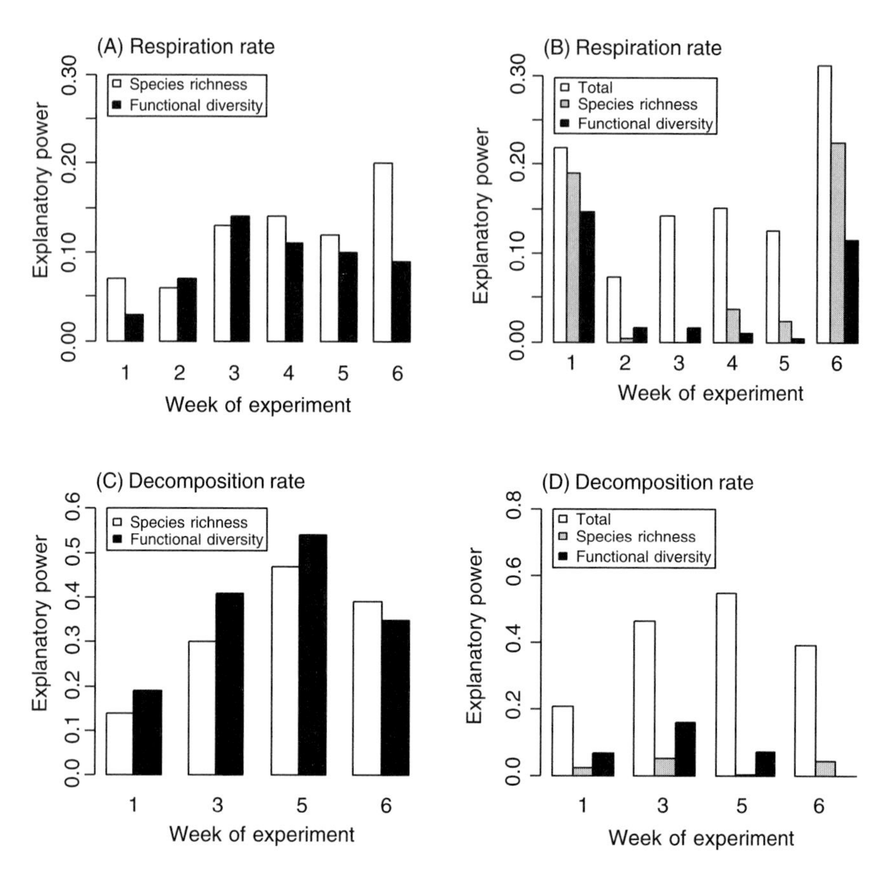

FIGURE 5 | The explanatory power (r^2) of species richness and functional diversity for respiration rates (**A**) and decomposition rates (**B**) when each was the sole explanatory variable. In (**C**) and (**D**) total explanatory power is the proportion of the variation in respiration rates (**C**) and decomposition rates (**D**) that is explained by species richness and functional diversity combined. The other bars represent the reduction in explanatory power cause by removing species richness or functional diversity from the combined explanatory power. While all of the combined models explain significant variation in ecosystem functioning, the unique explanatory power of species richness or functional diversity is never significant at $\alpha = 0.05$.

Modelling decomposition rate as a function of only FD results in a positive relationship that was significant at $\alpha = 0.01$ during all weeks analyzed (see Figures 4B and 5C). Using both species richness and functional diversity resulted in a total explanatory power that varied from about 20% to 60% across the different weeks of the experiment (see Figure 5D). The total explanatory power of the model was significant during all weeks of the experiment. The loss of explanatory power caused by removing species richness was significant at $\alpha = 0.05$ during only the third week. The loss of explanatory power caused by removing FD was significant at $\alpha = 0.05$ during the first week and at $\alpha = 0.001$ during the third and fifth week.

DISCUSSION AND CONCLUSIONS

The measure of functional diversity used above (Petchey and Gaston, 2002b) often explained a significant amount of variation in ecosystem functioning, perhaps because it provides a link between the resource use patterns of the species and ecosystem level processes. It was, however, not consistently better at explaining variation in ecosystem processes than species richness, in contrast to the results of analyses across single trophic level communities that vary in diversity (Petchey et al., 2004a). The combined explanatory power of species richness and functional diversity was rarely significantly greater than the explanatory power of either alone, indicating that both diversity measures explain similar components of the variation in ecosystem processes across communities. This is unsurprising given the strong correlation between species richness and FD. Functional diversity explained more of the variance in decomposition rate than in respiration rate, suggesting that a single measure of functional diversity will not be able to explain or predict all ecosystem processes. Different information about the species and/or different measures of diversity may be required for different functions, or diversity may have less functional significance for some ecosystem processes compared to others.

A number of hypothesis could explain the inability of functional diversity to explain ecosystem processes much better than species richness: (1) the incidence matrix may not contain the appropriate information; (2) analytical limitation caused by the experimental design; and (3) population and community dynamics make functional diversity a non-predictive concept in the context of food webs.

1. The incidence matrix contained the information used to calculate functional diversity. If this information is incomplete or inappropriate, no measure of functional diversity will link species and ecosystems. Binary information about feeding links is obviously incomplete: species feed more on some species than others. Though we had no information about these differences, if known they could easily be used to calculate functional diversity. Similarly, we characterized all autotrophs as using one resource, light, which is a great oversimplifi-

cation. Yet there must be some level of resource partitioning among autotrophs because some coexisted throughout the experiment. Again, this information could be included in the matrix from which functional diversity is derived. It is also possible that particular sets of species in a food web are critical for a particular ecosystem function, whereas others are not. For example, the unmanipulated bacteria in this system are the decomposers, so perhaps it is their diversity, and the diversity of their consumers that determine decomposition rate, whereas the diversity of autotrophs and herbivores might influence respiration. This suggests that the functional diversity of different sets of species in a food web could predict different ecosystem processes.

2. The experiment varied species richness and food web structure simultaneously, but not factorially, partly due to trophic dependencies (predators cannot persist without prey) and partly because this was not required to answer the primary question of the study. Consequently it is very difficult to separate potential effects of variation in species richness from effect of variation in trophic structure or functional diversity. A factorial manipulation of species richness and functional diversity would lead to more interpretable results and robust conclusions about the functional significance of diversity in food webs. Such an experiment would also allow more sensitive test of the influence of functional diversity and the importance of diversity within different trophic levels independently of the diversity within other trophic levels.
3. It seems valid to measure the functional diversity of food webs as previously suggested and use it as a descriptor of food web structure akin to species richness, trophic height, connectance, or even number of trophic species. However, is it sensible to assume that it will explain and predict ecosystem level processes?

Decades of research show that multiple factors determine the distribution of biomass and productivity in a food web (Polis, 1999). For example, theory and experiments clearly show that whether there is an odd or even number of trophic levels has a strong influence on whether there is a large or small standing biomass of autotrophs (Oksanen et al., 1981; Abrams, 1993; Polis, 1999). Presumably, such large changes in autotroph biomass would have large effects on ecosystem processes. However, whether there are an odd or even number of trophic levels is independent of the number of trophic levels (imagine a graph of number of trophic levels on the x-axis and zeros and ones on the y-axis to represent an odd or even number of trophic levels). Consequently, the number of trophic levels (or functional diversity) may not predict ecosystem functioning in a linear food chain or food webs with discrete trophic levels. However, real food webs show little evidence of discrete trophic levels (e.g., there is often a great deal of omnivory).

More research demonstrates the importance of the identity of trophic groups for ecosystem processes (Hairston and N.G. Hairston, 1993; Polis and Strong, 1996; Raffaelli et al., 2002), perhaps more so than the diversity of trophic strategies. We found that loss of top species (predators and herbivores) increases the primary productivity of aquatic protist communities in the laboratory (Petchey et al., 1999). Downing and Leibold (2002) conducted an experiment with diverse aquatic communities in outdoor mesocosms containing macrophytes, benthic grazers, and predators. Though all of the experimental communities contained all three manipulated trophic groups, they contained different species within those groups. The effect of these compositional differences had large effects on ecosystem processes, such as ecosystem-wide decomposition and respiration rates. Downing and Leibold concluded that trophic interactions and indirect effects are important in determining the response of the communities to changes in species composition within trophic groups.

Another example of why functional diversity may not predict food web functioning is provided by studies of interactions between trophic levels (Naeem et al., 2000a). These find evidence for 'complex effects' of diversity at two trophic levels that seem to occur because decomposers and producers simultaneously compete with each other and provide resources for each other (Harte and Kinzig, 1993; Loreau, 2001). Hence, there may be no easy way to predict producer biomass (an important determinant of ecosystem functioning) without knowing many details about the interactions between species. Similar results have been found in other systems, where manipulation of diversity at one trophic level has idiosyncratic effects on biomass at other trophic levels and ecosystem process rates (Wardle and Nicholson, 1996; Setälä, 2002).

There is good reason for trying to find a measure of functional diversity for food webs. It would allow prediction of effects of extinctions, and allow a wider understanding of the role of diversity in determining ecosystem level processes. Furthermore, many studies focus on the determinants of food web structure; we need to focus on translating that knowledge about determinants of structure into knowledge about the consequences of structure for functioning. Here, we transferred the concept of functional diversity directly from studies of single trophic levels to food webs to provide a measure of the functional diversity of food webs. It quantifies the diversity of trophic strategies employed by the species making up the food web. Consequently, it is an intrinsically interesting property of food webs. While FD explains significant variation in ecosystem functioning, it does little better than species richness. This forces the caveat that probabilistic processes associated with the design of the experiment could have caused the results (Huston, 1997). Other approaches that allow easy manipulation of food web structure without changing species richness would allow separation of the effects of species richness and functional diversity; modelling provides a possible setting for such a study.

The viewpoint presented here and illustrated by the re-analysis can been seen as an example of using food web structure to predict ecosystem functioning. The particular example is at one extreme of a continuum of possible strategies that represents the amount of information used to link between species and ecosystems. We asked whether only one property (functional diversity) of a food web explains and predicts ecosystem functioning. In addition, the single measure did not include species' abundances or interaction strengths, for example. The potential value of this extreme is that the product would be a broad generalization applicable to many different food webs. The cost may be lower explanatory and predictive power than if more information about the food web were used to predict functioning. There is no reason, however, to constrain the information used to predict functioning to this extent. Indeed, most ecologists would probably doubt whether any single property of a food web could predict ecosystem level properties. Perhaps predicting the functioning of food webs with any accuracy will require a middle ground on the continuum, where several different food web properties are used to predict functioning. It is also possible that functional diversity will be of limited use in combination with other properties. The other end of the continuum is to use extremely detailed knowledge about a specific food web to predict how it will function and respond to change. However, this extreme represents treating every food web as a special case and denying any general link between food web structure and function. Hopefully, the next decade will see development of general methods for predicting functioning from structure that explore the full range of this continuum.

ACKNOWLEDGMENTS

Joel Cohen, Andre De Roos, Dimitri Logofet, and Kevin Gaston provided valuable comments and help. Two anonymous reviewers provided excellent and stimulating comments on a previous draft of the manuscript. Jeremy Fox supplied the food web matrix in electronic format. All analyses were performed in R (the R Project for Statistical Computing) and Figure 2A was produced using Graphviz (open source graph drawing software from AT&T Labs). OP is a Royal Society University Research Fellow.

7.0 | TRACING PERTURBATION EFFECTS IN FOOD WEBS: THE POTENTIAL AND LIMITATION OF EXPERIMENTAL APPROACHES

Dave Raffaelli

Over the last decade food web ecology has become increasingly focused on environmental issues, like pollution, fragmentation, global change, and over-exploitation of natural resources (Polis and Winemiller, 1996; Baird et al., 2001). This is because the analysis of trophic interactions in food webs provides new ways to identify and quantify the overall consequences of direct and indirect effects of perturbations due to environmental stress and disturbance. The perturbation approach to understanding ecological dynamics, either through carefully controlled manipulations or by observing the effects of natural perturbations, such as disease epidemics, and man-made uncontrolled experiments, such as over-fishing, is beginning to provide major insights into how ecological systems might function. In particular, perturbations allow us to formally try to falsify hypotheses, by displacing systems from their normal operational bounds, adding rigour and persuasive power to explanations of other changes which follow the perturbation. Understanding the impacts of perturbations on food webs has long been a pre-occupation of, and a challenge to ecologists, and it remains a challenge today. Studies of the effects of perturbations on single species, or pairs of species, in isolation from the complex interaction network in which they operate are unlikely to reveal much about what might happen in nature, either to those species or to the system as a whole (see Culp et al., Chapter 7.1). However, whilst a food web or whole system approach is conceptually and philosophically more attractive, the complexities inherent in such an approach bring a further set of issues, many of which are difficult to resolve. This chapter outlines some of these issues and indicates how the chapters that follow have striven to address them.

Basic to the desire to move from the single to multi-species approach is the reasonable assumption that there are dependencies between species, so that perturbations of one species generate effects on others that propagate throughout the food web. A natural first step in trying to identify where in the food web those propagations are most likely to occur has been the construction of static pictures of binary webs depicting whom eats whom. However, even relatively simple binary web representations cannot be used to convincingly identify the likely pathways of effects propagation, given the large number of pathways and feedbacks as well as the fact that all linkages are given equal status. Nevertheless, there appears to be merit in documenting binary webs for the purposes of assessing perturbation effects, such as extinction, on general web properties such as connectance (see Harper-Smith et al., Chapter 7.5) and hence stability (Solé and Montoya, 2001; Dunne et al., 2002a; Williams et al., 2002).

One approach to identifying the likely linkages along which effects might be propagated is found within the energy flow framework for the initiatives within the International Biological Programme of the 1970s and 1980s (although this was not the sole motivation for that programme). Construction of whole-system flow diagrams and energy budgets allowed ecologists to identify those pathways and linkages which were likely to be important as well as the areas where more research was needed (Raffaelli, 2000). The empirical research underlying the energy flow approach also provides suggestions as to sensible parameter values for a suite of ecosystem-orientated models, such as Ecopath and related network analysis models (Christensen and Pauly, 1992; Ulanowicz, 1996) and exergy-orientated models (Jorgensen, 1998), allowing the investigation of perturbation effects at the system level. In addition, there has since the 1970s been a rapid and extensive development of multi-species models that have emerged from population and community ecology, exemplified by many of the papers in the present volume and specifically illustrated in the papers in the present chapter by Koen-Alonso and Yodzis (see Chapter 7.3) and Montoya et al. (see Chapter 7.2).

The energy flow and dynamic modelling approaches provide compelling evidence of the likely effects of perturbations within food webs, especially if extensive sensitivity analyses are carried out (see Koen-Alonso and Yodzis, Chapter 7.3), but the persuasive power of replicated manipulative field experiments has often proven to be much greater. Whilst the manipulative approach has a long history in ecology, Robert Paine's (1980) Tansley Lecture to the British Ecological Society marked something of a watershed in the food web community's appreciation of the relative value of connectedness (binary) webs, energy flow webs, and functional webs derived from experimental manipulation of the constituent species. Paine's analysis of a simple rocky shore grazer-algae food web showed that there is not always a clear match between linkages which are important in terms of material or energy flow and those revealed as functionally important by experimental disruption: linkages which carried much energy were functionally trivial, whilst linkages which had very little energy moving along them were functionally the most important for maintaining community structure and stability (Paine, 1980). This theme is revisited by Layman et al. (see Chapter 7.4) in a detailed and elegant analysis of a freshwater food web from which predatory fish have been removed. The exciting and novel aspect of their work is the application of stable isotope analysis to produce the quantitative linkages in the food web.

The persuasiveness of the experimental approach for assessing species interactions in food webs lays largely in its clear affinity with the Popperian hypothesis testing paradigm, the often clear-cut outcomes of the experiment, and the revelation of indirect effects which would not be apparent from most mathematical modelling treatments (however, see Montoya et al., Chapter 7.2, for an intriuging modelling approach). Despite their persuasiveness, experimental approaches to evaluating the effects of perturbations on food webs can be fraught with interpretational issues concerned with complexity, the need to move towards larger spatial and longer temporal scales, and the probable non-linearity of perturbation responses, so that short-term and long-term outcomes may be different (Raffaelli and Moller, 2000).

The spatial scales at which perturbation experiments are carried out is generally relatively small, usually because there is a need to trade off plot size against the number of replicates that can be handled efficiently or accommodated within a study area (Raffaelli and Moller, 2000). Whilst small-plot, highly replicated experiments yield elegant designs and are amenable to analysis using inferential statistics, society is usually interested in effects of perturbations which might occur at much larger, landscape scales. But highly replicated, large plot size experiments are uncommon in the literature and at the largest scales it may not be possible to have any replication at all, and often no reference (control) plots. Such designs only limit the persuasive power of the inferences drawn from the experiment if the effect size of the perturbation is small. For instance, if the effect of excluding lions from a single 100 km^2 plot leads to a 5% increase in their antelope prey, it will be difficult to convince other researchers that the effect was due to an absence of lions. If the effect is large in size, for instance a 500% increase in prey, or if the biology of the manipulated plot is contrary to the norm generally accepted by the ecological community, then even unreplicated experiments without control plots will be persuasive. This is indeed the case for many of the classic experiments in community ecology, many of which were not replicated and lacked controls, but which have lead to paradigms in mainstream ecology (Raffaelli and Moller, 2000). Nevertheless, large scale manipulations remain problematical for many ecologists because of their concerns over replication.

An additional aspect of large scale experiments is that there is often confusion about plot size and scale. Large plot sizes are only equivalent to large scale if the species (and system) of interest is relatively small. For instance, a 1m^2 plot for a rocky shore limpet may be equivalent to a 10,000 km^2 plot for a polar bear. Few experiments really explore the effects of increasing scale *per se* and those that do (Thrush et al., 1996; Thrush et al., 1997) indicate that outcomes are often scale-dependent.

The duration (temporal scale) of experiments is also an issue. Many experiments are run for quite short times, in marine intertidal systems typically weeks to months (Raffaelli and Moller, 2000). Whilst there are clear advantages in running short-term experiments (see Culp et al., Chapter 7.1), as well as acknowledged constraints on running experiments for long periods (Raffaelli and Moller, 2000), there may also be temporal scale effects similar to those of spatial scale, in that the outcome of the experiment is scale-dependent. If there are non-linear dynamics in the response variables, the short and long-term outcomes of a perturbation may be quantitatively and/or qualitatively different. In at least two long-term press perturbations (Paine, 1974; Barkai and McQuaid, 1988), prey size escapes eventually occurred which rendered those species invulnerable to predation once the perturbation was relaxed, and this locked the food web into an alternate stable state. Deciding on the appropriate temporal scale is thus important when designing food web perturbation experiments, but there are few guidelines available to ecologists, other than Yodzis' (1988) "rule-of-thumb" based on the generation times of all the species of interest in the web. When this was applied to a simple New Zealand food web, where estimates were required of the direct and indirect effects of rabbit removal on the abundance of an endangered reptile, the answer was quite sobering—at least 50 years—and the experiment was never carried out (Raffaelli and Moller, 2000). It seems likely that the majority of experiments designed to assess the effects of perturbations on real food webs do not run for sufficiently long for the long-term effects to be evaluated.

Finally, the majority of food web experiments deal with a single stressor or perturbation. By constructing experiments in this way the outcomes are easier to interpret, but in reality food webs will be subjected simultaneously or sequentially to a board range of stressors. Multiple stressors have been shown to act synergistically, antagonistically, or additively, depending on the system and stressor (Kennish et al., 2005), so that the outcomes of single stressor experiments cannot be used to reliably predict impacts on food webs when more than one stressor is present (see Culp et al., Chapter 7.1).

If experiments are too difficult to do routinely at the appropriate spatial and temporal scales and with the appropriate level of complexity, then how do we progress? One option is to reflect on the large scale "experiments" that have followed the arrival of arguably the most dramatic ecological perturbation of all—the arrival of humans in pristine ecosystems. There is much evidence (as well as contentious debate) that many food webs experienced large-scale changes on first contact with humans, from the mega-faunal overkills of the Mesolithic period (Flannery, 2001), to the more recent events in Australasia (Flannery, 1994) and in marine systems world-wide (Jackson et al., 2001). The wholesale removal of large-bodied consumers in these systems, both carnivores and herbivores, and the often simultaneous introduction of alien species such as rats, especially for the first contacts that occurred during imperial periods, has undoubtedly had huge effects on the food webs with which ecologists are working today. Unless these historical events are appreciated, and present-day food webs placed into context, it will be difficult to make sense of the dynamics and structure of those systems. Whilst it is difficult to convincingly re-construct food webs pre- and post-first contact so that the effects of removals and invasions can explored, there is consensus that consumers lower in the food chain, such as meso-predators, have been released by removal of larger predators. These kinds of ideas are taken up indirectly by Montoya et al. (see Chapter 7.1) and directly by Harper-Smith et al. (see Chapter 7.5), but their approaches are quite different. Montoya et al. develop a modelling approach to explore possible indirect effects in two well-documented aquatic food webs, whilst Harper-Smith et al. compare the properties of binary web descriptions affected to different extents by predatory fish introductions. These two chapters, together with the others, neatly illustrate the power and necessity of using a range of approaches—observation, experimentation, and modelling—for predicting the impacts of perturbations on the dynamics and structure of food webs.

7.1 | INSIGHT INTO POLLUTION EFFECTS IN COMPLEX RIVERINE HABITATS: A ROLE FOR FOOD WEB EXPERIMENTS

Joseph M. Culp, Nancy E. Glozier, Kevin J. Cash, and Donald J. Baird

Freshwater biomonitoring of the benthos dates back to the early twentieth century with the development of biological indicators to estimate pollution level (e.g., saprobity indices, Cairns and Pratt, 1993). Today, this indicator species approach is commonly used by regulatory agencies to produce aquatic guidelines based on the results of single-species laboratory-toxicity tests (Baird et al., 2001). However, single-species test results are rarely interpreted in an ecological context and thus are poor predictors of community response (Forbes and Forbes, 1994); therefore, their information may not be directly relevant to benthic community responses in aquatic ecosystems.

Another cornerstone of present day biomonitoring approaches is the use of field surveys to detect patterns of ecological response to stress. Multivariate methods such as RIVPACS (River Invertebrate Prediction and Classification System; Wright, 1995) are often employed to identify these patterns. However, this approach cannot easily link cause and effect. In addition, response variables measured in field surveys typically focus on how pollution modifies structural features of biological populations or communities (e.g., abundance of tolerant taxa, ratio metrics, diversity indices). We argue that in order to determine the causal mechanisms behind change resulting from complex anthropogenic stress regimes (e.g., effluent discharges), a broader food web perspective should be adopted to improve predictions of pollution effects on the long-term structure and function of ecosystems. In this paper we summarize examples of how food web experiments can improve understanding of pollution effects from pulp mill and metal mining effluents in complex river environments, and how an understanding of the food web aids interpretation of ecological responses to stressor gradients.

INVESTIGATION OF MODEL FOOD WEBS IN RIVERINE MESOCOSMS

Assessing pollution impacts in rivers is particularly challenging because flowing water environments receive multiple, interacting effluent discharges from cities and industries as well as non-point source inputs. Field biomonitoring in rivers is further hindered by uncertainties in estimating exposure (i.e., duration and concentration) to pollutants. In addition, spatial heterogeneity in the benthic environment frequently confounds environmental variables among field sites, making adequate treatment replication difficult (Glozier et al., 2002). In fact, expert review of the Canadian Environmental Effects Monitoring (EEM) program suggested that field assessments often were unsuccessful due to the confounding factors including multiple effluent discharges, habitat modifications from historical pollution effects, and a poor understanding of effluent exposure (Megraw et al., 1997).

We approach this problem by examining effects of pollutants on riverine food webs in mesocosms that simulate natural ecosystems. The approach emphasizes environmental realism with mesocosms situated alongside the study river, thereby improving the simulation of natural environmental conditions of physical (e.g., irradiance, temperature) and chemical water quality, as well as the ambient abundance of natural biota. The methodology can be statistically rigorous, allowing adequate replication of treatments and precise regulation of relevant variables such as effluent concentration (Culp et al., 2003a). Model food webs in field-based mesocosms also provide a much needed link between laboratory bioassays and field surveys. They allow questions ranging from the individual organism to the community to be posed, and chronic exposure treatments to be carried out at a smaller, more manageable scale than that of whole-ecosystem manipulations (Kimball and Levin, 1985; Shaw and Kennedy, 1996). Finally, this method can separate confounded effluents and environmental variables, an advantage that can help produce experimental tests to identify the causal mechanisms responsible for effects of effluents on benthic food webs (Clements and Kiffney, 1996).

EFFECTS OF PULP MILL EFFLUENT ON RIVER FOOD WEBS

Pulp mill effluent (PME) is a common pollution source in Canadian rivers as there are more than 100 PME discharges across the country. These effluents have the potential for both inhibitory (e.g., toxicity) and stimulatory (e.g., nutrient enhancement) effects on aquatic biota. PME is tightly regulated and wastewater treatment removes the majority of biochemical oxygen demand and toxicity. By applying in situ mesocosm technology to pollution problems in rivers across Canada, we have been able to decouple the responses of the food web to the potentially confounding effects of nutrients and contaminants, concluding that the common response of riverine food webs to this treated PME is that of nutrient enrichment (Culp et al., 2000a; Culp et al., 2000b; Culp et al., 2003a; Culp et al., 2004).

To illustrate the effects of PME on benthic communities in rivers, we focus on functional food webs from mesocosm experiments on the Saint John River, Canada; aspects of these studies were previously described in Culp et al. (2003a). For this food web analysis we assessed PME effects to functional feeding groups (FFG) of insects (i.e., insect feeding guilds). Aquatic insects were selected as the study assemblage because FFG membership of most insects is well described (Merritt and Cummins, 1996), insects dominated benthic invertebrate biomass of the food webs, and this assemblage comprised over 90% of benthic invertebrate densities. Insect taxa in the benthic communities and their FFGs are listed in Table 1. When trophic interactions of insect taxa consisted of two or more FFGs, density for the taxon was divided equally between or among FFGs.

The enrichment effect of PME from increased nutrients and detrital flocculants produced bottom-up responses in the food web that included increased algal biomass (Culp et al., 2003a), greater total insect numbers (i.e., benthic larvae plus emerged insects), and increased abundance of key FFGs (Figure 1A). PME enrichment also produced significantly higher insect emergence within all FFGs (except piercers)

Table 1. List of taxa and their functional feeding group assignments for benthic insect food webs of mesocosm studies that assessed pulp mill (PME) and metal mining (MME) effluent effects. Functional feeding group abbreviations: collector-filterers (CF), collector-gatherers (CG), scrapers (SC), shredders (SH), piercers (PI) and predators (PR). Presence and absence of taxa are indicated by + and –, respectively.

TAXON		Feeding Group	PME Web	MME Web
EPHEMEROPTERA				
Baetidae	*Acentrella*	CG	+	–
	Acerpenna	CG	+	+
	Baetis	CG, SC	+	+
	Callibaetis	CG	+	+
	Centroptilum	CG, SC	+	–
	Procleon	CG, SC	+	+
Baetiscidae	*Baetisca*	CG, SC	+	–
Caenidae	*Caenis*	CG, SC	+	–
Ephemeridae	*Ephemera*	CG, PR, CF	+	+
Ephemerellidae	*Drunella lata*	SC	+	–
	Drunella walkeri	SC	+	–
	Ephemerella	CG, SC	–	+
	Eurylophella	CG	+	+
	Serratella	CG	+	–
	early instars	CG, SC	+	–
Heptageniidae	*Epeorus*	CG, SC	–	+
	Leuctrocuta hebe	SC, CG	+	+
	Stenacron	CG, SC	–	+
	Stenonema	SC, CG	+	+
	early instars	SC, CG	+	+
Isonychidae	*Isonychia*	CF, PR	+	–
Leptophlebiidae	*Habrophlebia vibrans*	CG, SC	–	+
	Paraleptophlebia	CG	+	+
Potamanthidae	*Anthopotamus distinctus*	CF	+	–
Tricorythidae	*Tricorythodes*	CG	+	–
Leptophlebiidae	early instars	CG, SC	–	+
PLECOPTERA				
Capniidae	*Paracapnia*	SH	+	+
	early instars	SH	–	+
Chloroperlidae	*Haploperla*	PR	+	–
	Sweltsa	PR	+	+
	early instars	PR, SC, CG	–	+
Leuctridae	*Leuctra*	SH	+	+
	early instars	SH	–	+

Table 1. Cont'd

TAXON		Feeding Group	PME Web	MME Web
Perlidae	*Agnetina*	PR	+	+
	Acroneuria	PR	–	+
	Paragnetina/ Attaneuria	PR	–	+
	early instars	PR	–	+
Perlodidae	early instars	PR, CG, SC	–	+
Taeniopterigidae	*Taeniopteryx*	SH	+	+
TRICHOPTERA				
Brachycentridae	*Brachycentrus*	CF, SC	+	–
Glossosomatidae	*Agapetus*	SC,CG	+	–
	Glossosoma	SC	–	+
	Protoptila	SC	+	–
Hydroptilidae	*Hydroptila*	PI, SC	+	+
	Oxyethira	PI, CG	+	+
Hydropsychidae	*Cheumatopsyche*	CF	+	–
	Hydropsyche	CF	+	+
	early instars	CF, PR	+	–
Lepidostomatidae	*Lepidostoma*	SH	+	+
Leptoceridae	*Ceraclea*	CG, SH	–	+
	Mystacides	CG, SH	–	+
Limnephilidae	*Hydatophylax*	SH, CG	+	–
Odontoceridae	*Psilotreta*	SC, CG	–	+
Philopotamidae	*Chimarra*	CF	+	–
	Dolophilodes	CF	–	+
Phryganeidae	*Oligostomis*	PR, SH	–	+
Polycentropodidae	*Paranyctiophylax*	PR, SH, CF	–	+
	Polycentropus	PR, SH, CF	–	+
Psychomyiidae	*Psychomyia*	CG, SC	+	–
Rhyacophilidae	*Rhyacophila*	PR, SC, CG, SH	–	+
Uenoidae	*Neophylax*	SC	+	–
ODONATA				
Aeshnidae	*Aeshna eremita*	PR	–	+
Cordulegastridae	*Cordulegaster diastatops*	PR	–	+
Gomphidae	*Ophiogomphus rupinsulensis*	PR	+	+
	Gomphus	PR	+	–
	Lanthus parvulus	PR	–	+
Calopterygidae	*Calopteryx*	PR	–	+
COLEOPTERA				
Curculionidae	early instars	SH	–	+
Dytiscidae	*Oreodytes*	PR	+	–

Table 1. Cont'd

TAXON		Feeding Group	PME Web	MME Web
Elmidae	*Dubiraphia*	SC, CG	+	–
	Promoresia	CG, SC	+	+
	Optioservus	SC, CG	+	+
	Oulimnius	CG, SC	–	+
	Stenelmis	SC, CG	+	–
	early instars	SC, CG	+	+
Gyrinidae	*Dineutus*	PR	+	–
Psephenidae	*Ectopria*	SC	–	+
MEGALOPTERA				
Corydalidae	*Nigronia*	PR	–	+
Sialidae	*Sialis*	PR	–	+
HEMIPTERA				
Veliidae	*Rhagovelia*	PR	–	+
DIPTERA				
Athericidae	*Atherix*	PR	+	+
Ceratopogonidae	*Bezzia / Palpomyia*	PR	+	+
Empididae	*Chelifera*	PR, CG	–	+
	Hemerodromia	PR, CG	+	+
Simuliidae	*Simulium*	CF	+	+
Tabanidae	*Chrysops*	PR	–	+
Tipulidae	*Antocha*	CG	+	+
	Dicranota	PR	+	+
	Hexatoma	PR	+	+
	Brachycera	PR	–	+
Chironomidae				
Tanypodinae	*Ablabesmyia*	PR, CG	+	+
	Labrundinia	PR	–	+
	Monopelopia	PR	–	+
	Natarsia	PR	–	+
	Nilotanypus	PR	+	+
	Trissopelopia	PR	–	+
	Thienemannimyia spp	PR	+	+
Chironomini	*Chironomus*	CG, SH	+	+
	Cryptochironomus	PR	+	+
	Demicrypto-chironomus	CG	+	–
	Dicrotendipes	CG, CF	+	+
	Glyptotendipes	SH, CF, CG	+	–
	Nilothouma	CG	+	+
	Microtendipes	CF, CG	+	+
	Parachironomus	SH, CG	+	–
	Phaenopsectra	SH, CG	+	–

(Continued)

Table 1. Cont'd

TAXON		Feeding Group	PME Web	MME Web
	Polypedilum	SH, CG, PR	+	+
	Stenochironomus	CG, SH	–	+
	Xenochironomus xenolabis	PR	–	+
	early instars	CG	+	+
Pseudochironomini	*Pseudochironomus*	CG	+	–
Tanytarsini	*Cladotanytarsus*	CG, CF	+	–
	Micropsectra	CG	–	+
	Paratanytarsus	CF, CG	+	+
	Rheotanytarsus	CF, CG	+	+
	Stempellina	CG	–	+
	Stempellinella	CG	–	+
	Sublettea	CF, CG	+	–
	Tanytarsus	CF, CG	+	+
	early instars	CF, CG	–	+
Orthocladiinae	*Brillia*	CF, CG	+	–
	Cricotopus / Orthocladius	SH, CG	+	+
	Corynoneura	CG	+	+
	Diplocladius	CG	–	+
	Eukiefferiella	CG, SC, PR	+	+
	Heterotrissocladius	CG	+	–
	Krenosmittia	CG	+	–
	Lopescladius	CG	–	+
	Parakiefferiella	CG	+	–
	Parametriocnemus	CG	+	+
	Paratrissocladius	CG	+	–
	Psectrocladius	CG,SH	+	+
	Rheocricotopus	CG, SH	+	+
	Synorthocladius	CG, SC	–	+
	Thienemanniella	CG	+	+
	Tvetenia	CG	+	+
	Xylotopus par	CG	–	+
	early instars	CG	+	+
Diamesinae	*Pagastia*	CG, SC	+	+
	Potthastia gaedii	CG, SC	+	+

(see Figure 1B), suggesting that bottom-up effects accelerated growth and development, decreasing generation time. We noted considerable algal mat sloughing during the experiments as algal productivity increased which, combined with detrital flocculants from PME, augmented the availability of detritus (Figure 2). This increase in detritus is

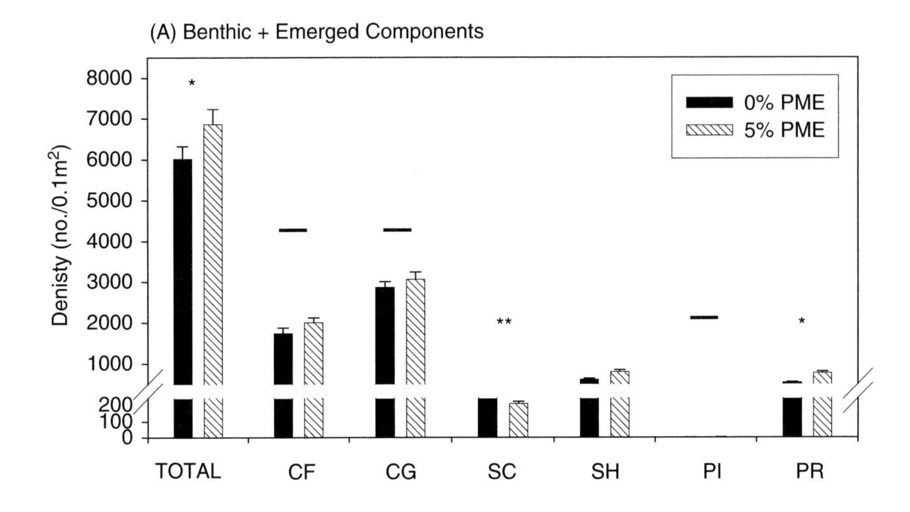

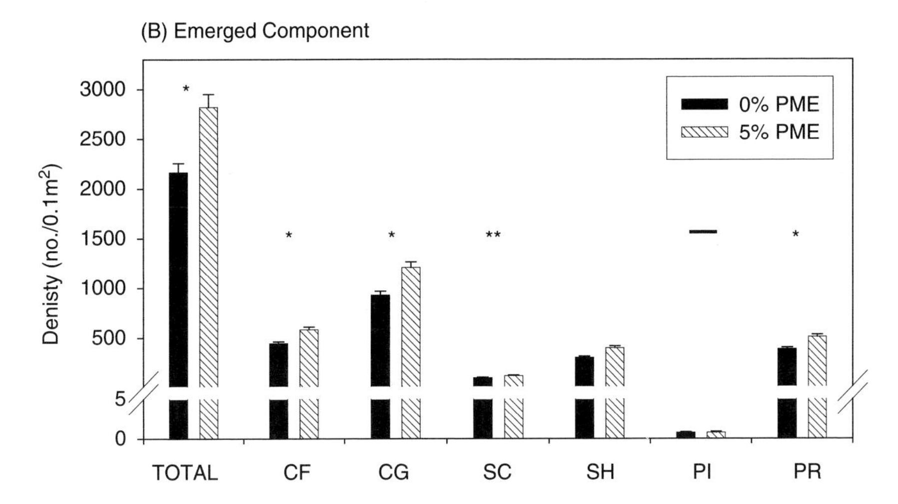

FIGURE 1 | Mean densities (± SE) in the 0% and 5% pulp mill effluent treatments of total insects and their distribution among functional feeding groups for (**A**) the combined sum of benthic larvae and emerged insects, and (**B**) emerged insects only. Functional feeding group abbreviations: collector-filterers (CF), collector-gatherers (CG), scrapers (SC), shredders (SH), piercers (PI) and predators (PR). See text for detailed explanation of changes in the functional food web. (*Indicates significant difference between treatments at $p < 0.05$; – indicates no difference.)

likely responsible for the more rapid growth of detritivore FFGs such as collector-gatherers and shredders.

A striking result was the significant increase in the abundance of invertebrate predators, particularly small-bodied chironomids which reside within the algal mats. Since taxonomic richness was similar in reference

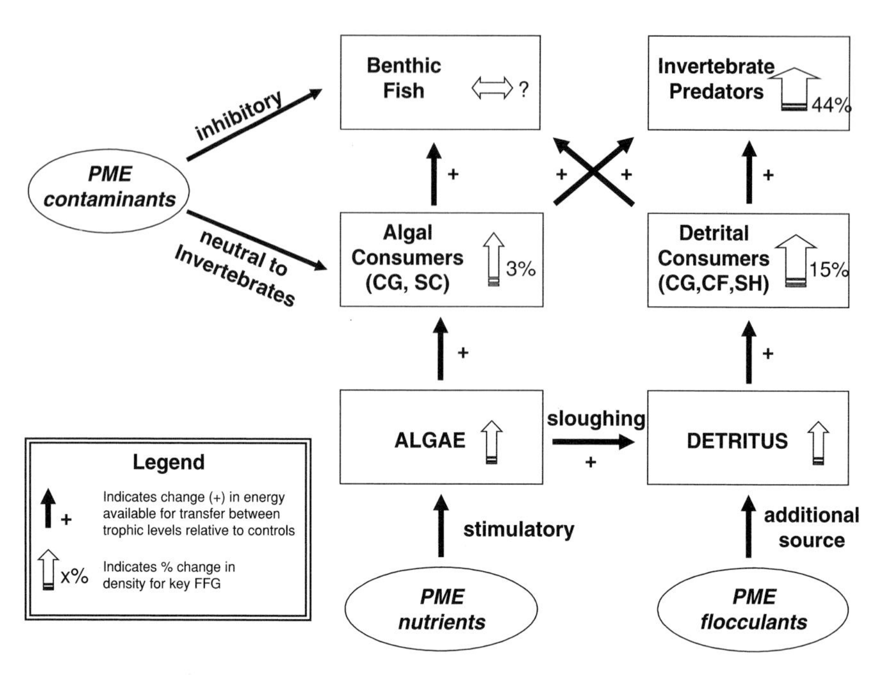

FIGURE 2 | Effects model illustrating the pathways through which a 5% dilution of pulp mill effluent (PME) modifies food-web interactions in large-river mesocosms, relative to controls. Functional feeding group abbreviations: scrapers (SC), collector-gatherers (CG), collector-filterers (CF), shredders (SH). See text for detailed explanation of changes in the functional food web.

and PME food webs, the increase in predators was likely a numerical response to increased prey abundance. We did not examine responses of benthic fish, but common benthic fish predators (e.g., sculpin) may respond similarly to PME enrichment of nutrient-enhanced food as suggested by increases in condition, gonad size, and egg mass downstream of PME discharge (Gibbons et al., 1998). Thus, while PME enrichment results in a bottom-up increase of most FFGs, the changes in abundance may be attenuated by increased predation losses indicating a fundamental change in food web structure as a result of effluent discharge. Finally, although PME contaminants can negatively affect reproduction of fish by disrupting endocrine pathways, our experiments with model food webs indicates that at environmentally realistic, low concentrations of PME, direct contaminant effects on invertebrates could not be detected.

By exploring the impacts of PME on food webs in mesocosms we developed an improved mechanistic understanding of how environmentally realistic, low concentrations of treated PME (< 5% v/v) modify community structure, increase the importance of detrital pathways and, ultimately, shift the metabolism of large rivers. These applied experiments can also make important contributions to general food web theory in river ecosystems. For example, our results suggest an increased importance of small-bodied predators under conditions of PME discharge likely because thicker algal mats provide greater microhabitat complexity and increased prey abundance. Furthermore, we have been able to describe a conceptual model of benthic community changes across a nutrient-contaminant gradient by linking food web experiment results to field observations at reference and impact sites (see Culp et al., 2000b, and Lowell et al., 2000, for details). This research program also provided scientific justification to incorporate mesocosm food web studies into environmental-effects monitoring programs, and to use the results obtained to inform Federal regulations (Dubé et al., 2002).

EFFECTS OF METAL MINE EFFLUENT ON RIVER FOOD WEBS

Food webs in mesocosms can be an important component of integrated monitoring strategies for riverine food webs when there is a need to evaluate different ecological risk scenarios. For example, we examined the effect of metal mining effluent (MME) using the scenario of predicted MME discharge following mine decommissioning (Dubé et al., 2002; Culp et al., 2003b). We asked the question, "Do food webs recover to local reference conditions after mine decommissioning?" Experimental treatments assessed the effects of 0%, 20%, and 80% (v/v) of a metal mining effluent on the benthic food web (algae to fish) of the Little River, New Brunswick, Canada, over a 24-day period during August and September 2000. Treatment levels included existing MME discharge (80%) and the predicted discharge scenario upon mine closure (20%). Mesocosm set-up and sampling of the food webs followed the methods outlined in Culp et al. (2003a), except that we included treatments that examined effects on two-level (algae/detritus→primary consumers) and three-level food webs (algae/detritus→primary consumers→sculpin). Measurement endpoints included algal biomass, invertebrate community composition and insect emergence for the two-level food web experiment. The three-level food web experiment focused on measuring the survival and growth of young-of-the-year slimy sculpin (*Cottus cognatus*). Sculpin were counted, weighed and the total length measured at the end of the experiment. Weekly grab samples of reference water and the dilution series (20%, 80% MME) were collected and analyzed for nutrients, general ions, and metals.

The concentration of nutrients and most metals was significantly higher in MME treatments with the highest levels occurring in the 80% treatment (Culp et al., 2003b). Mean algal biomass significantly increased in both MME treatments relative to reference streams with the mean values (±SE) for the 0%, 20% and 80% MME being 23.7 ± 3.4, 72.4 ± 9.6, and 131.2 ± 18.9 mg/m^2, respectively. Benthic invertebrate community composition and abundance were strongly affected by 80% MME, and less so by 20% MME. Furthermore, familial richness was not changed by exposure to 20% MME, but 80% MME significantly decreased invertebrate biodiversity with losses of several mayfly, stonefly and caddisfly families. Over the 26-day experiment, sculpins exposed to 0% or 20% MME were larger than individuals exposed to 80% MME and only 25% of the 80% MME fish survived (Dubé et al., unpublished manuscript).

We investigated the effects of MME on benthic food web structure by examining abundance changes in FFGs (Table 1) of benthic insects. Total insect abundance (i.e., emerged plus larval insect abundance) decreased under MME exposures of 20% and 80% (Figure 3A). While all FFGs had lower densities in 80% MME, this trend was not significant for several FFGs in 20% MME. In contrast, the emergence of all FFGs (except predators) and, consequently, total insect emergence increased in 20% MME (see Figure 3B). The trend of higher collector-filterer, collector-gatherer, scraper and shredder emergence in 20% MME is largely attributable to greater abundances of chironomid midges at the lower MME exposure. Chironomids are more tolerant of metal exposure than other aquatic insects such as mayflies (de Haas et al., 2002), and they appeared to respond positively to the thicker algal mats produced by bottom-up enrichment. Increased algal biomass also led to greater algal sloughing and input to detrital pathways as in the PME results, which likely contributed to the bottom-up abundance response of midges.

Unlike the positive response of midges to increased algal biomass, most algal and detrital consumers were unable to benefit from higher food availability under either MME exposure (Figure 4). In particular, abundances of mayfly collector-gatherers and scrapers were reduced as result of MME exposure. Because nutrients were saturated in both MME treatments, we hypothesize that the increased algal biomass in the MME treatments resulted directly from MME nutrient enrichment and, indirectly, from the release from top-down grazer control as grazer abundance decreased (see Figure 4). Niyogi et al. (2002) also found that MME reduced densities of algal consumers, leading to top-down release of grazing pressure and increased algal biomass. We predict that the pro-

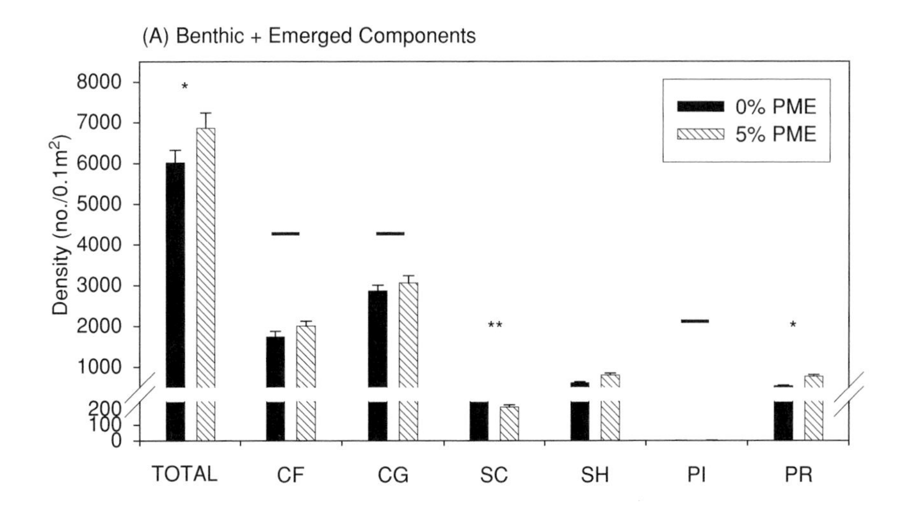

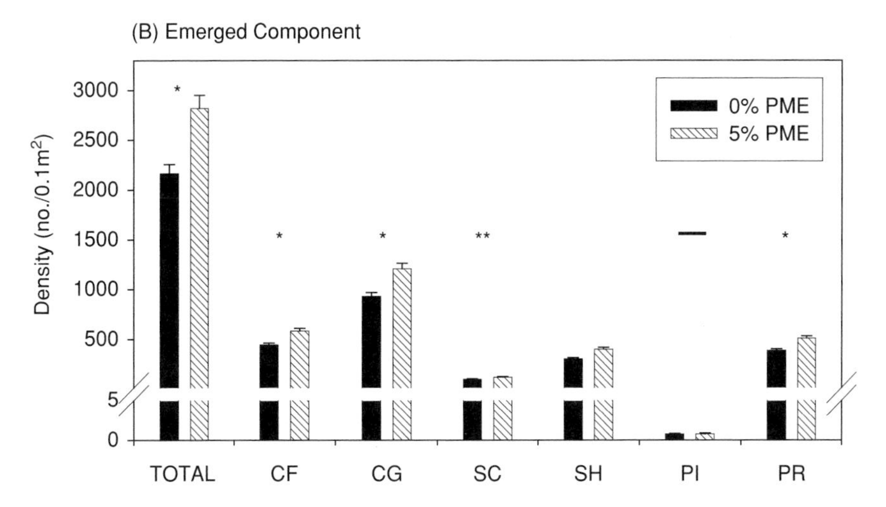

FIGURE 3 | Mean densities (± SE) in the 0%, 20% and 80% metal mill effluent treatments of total insects and their distribution among functional feeding groups for (**A**) the combined sum of benthic larvae and emerged insects, and (**B**) emerged insects only. Functional feeding group abbreviations: collector-filterers (CF), collector-gatherers (CG), scrapers (SC), shredders (SH), piercers (PI) and predators (PR). See text for detailed explanation of changes in the functional food web. (*Indicates significant difference among treatments at $p < 0.05$; – indicates no difference.)

posed MME concentration after mine closure will lead to enriched algal communities, and invertebrate communities dominated by metal-tolerant taxa such as chironomid midges. In addition, the abundances of pollution intolerant taxa (e.g., mayfly collector-gatherers and scrapers) will likely be reduced with fewer species compared to reference areas.

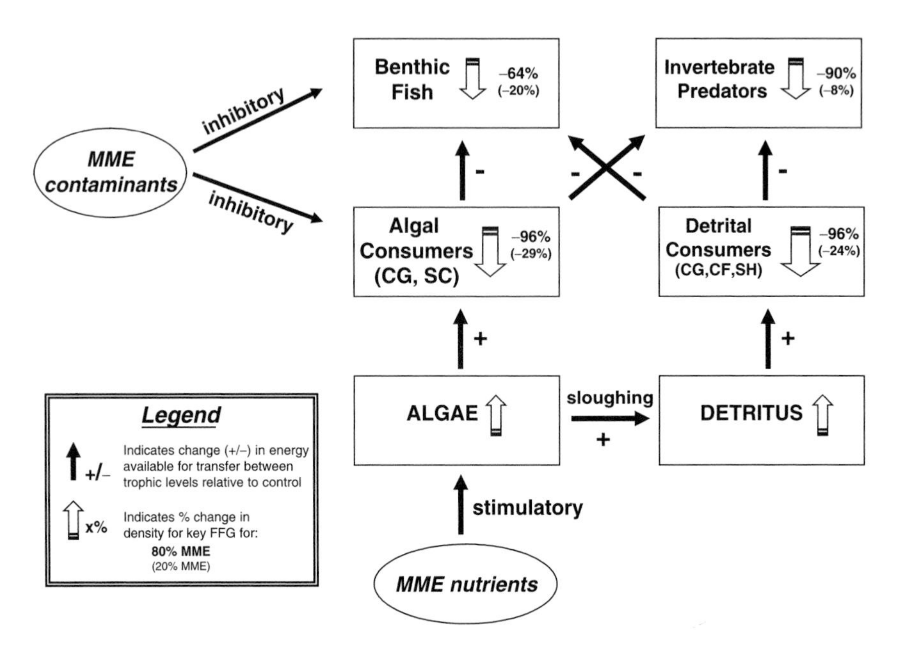

FIGURE 4 | Effects model illustrating the pathways through which 20% and 80% metal mining effluent (MME) modifies food-web interactions in an oligotrophic river relative to reference benthic communities. Functional feeding group abbreviations: scrapers (SC), collector-gatherers (CG), collector-filterers (CF), shredders (SH). See text for detailed explanation of changes in the functional food web.

DISCUSSION AND CONCLUSIONS

A criticism often leveled at mesocosm studies focuses on whether the small spatial-scale of the mesocosms is appropriate for assessing large-scale impacts on biodiversity and river ecosystem function. Schindler (1998) suggests that mesocosm experiments may not accurately predict ecosystem functioning at larger spatial scales because ecosystem dynamics are not fully represented, but this equally applies to other spatial scales (e.g., whole-ecosystem experiments). Nevertheless, river ecologists are faced with the dilemma that large-scale experiments, which incorporate contaminant or nutrient stressors, are not feasible on ethical and practical grounds because they may cause severe acute and chronic effects on riverine communities. The solution of using small-scale mesocosms to study food web responses to stressors at a manageable spatial-scale necessarily sets limits on the duration of experiments. This temporal limitation has two advantages: First, the many logistic constraints associated with manipulating pollution gradients in a flowing-water environment (e.g., cost, seasonality, complex life histories of biota) dictate shorter experimental periods so that the pollutant remains the key experimental factor. Second, experimenters must restrict the population processes of emigration and immigration in order to enhance the ability to detect the pollution signal. Reviews detailed in Lamberti and Steinman (1993) provide a thorough analysis of the potential deficiencies in flowing water mesocosm approaches. Clearly, in situmesocosm studies can provide important information on causal mechanisms of pollution effects including an indication of which species and trophic pathways are least tolerant of pollution stress (Clements and Kiffney, 1996; Culp et al., 2000a). Results must be interpreted with the consideration that they cannot predict changes in ecological process that may occur through long term changes in habitat (as a result of stressor pollution) or stressor-related changes in the genetic structure of populations within the community. In our approach we attempt to broaden the conclusions from mesocosm experiments through linkage to field measurements during the experiments and the use of formal weight-of-evidence postulates. The weight-of-evidence approach also serves to integrate data at a variety of different spatial and temporal scales (Lowell et al., 2000).

We strongly advocate the use of in situ food web experiments for ecological assessment. In complex environments this approach can elucidate dose-response relationships between components of the model food web and single or multiple stressors better than standard field observations. Food web experiments are not often used in a regulatory setting; however, their use can improve ecological risk assessments, help to develop ecologically relevant environmental guidelines, and form the basis of a diagnostic methodology to tease apart the effects of multiple stressors. In addition, the use of field-deployed mesocosms allows ecologists to examine the relationship between changes in biodiversity and function across stressor gradients. Through careful application of food web experiments we suggest that important progress will be made in understanding river ecosystem functions in response to anthropogenic threats. The results of ecotoxicological experiments in model food webs can provide a wealth of information on possible species interactions within a community, including knowledge of functional redundancy of species and various indirect effects. Thus, these more applied ecological experiments can make important contributions to more general food web theory.

ACKNOWLEDGMENTS

Financial support for this research was provided by the Fraser River Action Plan, Northern River Basins Study, Northern River Ecosystem Initiative, the National Water Research Institute (Environment Canada), the Toxic Substances Research Initiative Project No. 196, and a NSERC Discovery Grant to JMC. The manuscript benefited from constructive criticisms by two anonymous reviewers.

7.2 | PERTURBATIONS AND INDIRECT EFFECTS IN COMPLEX FOOD WEBS

José M. Montoya, Mark C. Emmerson, Ricard V. Solé, and Guy Woodward

Why are several common species of birds disappearing rapidly and mysteriously from Barro Colorado Island, for no readily apparent reason? Could there be a relationship with earlier extinctions of large predators, including jaguars and pumas? The fragmented habitat resulting from the building of the Panama Canal created an island that was too small to sustain large predators. These were the first species that suffered extinctions, leading to population increases of their prey species. These mesopredators fed on the eggs and young of ground nesting birds and their increase in numbers was sufficient to wipe out many bird populations (Emmons, 1987; Terborgh, 1992). This rather dramatic example of a trophic cascade illustrates that indirect interactions (i.e., those occurring when the effect of one species on another is mediated by intermediate species) are an important cause of many recent and historical extinctions (Jackson et al., 2001). Ignoring indirect effects may lead to a misunderstanding of the causes and consequences of perturbations. Identifying the causes and predicting the effects of disturbances in complex ecosystems is one of the major challenges faced by contemporary ecology in the light of the current global biodiversity crisis. This challenge raises three questions that we address here. First, is it possible to generalize and predict the effects of disturbances in complex communities considering both direct and indirect interactions among species? More specifically, are there any particular species' traits that predispose species to propagate disturbances throughout the food web, and how important are indirect effects relative to direct effects?

INDIRECT EFFECTS AND THE INVERSE COMMUNITY MATRIX

The general perception of indirect effects is that, given the complexity of real food webs, they are ubiquitous and likely to be important for community structure, and that they can also be highly unpredictable, leading to unanticipated results compared to predictions based on direct interactions (Sih et al., 1985; Sih et al., 1998). The presence of indirect effects has long been recognized (Darwin, 1859; Camerano, 1880; Elton, 1927; see Wootton, 2002 for review). More recent studies have shown that indirect interactions can be as important as direct interactions in their effects on the population abundances and traits (e.g., behavior, morphology) of other species (Garrity and Levings, 1981; Abrams, 1992; Schoener, 1993; Wootton, 1994c; Menge, 1995; Schmitz, 1997; Wootton, 2002), and that they can propagate as fast or faster than direct effects through the food web (Dungan, 1986; Abrams et al., 1996; Menge, 1997). In Menge's survey on marine rocky intertidal communities, for example, indirect effects accounted for 40% of changes in community structure resulting from species' manipulations (Menge, 1995). The food webs analyzed in Menge's survey and other studies rarely comprised more than 10 species, and they are all module sub-webs nested within the wider community food web. The reason for using such small modules is because it is logistically impossible to measure the strength of every interaction for all species pairs in an entire food web (see Berlow et al., 2004, for a review). Attempts to parameterize larger food webs by randomly assigning interaction coefficients between species supported the overwhelming importance of indirect effects (Yodzis, 1988; Abrams et al., 1996; Yodzis, 2000). However, the results were extremely sensitive to the values of the parameters defining the strength of each interaction. This lead to the conclusion that it is difficult to predict the effects of perturbations, because they are highly indeterminate, especially when we lack precise information on the range of interaction strengths (Pimm, 1993; Abrams et al., 1996; Yodzis, 1996 and references therein) or other relevant parameters (Schaffer, 1981; Solé and Bascompte, unpublished manuscript).

New empirical food web data with field measurements or biologically plausible estimates of interaction strengths provide a more precise avenue for exploring the effects of disturbances via direct and indirect effects. These novel measures of interaction strength also allow us to explore the effects of species traits on the probability of perturbations to propagate throughout a community. Here we will use data from two well-resolved aquatic food webs: the Ythan Estuary and Broadstone Stream food webs. In the Broadstone web (Figure 1A), interaction strengths are obtained by means of analyses of predators' gut contents (after Woodward and Hildrew, 2002a; Woodward et al., unpublished manuscript). In the Ythan (see Figure 1B), they are estimated from predator and prey body sizes, in agreement with recent empirical and theoretical studies showing that interaction strength scales with predator-prey body size ratio (Emmerson and Raffaelli, 2004b; see Emmerson et al., Chapter 4.3). In both cases,

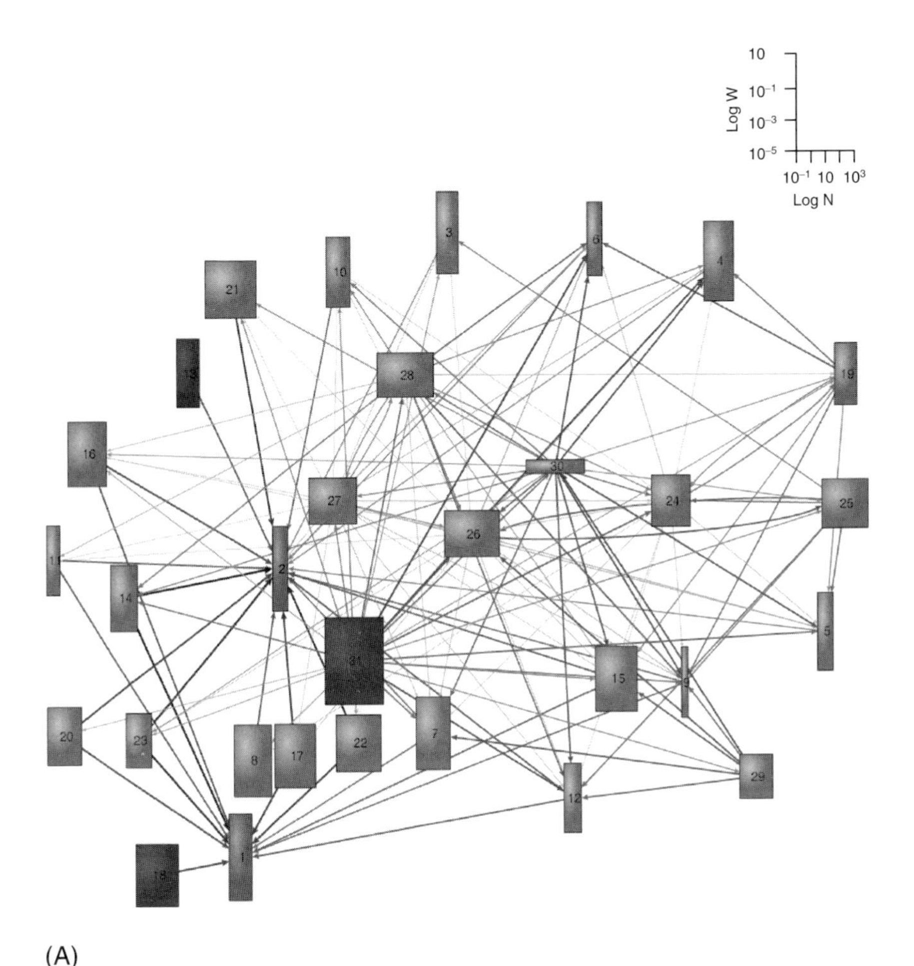

FIGURE 1 | Representation of Broadstone (**A**) and Ythan (**B**) food webs. Arrows point from predators to prey, and line thickness and darkness correspond to predator per capita interaction strengths (a_{ij}) in the community matrix. Rectangles correspond to species (most of them taxonomic species), and green, blue and red correspond to basal, intermediate and top species respectively. The base of the rectangle corresponds to species' numerical abundance (number/m^2), and its height to body size (mg of mean individual or mean adult, for the Broadstone and Ythan respectively). (See also color insert.)

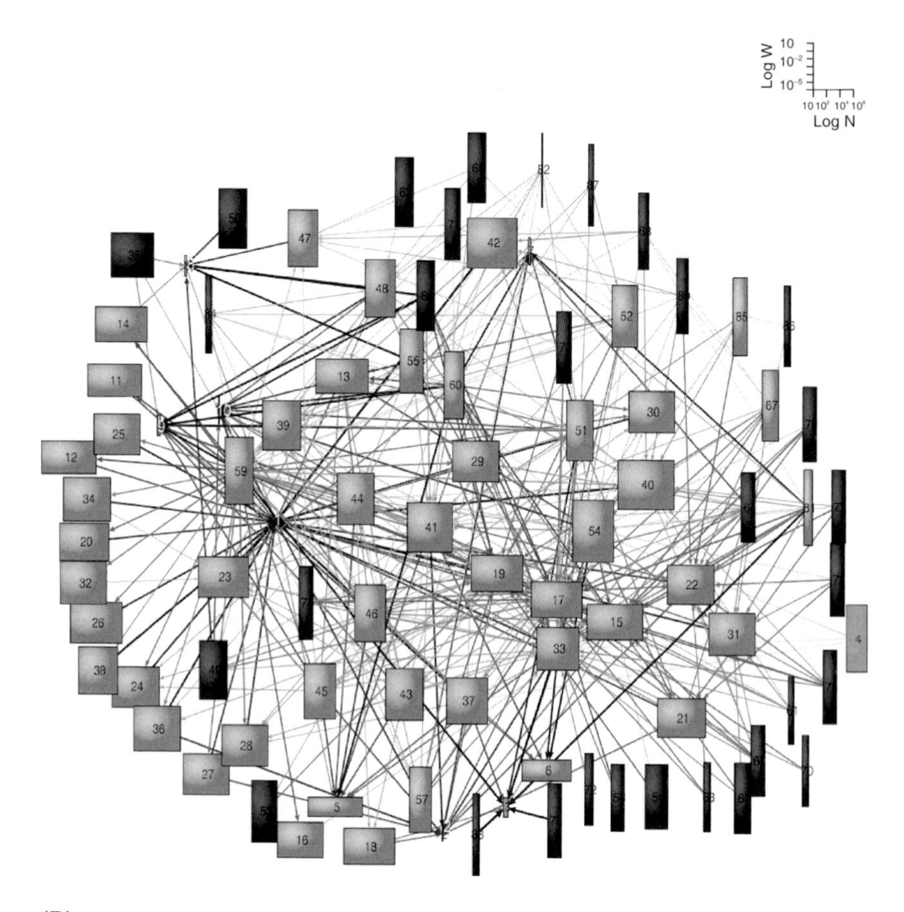

(B)

FIGURE 1 | *Cont'd* All data are log-transformed, and the scales for abundance and body size are indicated. For the Ythan web, we do not have direct measurements of some species abundances, so we have estimated them from their body size using observed allometric relationship between abundance (X) and body size (W) for the species whose abundance was already known: $X = 177.51^{*}W^{-1.185}$ (r^2=0,84; $p<0.0001$) (after Leaper and Raffaelli, 1999; Emmerson and Raffaelli, 2004b). This type of representation might be very helpful for visualizing the temporal variability in food webs (e.g., exploring how abundances, body sizes, or interaction strengths change through time), or the effects of the introduction and extinctions of species in complex food webs. Figures have been generated using the program Pajek. (See also color insert.)

interaction strengths are the values of the elements a_{ij} of the community matrix A and represent the per capita effect of species on each other's dynamics near equilibrium and are related with the elements of the Jacobian matrix c_{ij} through $c_{ij} = a_{ij}^{*} X_i$, where X_i is the population of species i (May, 1973; Neutel et al., 2002). Elements of the inverse of the community matrix (i.e., $-(a_{ij})^{-1}$) summarize the net effect of a small change in the population size of species j on the equilibrium density of species i due to the direct linkage with species i and all possible indirect pathways through which species i and j are connected (Levins, 1974; Bender et al., 1984; Yodzis, 1988, 1996 and references therein). Previous studies have shown that predictions based on the inverse community matrix reflect with high accuracy the observations from manipulative experiments (Stone and Roberts, 1991; Schmitz, 1997), supporting the view that the inverse community matrix can be a useful theoretical benchmark to analyze consequences of perturbations in natural communities.

We use a Lotka-Volterra model, defined as $dX_i/dt = \varphi(\vec{\boldsymbol{X}})$, where X_i indicates the population of the i-th species (i= 1,2,...,S) and $\varphi(\vec{\boldsymbol{X}}) = X_i$ $\epsilon_i + \sum_{j=1}^{S} a_{ij} X_j(t)$, where $a_{ij > 0}$ for $i > j$, and $_a ij < 0$ for $i<j$, and the equilibrium point $\vec{X}^* = (X_1^*, \ldots, X_s^*)$. A press disturbance is introduced as a small and sustained increment in the population of species k. Specifically, we consider $dX_i/dt = \varphi(\vec{\boldsymbol{X}}) + I_k$, where I_k is small ($I_k << X_{k^*}$), and its units are number of individuals/time, in contrast with a pulse disturbance, where the units would be number of individuals. Hence, the net effect on species i resulting from a perturbation on species k is given by the element of the inverse community matrix $(a_{ik})^{-1} = -\partial X_i^* / \partial I_k$. Here we do not differentiate amongst the types of indirect effects involved in each pathway between two species, but rather, we consider the resulting net effect of one species on another given all possible interaction pathways between them.

SPECIES' TRAITS AND DISTURBANCE PROPAGATION

Can we describe the net effect of each species on the rest of the food web (i.e., its net food-web effect), and is it possible to determine whether these effects are related to any characteristic of the disturbed species? We focused on three principal descriptors that drive many of the key dynamical and evolutionary processes governing community organization: numerical abundance, body size, and linkage density (considering both ingoing and outgoing links) (May, 1975; Peters, 1983; Cohen et al., 2003).

We calculated the mean net effect of species i on the web using the absolute values of the inverse community matrix, i.e., $|(a_{ik})^{-1}|$, k=1,...,S (in other words, focusing in the columns of A^{-1}). For the Ythan food web, abundance, body size, and linkage density of species are all correlated with their net effect on the community, while in Broadstone, linkage density is the only variable that exhibits a significant correlation (Figure 2).

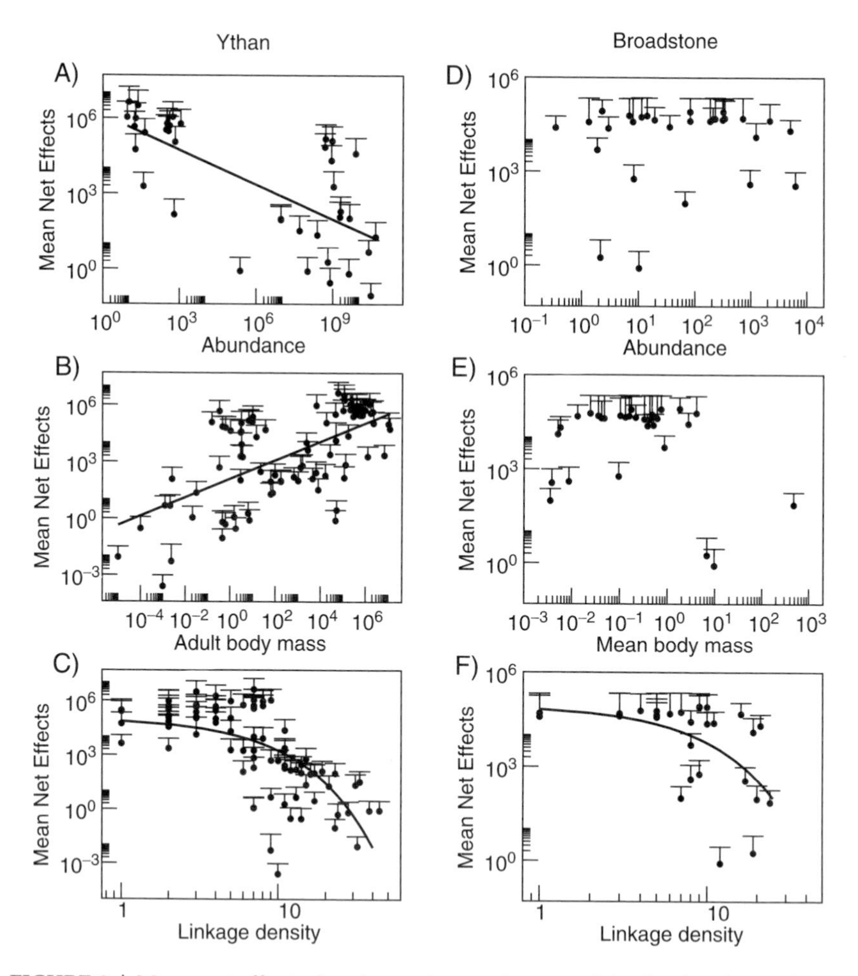

FIGURE 2 | Mean net effect of each species on the rest of the food web in relation to three community descriptors for the Ythan (a-c) and Broadstone (d-e) webs. Only significant fittings are drawn. **A,** Data on abundance (no individuals/m^2) are incomplete; power-law fitting, slope = −0.458, r^2 = 0.48, $P < 0.0001$. **B,** Mean adult body size (mg); power-law fitting, slope = 0.489, r^2 = 0,37, $P < 0.0001$. **C,** Linkage density considering ingoing and outgoing links; exponential fitting, slope = −0.41, r^2 = 0.45, $P < 0.0001$. **D,** Units of abundance (no individuals/m^2). **E,** Body size as mean body size across life-history (mg). **F,** Linkage density as in (C); exponential fitting: slope = −0.272, r^2 = 0.29, $P < 0.001$.

Allometric power relationships between body size and abundance are well documented for many different guilds and assemblages within natural communities and, more recently, for whole food webs, including Tuesday Lake (Cohen et al., 2003; Jonsson et al., 2004), the Ythan Estuary and Broadstone Stream (Emmerson et al., Chapter 4.3; Woodward et al., unpublished manuscript). These relationships show that species' numerical abundance decays with species' body-size in power-law fashion, with exponents 0.9-1.2 (see Cohen, Chapter 4.1, for more details). It follows, given the negative relationship between body-size and abundance, that if big species have large net effects, less abundant species will also have large effects. Trophic level or status also tends to increase with body-size (Elton, 1927; Cohen et al., 1993a), and the largest species in a web is often also the top predator, as is true for both the Ythan and Broadstone webs. Large predators are known or suspected to have important community and ecosystem-level effects (see Estes, 1995, Terborgh et al., 1999, for reviews), especially in aquatic systems, where top-down control appears to be particularly prevalent (Strong, 1992), and important indirect effects such as trophic cascades and keystone predation are often observed (Paine, 1969, 1974; Carpenter and Kitchell, 1993a). Our results for Ythan support this view, where we found that most bird species and other terrestrial predators have large net effects in the food web; that is, a small disturbance on these predators triggers large fluctuations on the equilibrium populations of other species within the web. We are aware of at least one other empirical-based study consistent with our results, which has shown that disturbances affecting predatory birds (eventually driving them to extinction) have large effects on the stability of aquatic communities (Wootton, 2001).

However, disturbances affecting large predators in Broadstone Stream did not have large consequences on other species. This is perhaps because of redundancy in the web. In Broadstone, the diet of each predator species is a nested subset of that of the next largest predator, so considerable trophic equivalence among similar-sized species is common (Woodward and Hildrew, 2002a). Species of comparable size may, therefore, compensate following perturbations. This mechanism has been demonstrated in a four species predator-prey subweb (Speirs et al., 2000), and these compensatory responses could account for the long-term persistence of the community (Woodward et al., 2002). An alternative, not exclusive, explanation is that the largest predators present in the Broadstone Stream (dragonfly larvae) are only as large as the small intermediate meso-predators in the Ythan Estuary food web. If we were to truncate the Ythan Estuary body size distribution so that it spanned the same range as in the Broadstone web we might expect to find that the correlations disappear (we have not done it, but this effect is likely to occur). For instance, fish are absent from the Broadstone web, because the stream is naturally acidic. Predatory fish, which are considerably larger predators than the dragonflies at the top of the Broadstone web, can have very powerful effects in freshwater food webs, including inducing strong cascading effects in certain instances (Power, 1990; Woodward and Hildrew, 2002b). The different results between Ythan and Broadstone are not likely to be a function of the way the two webs were generated. Indeed, we have estimated interaction strengths for the Broadstone web using the same approach used for the Ythan web (i.e., predator–prey body-size ratios), and such differences remained.

It is not immediately obvious why poorly connected species have considerably stronger effects than highly-connected species, a result that is consistent for both food webs. Such a counterintuitive result suggests that species with many connections may be acting as buffers, so that press perturbations affecting them might be attenuated along the multiple interaction pathways that connect them to other species within the food web. This would indicate that stability is lower when the net effect of one species on another is small, and we can say that "fewer links lowers stability." We can call this type of stability community-level stability. Notably, one of the main conclusions of MacArthur's seminal work is that *restricted diet lowers stability* (MacArthur, 1955), thus defining stability as fluctuations in the population density of focal species. We can call this type of stability species-level stability. If this is also true, then community-level and species-level stability should behave in similar ways, at least in relation to the number of links per species. This then raises the intriguing possibility that a similar mechanism is at work in both scenarios. An alternative possibility is that this relationship arises because species that have only a few links may have strong interactions, as has been suggested by several authors (Margalef, 1968; May, 1972; Borvall et al., 2000), so that perturbations might not be dampened through many interactions (McCann et al., 1998; McCann, 2000). However, contrary to this latter suggestion, we found that for both of our empirical food webs, the mean and median interaction strengths of each species had no relationship with linkage density.

DIRECT VS. INDIRECT EFFECTS

So far we have dealt solely with net effects, that is, the sum of the direct and indirect effects between two particular species. But, is it possible to address the importance of indirect effects relative to direct effects in disturbance experiments? In natural communities, multiple direct and indirect pathways are operating simultaneously (and these interactions might be synergistic or antagonistic). The most common approach to identify indirect effects is by combining results from field manipulations with knowledge of the natural history of the species involved, then constructing the different sequences of direct and indirect effects connecting each pair of species (i.e., interaction chains) (Wootton, 1994b, c; Menge, 1995; Montoya et al., 2003). However, it is not feasible to construct all interaction chains in large food webs, as there are simply too many possible permutations.

We suggest an alternative and simpler technique, based on the sign structure of the community matrix A (where only predator-prey interactions are represented) compared with the sign structure of the inverse community matrix A^{-1}. The direct effect of a predator on its prey (a_{ij}) is always negative, and vice versa. However, the net effect of the predator on its prey ($-(a_{ij})^{-1}$), considering every interaction pathway connecting them, can be either positive or negative. When a_{ij} and $-(a_{ij})^{-1}$ have different signs, we can consider indirect effects to be very important relative to direct effects, because they are reversing the net effect of a predator on its prey (a similar rationale is used in Higashi and Nakajima, 1995). An illustrative example is given by keystone predation illustrated in Figure 3A). Here, predator 1 feeds both in 2 and 3, $a_{12} > a_{13}$, $a_{12} < 0$, and $a_{13} < 0$. Species 3 is a competitively superior species that would outcompete species 2 in the absence of predation by 1. It is possible to evaluate the sign of an indirect interaction between two species by simply multiplying the elements of A connecting the species through an interaction chain. Hence, the indirect effect of 1 on 2 mediated by species 3 is positive through the interaction chain 1-3-2: (a_{12}) (a_{32}) = (−)(−) = (+). The inverse community matrix shows that this indirect effect has a greater influence on the net effect of species 1 on 2 given by $-(a_{12})^{-1}$, so that the sign of the direct effect and the sign of the net effect of predator 1 on prey 2 have opposite signs (i.e., $\text{sign}(a_{12}) \neq \text{sign}(-(a_{12})^{-1})$).

In around 40% of predator–prey interactions, we found that predators have a positive (beneficial) net effect on their prey species. A similar percentage is observed in prey–predator interactions, such that prey species have a negative (detrimental) net effect on their predators. This percentage is the same for both food webs and consequently we can conclude that in 40% of species pairs, indirect effects are very important for the outcome of press perturbations. Intriguingly, Menge has also found that indirect effects accounted for 40% of changes in the structure of 23 rocky intertidal communities resulting from manipulations (Menge, 1995). However, our technique underestimates the relative importance of indirect effects in those cases where indirect interactions between a predator

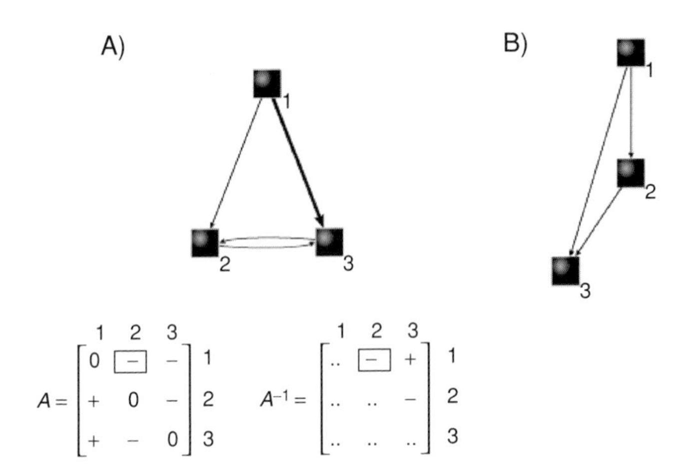

FIGURE 3 | Examples of food web modules where equilibrium combination of direct and indirect effects has different net outcomes. **A** corresponds to keystone predation, and (**B**) to intraguild predation. Arrows correspond to feeding links pointing from predator to prey, except links between 2 and 3 in (**A**) where they correspond to competitive interactions. In (**A**), *A* is the community matrix where the element a_{ij} is the direct effect of species *j* on *i* (measured as per capita interaction strength), and A^{-1} is the inverse community matrix, where $-(a_{ij})^{-1}$ is the net effect of *i* on *j* (we only represent relevant signs for the example). For determining the importance of indirect effects, we compare the sign structure of the community matrix A and the inverse community matrix A^{-1} (see text for further details)

and its prey have the same sign as their direct interaction. This is the case, for example, in some forms of intraguild predation. In Figure 3B, species 1 has a negative direct effect on species 2 measured by a_{12}, and also a negative indirect effect mediated by species 3, because 1 is feeding also on a prey of 2 (Polis and Holt, 1992). In this case, the comparative analysis of sign structure on A and A^{-1} does not consider whether this indirect effect has a large influence on the net effect of species 1 on 2.

UNCERTAIN NET EFFECTS OF DISTURBANCES?

Two types of uncertainties are associated with perturbations and indirect effects. First, predictions that are based on the interaction between direct and indirect effects may yield unexpected results when compared with predictions based on pairwise, direct interactions (Abrams, 1992). Second, there may be a lack of reliable and precise ranges of interaction parameters between species (Yodzis, 1988; Pimm, 1993; Yodzis, 1996).The first type of uncertainty is inherent to ecologically complex systems, and it can be system-specific. This is especially clear when indirect effects arise when a species modifies the interaction between other species (i.e., trait-mediated indirect effects). For example, when prey species are consumed by multiple predators, behavioural interactions between predators often lead to lower predation rates on these prey (Sih et al., 1998, and references therein). This risk-reduction phenomenon in the presence of multiple predators due to indirect interactions is very difficult to determine *a priori*. However, in other cases we can predict the net effect of one species on another, even when this net effect is opposite to its direct effect, as we have seen in the case of keystone predation. In the Ythan and Broadstone webs keystone predation may account for up to about 40% of the net positive effects of predators on prey species. The prediction of the net effect is possible in this case because we have a theoretical basis for the type of indirect effect involved, keystone predation (Polis and Holt, 1992), and because we assume trait-mediated effects are not playing a relevant role. Therefore, our analyses could potentially be more uncertain if we introduce morphological or behavioural indirect effects.

The second type of uncertainty can be eliminated if we have precise parameter estimates. For both food webs we lack information on self-damping terms a_{ii} (e.g., intraspecific competition or cannibalism), and for the Ythan web, interaction strengths are estimated from predator–prey body size ratios, assuming biologically reasonable scaling relationships (see Emmerson et al., Chapter 4.3, for further details). We have explored the robustness of our results using different a_{ii}, ranging from -10^{-3} to -10^{-6} for both webs, and different scaling exponents between interaction strengths and predator–prey body-size ratios for Ythan, generating a universe of community matrices with their correspondent inverses (generating a feasible parameter space for a population of community matrices and their inverses). We observed (described in detail in Montoya et al., unpublished manuscript) that results were very robust to the selection of community matrix. This suggests that once the patterning of predator–prey interaction strengths is known or properly estimated, the indeterminacy of ecological interactions is much lower than previously reported (Yodzis, 1988, 2000).

DISCUSSION AND CONCLUSIONS

Knowing how nature works, and predicting its behavior, are two basic aims of scientific inquiry. Predictive ability is especially important to unite the fields of ecology and conservation biology. In this chapter we attempt to define a predictive framework of press perturbations in empirical food webs considering direct and indirect density-mediated effects, and in so doing we hope to contribute to the integration of these two fields. We have shown that:

1. Some general and relatively "easy to measure" species' trait-level characteristics are related to (and might be the cause of) the net effect of species in the whole food web.
2. Indirect effects are very important for the propagation of disturbances, and this importance can be quantified by means of comparisons between the sign structure of the community matrix and its inverse.
3. Uncertainty on the net effects of disturbances can be reduced markedly if we know or we have plausible ranges of predator–prey interaction strengths.

However, these patterns and observations cannot be classified as general since they are based on a sample size of only two food webs. There is a pressing need for further investigation to determine their cause for more high quality, well resolved, food web data detailing body size, abundance, food web structure, and interaction strength.

ACKNOWLEDGMENTS

This work has been supported by a grant Comunidad de Madrid FPI-4506/2000 (JMM), ESF-project InterAct (JMM), and Spanish Research Council FIS2004-05422 (RVS).

7.3 | DEALING WITH MODEL UNCERTAINTY IN TROPHODYNAMIC MODELS: A PATAGONIAN EXAMPLE

Mariano Koen-Alonso and Peter Yodzis

Model uncertainty has two main sources, the mathematical structure of the model and the parameter values. Although elements of uncertainty are present in every mathematical model, the complexity and nonlinearity of food web models (and the actual systems they try to mimic) make them especially vulnerable.

In a sense, this very vulnerability has always been an underlying force fuelling the development of food web theory. Our lack of knowledge about how certain processes should be mathematically described and how much detail is necessary, has always been pushing the envelope towards new insights (e.g., random versus realistic food webs, functional responses and system stability, static versus dynamic food web structure). However, assessing uncertainty acquires importance in its own right when we address practical applications (Harwood and Stokes, 2003). The incorporation of uncertainty in applied ecosystem models has been recognized as one of the major challenges for implementing ecological approaches to the management of marine resources (deYoung et al., 2004).

Therefore, before believing or using the output of any model, we need to assess the impact of the model formulation on our conclusions. Although this proposition always applies, if we use admittedly incomplete depictions of a real system, like simplified food web models, it becomes essential.

Most applied studies pay some attention to parameter values (e.g., sensitivity analysis), but the influence of functional form is rarely touched upon (Fulton et al., 2003). Yodzis (1988, 1998, 2000) has addressed these issues in the context of local (near equilibrium) trophodynamic models. One of our goals here is to carry this analysis forward to the global arena (far from equilibrium). Another goal is to see whether, and how, food web theory can address practical problems in real systems.

As a first step, we have formulated a simplified food web for the marine ecosystem of Patagonia, off the south coast of Argentina. As we will see, our two goals complement one another—especially, the practical modelling needs the functional uncertainty analysis.

THE MODELLING FRAMEWORK

We derive our basic model structure and parameterization from bioenergetic and allometric principles. This enables setting physiologically meaningful values for some parameters, constrains others to be proportions, and reduces the number of parameters to be estimated (Yodzis and Innes, 1992).

The Basic Equations

Within this framework, we build models using two types of equations (Yodzis, 1998). In case species i is a basal, we use

$$\frac{dB_i}{dt} = \left\{ r_i \frac{B_i}{K_i} \left[K_i - B_i - \sum_c \alpha_{ic} B_c \right] \right\} - \sum_p B_p F_{ip} - m_i B_i - H_i \qquad (1)$$

If species i is a consumer, we use

$$\frac{dB_i}{dt} = B_i \left\{ -T_i + \sum_j e_{ji} F_{ji} \right\} - \sum_p B_p F_{ip} - m_i B_i - \mu_i B_i^{v_i} - H_i \qquad (2)$$

In these equations, the subindex i corresponds to the species whose dynamics is being described (the focal species), c indicates a competitor of i, p indicates a predator of i, and j indicates a prey of i. These subindices are used to identify a species or to indicate a summation over the set it represents (e.g., summation over c implies the summation over all competitors of the focal species i).

The parameters r_i and T_i are the intrinsic production/biomass ratio and the mass-specific respiration rate of species i, and are derived from

$r_i = f_{ri} a_{ri} w_i^{-0.25}$ and $T_i = a_{Ti} w_i^{-0.25}$ where f_{ri} is a proportion ($0 < f_{ri} < 1$), a_{ri} and a_{Ti} are allometric coefficients which depend on the metabolic type of species i (endotherms, vertebrate ectotherms, invertebrates, or phytoplankton), and w_i is the mean individual biomass of species i (Yodzis and Innes, 1992). B_i is the biomass of the focal species i (similarly B_c is the biomass of a competitor and B_p is the biomass of a predator of i; for the sake of brevity we will not repeat these obvious details in the subsequent definitions), K_i is the carrying capacity of i, α_{ic} is the competition coefficient between the species i and a competitor species c, F_{ji} is the functional response of i when preying on j, e_{ji} is the assimilation efficiency of i when eating j, m_i is the "other mortality" rate of i and represents the losses due to predation by species not included in the model, the term $\mu_i B_i^{v_i}$ represents density-dependent mortality (μ_i and v_i are positive constants), and H_i is the harvest rate of i.

The Functional Response

The functional response is a core feature in any trophodynamic model and describes the consumption rate of a given prey by a given predator (Holling, 1959b, 1965; Beddington, 1975; DeAngelis et al., 1975; Yodzis, 1989, 1994; Jeschke et al., 2002; Gentleman et al., 2003; Murdoch et al., 2003; Turchin, 2003). It can be a function of prey densities only (*laissez-faire* functional responses), or it can also depend on predator density (predator interference or facilitation), and/or other plausible variables (e.g., competitors densities, temperature, age, condition, etc.). This ample spectrum of possibilities provides plenty of room for a quite diverse set of dynamic behaviors. However, we typically have very little evidence to discriminate among alternative mathematical formulations.

Using time budget reasoning and simple foraging theory within a multispecies scenario (i.e., derived from basic principles), a fairly general way of writing the functional response for predator i when consuming a specific prey k (F_{ki}) is:

$$F_{ki} = \frac{C_{ki}}{1 + \sum_j h_{ji} C_{ji}} \qquad (3)$$

where the subindex j indicates any prey of predator i, including the focal prey k, and the summation over j implies a summation over all prey of predator i, C_{ji} is the capture rate (amount of prey j captured by predator i per unit of searching time), and h_{ji} is the handling time for prey j. The inverse of this handling time is the asymptotic consumption rate of prey j (J_{ji}) when density of j tends to ∞.

Assuming different forms (and functional dependencies) for C_{ji} we can derive different functional responses. For example, if for all j, $C_{ji} = s_{ji} B_j$, where s_{ji} is a positive constant, we obtain the classic Holling Type II functional response; or if $C_{ji} = \frac{s_{ji} B_j}{1 + q_i B_i}$, where s_{ji} and q_i are positive constants, we get the Beddington functional response (Table 1 contains more examples). Although equation (3) is not fully general (i.e. there are functional responses that cannot be derived from it), it provides enough generality to give a common ground for many widely used functional responses (Yodzis, 1994).

If for simplicity we assume that J_{ji} depends only on the predator, which is also consistent with assuming a digestion limited functional response (i.e. a predator-dependent digestive pause) (Jeschke et al., 2002), we can rewrite (3) as:

$$F_{ki} = \frac{J_i C_{ki}}{J_i + \sum_j C_{ji}} \tag{4}$$

In this way, we can use the bioenergetic-allometric framework to estimate J_i as $J_i = f_{Ji}\, a_{Ji}\, w_i^{-0.25}$ where f_{Ji} is again a proportion and a_{Ji} is another allometric coefficient (Yodzis and Innes, 1992).

THE STUDY CASE: THE PATAGONIA SYSTEM

We have a good knowledge of the exploitation history of the Patagonian marine community. The currently most abundant marine mammal, the southern sea lion (*Otaria flavescens*), was harvested between 1920 and 1960 and its population was drastically reduced (Crespo and Pedraza, 1991). Even though it has not reached pre-harvest levels yet, this population is clearly recovering (Reyes et al., 1999; Dans et al., 2004). In the 1970s the hake (*Merluccius hubbsi*) fishery developed in Patagonia, followed in the 1980s by the squid (*Illex argentinus*) fishery. During the 1990s the Patagonian hake stock virtually collapsed (Aubone et al., 1999).

The anchovy (*Engraulis anchoita*) is the main forage fish in this system, and together with squid and hake are the core species of the Patagonia marine food web (Angelescu and Prenski, 1987; Prenski and Angelescu, 1993, Dans et al., 2003). The hake has anchovy and squid, together with zooplankton, as main food items, while zooplankton is the principal food resource for both anchovy and squid (Angelescu, 1982; Ivanovic and Brunetti, 1994). Sea lions prey on these species, but most heavily on hake and squid (Koen-Alonso et al., 2000).

The Patagonia Model

We used time series of population biomass and annual catches for squid, anchovy, hake, and sea lion to build a simple global trophodynamic model (Figure 1) based on equations (1) and (2) (Koen-Alonso and Yodzis, 2005).

Squid, anchovy and, to a lesser extent, hake feed on zooplankton, but the information available on zooplankton is too scarce and sparse to enable modelling that component. Furthermore, zooplankton is an aggregated group composed of many very different species. Therefore, we treated squid and anchovy as basal species. Since squid and anchovy are relying on similar resources, there is room for competition between them. The competition terms in equation (1) take this into account.

Another modelling feature that we incorporated is the provision of alternative prey for the consumer species. Typically, if equation (4) is used to represent the functional response, all the consumption will be allocated among the specified prey. However, if we are not considering

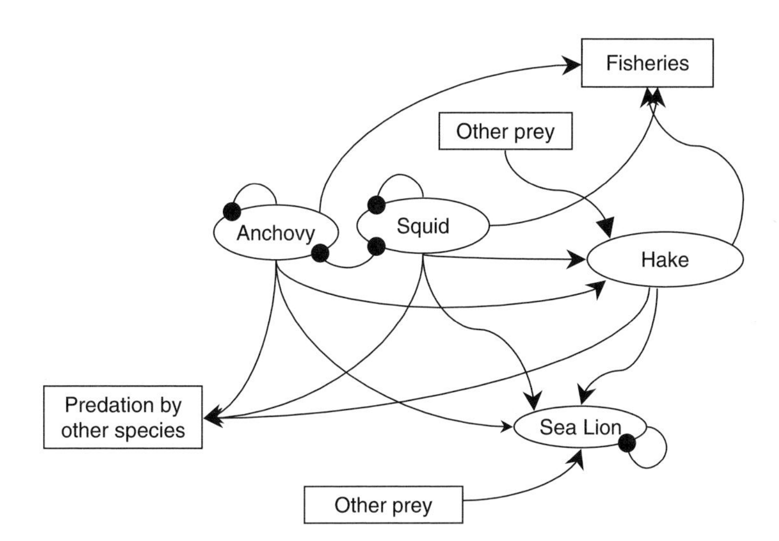

FIGURE 1 | General sketch for the Patagonia models. The ovals indicate dynamic equations while the boxes correspond to constant effects. Squid and anchovy were modeled as basal species [equation (1)], while hake and sea lion were modeled as consumers [equation (2)]. When fitting the models, the annual catches were used as harvest rate H_i and were annually updated. The double negative link between anchovy and squid indicates competition, the self-negative loops in the squid, anchovy, and sea lion correspond to density-dependent mortality, and predation by other species was represented by a linear mortality term [see equations (1) and (2)]. The other prey effects were represented by an additional and constant prey biomass term in the functional responses of the consumer species.

the entire food web, is clear that some of the consumption will be on species not modeled. If we do not take this into account, we will overestimate the consumptions of the prey in the model. We addressed this issue in the simplest way possible. We added one additional prey source in the functional response of each consumer. They were modeled as a constant prey biomass always available to the consumer, and treated as additional parameters.

The parameters in the model were estimated by maximum likelihood, and assuming multiplicative lognormal errors (Pascual and Kareiva, 1996; Hilborn and Mangel, 1997). The actual fitting involved the minimization of the negative log-likelihood using a simulated annealing-based minimization algorithm developed by us (Fortran 77 subroutine available upon request), and the system of ordinary differential equations was solved using the Gear's stiff method (Visual Numerics, 1997). Although population time series start in 1970, the sea lion harvest took place between 1920–1960 (Figure 2). Therefore, the starting point was set before that time and the catch information was used to annually update the harvest rates in the model during the fitting process.

Model Uncertainty in the Patagonia Model

Fitting a global trophodynamic model to time series requires full specification of the form of the functional response (Yodzis, 2001). Therefore, to explore the associated structural uncertainty, we need to fit the model with alternative functional responses.

We selected five mathematical forms for the functional response (see Table 1), and we fitted the otherwise identical model with each of them. These forms covered a fairly large range of possibilities. The multispecies Holling Type II with predator interference (see Table 1) is a general formulation that can produce both the Beddington functional response (Beddington, 1975; DeAngelis et al., 1975; Yodzis, 1994) or the classical ratio-dependent functional response (Açacaya et al., 1995; Jost and Arditi, 2000); the generalized *laissez-faire* Holling allows Type II and Type III shapes, the frequency-dependent predation formulation is a *laissez-faire* Type III which collapses to a Type II when only one prey is present, the Evans functional response (Evans and Garcon, 1997) combines linear and non-linear effects on the capture rate C_{ki}, and the Ecosim formulation (Walters et al., 1997; Walters and Kitchell, 2001) is essentially a linear functional response (without upper limit) with predator interference. This last functional response was considered due to its extensive use as part of the widespread package Ecopath with Ecosim (Walters et al., 1997).

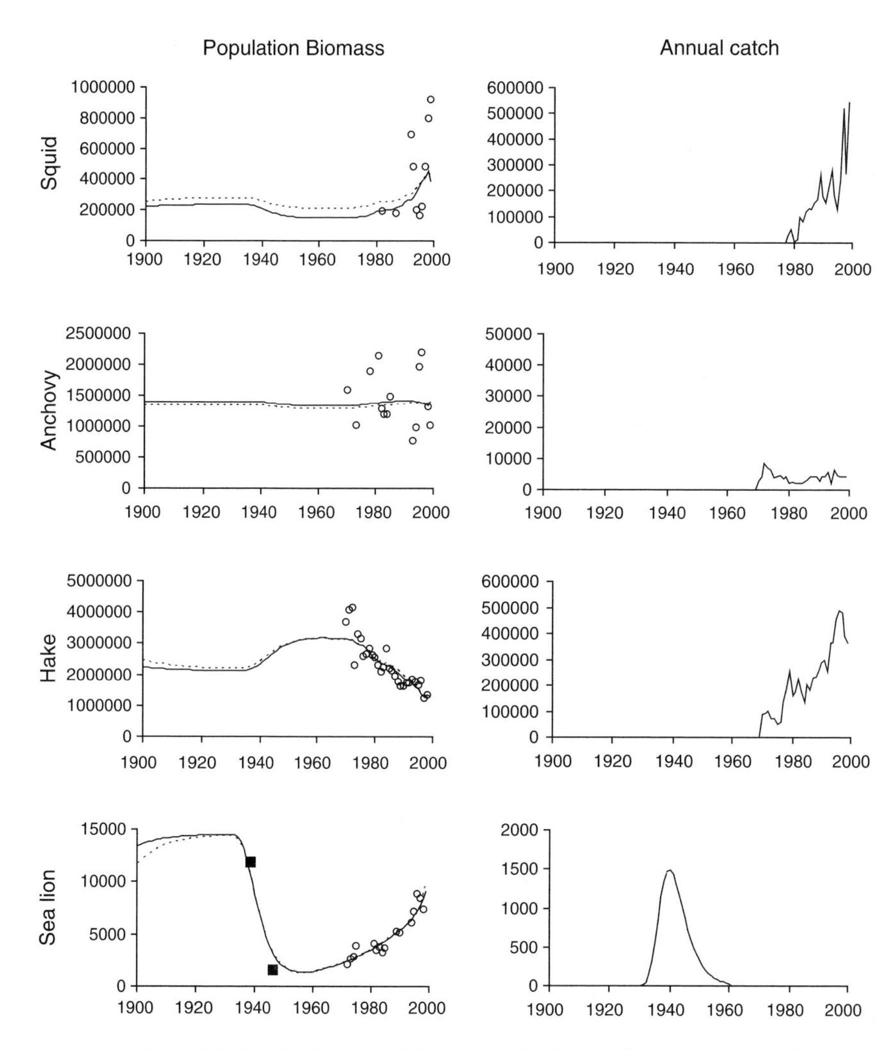

FIGURE 2 | Models fits. Left panels: Maximum likelihood fits for the Evans (continuous line) and generalized Holling (dotted line) Patagonia models. The data used for fitting are indicated by empty circles. The two black squares in the sea lion panel indicate early data which, due to methodological differences with recent surveys, were not used for fitting. Considering this, the close agreement with model predictions is remarkable. Right panels: Annual catches used to fit the models. The sea lion catches are derived from a fitting of the cumulative catch curve (the original records do not provide annual catches). All biomasses and catches are in metric tonnes.

After fitting the models, we used the Akaike's information criterion corrected for sample size (AIC_c) to rank and select those models that had comparatively enough empirical support to deserve further analysis (Burnham and Anderson, 2002). Only the Evans and the generalized Holling models were worthy of such consideration (see Table 1, Figure 2).

Parameter uncertainty was then evaluated by applying the sampling-importance-resampling (*SIR*) algorithm (Gelman et al., 2004), and considering uniform prior distributions for all parameters. The range for the priors was defined by $\theta_{MLE} \pm 0.5\theta_{MLE}$, where θ_{MLE} is the maximum likelihood estimator of parameter θ. This approach allows a good exploration of parameter space in the neighborhood of θ_{MLE}, and produces an approximation of the posterior distributions of the parameters in that region. It also allows building empirical 95th percentile ranges for the parameters, which can be considered a first approximation to actual confidence limits.

Exploration of Model Behavior

The procedures described in the previous section allowed us to select two plausible structural models, and to generate distributions for their parameters. Using this information, we can explore how these models behave under different scenarios and evaluate the similarity in their responses. Generally speaking, we can distinguish between two approaches for this exploratory analysis. The first one is narrower in scope but provides a thorough exploration and is best suited for specific and concrete questions (e.g., evaluating a particular management proposal). The second one is broader in scope but provides a less detailed picture (e.g., a general overview of model behavior).

The first approach requires a very specific description of the scenario to be explored (e.g., specific catch levels and/or harvest rates for all the species in the model). Within a precisely defined scenario, the dynamic of the modeled system can be simulated and Monte Carlo techniques implemented. The distributions obtained from the *SIR* algorithm can be used for sampling parameter values. Within each model this procedure will provide insight on the effect of parameter uncertainty, while the comparison of results between models will allow the evaluation of the structural uncertainty. If we want to give one step further, we can also use model averaging techniques to generate a compound answer (Burnham and Anderson, 2002; Gelman et al., 2004).

Given our goals, the second approach is the one that we actually followed. Instead of simulations, we relied on continuation and bifurcation analysis (Doedel et al., 1991a, b; Doedel et al., 1998). In this way, we can track how the attractor of the system changes as we vary one or more parameters (the bifurcation parameters). In this case we defined the harvest rates as $H_i = h_i B_i$ where h_i is the harvesting mortality (or fishing mortality if you prefer), and used the harvesting mortality h_i as bifurcation parameter. Although this approach allows building a general picture

Table 1. Functional responses used to build alternative Patagonia trophodynamic models, and corrected Akaike's information criteria corresponding to their best fits.

Functional response	Equations (expressed in terms of a focal prey k)		AIC_c
	Capture rate C_{ki}	Functional response (eq. 4) $F_{ki} = J_i \frac{C_{ki}}{J_i + \sum_j C_{ji}}$	
Multispecies Holling Type II with predator interference (based on ideas from Yodzis, 1994)	$C_{ki} = \frac{a_{ki} B_k}{\left(U_i + B_i^{u_i}\right)}$	$F_{ki} = J_i \frac{a_{ki} B_k}{J_i\left(U_i + B_i^{u_i}\right) + \sum_j a_{ji} B_j}$	2011.7
Multispecies generalized *laissez-faire* Holling	$C_{ki} = a_{ki} B_k^{bi}$	$F_{ki} = J_i \frac{a_{ki} B_k^{b_i}}{J_i + \sum_j a_{ji} B_j^{b_i}}$	1965.7
Frequency-dependent predation (Yodzis, unpublished)	$C_{ki} = \frac{p_{ki} B_k^{b_i}}{\sum_j p_{ji} B_j^{b_i}} a_{ki} B_k$	$F_{ki} = J_i \frac{p_{ki} a_{ki} B_k^{1+b_i}}{J_i \sum_j p_{ji} B_j^{b_i} + \sum_j p_{ji} a_{ji} B_j^{1+b_i}}$	2073.1
Evans (Evans and Garcon, 1995)	$C_{ki} = a_{ki} B_k (1 + b_i B_k)$	$F_{ki} = J_i \frac{a_{ki} B_k \left(1 + b_i B_k\right)}{J_i + \sum_j a_{ji} B_j \left(1 + b_i B_j\right)}$	1963.6
Ecosim (Walters et al., 1997; Walters and Kitchell, 2001)	This functional response cannot be cleanly obtained from eq. (4) without additional assumptions	$F_{ki} = \frac{a_{ki} v_{ki} B_k}{v'_{ki} + v_{ki} + a_{ki} B_i}$	2205.9

F_{ki}: functional response of predator i on prey k; B_k : biomass of the focal prey k; B_j : biomass of a prey j (the summation over j implies the summation over all prey of i, including the focal prey k); B_i : biomass of predator i; J_i : asymptotic consumption of predator i. In the corresponding equations, a_{ki}, p_{ki}, v_{ki}, v_{ki}', U_i, u_i, and b_i are positive constants. In the frequency-dependent predation functional response the $\sum_j p_{ji} = 1$. When appropriate, remember that the subindices ki are a short-hand for $j=k,i$ where j indicates a generic prey, and k indicates a specific (focal) prey of predator i.

of the expected changes in the attractor of the system, incorporating parameter uncertainty is not that simple due to the difficulties in automatizing the bifurcation analysis.

Therefore, in this case we assessed the effects of parameter uncertainty as follows. We used the 95th percentiles of the estimated posteriors of the parameters to define a 95th percentile region in parameter space. Within that space, we selected four parameter sets with the highest negative log-likelihoods (i.e., the worst parameter sets). The idea was to find "extreme" parameter sets within an approximate 95% confidence region. We repeated the bifurcation analysis for these "extreme" sets, hoping that they will reflect the boundaries of the expected behaviors of these models under exploitation, or at least provide some insight about their uncertainty.

This approach does not provide a comprehensive exploration of parameter space; it is possible that less "extreme" parameter sets can generate even more "extreme" dynamics. However, we are not relying here on detecting the highest discrepancy in model behavior between parameter sets, but the lack of it. We are assuming that is very unlikely that a given qualitative behavior which is present for both the most likely parameters (the *MLE* parameters) and the less likely ones (the "extreme" parameters) will change dramatically in between. Only if all bifurcation analyses showed a common general pattern, we considered the predicted model behavior to be a robust result.

Some Results From The Patagonia Model

Although five models were fitted, we were able of selecting just two for further analysis. These models corresponded to the Evans and generalized Holling functional responses.

In both cases, the estimated parameters rendered Type III shaped functional responses and the fits were almost identical; even showing a similar increase in hake biomass as a consequence of the sea lion decline (see Figure 2). This result is particularly interesting, because a common assumption in single-species biomass-production fishery models is that the target species was at equilibrium before the beginning of the fishery (the "virgin biomass"), and this biomass is usually equated to the target species carrying capacity. Here, making such assumption can lead us to (erroneously) believe that the Patagonia system can sustain a much higher hake biomass than what it actually can (at least according to our models). A misconception like this can have serious consequences.

Although the fits of these models may suggest similar responses to harvesting, the exploration of their behavior under different exploitation scenarios actually showed some important divergences. For example, the harvesting of squid produced an unexpected increase in squid equilibrium biomass for intermediate levels of harvesting (including hysteresis behavior) for the Evans model, whilst the generalized Holling model showed a monotonic reduction of the squid equilibrium biomass until the extinction point (Figure 3). Therefore, the expected behavior under exploitation is obviously model-dependent. We cannot simply trust a single model configuration. The examination of the bifurcation pattern in the "extreme" parameter sets of the Evans model did not show the non-monotonic response of squid to its own harvesting (see Figure 3). Therefore, although the non-monotonic response is technically the most likely behavior for the Evans model, cannot be considered a robust prediction from it.

However, not all results were blurry. For example, both models showed a similar bifurcation pattern when harvesting hake. These

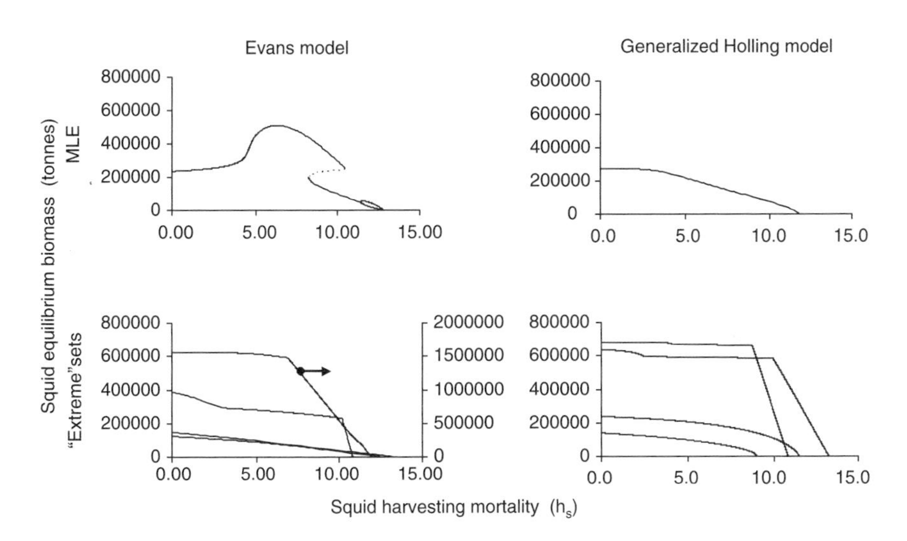

FIGURE 3 | Squid equilibrium biomass as a function of squid harvesting mortality (h_s) (bifurcation diagrams) for the Evans and generalized Holling models. The first row of diagrams correspond to the best fits (*MLE* parameters) and the second to the "extreme" parameter sets (see details in the text). All analyses started with the system at equilibrium and no harvest, and the squid harvesting mortality (h_s) was increased from zero until reaching squid extinction. All attractors (equilibrium points or limit cycles) are stable unless indicated with dotted lines. For the MLE diagram of the Evans model (top-left), the splitting of the single equilibrium in two branches indicates a Hopf bifurcation, where the two branches indicate the amplitude of the cycles. The arrow in one bifurcation diagram of the "extreme" sets of the Evans model (bottom-left) indicates that it should be read on the secondary y-axis.

analyses indicated that the equilibrium biomass of anchovy goes down when the hake harvesting mortality increases (i.e., harvesting the predator has a negative effects on its prey) (Figure 4). In this case, the bifurcation analyses of the "extreme" sets indicated that the equilibrium biomass of anchovy, though showing some increases at lower levels of hake exploitation, always goes down dramatically when the hake harvesting mortality becomes larger (see Figure 4). This result suggests the anchovy typically goes extinct before the hake, which is the species being harvested. There is a pretty robust indication here that an intense hake fishery may have an overall negative effect on anchovy.

DISCUSSION AND CONCLUSIONS

Model predictions of the responses of stocks to changes in harvest rates, both in local and global trophodynamic models, show consider-

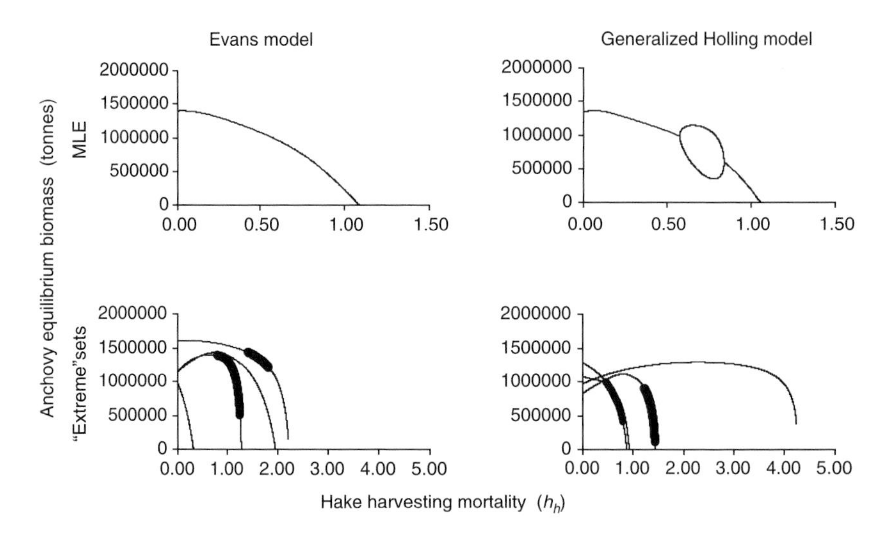

FIGURE 4 | Anchovy equilibrium biomass as a function of hake harvesting mortality (h_s) (bifurcation diagrams) for the Evans and generalized Holling models. The first row of diagrams correspond to the best fits (*MLE* parameters) and the second to the "extreme" parameter sets (see details in the text). All analyses started with the system at equilibrium and no harvest, and the hake harvesting mortality (h_h) was increased from zero until reaching hake extinction. All attractors (equilibrium points and limit cycles) are stable. For the MLE diagram of the generalized Holling model (top-right), the blob indicates an stable periodic attractor bounded by Hopf bifurcations, and the two branches indicate the amplitude of the cycles. For the "extreme" sets, the thick lines (actually very close filled circles) correspond to stable periodic attractors bounded by Hopf bifurcations. These last diagrams provide no information about the amplitude of the cycles.

able variation with respect to adjustments of parameters and of functional forms (Yodzis, 1988, 1998, 2000, this chapter). Unfortunately, some predictions are not robust; fortunately and importantly, others are. Therefore, trophodynamic modelling has the potential to be of practical use, but its application requires care. For example, the one functional response that is actually widely used for marine systems, the Ecosim form, performed for the Patagonia system far worse than the others (see Table 1). Also, even though the two selected models were virtually identical in terms of their general shape and goodness-of-fit, their predictions were significantly different for some exploitation scenarios. This clearly indicates that a single model structure cannot be considered *per se* a reliable source for making inferences about the modeled system.

The Patagonia model also provides other insights. The effects of fishing impacts on a community not only can contradict our initial expectations, but they also can reverse their signs depending on the intensity of the exploitation (see Figure 3). Non-monotonic responses to perturbations have been theoretically suggested (Yodzis, 1996), and recently found in simulated predator–prey systems for the very species being harvested (i.e., for some exploitation levels the effect of fishing is an increase in the biomass of the exploited species) (Matsuda and Abrams, 2004). Our results constitute the first applied example where a plausible solution shows this kind of behavior.

The inclusion of competition among our "basal" species reflects the real system, and preliminary work where it was excluded, indicated that is essential for the good performance of the model. Competition among basal species, whether surrogates as here or real ones, is likely very common in nature, but many food web models still ignore this interaction (Berlow et al., 2004). Although more complex ways of including competition among basal species are currently being studied (see Brose et al., Chapter 2.1), our simpler implementation allows considering it without significantly increasing model complexity. In either case, it appears that it might be better routinely to include this interaction than routinely to exclude it from food web dynamic models. It is also worth mentioning that the best fits of the selected Patagonia models shared *laissez-faire* Type III functional responses. Although the presence of prey refuges, predator learning, and prey switching are traditional arguments for explaining Type III shapes (Murdoch, 1969, 1973; Chesson, 1983), and it is well known that this functional response enhances stability (Oaten and Murdoch, 1975a; Murdoch et al., 2003), there is an apparent consensus among ecologists that Holling Type II is the most plausible/general shape for the functional response (Jeschke et al., 2002).

However, adaptive feeding and prey-switching, by reconfiguring the trophic structure of the food web along time, have profound effects on system features like connectance, flexibility, and population persistence (see Kondoh, Chapter 3.3). Coupling and decoupling of consumer-resource systems in space and time, both at the species and/or food web module levels (weak-link and weak sub-system interaction effects) have been proposed as key processes in the generation and regulation of ecosystem stability (McCann et al., 1998; McCann, 2000; see McCann et al., Chapter 2.4). These mechanisms can generate Type III shaped functional responses (McCann, 2000; McCann et al., Chapter 2.4).

From an individual perspective, feeding and the ways in which it varies in space and time, are mainly behavioral processes. It is also plausible that this individual behavior might be better described by a Type II shaped functional response. However, most food web models typically integrate (at different degrees), individual responses over space, time, and population structure. Therefore, their functional responses should also integrate many of the feeding behavior features, and the Type III shape appears to be a consistent phenomenological emergent from many plausible ecological mechanisms. Under this light, the observed agreement between current ideas in food web theory and our empirical results suggests that Type III functional responses might be a better standard choice for simple food web dynamic models than the customary Type II.

ACKNOWLEDGMENTS

Ideas, contributions, and data from Enrique (Kike) Crespo, Silvana Dans, and Susana Pedraza were the foundations of the path that led us to this point. Dave Vasseur provided stimulating feedback in both the ecology of this problem and the development of the computer code. Geoff Evans not only brought to our attention his functional response, he also became an insightful, sharp and extremely helpful devil's advocate. Discussions with John T. Anderson, Peter Shelton, Paul Fanning, and Alida Bundy also helped us to improve this work. Norma Brunetti, Susana García de la Rosa, Mario Lasta, Marcelo Pérez, Bruno L. Prenski, Felisa Sánchez, and Ramiro Sánchez generously shared their time, data, and experience. Comments from two anonymous reviewers gave us the opportunity for clarifying many points. Financial and/or logistic support was provided by the National Research Council (CONICET), the Institute for Fisheries Research and Development (INIDEP), and the University of Patagonia (UNP) in Argentina; the Natural Sciences and Engineering Research Council (NSERC) and the Department of Fisheries and Oceans (DFO) in Canada.

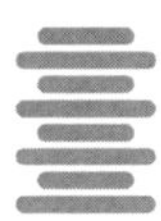

7.4 | DESCRIBING A SPECIES-RICH RIVER FOOD WEB USING STABLE ISOTOPES, STOMACH CONTENTS, AND FUNCTIONAL EXPERIMENTS

Craig A. Layman, Kirk O. Winemiller, D. Albrey Arrington

Complexity and variability of ecosystems, especially in the tropics (e.g., rain forests, coral reefs, freshwater lotic systems), render obsolete simple solutions to describe how humans impact food webs. Yet this understanding is essential to help stem biodiversity loss and assess community- and ecosystem-level responses to human-induced habitat change (Chapin et al., 1998; McCann, 2000). Human impacts are especially difficult to assess in complex webs due to indirect (Wootton, 1993), diffuse (Yodzis, 2000), and emergent (Sih et al., 1998) effects. In this context, characterization of trophic dynamics (i.e., food web structure, energy flow) using multiple, complementary methodologies is most likely to lead to an understanding sufficient to derive useful predictions regarding effects of human-induced perturbations.

We describe research conducted in a species-rich, floodplain river in Venezuela, the Cinaruco, where we have employed three methodologies to describe different aspects of food web structure: (1) stable isotope analyses, (2) stomach contents analyses, and (3) functional experiments. Because of the recent emphasis on stable isotopes in food web research (Post, 2002a; Vanderklift and Ponsard, 2003, and references therein), we devote much of this chapter to describing how we have used stable isotope ratios to characterize food web structure in the Cinaruco River. The importance of using empirically-derived stomach contents analysis has increasingly been acknowledged, and together stable isotope ratio and stomach contents analysis can provide detailed descriptions of energy flow through food webs. These approaches, however, do not reveal functional roles of organisms in the web (Paine, 1980). To illustrate, we compare results of experimental exclusion of large piscivores with results derived from stable isotope and stomach contents analysis of the same piscivores. Each of these methodologies may lead to different predictions regarding human impacts on food web structure, and we provide an example by examining how illegal commercial netters may affect the Cinaruco River food web through removal of large piscivore species.

METHODS

Study Site

The Cinaruco River is a moderate blackwater (sensu Goulding, 1980) floodplain river in southwestern Venezuela (6° 32´ N, 67° 24´ W). The river has a forested riparian zone, but open grassland dominates the drainage basin (ca. 10,000 km^2). Hydrology is strongly seasonal, with the river water level fluctuating more than five meters annually. In the wet season (May to October) the riparian forest is flooded, and dispersal of organisms is extensive (Lowe-McConnell, 1987). Rapidly falling water levels in the river from November to January increases fish densities, leading to intense biotic interactions (e.g., predation) during the dry season (February-April) (Winemiller and Jepsen, 1998; Layman and Winemiller, 2004). Maximum width of the main channel during the dry season is 50–200 meters. The river supports an extremely diverse fish assemblage (>280 species) with taxa representing a wide range of ecological attributes and life history strategies (Jepsen et al., 1997; Winemiller et al., 1997; Arrington and Winemiller, 2003; Winemiller and Jepsen, 2004).

Isotopic Analyses

Samples were collected 1999–2003 from a 25-kilometer stretch of the Cinaruco River. In this chapter, we discuss a sub-set of fish samples from an extensive isotopic ratio data base (n > 1,900 individual samples).

Collection, preservation, and preparation follow Arrington and Winemiller (2002). Briefly, fishes were collected using nets and hook and line, and represent a random sub-sample from the community. Immediately following capture, approximately 2 grams of dorsal muscle was removed and covered with 20 grams of non-iodized table salt. In the laboratory, salt-preserved samples were rinsed in distilled water, soaked in distilled water for 4 hours, and dried at 60°C for 48 hours. Once dry, samples were ground to a fine powder using a mortar and pestle and loaded into tin capsules. Samples were analyzed at the Stable Isotope Laboratory at the University of Georgia's Institute of Ecology for determination of percent carbon, percent nitrogen, and stable isotope ratios. Stable isotope values are reported using δ (delta) notation where:

$$\delta^{15}C \text{ or } \delta^{15}N = ([R_{sample} / R_{standard}] - 1) \times 1000 \quad (1)$$

where R is $^{13}C{:}^{12}C$ or $^{15}N{:}^{14}N$. Working standards were bovine (n = 49, $\delta^{13}C$ = –22.11%, SD = 0.06%, 48.8% C, $\delta^{15}N$ = 7.47, SD = 0.07%, 10.0% N) and poplar (n = 81, $\delta^{13}C$ = –27.34%, SD = 0.10%, 48.1% C, $\delta^{15}N$ = –2.47, SD = 0.16°/%, 2.7% N). We did not perform lipid extraction prior to stable isotopic analysis of samples because the range of lipid content among samples was relatively small (Arrington, *unpublished data*), and there was typically no correlation between the relative amount of lipid in a sample and the isotopic signature of either carbon (e.g., *Semaprochilodus kneri*, n = 91, R^2 = 0.015; P = 0.25) or nitrogen (e.g., *S. kneri*, n = 89, R^2 = 0.012; P = 0.31).

A common use of stable isotope ratios is to estimate the trophic position of secondary consumers using $\delta^{15}N$, and a crucial step in this calculation is determination of baseline $\delta^{15}N$. It is now widely acknowledged that primary consumers provide the best baseline because they integrate temporal and spatial variation in isotopic signatures of basal resources (Cabana and Rasmussen, 1996; Vander Zanden and Rasmussen, 1999; Post, 2002a, b). When consumers acquire nitrogen from more than one food web module (sensu Holt, 1997, e.g., the littoral and pelagic food web in a lake), the following model is typically used to assess trophic position of a secondary consumer (S):

$$\text{Trophic position} = \lambda + (\delta^{15}N_S - [\delta^{15}N_{B1} \times \alpha + \delta^{15}N_{B2} \times (1 - \alpha)])/\Delta \quad (2)$$

where λ is the trophic position of the organism(s) used to estimate baseline values (in this study λ = 2, the trophic position of primary consumer taxa), S is the consumer in question, B1 and B2 are the two baseline taxa, α the proportion of nitrogen in the consumer derived from the food web module which B1 represents, and Δ is the enrichment in $\delta^{15}N$ per trophic level. If it is assumed that nitrogen and carbon move through a food web in similar fashion, α can be estimated as:

$$\alpha = (\delta^{13}C_S - \delta^{13}C_{B2}) / (\delta^{13}C_{B1} - \delta^{13}C_{B2}) \quad (3)$$

where the consumer taxa of interest is designated by S and baseline taxa are designated by B.

A species used to estimate baseline should: (1) temporally integrate isotopic changes at a scale near that of the secondary consumer of interest, and (2) capture spatial variability of resources supporting the secondary consumer (Post, 2002b). More generally, baseline taxa should integrate (temporally and spatially) the main sources of basal production that support consumers, thereby representing the specific energy sources for the overall food web. In tropical lotic waters, fish often occupy foraging niches that in temperate aquatic systems are dominated by invertebrates (e.g., processors of particulate organic matter) (Winemiller, 1990; Flecker, 1992; Winemiller and Jepsen, 1998), and thus detritivorous and herbivorous fishes may serve as ideal baseline organisms. Tissue turnover rates in invertebrates are much higher than that of fish (Fry and Arnold, 1982; Peters, 1983; Hesslein et al., 1993), and thus using fish to characterize the isotopic baseline in order to estimate trophic position of secondary consumers (i.e., typically other fish species) provides a better match of temporal integration of resources. Fish also are more vagile than most invertebrates, providing a spatial integration of basal production sources. This integration is especially important in floodplain rivers that have a high-level of habitat heterogeneity (Arrington, 2002; Hoeinghaus et al., 2003a) and numerous basal resources (Hamilton and Lewis, 1992; Forsberg et al., 1993; Lewis et al., 2001; Bunn et al., 2003).

Based on these considerations we identified two species, which integrate the two dominant source pools of primary production in the Cinaruco River: (1) *Semaprochilodus kneri* (Characiformes: Prochilodontidae) for autochthonous algal/detrital resources, and (2) *Metynnis hypsauchen* (Characiformes: Characidae) for allocthonous C_3 (and to a lesser extent C_4) plant material (see also Hamilton and Lewis, 1992; Lewis et al., 2001). *S. kneri*, similar to other prochilodontids, feed on a mixture of algae and detritus (Bowen, 1983; Bowen et al., 1984; Goulding et al., 1988), integrating *in situ* benthic algae and periphyton production, as well as detrital matter derived from macroalgae, periphyton, macrophytes, and phytoplankton. The $\delta^{15}N$ of *S. kneri* is representative of $\delta^{15}N$ of other algivore/detritivore taxa in the river (e.g., the con-generic *S. laticeps*, loracariids, and curimatids, *n* of taxa = 15, *unpublished data*; see also Jepsen, 1999). Allochthonous production may also be important in tropical floodplain rivers, especially since numerous taxa consume parts of terrestrial plants (Gottsberger, 1978; Goulding, 1980; Hamilton and Lewis, 1987; Goulding et al., 1988). For example, in the blackwater Río Negro, 79 fish species were found to consume forest fruit and seeds (Goulding et al., 1988). *M. hypsauchen* feeds primarily on terrestrial plant material, including leaves, fruits, and seeds (Layman, *unpublished data*), and its isotope signature reflects a combination of C_3 and C_4 plants, weighted heavily toward to the former source (Jepsen, 1999).

Stomach Contents Analyses

From 1999–2003, piscivores were collected from the Cinaruco River by gill netting and hook and line. Stomachs were analyzed using one of two methods. *Cichla* spp. were examined by pressing down the posterior region of the tongue and pushing gently on the fish's stomach while holding the fish in a head-down position (Layman and Winemiller, 2004). Other piscivore species were euthanized, and stomachs removed for examination. All prey items were identified to the lowest taxonomic level possible, measured (SL ± 1.0 mm), and quantified volumetrically.

Experimental Manipulations

A large-bodied fish exclusion experiment conducted in the dry season (January–March) of 2001 evaluated, at a relatively large-scale (approximately 500 m^2), whether prey altered their spatial distribution in the absence of large-bodied piscivores. Exclosures were made of poultry wire (mesh 2.5 cm) which allowed most prey taxa to move freely in and out of experimental areas, while preventing entry of large-bodied piscivores. Control areas were of an equivalent size, but gaps in the wire allowed passage of all fishes. After two weeks, fish assemblages in experimental areas were sampled, and abundance of fishes were used to assess potential community-level responses to piscivore exclusion. Complete methods can be found in Layman and Winemiller (2004).

DISCUSSION AND CONCLUSIONS

Isotopic Analyses

What do $\delta^{13}C$ and $\delta^{15}N$ tell us about a secondary consumer, *Cichla temensis*, one of the most common piscivores in the Cinaruco River (Jepsen et al., 1997; Winemiller et al., 1997)? Since the trophic fractionation of $\delta^{13}C$ is assumed to be close to zero (Rounick and Winterbourn, 1986; Peterson and Fry, 1987; Post, 2002b), $\delta^{13}C$ suggests that *C. temensis* assimilates a large proportion of its energy from the autochthonous food web module (Figure 1), perhaps via *S. kneri* (Winemiller and Jepsen, 1998; Winemiller and Jepsen, 2004). The variability in $\delta^{13}C$ values observed among individual *C. temensis* indicates that a diversity of prey types support individuals of this top predator species, and emphasizes the importance of large sample sizes to characterize this variability. Significant overlap in $\delta^{13}C$ of basal resources in tropical floodplain rivers (Hamilton and Lewis, 1992; Lewis et al., 2001) complicates determination of resources supporting upper trophic levels. Other sources of information, such as additional isotopes (Wainwright et al., 2000; Hsieh et al., 2002; Currin et al., 2003; Litvin and Weinstein, 2003) or stomach

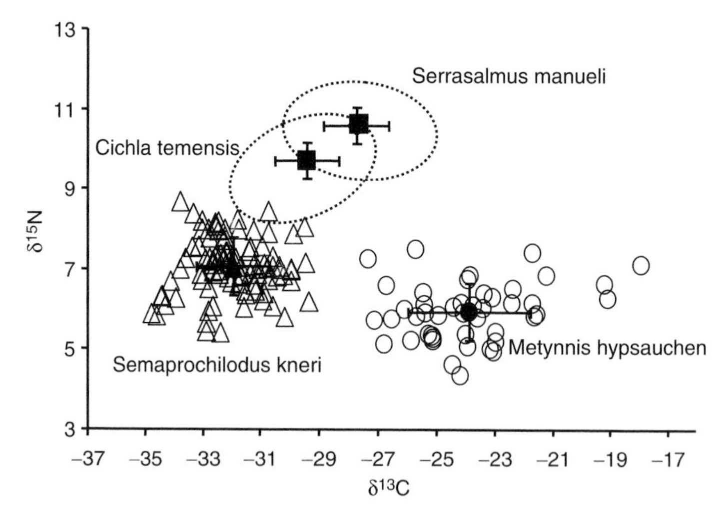

FIGURE 1 | Biplot of carbon and nitrogen stable isotope signatures of individual *Metynnis hypsauchen* (circles; $n = 45$), and *Semaprochilodus kneri* (triangles; $n = 95$). The two ellipses represents ninety-five percent confidence intervals of *Cichla temensis* ($n = 184$) and *Serrasalmus manueli* ($n = 148$) stable isotope signatures. Open symbols are values for individual specimens, and filled symbols means of all individuals of the species (± 1 SD).

contents, are needed to further identify the importance of specific resources to secondary consumers.

Ratios of stable isotopes indicate that large-bodied piscivores feed relatively low in the food web. The piranha *Serrasalmus manueli* occupied the highest estimated mean trophic position in the web, 3.6 (using a trophic level fractionation, Δ, of 2.54, see Vanderklift and Ponsard, 2003), with individuals varying from 2.9–4.4 (trophic position 1 = primary producers). Trophic position of individual *C. temensis* ranged from 2.8 to 3.7 (mean 3.2), suggesting *C. temensis* predominantly feed on organisms at trophic position 2 (herbivore or detritivores). A predator's foraging strategy is optimal when it exploits the thermodynamically richest resource possible (MacArthur and Pianka, 1966; Hastings and Conrad, 1979). By selectively feeding on herbivorous (e.g., *M. hypsauchen, Myleus* spp.) and algivorous/detritivorous (e.g., *S. kneri,* hemiodontids) taxa, piscivores in the Cinaruco River exploit short, productive food chains. Lewis et al. (2001) describe a similarly "compressed" food web for the Orinoco River, which may explain why secondary fish production is so high in these floodplain rivers.

Estimation of trophic position is especially sensitive to two factors. First, fractionation may vary with taxonomy (DeNiro and Epstein, 1981; Macko et al., 1982; Minagawa and Wada, 1984; Hobson and Clark, 1992; Vanderklift and Ponsard, 2003), form of nitrogen excretion (Vanderklift and Ponsard, 2003), quality of food (DeNiro and Epstein, 1981; Hobson, 1993; Hobson et al., 1993; Webb et al., 1998; Adams and Sterner, 2000), and turnover time of tissues (Schmidt et al., 2003). Thus, a given trophic transfer may vary substantially from the mean fractionation value employed (Post, 2002b). Reliance on a single assumed value of Δ is the weakest component of stable isotope use to estimate trophic position (Vander Zanden and Rasmussen, 2001; Phillips and Koch, 2002), and this is especially true in species-rich systems where community members vary greatly in taxonomy, body size, diet, excretion, etc. Second, it is possible that the excessive enrichment of *S. kneri* $\delta^{15}N$ values, relative to sources of primary production, may have resulted in an underestimation of trophic position of secondary consumers. For example, $\delta^{15}N$ of filamentous algae was found to be 1.3 ± 0.9 and $\delta^{15}N$ of *S. kneri* 7.1 ± 0.7 (see Figure 1). Such measures represent a greater enrichment in *S. kneri* than would be expected even when using the largest estimated mean values of Δ (Post, 2002b). Enriched primary consumer $\delta^{15}N$ values (relative to expected values based on temporally and spatially integrated $\delta^{15}N$ of primary producers) also have been found in temperate lakes (Post, 2002b) and a river in the Andean piedmont of Venezuela (Peter McIntyre, personal communication). Hamilton and Lewis (1992) showed that algae can have a significantly different isotopic signature from bulk algal/detrital samples, suggesting that the heterogeneous nature of detritus (and the microbial community associated with it) may result in *S. kneri* assimilating organic matter enriched in $\delta^{15}N$ relative to isolated algae samples. *All* of the resources integrated by consumers must be carefully considered when determining baseline $\delta^{15}N$.

Stomach Contents Analyses

Stable isotope ratios are most informative when combined with stomach contents analysis. This is especially important in species-rich systems where predators may consume diverse prey items, and it is impossible to identify particular predator–prey interactions from stable isotopes ratios alone. Table 1 illustrates the diversity of prey taxa consumed by the seven most common large-bodied piscivores in the Cinaruco River. Dietary breadth was high, yet these are likely conserva-

Table 1. Stomach contents analysis of the seven most common large piscivore species of the Cinaruco River. Prey taxa were identified to the species or genus level. The last column gives the proportion of identified taxa (i.e., excluding contents of unknown identity) of diet contents (by volume) that were *Moenkhausia* af. *lepidura*, the prey fish species that responded most significantly in large-bodied piscivore exclusion experiments.

Piscivore Species (number of individuals examined)	Number of Unique Prey Taxa	% Stomach Contents *Moenkhausia* af. *lepidura*
Boulengerella cuvieri ($n = 257$)	17	2.7%
Boulengerella lucius ($n = 272$)	11	3.5%
Cichla intermedia ($n = 97$)	10	0
Cichla orinocensis ($n = 703$)	29	0
Cichla temensis ($n = 1,159$)	40	<1%
Hydrolycus armatus ($n = 406$)	24	0
Serrasalmus manueli ($n = 668$)	31	0

tive estimates of the diversity of food resources because: (1) the number of unique prey items was significantly related to number of stomachs examined for each piscivore species ($R^2 = 0.92$, $p < 0.0001$; *n* of piscivore stomachs examined >5,000), and (2) the high frequency of empty stomachs for many top predators (Arrington and Winemiller, 2002). *C. temensis* were found to consume 40 unique prey taxa characterized by substantial variation in body size, morphology, trophic position, and life history strategy. *S. kneri* made up 41% (by volume) of the total diet of *C. temensis*, and reliance on *S. kneri* as a food source is consistent with $\delta^{15}N$ values suggesting *C. temensis* feeds relatively low in the food web. No other prey item accounted for more than 7% of diet contents. Other important components of *C. temensis* diet included hemiodontids (algivore/detritivores), juvenile *Cichla* spp. (piscivorous), *Brycon* spp. (omnivorous), other cichlid taxa (insectivorous/piscivorous), and various small characiform species (insectivorous/omnivorous).

Although integration across time and space is desirable when selecting taxa to serve as baseline $\delta^{15}N$ indicators, specific temporal or spatial feeding patterns may not be apparent using stable isotope ratios without associated stomach contents data. For example, in November (when the river water level typically begins to drop) *S. kneri* comprised 59.8% of the identifiable prey items found in the stomachs of *C. temensis*, with a guild of small characid species (e.g., *Bryconops caudomaculatus*, *Moenkhausia* spp., *Hemigrammus* spp.) comprising only 5.8% of their diet. At the end of the dry season (April), no *S. kneri* were identified in *C. temensis* stomachs, and 26.7% of *C. temensis* stomach contents were characids. Likewise, piscivores undergo ontogenic diet shifts (Werner and Gilliam, 1984; Winemiller, 1989c; Jepsen et al., 1997; Post, 2003), necessitating stable isotope analyses across size-groups or age classes. Opportunistic feeding likely contributes to variability in $\delta^{13}C$ and $\delta^{15}N$ of Cinaruco piscivores (e.g., see *C. temensis* in Figure 1). Prey species are not homogeneously distributed among habitats (Arrington, 2002; Hoeinghaus et al., 2003a; Layman and Winemiller, 2004; Layman et al., 2005), and piscivores may feed on different taxa depending on foraging location. Moreover, some individual piscivores may move long distances and others remain restricted to one area (Hoeinghaus et al., 2003b), which also affects prey species consumed.

Energy Flow and Functional Food Webs

Food webs developed using stable isotopes or stomach contents can depict the flow of energy in an ecosystem, but do not allow for determination of "functional" roles of organisms (Paine, 1980; Polis, 1991; Polis and Strong, 1996). Field experiments are often used to examine the effect that one organism has on any defined attribute, such as a community (e.g., abundance of other species) or ecosystem (e.g., primary production) level parameter. In the Cinaruco River, both stomach contents and isotopic analyses suggested relatively low trophic position of the piscivores *Cichla temensis* and *Serrasalmus manueli*, but these methodologies do not reveal what functional effect these species have in the food web.

In a large-scale experiment (approximately 500 square meters), we evaluated the effect of large-bodied piscivore exclusion on spatial distribution of prey. After two weeks, abundance of fishes in a specific size range (40–110 millimeters) was significantly higher in exclusion areas, suggesting a size-based behavioral response by prey (Layman and Winemiller, 2004). The most commonly collected taxa in this size range, *Moenkhausia* af. *lepidura*, was significantly more abundant in exclusion than control areas (Kruskal-Wallis, $P = 0.002$). This prey species, however, comprises a relatively small proportion of diets of excluded piscivorous fishes (see Table 1; based on complete stomach contents data set across all seasons and sampling sites). Although this experiment examined only one component of a community-level functional response (effect of predators on prey), it demonstrates how experimental manipulations can provide insights not ascertainable from either stable isotopes or stomach contents analyses.

Food Web Perturbations and Predictive Models

A central goal of food web research is to produce predictive models, especially those that can be used to assess human-induced impacts on food web structure (see Winemiller and Layman, Chapter 1.2). One of the most acute environmental problems in Venezuela is overexploitation of fisheries (Rodríguez, 2000). In the Cinaruco River, illegal commercial netting is a relatively new development, but is becoming increasingly intense (Hoeinghaus et al., 2003b). Populations of many large-bodied fish taxa, including *C. temensis* and *S. kneri*, are in decline, which will likely lead to shifts in food web structure. Employing both energy flow (stable isotope and stomach contents analyses) and functional/experimental approaches allow the most comprehensive assessment of effects induced by this netting activity.

For example, stable isotope and stomach contents analyses both suggest that netting may serve to *increase* food chain lengths in the system. A decline in populations of *S. kneri*, one of the most important prey of large-bodied piscivores such as *C. temensis*, may cause piscivores to consume larger quantities of omnivorous and invertivorous taxa, thereby increasing mean food chain length. Further, removal of *C. temensis*, the primary predator of *S. kneri*, results in a shift in the community to piscivore taxa that feed at higher trophic levels (as is reflected by their $\delta^{15}N$ values (Jepsen and Winemiller, 2002). Consequently, we hypothesize that average food chain length (sensu Post, 2002a) will increase with increased commercial harvest (Figure 2). This hypothesis differs from the conventional belief that "fishing down food webs" (sequentially removing top predators from food webs) results in decreased average food chain length (Pauly et al. 1998). Such an approach to analyzing human impacts based on realized measures of trophic position (typically with stable isotopes) are becoming increasingly common (France et al., 1998; Pauly et al., 1998; Jennings et al., 2001; Jennings et al., 2002).

Field experiments suggest additional effects of piscivore removal by commercial netters. Piscivore exclusion on sand bank habitats results in increased abundance of prey fishes in the size range most commonly consumed by piscivorous fishes, likely a behavioral response to piscivore exclusion (Layman and Winemiller, 2004). Subsequent work has shown this pattern to hold at the landscape scale, as distinct size-based differences in fish assemblage structure between netted and un-netted lagoons that correspond closely to those at the experimental scale (Layman et al., unpublished manuscript). Two small characins, *Moenkhausia* sp. af. *lepidura* and *Bryconops caudomaculatus*, displayed

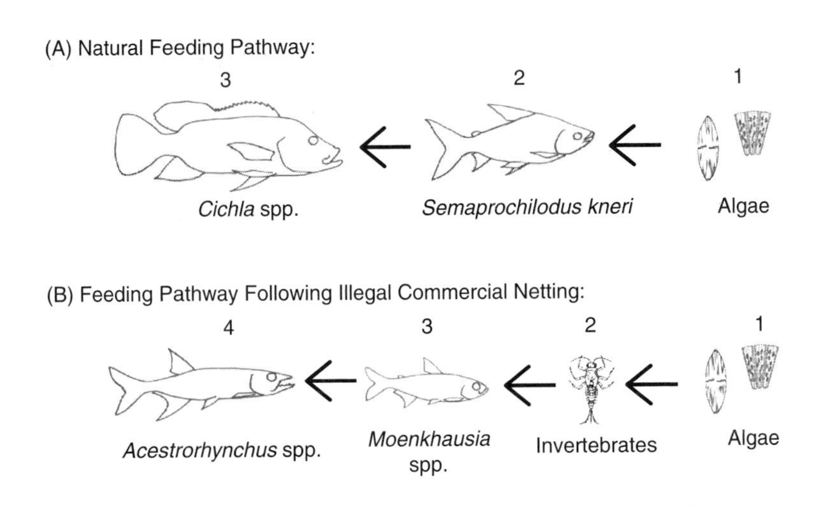

FIGURE 2 | Hypotheses of primary paths of energy flow in the Cinaruco River: (A) natural feeding pathway; (B) feeding pathway following removal of *Cichla* spp. and *Semaprochilodus kneri* by illegal commercial net fishermen. Numbers represent estimated trophic position.

the most significant responses to piscivore exclusion, and also had the greatest difference in abundance between netted and un-netted lagoons. These are among the species we expect will increase in abundance if commercial netting activity continues to increase, a prediction that could not be derived from energy flow methodologies alone.

The questions of interest in any food web study will determine the methodologies that must be employed in that particular instance. But in most food webs, particularly species-rich systems, even seemingly simple questions may be impossible to answer adequately without multiple research techniques. When moving from abstraction to prediction, research paths may lead in different directions, but going down each may be necessary to provide the information necessary to assess human impacts on food webs.

ACKNOWLEDGMENTS

Don Taphorn of Universidad de Los Llanos Occidentales (UNELLEZ) assisted with logistics and fish identification. We thank the numerous people involved with field collections, including David Hoeinghaus, Carmen Montaña, Alexis Medina, and Pablo Medina. This work was funded by an U.S. Environmental Protection Agency Science to Achieve Results Graduate Fellowship, William "Bill" Baab Conservation Fellowship, Texas A&M Regents and Tom Slick fellowships, a Fulbright Fellowship, and National Science Foundation grants DEB 0107456 and DEB 0089834.

7.5 | COMMUNICATING ECOLOGY THROUGH FOOD WEBS: VISUALIZING AND QUANTIFYING THE EFFECTS OF STOCKING ALPINE LAKES WITH TROUT

Sarah Harper-Smith, Eric L. Berlow, Roland A. Knapp, Richard J. Williams, and Neo D. Martinez

Many chapters of this volume document important advances and opportunities in food web research that have arisen since Polis and Winemiller's (1996) book on food webs a decade ago. Over the past two decades, the contribution of structural food webs (binary maps of who eats whom) to general ecological theory has been hotly debated (Paine, 1988; Polis, 1991; Banasek-Richter et al., 2004). However, in this debate, one fundamentally important contribution of structural food webs is often overlooked, namely the very *concept* of a food web as a means to intuitively and synthetically visualize and communicate about complex interconnections among species in natural communities. Biology teachers widely use the food web concept as a primary vehicle for teaching students of all ages about the fundamental concept of interdependency, interspecific interactions, and indirect effects of species extirpations and invasions (Barman and Mayer, 1994; Ford and Smith, 1994).

Given this pervasive use of structural food webs as a teaching tool, it is perhaps surprising that they are rarely used in resource management to communicate about system-level impacts of human activities. Despite calls for more holistic approaches to 'ecosystem management,' a single species focus remains the norm. Broader considerations may include physical or biotic 'habitat' protection, but rarely do they explicitly incorporate fundamental species interactions, such as feeding relations. We propose that, independent of any theoretical debate, structural food webs are a powerful data visualization and communication tool for natural resource management and conservation. In addition to qualitatively describing multivariate community changes in an intuitively tractable form, quantitative changes in web structural properties may offer novel insights into potential ecological consequences of these changes. For example, if a human impact simplifies food web structure by reducing the realized fraction of all possible links among species, or connectance (*C*), it may make a community more vulnerable to cascading extinctions following species loss (Dunne et al., 2002a).

A CASE STUDY

The introduction of salmonid fishes into mountain lakes of the Sierra Nevada of California is an ideal case study to explore the potential value of simple, structural food webs for visualizing and communicating about the community-level impacts of an introduced species. More than 99% of the approximately 10,000 high elevation lakes (> 3000 m) in this approximately 5 million hectare range were historically fishless as a result of numerous barriers to upstream movement by fishes (Bahls, 1992; Knapp, 1996). Fish stocking began in the mid-1800s to provide recreational fishing opportunities, and over the course of the next century trout were introduced into 80%–95% of the lakes (Bahls, 1992; Knapp, 1996; Knapp and Matthews, 2000; Schindler et al., 2001). The most commonly introduced species were rainbow trout (*Oncorhynchus mykiss*), golden trout (*O. m. aguabonita*), and brook trout (*Salvelinus fontinalis*).

Trout are voracious generalist predators and, as in other aquatic systems throughout western North America, they have had a dramatic impact on the aquatic fauna. Many of these impacts appear to be directly trophically mediated (Vander Zanden et al., 2000; Knapp et al., 2001; Beisner et al., 2003; Vander Zanden et al., 2003). Recently, the improved understanding of the impacts of fish introductions on Sierra Nevada lake ecosystems has resulted in increased scrutiny of fish stocking practices. However, in the development of new stocking management strategies, community-level impacts of trout have generally been ignored and all focus has been directed at a single 'charismatic' species, the mountain yellow-legged frog (*Rana muscosa*) whose distribution and abundance is strongly influenced by introduced fish (Bradford, 1989; Knapp and Matthews, 2000; Knapp et al., 2001; Vredenburg, 2004). As a consequence of strong predation by trout on *R. muscosa* tadpoles and adults, this species is now extirpated from 50%–80% of its historic localities (Bradford et al., 1994; Drost and Fellers, 1996; Jennings, 1996). Consequently, the U.S. Fish and Wildlife Service recently determined that the listing of *R. muscosa* as "endangered" under the U.S. Endangered Species Act is warranted (Federal Register, 2003). Due to the increasingly precarious status of *R. muscosa*, the California Department of Fish and Game recently halted most stocking of non-native trout in high elevation Sierra Nevada lakes and efforts to restore some lakes to their naturally fishless condition are underway (Knapp and Matthews, 1998; Milliron, 1999; Vredenburg, 2004). This single-species focus was likely due, in part, to the ease of describing the fish-frog interaction compared to the considerably more complicated community-level changes that accompany fish introductions.

Using existing lake community data (Knapp et al., 2001, and Knapp, unpublished data), we explored the use of basic structural food webs to visualize the multivariate community level impacts of introduced trout in a compact, intuitively tractable form. We constructed food webs representative of four different lake categories to illustrate both the dramatic community changes due to humans adding fishes to lakes and the recovery trajectory of these stocked lakes after stocking was terminated and fish populations disappeared: (1) Fish present (*F*), (2) No fish for 5–10 years ($NF \leq 10$), (3) No fish for >10 years ($NF > 10$), (4) Never any fish ($NF\infty$). Prior to our analysis, Knapp et al. (2001) documented clear differences among these lake categories in the number of commonly occurring species. Many food web characteristics are known to be "scale-dependent" in that they systematically depend on the number of species in the web (Martinez, 1993b). Therefore, we used a model of food web structure (hereafter, the "Niche Model") that uses species number as an input and helps to normalize comparisons by accounting for systematic scale-dependencies.

METHODS

We used lake faunal community composition data from a survey of approximately 200 Sierra Nevada lakes in a 1400-km^2 study area that included portions of the John Muir Wilderness (Inyo and Sierra National Forests) and Kings Canyon National Park. A map and descriptions of the study area and sampling methods are detailed in Knapp and Matthews (2000) and Knapp et al. (2001). Briefly, the lakes were generally small (0.5–10 ha surface area), > 3 m in maximum depth, and located in the subalpine and alpine zones (elevation 2870–3600 m). Lakes throughout the study area are generally oligotrophic, species-poor, and cold (maximum temperature < 17° C, and ice-free for only 4 mo/yr (Melack et al., 1985; Bradford et al., 1994). They are similar in physical and chemical characteristics due to their common glacial origin and their location in watersheds dominated by intrusive igneous bedrock (California Division of Mines and Geology, 1958; Melack et al., 1985).

Data on the occurrence of amphibian, reptile, benthic macroinvertebrate, and zooplankton taxa were first compiled across all lakes sampled to create a speculative 'meta-web,' that is, a map of all potential feeding interactions for all taxa observed in the entire study area that could reasonably occur if all taxa co-existed in the same lake. Only one amphibian (*R. muscosa*) and one reptile (the mountain garter snake, *Thamnophis elegans elegans*) were commonly observed during lake surveys. All benthic macroinvertebrate taxa were identified to genus using Merritt and Cummins (1984) except mites and oligochaetes, which were identified to order level. Following identification at the genus level, several genera of dytiscid beetles were grouped into the tribe Hydroporini. Similarly, two corixid genera were grouped into the family Corixidae. Zooplankton crustaceans were identified using Edmondson (1959) and Pennak (1989), and rotifers were identified using Stemberger (1979). Crustaceans and rotifers were identified to species except for cyclopoid copepods, which were identified to family (Cyclopoda), and the rotifers *Kellicottia, Keratella,* and *Polyarthra,* which were identified to genus. Basal species were highly aggregated into groups called 'benthic algae,' 'phytoplankton,' and 'detritus.'

From this meta-web, we constructed a representative food web for each lake category by compiling species lists across multiple lake samples (Table 1). A taxon was included in a web for a given lake category if the taxon was observed in at least 20% of the surveyed lakes of that type (Table 2). Other rare taxa, observed in less than 20% of the lakes in a category, were excluded from the representative food web. Six representative webs were created: (1) Fish present (*F*), (2) No fish for 5–10 years

Table 1. Number of lakes sampled for benthic and planktonic fauna in each lake category. Lakes were selected at random from a larger sample of 533 total lakes. Lakes selected for zooplankton analysis were not necessarily selected for benthic macroinvertebrate sampling and vice versa; both zooplankton and benthic invertebrate samples were processed for 65% of selected lakes.

Lake Category	Benthic Macroinvertebrates	Zooplankton
Never Fish (NF∞)	67	62
Fish Present (F)	100	89
No Fish (NF ≤10)	5	6
No Fish (NF >10)	17	20

(*NF* ≤ *10*), (3) No fish for >10 years (*NF* >*10*), (4) Never any fish (*NF*∞), and (5) a hypothetical food web: *NF*∞ with fish added (*NF* + *F*). The latter incorporated fish and their feeding links into the *NF*∞ web to depict a hypothetical food web just after fish are originally introduced and before other species are lost.

Potential feeding links for a speculative meta-web of all taxa included in any lake category were assigned one of three certainty levels. "Certain" feeding interactions were observed either directly or in gut contents by R. Knapp or reported in a peer-reviewed published account. "Probable" feeding interactions were based on direct observations or peer-reviewed published accounts for similar or analogous taxa. "Possible" feeding interactions were those that were likely based on available natural history information, but there were no direct observations or peer-reviewed published accounts of predation for these or analogous taxa. In addition to direct observations and gut contents analyses, we extracted potential feeding relations from a combination of taxonomic keys and texts (La Rivers, 1962; Usinger, 1963; Thorpe and Covich, 1991), expert knowledge, published aquatic food webs (Martinez, 1991), the U.S. Environmental Protection Agency Rapid Bioassessment Protocol's functional feeding group listing (Barbour et al., 1999) and scholarly websites (Knapp, 2003). All three levels of link certainty were included for the webs presented, and thus indicate potential links based on the best available information. In all cases, all life stages of a taxon were included in one node. Thus a feeding link between two nodes indicates that at least one life stage of the consumer feeds upon at least one life stage of the prey. For a copy of the entire feeding interaction matrix for the meta-web with certainty levels for each link, please contact E. L. Berlow (eberlow@ucsd.edu).

Table 2. List of common taxa and associated trophic species groupings for each of the five lake categories. A taxon was included in a given web if it was present in at least 20% of the lakes sampled for that lake category. 'NS' = never stocked; 'S' = stocked fish present, 'SNF <10' = stocked in the past and has been fishless for less than 10 years, 'SNF >10' = stocked in the past and has been fishless for more than 10 years, 'NSF' = never stocked with fish added. Taxa with the same number in a given column are in the same trophic species group for that lake category. Note that some species change trophic species group identity among lake categories. *Multiple trout species include: golden trout (*Oncorhynchus mykiss aguabonita*), rainbow trout (*Oncorhynchus mykiss*), brook trout (*Salvelinus fontinalis*), and brown trout (*Salmo trutta*).

	Species	Common Name	NS∞	F	NF≤10	NF>10	NSF
Reptile	*Thamnophis elegans elegans*	mountain gartersnake	1	–	–	1	1
Amphibian	*Rana muscosa*	mountain yellow-legged frog	2	–	–	2	2
Trout spp	*Multiple trout species**	trout	–	3	–	–	3
Benthic	*Ameletus edmundsi*	mayfly	4	–	4	4	4
Macroinvertebrates	*Callibaetis ferrugineus*	mayfly	4	–	4	4	4
	Agabus spp.	predaceous diving beetle	5	–	–	–	5
	Corixidae	water boatmen	6	–	6	6	6
	Culex spp.	mosquito	–	7	–	–	–
	Hydroporini	water beetle	8	8	8	8	8
	Hesperophylax spp.	caddisfly	9	–	9	9	9
	Desmona spp.	caddisfly	9	–	9	9	9
	Limnephilus spp.	caddisfly	–	–	–	9	–
	Polycentropus spp.	caddisfly	11	–	11	–	11

	Psychoglypha spp.	caddisfly	9	9	9	9	9
	Oligochaeta	aquatic worm	12	12	12	12	12
	Sialis spp.	alderflies	13	13	13	13	13
	Pisidium spp.	freshwater pea clam	14	14	14	14	14
	Acari	mites	–	15	–	–	–
	Chironomidae	midges	16	16	16	16	16
Zooplankton	*Hesperodiaptomus shoshone*	copepod	17	–	17	17	17
	Daphnia midden-dorffiana	water flea	18	–	20	18	18
	Leptodiaptomus signicauda	copepod	19	19	20	19	19
	Daphnia rosea	water flea	–	19	20	19	–
	Kellicottia spp.	rotifer	21	–	21	21	21
	Keratella quadrata	rotifer	21	21	21	21	21
	Keratella spp.	rotifer	21	21	21	21	21
	Polyartha spp.	rotifer	21	21	21	21	21
Producers	Benthic Algae	benthic algae	22	22	22	22	22
	Detritus	dead organic matter	23	23	23	23	23
	Phytoplankton	pelagic algae	24	24	24	24	24

Thus, the webs presented here were based on feeding information compiled for all species found in the study area. They are meant to characterize potential food webs for the most common species (i.e., ≥ 20% occurrence) observed in each lake category. While the taxonomic resolution was not equal across all groups (see Table 2), and while some uncertainties exist for particular links, we assumed that any potential biases of this approach were similar across all lake categories. Thus, the consistency of the approach facilitates comparisons among the different web types within this one ecosystem.

For each lake category, taxa were grouped into 'trophic species,' or groups of one or more taxa that have identical predators and prey (see Table 2; Williams and Martinez, 2002). All food web visualizations and network measures presented here are based on these functional groupings. We initially focused on two fundamental, orthogonal network measures of food webs: the number of trophic species (***S***) and connectance (***C***), which is the realized fraction of all possible links (***L***) among the species, or L/S^2 (Martinez, 1993b).

Other food web properties systematically depend upon ***S*** and ***C*** (Williams and Martinez, 2000; Camacho et al., 2002a, b; Williams and Martinez, 2002). In order to normalize for this dependence, we used the Niche Model (Williams and Martinez, 2000; Camacho et al., 2002a, b) to evaluate the degree to which differences in web structure among lake categories could be explained by factors other than changes in ***S*** and ***C***. The Niche Model accepts ***S*** and ***C*** as input parameters and generates food webs similar in structure to empirical food webs with the same ***S*** and ***C*** (Williams and Martinez, 2000; Dunne et al., 2004). Therefore, the model forms a consistent benchmark for detecting more subtle changes in food-web architecture that are not simply expected results from changes in ***S*** and ***C***.

The structure of each of the four empirical webs (*F*, *NF* ≤ 10, *NF* >10, *NF*∞) was visualized and statistically compared against both the Niche Model and a completely random network model (Williams and Martinez, 2000). The random model assigns each of S^2 possible links with equal probability, ***C***. This provides a null model of the statistical universe of similarly sized and connected networks but free of any biological constraint. The Niche Model, on the other hand, constrains link structure by three rules. The first places species on a community niche dimension by assigning each of ***S*** species a uniformly random "niche value" (n_i) between 0 and 1. The second assigns each ***i***th species a beta-distributed feeding range (r_i) between 0 and 1 with a mean of ***C***. The third places the species ***i***'s range by placing the range's uniformly between $r_i/2$ and n_i. This last rule places establishes a relaxed trophic hierarchy where species tend to feed on those to their 'left' on the niche axis but allows cannibalism and looping (see Williams and Martinez, 2000, for details). The species with the lowest n_i is forced to be a basal species with no prey as are all other species that happen to have no species within their feeding range. Each model was run 1,000 times for each set of inputs. The mean and standard deviation of each of 12 properties was measured for each set of 1,000 webs. A model's fit to a given empirical property was considered good if the latter was within two standard deviations of the model's mean. 'Standardized model errors' for each web property measure the empirical property minus the model mean divided by the model standard deviation (Williams and Martinez, 2000).

Consistent with other studies, the Niche Model consistently gave a much better fit to the empirical data than the completely random network model for all lake categories, thus we only present data for the former. We focus on the Niche Model as a means to account for expected changes in web topology due simply to changes in ***S*** and ***C***, and thus to better assess whether fish-altered web structures differed significantly from what would be expected for a typical food web of similar size and complexity.

RESULTS

The NF∞ web was the most species rich and structurally complex of any of the lake types, with 28 common taxa that were grouped into 19 trophic species (Figure 1, Table 3). The Niche Model fit 15 of the 17 food web properties well, while the random model only fit 8. This suggests that the NF∞ web structure is not random but similar to other previously observed, nonrandom webs well fit by the Niche Model. One visually striking difference between the NF∞ and F webs is the reduced complexity of aquatic communities with fish present (see F, Figure 1). Approximately one third of the common NF∞ taxa are absent in the F webs, which are dominated by small or visually inconspicuous fauna that escape predation by fish (Knapp et al. 2001). The loss of functional diversity in terms of trophic species (26%) was slightly less than loss of taxonomic groups (29%, see Table 3). While 11 taxa are less common in F webs, two taxa are more common (i.e., found in 20% or more of F webs): mosquitoes (*Culex* spp.) and mites (Acari) (see Table 2, Figure 1).

Beyond the loss of functional and taxonomic diversity, structural complexity was also lost as reflected in the 24% decline in C from 0.202 (NF∞) to 0.153 (F). These combined changes greatly simplified the structural complexity in F webs. For example, other web properties indicative of

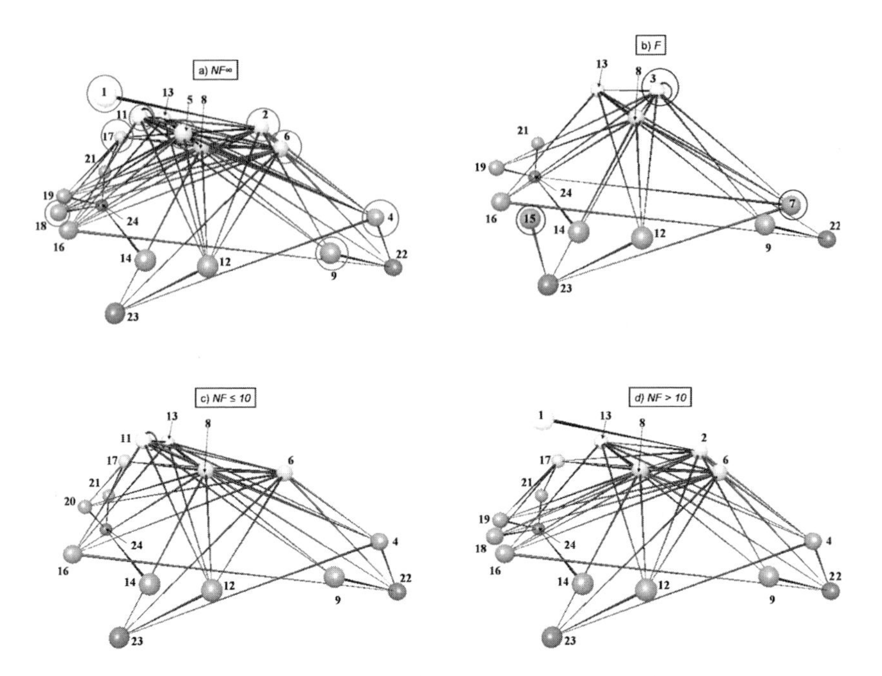

FIGURE 1 | Food webs for the four lake categories: (a) "*NF∞*" = Never any Fish, (b) "*F*" = Fish present, (c) "*NF* ≤ *10*" = No Fish ≤ 10 years, (d) "*NF* > *10*" = No Fish > 10 years. Each node is a separate trophic species which may consist of one or more taxa that share all the same predators and prey. Webs are oriented so that the basal species are at the bottom in dark orange and the top predators are at the top in yellow. Link shape indicates the direction of the feeding relation, with the wider end being the predator and the narrow end the prey. If two species eat each other, the link is narrowest in the middle. Cannibalistic links are indicated by a loop returning to the same node. In the *NF∞* web, red circles highlight the trophic species that are lost in the *F* web. In the *F* web, blue circles indicate new trophic species relative to the 'Never Stocked' web. For a list of taxa and trophic species, see Table 2. (See also color insert.)

structural complexity in the NF∞ web decline more than 25% in the F web (see Table 3). These include: the fraction of intermediate species with both predators and prey, the number of food chains, the mean and standard deviation of food-chain lengths, the mean and variance of prey-averaged trophic level (Williams and Martinez, 2004b), the fraction of species within loops (e.g., species A eats B which eats A), and fraction of clusters (modules where at least one of two species that share a predator or prey also eat each other). This latter "clustering coefficient" closely associated with "small world" network topology (Watts and Strogatz, 1998; Dunne et al., 2002b) declines by more than 40% in the F web compared to the NF∞ condition. While NF∞ lakes exhibit relatively distinct benthic and pelagic modules, in the smaller F webs these modules are connected by generalist fish consuming both pelagic and benthic taxa (see Figure 1, Table 2). Consistent with other studies (Solé and Montoya, 2001; Williams et al., 2002), one small-world property that remains relatively constant across all the webs is the "characteristic path length," or mean shortest distance between any two species, which is approximately two degrees of separation (see Table 3).

The NF $\leq$ 10 and NF > 10 webs lie relatively neatly on the spectrum from simple F webs to complex NF∞ webs with NF > 10 being most similar to the NF∞ web (see Figure 1, Table 3). While 14 properties of the F web are at least 25% different from those in the NF∞ web, the NF $\leq$ 10 and NF > 10 webs had only 9 and 2 properties, respectively, that differed $\geq$ 25% from the NF∞ web (see Table 3). The mean absolute change in web properties relative to the NF∞ condition declines from 31% (F vs. NF∞) to 19% (NF $\leq$ 10 vs. NF∞) to 12% (NF >10 vs. NF∞). Lakes that have been fishless for > 10 years recover both the mountain yellow-legged frog and the snakes that feed on them (see Figure 1, Table 2).

The Niche Model fits each of the webs quite well ranging from fitting all properties of the F web to fitting all but one property of the NF $\leq$ 10 and NF > 10 webs to fitting all but two properties of the NF∞ web (see Table 3). Not only are the normalized errors within the range determined as a good fit, but the mean absolute value of Niche Model errors is very close to the ideal of 1 standard deviation (Williams and Martinez, 2000) for the F (0.86), NF $\leq$ 10 (1.07), NF > 10 (0.93) and NF∞ (1.00) webs. In contrast, the mean absolute random model errors increase from F (1.69) through NF$\leq$ 10 (3.69) and NF > 10 (3.52) to NF∞ (5.95), confirming the random model's poorer fit to the data. The random model's comparatively good fit to the F web is likely due to its extremely small size (14 nodes).

The NF + F web illustrates a hypothetical situation where trout are recently added to a NF∞ web (Figure 2). In this web, seven of the eight trophic species that are lost in the F webs are directly fed upon by trout (see Figure 2a). The only other species that disappears is the mountain garter snake (*Thanmophus elegans elegans*), which feeds on the frogs that the fishes extirpate. These data help visualize one proposed indirect effect of fish on snakes: snakes consume frogs and disappear when frogs are driven locally extinct by fish.

DISCUSSION AND CONCLUSIONS

Basic binary structural food webs provided a compact, visually accessible description of the dramatic, multivariate, community-level changes (and recovery) of alpine lakes in response to introduced trout. Stocked

Table 3. Food web properties for each lake category that were either empirically observed ('Empirical') or randomly generated using the Niche Model for the same size and connectance as the empirical web in that category. The 'Empirical % Change from NS' for a given property is the difference in the empirical value between that lake type and the *NF* condition, standardized by the *NF* value. Numbers italicized in boldface indicate where the difference was ≥ 25% of the *NF* condition. The 'Standardized Niche Model Error' is the number of model standard deviations that the empirical value differs from the model mean (n = 1000). Standardized Model Errors ≥ 2 are italicized in boldface and indicate where the model mean was considered significantly different than the empirical value for that lake type. Postive (and negative) values indicate that the model over-(or under-) estimated the empirical value. Definitions for the web properties are as follows: '%Top', '%Intermed', and '%Basal' = the percent of all species in the top, intermediate, and basal trophic levels; 'GenStdDev' = the standard deviation of 'Generality' which for each species is (# of prey)/(L/S); 'VulStdDev' = the standard deviation of 'vulnerability' which for each species is (# of predators)/(L/S); Connectance' = Fraction of all possible links that are realized; 'Links/Species' = Average number of links per node; 'Link Stdev' = Standard deviation of the number of links per node; '# Chains' = Number of food chains in the web (i.e., paths from any species to a basal species); 'Chain Len' = Average food chain length (i.e., average number of links from any species to a basal species); 'Tro Lev' = mean prey-averaged trophic level of the web; 'TroLev StDev' = Standard deviation of the mean prey-averaged trophic level; 'Loop' = Fraction of species that are part of a trophic loop (e.g. A eats B eats A); 'Cannibal' = Fraction of species that are cannibals; 'ConsumerMeanTL' = Mean prey-averaged trophic level of all consumers (i.e., does not include basal species); 'Cluster Coefficient' = Average fraction of pairs of species one link away from a species that are also linked to each other (Dunne et al 2002); and 'Characteristic Path Length' = Average shortest path length between all pairs of species; generality = (# of prey)/(L/S), and vulnerability = (# of predators)/(L/S).

	Never Fish			Fish				No Fish ≤ 10 yrs				No Fish > 10 yrs			
Food Web Property	Empirical	Niche Mean	Standardized Niche Model Error	Empirical	Empirical % Change from NF	Niche Mean	Standardized Niche Model Error	Empirical	Empirical % Change from NF	Niche Mean	Standardized Niche Model Error	Empirical	Empirical % Change from NF	Niche Mean	Standardized Niche Model Error
# Taxonomic Species	28	–	–	20	***–29***	–	–	23	–18	–	–	28	0	–	–
# Trophic Species	19	19	–	14	***–26***	14	–	15	–21	15	–	17	–11	17	–

Connectance	0.20	0.20	–	0.15	*–24*	0.15	–	0.19	–5	0.19	–	0.17	–16	0.17	–
%Top	10.53	8.02	–0.33	14.29	*36*	13.59	–0.07	6.67	*–37*	10.35	0.41	11.76	12	10.95	–0.09
%Intermed	73.84	73.16	–0.06	64.29	–13	59.79	–0.37	73.33	–1	67.15	–0.57	70.59	–4	65.96	–0.43
%Basal	15.79	18.82	0.48	24.43	*55*	26.62	0.64	20.00	*27*	22.50	0.33	17.65	12	23.08	0.52
GenStDev	1.15	0.97	–1.36	1.31	14	1.05	–1.44	1.20	4	1.00	–1.23	1.27	11	1.02	–1.62
VulStDev	0.50	0.61	0.81	0.30	*–39*	0.67	0.37	0.45	–9	0.63	1.20	0.56	12	0.64	0.54
Links/Species	3.84	3.86	0.61	2.14	*–44*	2.19	0.38	2.87	*–25*	2.89	0.15	2.88	*–25*	2.90	1.10
LinkStdev	0.63	0.47	*–2.24*	0.63	0	0.52	–1.21	0.62	–2	0.49	1.60	0.56	–11	0.51	–1.71
# Chains	280	–	–	44	*–84*	–	–	68	*–76*	–	–	137	*–51*	–	–
ChainLen	4.26	5.29	1.27	3.02	*–29*	3.81	1.20	3.21	*–25*	4.39	1.61	3.85	–10	4.50	0.83
ChainLen StdDev	1.13	1.39	1.34	0.73	*–35*	1.13	1.68	0.80	*–29*	1.24	*2.07*	1.05	–7	1.28	0.98
LogChain Count	2.45	2.78	1.21	1.64	*–33*	1.82	1.03	1.83	*–25*	2.17	1.61	2.14	–13	2.27	0.61
TroLev	2.54	3.35	0.86	2.03	–20	2.66	0.93	2.25	–11	3.00	1.00	2.32	–9	2.93	0.85
TroLevStdDev	1.02	1.45	0.81	0.74	*–28*	1.25	1.17	0.90	–12	1.35	1.07	0.95	–7	1.33	0.95
Loop	0.37	0.28	–0.36	0.21	*–42*	0.07	–0.94	0.27	*–28*	0.17	–0.45	0.29	–20	0.15	–0.69
Cannibal	0.05	0.24	*2.13*	0.07	*36*	0.15	0.91	0.07	*27*	0.21	1.56	0.00	0	0.18	*2.24*
Consumer MeanTL	2.82	3.8[illegible]	1.02	2.31	–18	3.22	1.25	2.56	–9	3.54	1.24	2.60	–8	3.48	1.16
Characteristic Path Length	1.88	1.71	–1.69	2.03	8	1.96	–0.37	2.03	8	1.78	–1.85	1.96	4	1.84	–0.95
Cluster Coefficient	0.29	0.31	0.34	0.19	*–36*	0.23	0.64	0.27	–6	0.29	0.25	0.24	–19	0.26	0.49

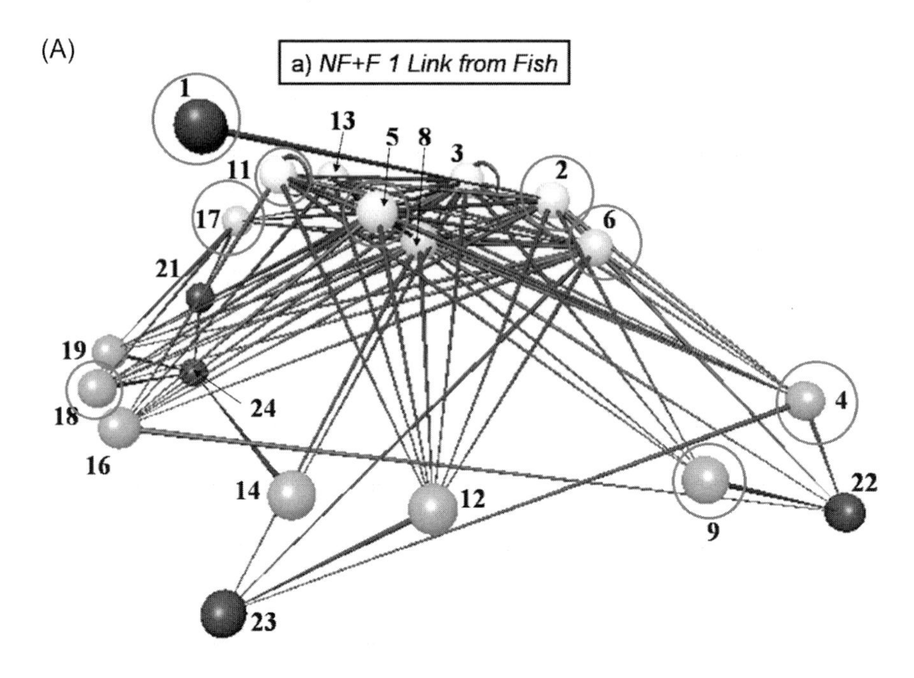

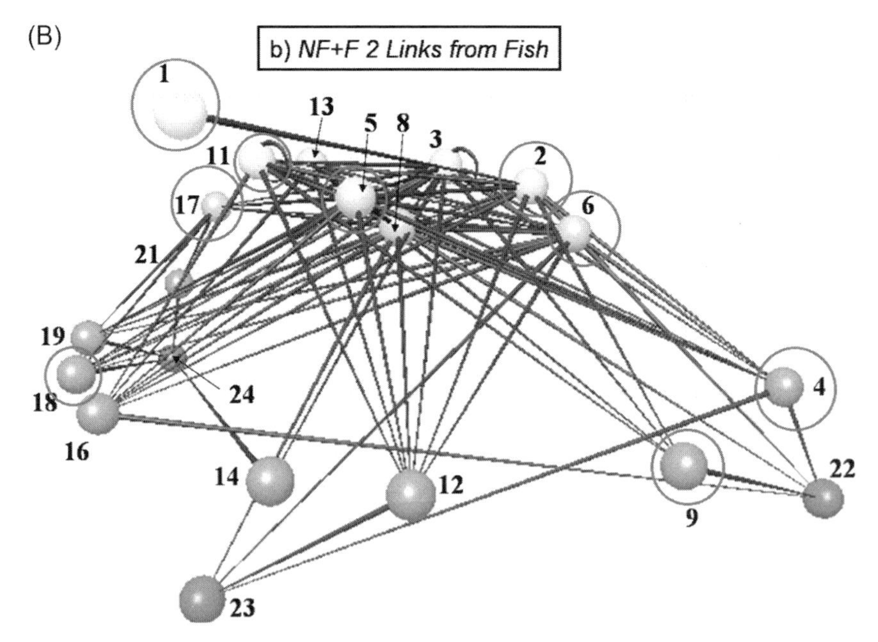

FIGURE 2 | A hypothetical *NF+F* web showing the *NF*∞ condition with fish added. **A,** All direct feeding links from trout are highlighted in color, with all other links and nodes blackened. **B,** All nodes and links two feeding links away from trout are shown in color, which includes the entire web. Red circles indicate trophic species that are lost in lakes where fish are present. For a list of taxa and trophic species, see Table 2. (See also color insert.)

lakes with trout present had dramatically simplified webs compared to lakes that never had fish. Not only did *F* webs have fewer trophic and taxonomic species, but *F* webs also had 24% lower connectance (*C*) which may make *F* webs more sensitive to further extirpations. Dunne et al. (2002a) analysis suggests a threshold of approximately $C = 0.10$, below which further reductions in connectance greatly reduce a web's structural "robustness" to species deletions. "Robustness" was quantified as the mean fraction of species that had to be removed in order to result in a total loss of ≥ 50% of the species (i.e., primary species removals plus secondary extinctions due to loss of all resource species). In this structural analysis, secondary extinctions occur when a species loses all of its prey. Thus, the observed change in *C* in the Sierra lake webs from 0.20 ($NF\infty$) to 0.15 (*F*) suggests that the fish introductions moved lake webs closer to this critical structural threshold where losing a small fraction of species could lead to many secondary extinctions. Direct species deletion simulations (Dunne et al., 2002a) of specific lake webs described here would help test these predictions.

The consistently good fit of the Niche Model webs to each category of lake webs suggests that most of the changes in web structure due to the introduction of (and recovery from) fish could be explained by changes in web size (*S*) and connectance (*C*) alone. Thus, despite dramatic effects of fish on community structure (Knapp et al., 2001; see Figure 1), none of the webs differed significantly from what would be expected for a typical food web of similar size and complexity. The comparatively good fit of the random model to the simplest fish-impacted webs (*F* webs with 14 trophic species) is probably due to the higher variability among random webs and the smaller differences between niche and random models in webs with few species.

In this system, we propose that many of the community-level impacts of trout could have been anticipated with a simple knowledge of trophic relationships (e.g., see Figure 2). As in other aquatic systems, most of the faunal changes attributable to fish appear to be due to the direct effects of consumption (Beisner et al., 2003). A hypothetical map of potential, direct trophic links from trout to the fauna of a pristine, never stocked lake (see Figure 2a) would have concisely demonstrated the likelihood of pervasive community impacts. We stress that these maps are relatively easy to construct, particularly by resource agency specialists who have an intimate knowledge of the natural history of their system, provided two conditions: (1) They are explicitly acknowledged as works in progress, and the relative certainty of each link is clearly specified; and (2) They are used primarily for comparisons *within* one system over space, time, or other conditions (e.g., invaded or not). The first condition stresses the heuristic value of the process by considering each web to be an operating hypothesis that helps target further empirical or theoretical investigations. The latter condition facilitates comparisons by minimizing variation among webs due to link uncertainties or inconsistent levels of resolution within a web (e.g., top predators identified to species level and all phytoplankton lumped into one group).

Is this situation unique? It has been well documented that species dynamics are difficult to infer from a simple knowledge of binary link structure alone (Paine, 1980, 1988). Similarly, the presence and relative abundances of species in a community are often regulated by non-trophic processes such as recruitment, abiotic conditions, disturbance, facilitations, and interference competition (Connell, 1983; Schoener, 1983; Sousa, 1984; Menge and Sutherland, 1987; Callaway et al., 2002). However, our case study follows others (Mitchell and Power, 2003; Sinclair et al., 2003) in suggesting there are many situations where species' mortality and impacts are predictably driven by overwhelmingly strong trophic interactions, and in these situations very basic binary information about food web structure is informative. It would be very interesting to see if this binary approach applies as effectively to the many systems that have experienced invasions of new predators and predictable, strong, trophically driven community changes (Lodge et al., 1994; Vander Zanden et al., 2000; Schindler et al., 2001; Woodward and Hildrew, 2001; Courchamp et al., 2003).

Even in communities where population dynamics are not necessarily regulated by trophic interactions, food webs can help visualize and communicate about changes in community composition and the relationships among species. Simple networks of potential links are easy to construct, encourage system level thinking, suggest testable hypotheses about the species impacts, and in the future could include other types of non-trophic links and interdependencies known to be important for that community. Scientific visualization of such networks can help us understand and communicate both changes and fundamental regularities in the structure of complex ecological networks for both basic and applied purposes.

ACKNOWLEDGMENTS

We thank Jennifer Dunne, Dave Herbst, Jacques Finlay, and Phil Pister for advice on constructing the food webs. REU interns at the U.C. White Mountain Research Station provided helpful discussions, feedback, and moral support. This project was funded by a NSF-Research Experience for Undergraduates grant #0139633 to E.L.B and the U.C. White Mountain Research Station, NSF grant #0342332 to N.D.M, and NSF grant #DEB-0075448 to R. Knapp and O. Sarnelle.

8.0 | PREFACE: THEMATIC REVIEWS

Peter C. de Ruiter, John C. Moore, and Volkmar Wolters

The four chapters brought together in this section emerged from the food web symposium (Schloss Rauischolzhausen, Germany, November 2003; see Chapter 1.0) and the discussions that followed. In total, six of these discussion papers became part of this book. Two of them became part of one of the foregoing sections (i.e., the one regarding body-size (see Chapter 4.4)) and the one on modeling food webs and nutrient cycling (see Chapter 5.6). The other four chapters discuss the issues of complexity–stability in food webs, detail and resolution in measuring properties of food webs, developments in aquatic food web research, and spatial aspects of food webs. The primary aim of these papers is to provide an up-to-date review of the issue and give a perspective for future research. Together they encapsulate the change in our thinking about food webs in general and our progression towards the concept of dynamic food webs.

8.1 | HOW DO COMPLEX FOOD WEBS PERSIST IN NATURE?

Anthony I. Dell, Giorgos D. Kokkoris, Carolin Banašek-Richter, Louis-Félix Bersier, Jennifer A. Dunne, Michio Kondoh, Tamara N. Romanuk, and Neo D. Martinez

THE COMPLEXITY–STABILITY RELATIONSHIP

Natural ecological communities are composed of a large and often indeterminate number of taxonomic species that trophically interact in myriad ways. Food webs describe the networks of these relations. While the population dynamics of individual species are often highly variable (Bjornstad and Grenfell, 2001), the overall structure of the trophic relations of the community, its food web, is comparatively more stable as they exhibit remarkably consistent patterns (Martinez, 1993b, 1994; Warren, 1994; Camacho et al., 2002a; Garlaschelli et al., 2003) and follow surprisingly consistent rules (Williams and Martinez, 2000; Camacho et al., 2002b; Cattin et al., 2004). This consistency combined with population variability makes natural food webs both rather dynamically and structurally complex and also somewhat stable over ecological time. For the most part theory has been unable to explain these high levels of complexity in terms of diversity and number of trophic relations because these elements are traditionally thought to decrease stability (May, 1973) and population persistence (Brose et al., 2003; Williams and Martinez, 2004c) in modeled communities. This disparity between real patterns and those predicted by theory has been one of most pressing issues facing ecologists for the past few decades. If the mechanisms driving the trophic dynamics of natural communities are to be understood, this paradox needs to be resolved and a robust theoretical framework needs to be developed that adequately explains the persistence of complex food webs in a way that is consistent with high quality empirical data. Identification of the mechanisms or "devious strategies" (May, 1973) that permit the persistence of complex food webs would be a valuable discovery for community ecology and would resolve a major paradigm within the complexity–stability debate (Brose et al., 2003). This chapter broadly outlines the current state of the complexity–stability relationship in food webs, the different approaches used to examine this issue, our current understanding of the mechanisms that appear to stabilize complex natural food webs and highlights some for the most promising research directions for future focus.

FOOD WEBS, STABILITY AND COMPLEXITY

A food web is a simplified model of the trophic structure and dynamics of a community that is used to derive and test food web theory. Therefore, empirical and conceptual limitations of food webs can limit our understanding of the trophic dynamics of natural communities. A prime example is the failure of the food web community to adopt a universal convention that defines a food web and details precisely how one should be constructed—problems that stem from this are well recognized (Winemiller, 1990; Martinez, 1991; Polis, 1991; Cohen et al., 1993b; Martinez, 1993a; Deb, 1997; Goldwasser and Roughgarden, 1997; Martinez et al., 1999; Borer et al., 2002). Proponents of food web theory argue that while such problems are important, they can be rigorously addressed so that general robust conclusions can still be made about the underlying mechanisms that structure ecological communities. Until general conventions are better developed and adopted, researchers need to describe in detail the precise way in which each food web was constructed to avoid confounding food web research—this involves describing in detail aspects of web construction and the use of a well-defined and tractable vocabulary to both avoid confusion and facilitate research. We therefore briefly introduce the concepts of complexity and stability.

Understanding Complexity and Stability

The complexity–stability paradigm was formalized by MacArthur (1955) who expressed stability as the effect of a species with 'abnormal' abundance on the abundance of other species (the community was stable if the effect was small). Citing Odum (1953), MacArthur considered the 'amount of choice which the energy has in following the paths up through the food web' as a direct measure of stability. This 'amount of choice,' a static feature of web topology, is more synonymous with what is now called complexity (its past usage as a measure of stability illustrates how both concepts are closely entwined). Because complexity refers to a static description of food webs while stability is a characteristic of their dynamical behavior, we describe both concepts separately.

By defining complexity as the amount of choice energy has when flowing through a food web it is clear that complexity is a function of both the number of species S and the number of trophic links L. Complex food webs contain many species and many interactions. A classic measure of complexity is connectance and while many different expressions have been used for the number of possible interactions (Warren, 1994) the most commonly accepted is L/S2 which counts all possible inter- and intra-specific interactions (Martinez, 1992). Importantly, though, connectance reflects a proportion of links and therefore does not fully account for the number of species in a web. A measure incorporating both the proportion of links C and species richness S is simply their product SC, which simplifies to L/S (a quantity called 'link density'). Link density is perhaps the measure that best corresponds to the essence of MacArthur's 'amount of choice.'

While a conceptual appreciation of stability in the context of complex food webs appears straightforward, measurement of stability in dynamic systems is more difficult, not least because it is often ill-defined (Pimm, 1982; McCann, 2000). McCann (2000), in a excellent review on this topic, grouped definitions of stability into two broad categories—those based on a system's ability to defy change and those based on the system's dynamic stability. MacArthur's (1955) definition of stability centers on the ability of the system to withstand 'abnormal' abundance of one of its species (and would therefore fall into the first category) whereas the influential efforts of May (1972), who used local stability analysis to examine the dynamics of randomly constructed food webs with randomly assigned interaction strengths in the vicinity of its equilibrium, refers to the stability of the system as a whole (the second category). Recently, the problem of species loss in a food web has been examined through the study of secondary extinctions; that is, the cascade of extinctions caused by the removal of a species (Dunne et al., 2002a). Such a measure of the robustness of food webs is intriguing because it does not rely on dynamical models, but only on the architecture of the system. This methodology, interestingly enough, is in close agreement with MacArthur's (1955) original concept of stability.

APPROACHES FOR EXPLORING THE COMPLEXITY–STABILITY RELATIONSHIP

It is somewhat paradoxical that the recent escalation in diversity of approaches used to explore the stability–complexity relationship is on one hand the cause of much contention within the food web community and on the other appears directly responsible for generating so much recent advancement on this issue. Evidence to date indicates the significant role of multiple mechanisms in allowing the persistence of complex food webs (see later discussion), suggesting that a synthetic approach employing an array of investigatory techniques is most likely to provide the intellectual breakthrough ecologists are hoping for. Empirical descriptions of real food webs are likely to always be open to improvement, but theory is invaluable in revealing new and unsuspected behaviors (Raffaelli, 2002) and subsequently confirming and further elucidating those mechanisms observed in real systems.

Topological Analysis of Complex Empirical Webs

The topological analysis of empirical webs has historically been a productive way to search for mechanisms that might stabilize real webs (MacArthur, 1955; Pimm, 1979; Schoenly and Cohen, 1991) and recent advances in the study of complex networks has renewed interest in this approach (Williams and Martinez, 2000; Solé and Montoya, 2001; Camacho et al., 2002a, b; Dunne et al., 2002a, b; Montoya and Solé, 2002; Allesina and Bodini, 2004). Topological studies assess food web properties using the number and distribution of connections among nodes in the food web (Borer et al., 2002). Understanding the topology of complex food webs is a key requirement in understanding their stability because a web's structure can directly affect its ability to tolerate extinction of species (Arii and Parrott, 2004). The past decade has seen considerable growth in the number of properties that describe food web topology (Bersier et al., 2002; Dunne et al., 2004) many of which directly describe aspects of web complexity and stability. Increasing the taxonomic resolution of empirical webs remains a major empirical focus with current well-resolved webs containing well over 100 trophic species (Dunne et al., 2002b). While it clear that the number of species and trophic interactions recorded in these webs are still below that in nature, determining the mechanisms that stabilize them will be a major step towards isolating those mechanisms that act in real communities.

Theoretical Analysis of Complex Webs

Since the seminal work of May (1973), much theoretical work has been devoted to the search for mechanisms that might allow complex model webs to persist. Lotka-Volterra type equations have long been used in the exploration of the complexity–stability relationship. Although historically productive, the use such linear approaches has been criticized as unrealistic by many authors. For example, linear shaped functional responses imply that a predator's capability is not restrained when prey is abundant (Armstrong and McGehee, 1976), which certainly cannot be true in real systems. The recent integration of non-linear dynamics into food web models has resulted in a substantial increase in the levels of complexity at which stable communities can be maintained (Drossel et al., 2004). The success of this approach, together with significant advances in computing technology that are a requirement of such techniques, has seen significantly advancement on this front. The works by Vandermeer et al. (2002), Huisman and Weissing (2000, 2001), Williams and Martinez (2004c), and Kondoh (2003a; see Chapter 3.3) are great examples of the power of non-linear dynamic approaches.

WHAT MECHANISMS MIGHT ALLOW COMPLEX FOOD WEBS TO PERSIST IN NATURE?

Real community food webs are complex and, at least at some level, persistent. Whilst a variety of mechanisms have been proposed, ecologists still lack a clear understanding of what processes maintain the trophic structure of natural communities. Still, much progress has been made on this front. It would be surprising if one mechanism was solely responsible for these patterns and evidence to date suggests a central role of the following mechanisms:

Food Web Topology Is Not Random

It is now universally accepted that food web structure is not random, but the precise nature of these patterns remains contentious (Havens, 1992; Martinez, 1993a). In general, non-random patterns in food web structure appear to be an emergent property of dynamical constraints on species interactions (Fox and McGrady-Steed, 2002; Montoya and Solé, 2002). Comparative studies of food webs from a wide variety of ecosystems have isolated several topological patterns that appear to apply to a wide range of webs from different ecosystems (Pimm, 1982; Pimm et al., 1991; Williams and Martinez, 2000). Four well-established patterns that appear to be critical to the stability of complex webs are that (i) connectance is scale-invariant, (ii) webs are characterized by short average path lengths between species, (iii) webs are short and fat, and (iv) omnivory is common.

Connectance Is Scale-Invariant

Early work suggested that connectance was scale-dependent (Schoenly et al., 1991; Havens, 1992), but re-analyses of these data (Martinez, 1993b; Murtaugh and Kollath, 1997) and of more complex empirical webs (Williams and Martinez, 2000) indicates that connectance is a scale-invariant property of food webs (i.e., it does not change systematically with the number of species). Connectance varies considerably in empirical food webs from about 3 to 32 percent (Dunne et al., 2002a, b), but on average is around 10 percent. Connectance is generally highest in empirical webs that have high proportions of intermediate and omnivorous species, two features that have been invoked as factors that increase the stability of natural food webs (Dunne et al., 2002a, b). However, theoretical studies often suggest a destabilizing effect of connectance on stability (May, 1973, Williams and Martinez, 2004b, c), but this is not always true (Dunne et al., 2002b; Fussmann and Heber, 2002; Kondoh, 2003a). While considerable attention is still required to determine how connectance affects food web stability, there is no doubt that the stability of natural food webs is, to some extent, influenced by the diversity of ways in which energy flows throughout it.

Short Average Path Lengths Between Species

As connectance increases, the mean distance between all nodes in a web also decreases. A recent analysis of seven highly resolved empirical webs (Williams et al., 2002) shows that species are, on average, only two links apart, with >95% of species typically within three links of each other. This suggests that real food webs could be sensitive to external forcing because local effects could permeate both rapidly and widely. Counter to this is the observation that real food webs are highly connected, have short path lengths, and are stable. Williams et al. (2002) suggest short path lengths may be a function of mechanisms associated with population dynamics and in particular the effects of weak interactions which can increase web stability and species coexistence (Warren, 1994; Berlow, 1999; McCann, 2000). For example, McCann et al. (1998) show that strong links are embedded in weak links and Neutel et al. (2002) show that strong links should not occur in the same interaction loop.

Webs Are Short and Fat

Numerous mechanisms have been invoked as to why food chain lengths are apparently quite short (Pimm, 1982), including organismal design or size constraints (Elton, 1927), available energy (Elton, 1927), disturbance (Power et al., 1996a), and ecosystem size (Post et al., 2000b). While a single mechanism does not appear to limit food chain length across ecosystems, the number of trophic transfers is ultimately limited by dissipation of energy both within and between trophic levels. One hypothesis that may account for the shortness of food chains relates to dynamical constraints associated with longer food chains—constraints that appear related to the productivity of the system (Moore et al., 1993). Pimm and Lawton (1977) showed that in a two to four trophic level system longer food chains had longer return times, suggesting they were less stable than chains of shorter length. Similar results have been observed in more complex models (DeAngelis et al., 1983; Carpenter et al., 1992; Moore et al., 1993) and in some simple real webs (Pimm and Kitching, 1987; Lawler and Morin, 1993) but these results are not universal (Sterner et al., 1997).

Omnivory Is Common

The prevalence of omnivory and its role in stabilizing or destabilizing food web dynamics has been a recurring focus of both empirical and theoretical research. Omnivory is usually defined as feeding at more than one trophic level, and is quantified by various ways of characterizing means and standard deviations of the length and sometimes strength of food chains leading from a consumer species to one or more basal taxa (Williams and Martinez, 2004b). Early food web stability modeling work, following May's (1973) equilibrium Lotka-Volterra community matrix approach, predicted that omnivory was likely to be uncommon in natural food webs, based on studies of 4-species modules (Pimm and Lawton, 1977, 1978). In particular, local food web stability depended on there being few omnivorous species, with any omnivores feeding on prey separated by only one trophic level. Initial surveys of empirical food web structure appeared to uphold this dynamical prediction (Pimm, 1982).

A prominent review of early food web research (Pimm et al., 1991) began to backpedal on earlier empirical claims by suggesting that omnivory is rare in "some webs" with "many exceptions." That same period of time saw the emergence of detailed empirical studies of complex food webs displaying high degrees of omnivory (Polis, 1991; see Polis and Strong, 1996 for review), with many researchers now suggesting that earlier data was severely flawed due to poor species resolution. Recent analyses of 18 relatively high quality terrestrial and aquatic community food webs (Dunne et al., 2004) show that percentages of omnivorous taxa range from 8% to 86%, with 13 of the webs displaying omnivory levels of 50% or greater.

Not surprisingly, modeling work also eventually shifted its tone, aided by a diversification of its approaches and definitions. Recent simple food web models (generally less than 10 taxa) suggest a stabilizing role of omnivory. Using a Lotka-Volterra approach, but exploring a different aspect of stability (the resistance of food webs to further extinction following species loss), Borvall et al. (2000) reported that omnivory appears to augment resistance when a herbivore is lost. Non-equilibrium approaches also suggest that omnivory stabilizes simple food webs by eliminating locally chaotic dynamics or positively bounding such dynamics further from zero (McCann and Hastings, 1997). Metacommunity structure analysis provides evidence that in food webs with high levels of omnivory, top species may persist at higher levels of habitat destruction (Melian and Bascompte, 2002). A purely structural analysis of "reliability flows" in small empirical food webs, an engineering stability concept which refers to the probability that sources are connected to sinks, showed that high levels of omnivory are favorable for reliable network flows (Jordan and Molnar, 1999).

Nonetheless, modeling in more specious systems has uncovered less evidence for either a positive or negative relationship between omnivory and aspects of stability. Dunne et al. (2002a), focusing on the network structure of 16 empirical food webs, examined rates of potential secondary extinctions due to primary species loss. Their simulations suggested that food web robustness to species loss (i.e., lower levels of secondary extinctions) increases with connectance but is unrelated to species richness and omnivory. In a modeling framework that integrates realistic food web structure and diversity with plausible non-linear, non-equilibrium dynamics, Williams and Martinez (2004c) found that omnivores that prefer prey at higher trophic levels lead to lower persistence of species. However, this behavior is unusual in empirical food webs (Williams and Martinez, 2004b). When omnivores prefer lower trophic level prey, there is little effect on food web persistence (Williams and Martinez, 2004c). In general, omnivory tends to be correlated with connectance (Dunne et al., 2004), and thus the teasing apart of a connectance versus omnivory effects, if indeed there is any difference, needs to be carefully considered.

The Importance of Weak Links

Variations in the strength of species interactions within food webs may also contribute to the stability of food webs over ecological time. In 1972, May bridged the gap between two, up to that point, largely separate approaches to food web analysis. By combining parameters pertaining to food web topology and derived from binary information on feeding interactions (S and C) with a measure of the strength of these interactions (average interaction strength) he was able to formulate stability criteria for model communities. His considerations are based on randomly assembled food webs as well as randomly distributed interaction strengths and predict increases in average interaction strength and/or complexity (i.e., S and/or C) to yield a decrease in stability for the given system. These results suggest that average interaction strength should be weak in species-rich, highly connected systems (May, 1973). In May's models, measures of interaction strength are the elements in a Jacobian matrix at equilibrium, which represent the direct effect of an individual of one species on the total population of another species at equilibrium (Wootton, 1997; Laska and Wootton, 1998). However, this is just one of several possible definitions of "interaction strength" within an ecological context (Wootton, 1997; Laska and Wootton, 1998).

Indeed, it is worth noting that the way in which empiricists deal with interaction strengths (with a focus predominantly on ecosystem response to disturbance) is very different from the per capita effects used in models. Depending on the concept adopted, the analysis of a food web can reveal differing of interaction strength, which may involve different interpretations of the same data set (Laska and Wootton, 1998; Berlow et al., 2004). Empirical work on real systems shows interaction strengths to be highly skewed towards many weak and a few strong interactions (Paine, 1992; Goldwasser and Roughgarden, 1993; Fagan and Hurd, 1994; Raffaelli and Hall, 1996; Wootton, 1997). This pattern is especially noteworthy because it contradicts the practice of drawing estimates of interaction strength from a uniform distribution to feed dynamic models for lack of better approximations, or of understanding the focus of many ecologists on putative "important" interactions to mean that most interactions in nature are strong.

But is the configuration of interaction strength important for questions of community stability? De Ruiter et al. (1995) linked the differing approaches by deriving values of the Jacobian matrices from empirical observations. Their results indicate that the patterning of interaction strengths is indeed essential to maintaining system stability (see also Yodzis, 1981a; Haydon, 2000) even though the strength of an interaction is not directly correlated with its impact on stability. What appears decisive is that "weak" interactions may be "strong" in reference to their stabilizing effects for the community as a whole (Paine, 1980; Hall and Raffaelli, 1993; Polis, 1994; Laska and Wootton, 1998; Kokkoris et al., 1999). More detailed work on the distribution of interaction strengths within food webs has shown that trophic loops are organized in such a way that long loops contain relatively many weak links and suggests that this patterning enhances food web stability because it reduces maximum 'loop weight' and thus reduces the amount of intraspecific interaction needed for matrix stability (Neutel et al., 2002). The results of a laboratory-based mesocosm study in turn imply that the patterning of interaction strengths in an ecosystem is determined by the body size distributions of its predators and prey (Emmerson and Raffaelli, 2004b).

In recent models that relate interaction strength to food-web dynamics, weak links appear to dampen strong and potentially oscillatory consumer-resource interactions (McCann et al., 1998) thus increasing food web persistence. The mechanisms responsible rely on a reduction in resource growth rates and consumer attack rates to stabilize consumer-resource interactions (Rosenzweig and MacArthur, 1963; McCann et al., 1998). Apparent competition occurs when a consumer species involved in a potentially oscillatory consumer-resource interaction trades off resource preference, thus mitigating the potentially critical interaction. The same effect is achieved by exploitative competition if a further consumer competing for the same resource is able to inhibit the growth rate of the shared resource. Lastly, food-chain predation describes the situation of top-down control when a top predator feeds on an intermediate species and thus indirectly restrains the interaction between the intermediate species and its resource (Hairston et al., 1960).

An additional effect of weak interactions is advocated by Berlow (1999) who experimentally assessed variability in the strength of these links. By magnifying spatio-temporal variation in natural communities "weak" interactions may be essential for the maintenance of diversity (or complexity), which in turn is expected to give rise to stability (McCann, 2000). A possible explanation for this relationship is that a system's stability depends on its capability to respond differentially to fluctuating conditions, which can increase as the system becomes more diverse (Naeem and Li, 1997; Naeem, 1998). Random distributions of interaction strength—as employed in May's models—cannot produce such structures (McCann, 2000).

The Dynamic Nature of Food Web Topology

Organisms may temporally change their diet and feeding activity through ontogenetic niche shift (Werner and Gilliam, 1984), behavioral flexibility (Stephens and Krebs, 1986), and evolutionary diet shift (MacArthur and Levins, 1964; Pimentel, 1971). Similarly, prey behavior such as anti-predation defense may influence a trophic interaction with the predator (Lima and Dill, 1996). Intra-population synchronization of such foraging-related behavior would activate or inactivate a trophic link at the population level. Furthermore, fluctuation in abiotic environmental conditions, which determine the possibility of prey-predator encounter, also leads to a temporal variation in food web architecture (Schoenly and Cohen, 1991; Tavares-Cromar and Williams, 1996). On top of this, trophic links can fluctuate at multiple time scales driven by different factors operating at different time scales. The time scale of the fluctuation would range from hours (Dell, unpublished data) to multiple years depending on the species that are linked and the precise mechanism involved. Such patterns hint towards a food web topology that is dynamic.

But is such temporal variation in food web architecture likely to affect population dynamics and the complexity–stability relationship in real ecological systems? Theory suggests a key factor is the relative time-scale of the linkage dynamics to that of population dynamics: if the linkage dynamics occurs at a time scale much longer than the population dynamics then food web topology can be considered, in effect, "constant" while the population level fluctuates. If true, then flexibility is less likely to have a major effect on complexity–stability relationship as it only defines the "static" food web architecture, which constrains the population dynamics (Pimm, 1991). In contrast, if the time scale of the focal linkage dynamics is comparable to that of population dynamics, the change in population level should be continuously influenced by the changing link strength (Abrams, 1982, 1984). In this situation there emerges the possibility that fluctuations qualitatively alter the complexity–stability relationship.

Among a number of factors that drive linkage dynamics, adaptation (population-level evolution, individual-level behavioral flexibility) has been intensively studied as a major factor that could alter the relationship between food-web architecture and population stability (Holling, 1959a; Abrams, 1982, 1984; Matsuda and Namba, 1991; Matsuda et al., 1993; Kondoh, 2003a; Takimoto, 2003). An important property of such an adaptation effect is that it creates a reciprocal interaction between population dynamics and linkage dynamics through the selection of links to be dependent on the relative population levels of interacting species. Predators tend to consume more abundant prey (foraging switch, Stephens and Krebs, 1986) and prey tends to avoid more abundant consumers (defense switch, Sih et al., 1998). It follows then that in the presence of adaptation an increase in trophic links can have very different effects on population stability. Competitive exclusion usually results in the sole persistence of the superior competitor when two or more predators share resources (Tilman, 1982) so the addition of more predators simply leads to more extinctions. However, in the presence of defence switch, a predator, whose (adaptive) prey has additional predators, is more likely to be "ignored" by the prey when its population level is low. An increase in predator species can therefore enhance the persistence of more predators (Tansky, 1978; Teramoto et al., 1979). A similar argument holds for a multiple-prey-one–predator system (Matsuda et al., 1993; Abrams and Matsuda, 1996; Matsuda et al., 1996). In the absence of adaptation such a system is unstable as all prey species other than the prey that is most tolerant to the predator become extinct due to the negative indirect effect between the prey species (apparent competition, Holt, 1977). A population of a prey species whose (adaptive) predator has other prey can be more stable when its population level is low as it is more likely to be "ignored" by the predator. Kondoh (2003a) directly showed that foraging adaptation reverses a classically negative complexity–stability relationship into a positive one in some cases (Brose et al., 2003).

A central unresolved topic here is the relationship between the relative time scale of population dynamics to that of linkage dynamics in model and real food webs. In considering how such processes affect the persistence of complex food webs, it seems essential to account for interspecific heterogeneity within each web. Because the time scale at which populations or trophic links fluctuate is likely to vary between species, so too will the time scale of linkage dynamics that are most influential to the population dynamics of each species. This suggests varying effects of linkage dynamics to the population dynamics of each species. It would be interesting to explore if this heterogeneity has any consistent effects on the relationship between topological flexibility and population dynamics or community maintenance.

8.2 | POPULATION DYNAMICS AND FOOD WEB STRUCTURE—PREDICTING MEASURABLE FOOD WEB PROPERTIES WITH MINIMAL DETAIL AND RESOLUTION

John L. Sabo, Beatrix E. Beisner, Eric L. Berlow, Kim Cuddington, Alan Hastings, Mariano Koen-Alonso, Giorgos D. Kokkoris, Kevin McCann, Carlos Melian, and John C. Moore

Real food webs exhibit both exquisite detail and exhaustive resolution (Paine, 1980; Schoener, 1989; Winemiller, 1990; Polis, 1991). This observation presents a considerable challenge for food web theory because both detail (e.g., parameters describing feeding relationships) and resolution (e.g., the number of state variables representing species) increase model complexity thereby reducing the analytical tractability of model systems and the clarity of their prediction. Traditionally, the field has carefully avoided this problem by building models that capture *either* detail *or* resolution. The divide between detail and resolution has in turn led to the formation of two very distinct schools of food web theory. The first school, which we call the "detail school," focuses on *population dynamics* and relies strongly on reductionism. This school simplifies food webs to a system of two to four focal interactions and examines the effects of variation at the level of the individual on the stability of population dynamics over time (Lotka, 1925; Volterra, 1926; Gause, 1934; Rosenzweig and MacArthur, 1963; Schoener, 1973; Holt, 1977; Yodzis and Innes, 1992; McCann et al., 1998). Here, details such as spatial refugia, size structure, diet specialization, predator satiation, and many other individual properties determine patterns of coexistence, the location of stable equilibria or the nature of variability in abundance patterns over time. The quintessential features of population dynamics models (i.e., the "detail school") are detailed mechanism at the individual or behavioral level and measurements in units of time.

The second school, which we call the "resolution school," focuses on regularities in the *food web structure* and the stability of non-random structures in more resolved, speciose food webs (Gardner and Ashby, 1970; May, 1972; DeAngelis, 1975; Cohen, 1977a; Yodzis, 1981a, 1982; Cohen and Newman, 1988; Cohen et al., 1990a; Martinez, 1991, 1992, 1993b; Moore et al., 1993; de Ruiter et al., 1995; Williams and Martinez, 2000). For example, numerous studies over the last three decades show how "regularities" in the structure of food webs determine their capacity to return to equilibrium following small perturbations (i.e., system resilience).

These regularities can be divided into two categories—macroscopic and microscopic. Classical examples of macroscopic descriptors are global connectance (Gardner and Ashby, 1970; May, 1972), the ratio predator to prey species (Cohen, 1977b; Cohen et al., 1990a) and the degree of omnivory (Pimm and Lawton, 1978; Williams and Martinez, 2000). By contrast, microscopic descriptors of food web structure include guilds (Root, 1967), blocks and modules (May, 1972), cliques and dominant cliques (Cohen, 1977a; Yodzis, 1982), compartments (Pimm, 1979), subwebs (Paine, 1980), block submatrices (Critchlow and Stearns, 1982), and complexes (Sugihara et al., 1989). In contrast to models of population dynamics advocated by the detail school, the resolution school focuses on the structure of feeding links between consumers and resources and the stability of this structure within more complete sets of species comprising a food web (but see Polis, 1991). Further, these maps occur in one place and at one time or in one place over different stages, representing cumulative data.

Clearly there are strengths and weaknesses to both approaches. The trade-off between detail at the level of the individual and resolution at the level of the whole system best exemplifies the compromises made by each. Models of *population dynamics* (hereafter, synonymous with the "detail school") typically focus only on a few (two to four) species but capture copious mechanistic detail when describing the interactions among these select few species. Conclusions from these models are then inferred to apply to more complex systems with higher numbers of species. However, few studies have identified empirical or formal rules for distilling a complex system to a smaller, more analytically tractable subset of interactions (Murdoch et al., 2002). By contrast, models of *food web structure* (hereafter, synonymous with the "resolution school") describe the binary relationships of qualitative feeding among a more complete set of species in a community. Resolution at the system level often comes at the cost of mechanistic detail at the level of the individual. Only recently have studies begun to integrate detail, temporal dynamics and topological resolution into a more cohesive theory of food webs (Neutel et al., 2002; Cohen et al., 2003; Kondoh, 2003a; Emmerson and Raffaelli, 2004b).

Empirical studies of food webs can also be distinguished along gradients of detail and resolution. For example, experimental studies of food chains (Estes et al., 1978; Oksanen et al., 1981; Carpenter and Kitchell, 1988; Power, 1990) simplify communities to a subset of strongly interacting species that presumably determine the dynamics of the larger system. These studies draw heavily from theories of population dynamics (Rosenzweig and MacArthur, 1963; Rosenzweig, 1969, 1971, 1973). By contrast, comparative studies of food web energetics characterize patterns of biomass and energy flow through more complete food webs (Lindeman, 1942; Margalef, 1963; Winemiller, 1990; Polis, 1991; Polis and Strong, 1996; Hall et al., 2000; Neutel et al., 2002). These studies find a more natural foundation in theories of food web structure (May, 1972; Cohen, 1977a; Pimm et al., 1991; de Ruiter et al., 1995; Williams and Martinez, 2000) though this foundation is not always upheld by empirical observation (Paine, 1980; Winemiller, 1989b; Polis, 1991).

Finally, despite the diversity of food web theory, only rarely do predictions from models map to the temporal or spatial scales of either type of empirical investigation (Berlow et al., 2004). Population dynamic models typically evaluate the effects of detail on *asymptotic behavior* (Hastings, 2004) occurring often over very long periods of time. Predictions at time scales of longer than a few years are out of reach of experimental food web studies executed on seasonal or short, multi-annual time scales. This is especially true for studies that include long-lived or slow growing species. More importantly, transient dynamics may play a stronger role in empirical systems than assumed by most theoretical studies (Higgins et al., 1997; Hastings, 2004). Similarly, more resolved models of food webs evaluate the effects of subsets of non-randomly interacting species on the resilience of the whole system in the face of small perturbations in the abundance of one or at most a small number of species. Often the return times predicted by models of food web structure are much longer than the typical empirical studies of food webs, grant cycles, or even lifetimes of investigators. Moreover, most empirical data used to test models of food web structure is static—a snap shot of the structure at one time in one place. Thus, empirically measured topologies may offer only a glimpse of transient rather than stable structure in time and space. In either case, current theory rarely provides predictions at time scales relevant to real webs observed over measurable time intervals.

In this chapter we argue that two major deficiencies of food web theory to date are (1) the almost mutually exclusive treatment of *detail* and *resolution* in food web models and (2) the lack of tangible timescales in mapping theory to empirical research on food webs. We highlight some recent advances in food web theory that represent the first steps towards integrating detail at the individual and population level with resolution at the whole system level and suggest that future progress depends on continued integration of these historically separate lines of investigation. Finally, we suggest a working conceptual model for integrating detail and resolution to make predictions about the links between population and whole system persistence. We argue that models of real food webs will only be valuable for more general empirical prediction if they include the *minimum* amount of detail *and* resolution necessary for capturing *measurable* properties of real food webs.

OSCILLATING THEORETICAL PERSPECTIVES—ROADBLOCKS OR GRIST FOR INTEGRATION?

Over the past century, interests have oscillated between "detail" (theories of population dynamics: Lotka, 1925; Volterra, 1926; Gause, 1934; Holling, 1959a; Hairston et al., 1960; Rosenzweig and MacArthur, 1963; Paine, 1966; Murdoch, 1969; Rosenzweig, 1969, 1973; Holt, 1977; Paine, 1980; Oksanen et al., 1981) and "resolution" (theories of food web structure: Lindeman, 1942; Odum, 1957; Patten, 1959; Gardner and Ashby, 1970; May, 1972, 1973; Cohen, 1977b; Pimm and Lawton, 1977; Lawton and Pimm, 1978; Rejmánek and Stary, 1979; Pimm, 1980a, b; Briand and Cohen, 1984; Briand and Cohen, 1987, 1989). This scientific divide has continued over the past 30 years. For example, studies of the structure of more resolved food webs inspired by classic studies of diversity and complexity (Maynard Smith, 1974) were in vogue from the late seventies through the late eighties (May, 1972; DeAngelis, 1975; Cohen, 1977a, b; Pimm and Lawton, 1977; DeAngelis et al., 1978; Pimm and Lawton, 1978; Pimm, 1979; Rejmánek and Stary, 1979; Pimm, 1980a, b; Pimm and Lawton, 1980; Yodzis, 1980, 1981a, b, 1982, 1984; Briand and Cohen, 1987, 1989; Yodzis, 1989). Interest in this approach waned in the late eighties as the field experienced a shift towards empirical research on population dynamics in simple food chains (Carpenter et al., 1985; Carpenter and Kitchell, 1988; Power, 1990, 1992), indirect effects (Schmitt, 1987; Wootton, 1994a; Wootton, 1994b, c) and interaction strength (Paine, 1992; Wootton, 1997). Most recently, ecology has seen a revitalized interest in more holistic studies of food web structure, catalyzed in large part by theory on self-organized networks imported from studies in physics and informatics (Albert et al., 2000; Dunne et al., 2002a, b; Neutel et al., 2002; Williams et al., 2002; Cohen et al., 2003). Many of the newer structural models have built on the empirical criticisms levied against older structural models.

Several of these studies provide evidence that the gap between detail and resolution in food web theory may be narrowing. First, mounting empirical evidence upholds a positive relationship between omnivory and stability (Polis, 1991; Diehl, 1992, 1995; Fagan, 1997). As a result, there has been increasing interest in understanding the role of weak interactions in conferring whole system stability (McCann et al., 1998; Berlow, 1999; Neutel et al., 2002). Empirical studies of microcosm communities suggest that population dynamics and nonrandom extinctions constrain feasible food web structures to those with constant connectance (Fox and McGrady-Steed, 2002). Finally, dynamic models of food webs now include more species (Yodzis, 2000) and focus specifically on how the prevalence of strong and weak interactions may increase the stability of small modules of three to four species by striking a balance between strongly oscillatory and more stable subsystems within these small modules (McCann et al., 1998). Taken together, these results may have implications for understanding the stability of more resolved food webs. For example, whole system stability may be influenced by the *arrangement* of stable and unstable groups of interacting species, the structural organization of weak and strong links, and the correlation structure of interaction coefficients within a community (de Ruiter et al., 1995; McCann et al., 1998; Kokkoris et al., 2002; Neutel et al., 2002). These studies suggest that a more synthetic theory of community organization will arise from further explorations of how population dynamics and link structure interact because all links, not just the strong ones, may regulate community-level dynamics.

BRIDGING THE DIVIDES BETWEEN DETAIL AND RESOLUTION

Despite recent advances in the integration of structural and dynamic approaches to food webs, few theoretical approaches satisfactorily combine detail and resolution to identify links between population dynamics and the stability of system structure. For example, does the strength of interactions between two to three interacting species depend on their location within a trophic network? Or conversely, do the details of population dynamics in one or a few modules change the structure of whole systems over time? We suggest that progress will come only by identifying strong empirical links between the details of population dynamics and the resolution of food web structure. Later we outline how the population and whole system persistence are linked through properties at the individual, internal system or external system level (Figure 1).

Persistence of Populations and Systems

Models of population dynamics and food web structure both attempt to understand *persistence* (see a in Figure 1). Population models analyze population persistence as a function of individual properties, while models of larger food webs evaluate the resilience of whole systems of interacting species as a function of observed structural properties of entire food webs. The analysis of persistence is achieved in both cases via the same methods—by evaluating the sign and magnitude of the real compo-

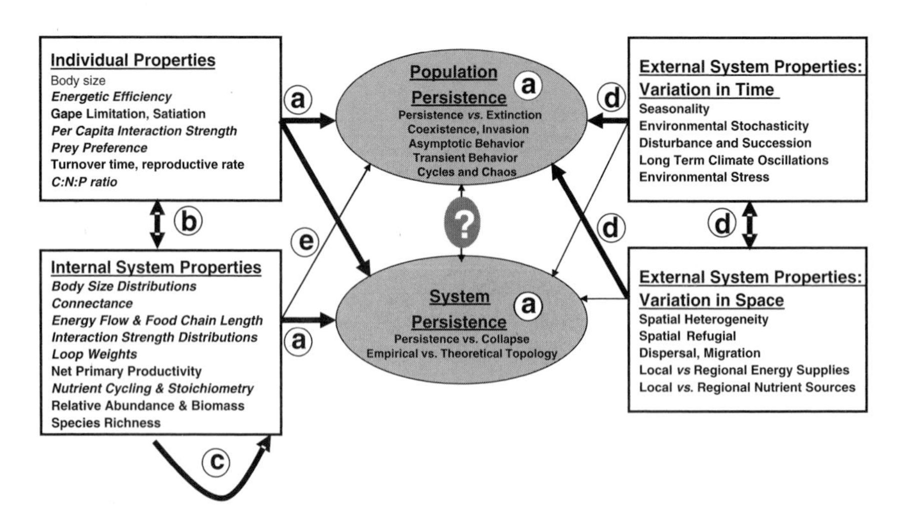

FIGURE 1 | Conceptual model linking population dynamics to food web structure.

nents of eigen values from the determinant of an $n \times n$ community matrix, where n is the number of species included in the model (May, 1973).

Scaling between Individual and System Properties

There is a great deal of *empirical* interest in understanding system properties by scaling up from individual properties (see b in Figure 1). Ecological efficiencies (Lindeman, 1942) for example, are at the heart of most arguments about food chain length (Schoener, 1989; Post et al., 2000b; Post, 2002a). Similarly, consumers can determine the body size distribution of resources (Brooks and Dodson, 1965), and gape limitation can determine the relative strength of top down control of resource biomass by consumers (Hairston and N.G. Hairston, 1993). Diet generality and prey preference patterns in concert with species diversity (a systems property) determine connectance (Williams and Martinez, 2000). Recent work suggests that body size, through its influence on metabolism and prey choice, may be one critical determinant of consumer-prey interaction strengths, biomass flows, and link structure (Cohen et al., 2003; Emmerson and Raffaelli, 2004b).

Control Pathways among System Properties

There has been a great deal of attention devoted to causal links between various system properties (see c in Figure 1). For example, studies of exploitation ecosystems (Oksanen et al., 1981; Oksanen, 1988; Persson et al., 1992; Schmitz, 1993; Wootton and Power, 1993; Power et al., 1996a) examine the effects of net primary productivity and food chain length on the relative biomass at individual trophic levels. Similarly, the numerous cause and effect relationships between ecosystem functions (e.g., net primary productivity) and species diversity are reviewed elsewhere in this volume.

External Effects on Structure

External system processes such as climate forcing, environmental stochasticity, spatial heterogeneity, and source sink dynamics all have well-known effects on population persistence (see d in Figure 1; Pulliam, 1988; Pulliam and Danielson, 1991; Ives, 1995; Tilman and Kareiva, 1997; Ives and Jansen, 1998; Ripa et al., 1998; Cuddington and Yodzis, 1999; Anderies and Beisner, 2000; Turchin et al., 2000; Ives et al., 2003; Kilpatrick and Ives, 2003). In some cases, the effects of both spatial and temporal variability are well understood (Kendall et al., 2000; Bjornstad and Grenfell, 2001). The effects of these external system properties are almost completely ignored by models of more resolved food webs (Winemiller, 1990).

System Effects on Dynamics

Very little theory in population dynamics relates system properties directly to population persistence (see e in Figure 1). By contrast, since most system properties (e.g., interaction strength distributions or loop weights) derive from individual properties (e.g., prey choice and feeding rates), structural studies at least indirectly rely on properties at multiple scales to understand structural persistence.

EXAMPLES OF INTEGRATING DETAIL AND RESOLUTION

As we have previously outlined, food web theory has oscillated between diparate schools promoting population-level detail or system-level resolution, and these oscillations were largely fueled by important empirical observations (Lindeman, 1942; Paine, 1980; Power, 1990; Polis, 1991; Paine, 1992; Hairston and N.G. Hairston, 1993; Polis and Strong, 1996). Yet population and system persistence are clearly linked in many ways through the relationships between individual and system properties (see Figure 1). Integration of these two types of empirical observations and theoretical approaches depends on combining the strengths of the two approaches—the scaling of the details of individual properties to whole system structure and understanding the constraints set by structure on the dynamics of smaller groups (modules) of species. We provide a few examples of how these two paradigms may be merged to construct productive and testable hypotheses.

Oscillatory Subsystems Scaled Up?

McCann et al. (1998) showed that the stability of small modules of species depends on a balance struck by oscillatory and stable "subsystems" within these modules. Here subsystems are really individual populations. From a network perspective, this result suggests that understanding the mean, variance, and relative location of stable modules (groups of three to four interacting species) within a food web network may provide insights into the stability of the network as a whole. Post et al. (2000b) examined at how food chain couplings influenced stability. Their work, motivated by the observation that mobile predators frequently link littoral and pelagic pathways in lake food webs, suggests that such coupling could be a potent stabilizing factor. Other systems show similar structure. For example, soil food web ecologists have argued that higher order, more mobile, generalist consumers couple the bacterial and fungal channels (see McCann et al., Chapter 2.4, for other examples), and this coupling has important implications for food web stability (Moore et al., 2003). Teng and McCann (2004) extended the Post et al. (2000b) analysis to consider an array of modules coupled by higher order consumers. The results suggest that oscillatory modules (three to seven species) coupled to stable modules (three to seven species) can act to dampen the dynamics of the potentially oscillatory system. These results argue that the weak interaction effect can be scaled up to the weak module or compartment effect.

Interaction Strength: Detail vs. Resolution

Interaction strengths among species, are both individual and system level properties (see Figure 1). As such they may be a unifying bridge between the "detail vs. resolution divide." A clear problem is the lack of consensus on how interaction strengths are defined and measured. One root of this problem is the difference between population and system level approaches (Berlow et al., 2004). Some interaction strengths specify individual or population-level traits (e.g., the Jacobian matrix element, maximum feeding rate, biomass flow along one link); while other interaction strengths are system-level traits (e.g., the impact of a change in one link or all links to and from a given species on the dynamics of other species). The second category includes the majority of field and microcosm perturbation experiments. Thus, in the first category individual interaction strengths are independent of the network, in the second category they are in theory inseparable from their network context. Empiricists and theoreticians with a population dynamics focus often argue that if one wishes to characterize the dynamic linkages among species, very few methods will be as accurate as direct manipulations of species. However, even when these studies focus on a single interaction, the outcome of press perturbations will always be a whole system response (Bender et al., 1984; Yodzis, 1988).

Ironically, empiricists and theoreticians with a system-level focus actually tend to use interaction strength measures that are more specifically an individual or population-level, rather than system-level, trait. These measures may have more easily quantified empirical proxies than those extracted from a laborious species perturbation experiment. For example, when characterizing interaction strengths for a large assemblage of species, it is possible to translate the partial derivatives of the inverse community matrices into measurable units in empirical settings (de Ruiter et al., 1995; Neutel et al., 2002). Other promising approaches for rapidly assessing feeding rates to parameterize dynamic food web models include the use of biomass abundance patterns, predator and prey body size ratios, and metabolic theory to estimate relevant parameter values (Yodzis and Innes, 1992; Neutel et al., 2002; Cohen et al., 2003; Emmerson and Raffaelli, 2004b; Williams and Martinez, 2004a).

Interaction strengths that are easily measured or estimated properties of individuals or populations, such as maximum feeding rates, may translate easily into model parameters, but they do not necessarily predict the strengths of dynamic linkages among populations (Paine, 1980; de Ruiter et al., 1995). Given these constraints and trade-offs, species manipulation experiments are unlikely candidates for directly parameterizing community models. Instead they may be better used as independent tests of models that were parameterized with more empirically accessible data. Thus, some critical ingredients for integrating detail and resolution concerning population dynamics in food webs include: (1) Translating individual or population-level interaction strength parameters of both detailed and resolved community models into units that are easily measured or estimated; (2) Using both detailed and resolved models to target specific species manipulations that would serve as critical system-level empirical tests of individual/population-level interaction strengths parameterized with other data; and (3) Using experimental manipulations to test both detailed and resolved models in order to determine the minimum level of detail and/or resolution necessary to capture key population dynamics within a whole-system context.

Switching, Omnivory, and Connectance

One of the most difficult details to integrate into more resolved models of food webs is the empirical reality of predator–prey behavior. Many behaviors have been routinely shown to have strong dynamic effects on the stability of more detailed models of population dynamics. These behaviors include non-linearities in the response of consumers to prey (functional responses, Holling, 1959a; Murdoch, 1973; Oaten and Murdoch, 1975b; Oaten and Murdoch, 1975a), predator interference and ratio dependence (DeAngelis et al., 1975; Free et al., 1977; Getz, 1984; Arditi and Ginzburg, 1989; Berryman, 1992), and seasonal, density- or frequency-dependent and ontogenetic shifts in diet (Murdoch, 1973; de Roos and Persson, 2002). For example, saturating (Type II) functional responses make even moderately sized structural webs unstable and thus, from the standpoint of theory, less likely to exist empirically. Sigmoidal functional responses alleviate this problem to a certain extent (Williams and Martinez, 2004a). The sigmoidal functional response is a potentially common emergent property of a variety of plausible ecological mechanisms including switching, prey refuges, predator learning and weak links (Murdoch, 1973; McCann, 2000; van Baalen et al., 2001). On the surface, this would seem to suggest that a potentially rich set of details could be simplified into a functional response that produces stable food web topologies.

Asymptotic functional responses suggest that predator per capita effects (i.e., interaction strength) and population effects (i.e., energy flow between prey and predator populations) may depend on prey density. Similarly, switching among alternate prey species by predators (seasonally or spatially) implies variability in patterns of food web structure or connectance. Thus, both switching and saturation imply that patterns of energy flow, interaction strength and link structure, all central to structural food web approaches, are not constant over time—a common assumption made when applying point estimates of these properties in more resolved models of food webs. Kondoh (2003a) showed that switching by adaptive foragers stabilized large and complex food webs similar in structure to those observed in nature. Though switching, ontogenetic diet shifts and even individual variability in diet preferences are well known behavioral attributes of a variety of predators, empirical measures of temporal variation in connectance, interaction strength or energy flow at the system level are rare. Taken together, these results suggest that empirical "snapshots" of food web structure may not be adequate benchmarks for testing models of large, more resolved food webs.

Clearly, multiple point estimates of the membership and feeding links within a single food web is not a trivial empirical endeavor. At a coarse scale, stable isotopes may prove useful in alleviating this problem by identifying time-integrated measures of energy flow from various source pools at either different trophic levels, from different habitats or from different energy channels (Post, 2002b). At a finer scale, diet information from multiple individuals (e.g., of different age classes and from different spatial locations) could be used to calculate the probabilities of consumption of different prey types and, thus, a less binary roadmap of feeding links within a food web.

Temporal Variability

Cycles, chaos, and stability are all deterministic concepts—measurements of the signal or "process" underlying population dynamics. In the field, we often see more noise than signal. Some of this noise is measurement error, while a good deal of it is stochastic environmental variability in recruitment or survival ("process error") unrelated to the underlying process of interest. Stochasticity is one of the most difficult divides between theory and data in food web science. Many models of population dynamics evaluate the effect of stochastic variation in abundance on the persistence of interacting species (Ives and Jansen, 1998; Ripa et al., 1998; Kilpatrick and Ives, 2003; Sabo, in press); however, very few of these insights have been integrated into a systems perspective. New work suggests that species interactions (process) can attenuate (Kilpatrick and Ives, 2003) or amplify (Sabo, in press) stochastic variation in abundance (process error) of key species in simple food webs. This new work suggests that empiricists may find more fruitful tests of theory by measuring second rather than first moments of abundance in (longer term) experiments.

Numerous new approaches provide rigorous statistical tools for quantifying process (e.g., interaction strength) in the face of process and measurement error (De Valpine and Hastings 2002; De Valpine, 2003; Ives et al., 2003; De Valpine 2004). For example, multivariate autoregressive models, or MARs, (Ives et al., 2003) use time series of abundance data from a set of interacting species to estimate interaction strength and the stability of the entire species assemblage. Stability can also be assessed under different environmental regimes or in response to a perturbation thus providing a tool for predicting the effects of environmental stress and disturbance on both population dynamics and food web structure.

Timescale Mismatch

Useful models should yield predictions that are tractable for testing by either field experiments or observational data. Models of population dynamics provide qualitative answers to two types of questions: (1) do the species described by the model coexist? (i.e., population persistence, see Figure 1), and if so, (2) how stable (or variable) are their abundances in time as a result of their species interaction? Predictions from the first of these questions are frequently tested by short-term experiments, most during the active season of a single year (Nisbet et al. 1997). Predictions of the latter type have rarely been tested in field settings, but have been more the realm of lab studies of cultured populations (Costantino et al., 1997; Cushing et al., 2001; Dennis et al., 2001). In both cases, predictions may be based on long-term (i.e., asymptotic) model behavior that ignores transient behavior and stochastic variation—both prevalent in real ecosystems.

By contrast, models of food web structure provide a qualitative answer to questions about the stability and thus the "persistence" (see Figure 1) of theoretical food webs of varying size (number of species), complexity (number of feeding links), and dynamics (relative distribution of weak and strong feeding links). One then asks if empirical webs (assumed to be stable by virtue of being observed) have similar properties to feasible theoretical webs. Large empirical food webs are rarely constructed in multiple sequential seasons or years (see previous discussion). In some of our best food webs, time may even be ignored (e.g., diets sampled in multiple seasons of the same year). As is the case with predictions from population dynamics, structural models also focus on asymptotic behavior (resilience) ignoring transient dynamics and stochastic variation.

Given the mismatch between the timescales of theory (asymptotic behavior) and data (seasonal collections or experiments), future models of food webs must either be refocused to make predictions at smaller timescales, or empiricists must learn to measure variability in the effects of their experiments or the membership of larger food webs across larger time scales. We acknowledge that neither is a trivial task; however, several recent approaches suggest some promise. For example, Nisbet et al. (1997) modeled *single season* interactions among three trophic levels in open ecosystems by substituting dispersal (immigration) for reproduction for higher trophic levels. Predictions from models on such short timescales are more appropriate for evaluation with data from short-term experiments. At higher levels of food web resolution, Multivariate Autoregressive Models (Ives et al. 2003) provide a promising tool for evaluating whole system stability, once more comprehensive time series data on larger sets of species are available.

DISCUSSION AND CONCLUSIONS

Approaches to food web theory have historically oscillated been the perspective of population-level detail or system-level resolution. Details include predator–prey behavior, temporal variability, spatial habitat structure, and links among food webs supported by alternate energy sources (detritus or primary production) or disparate habitats. Resolution includes the architecture of large ecological networks, including system-level patterns of energy mass flow, interaction coefficients, and species' body sizes and abundances. Each viewpoint has strengths and weaknesses. Neither viewpoint will ever provide an accurate caricature of a real food web without acknowledging the strengths of the countervailing view. We suggest that progress will come only by identifying strong empirical links between the details of population dynamics and the resolution of food web structure. Population and whole system persistence are linked through properties at individual, internal system, and external system levels (see Figure 1). We argue that models of real food webs will only be valuable for more general empirical prediction if they include the minimum amount of detail and resolution necessary for capturing measurable properties of real food webs. Model predictions should be testable with logistically feasible experiments (e.g., at seasonal timescales) or creative uses of non-experimental empirical data (e.g., longer-term properties of whole systems).

8.3 | CENTRAL ISSUES FOR AQUATIC FOOD WEBS: FROM CHEMICAL CUES TO WHOLE SYSTEM RESPONSES

Ursula M. Scharler, Florence D. Hulot, Donald J. Baird, Wyatt F. Cross, Joseph M. Culp, Craig A. Layman, Dave Raffaelli, Matthijs Vos, and Kirk O. Winemiller

Aquatic ecosystems worldwide provide important resources for human populations. Estuarine and marine habitats provide many essential ecosystem services, including climate regulation, yet currently these systems are impacted over large areas through over-exploitation and degradation (Alongi, 2002; McClanahan, 2002; Verity et al., 2002). In many countries, freshwater is a scarce resource that requires sensitive political engagement to address the social, economic and environmental problems arising from reductions in freshwater availability. Moreover, aquatic ecosystems represent an integrative picture of environmental impacts and management practices operating in the terrestrial hinterland (Beeton, 2002; Hall, 2002; Malmqvist and Rundle, 2002; Kennish et al., 2005). As a consequence, an ecosystem approach to analyse food webs of these systems is increasingly valued by managers and research funding agencies alike, but major challenges lie ahead in the application of food web research to real-world issues of ecosystem management.

The understanding of food web structure, function, dynamics, and complexity is the central theme of food web research. Within this context, there is an array of approaches aimed at answering the ultimate question: "What drives food web dynamics?" A major goal is to be able to predict food web behavior under external (e.g., climate, human exploitation, extensive nutrient input) and internal influences (e.g., population dynamics, feedback, self-organization), in order to manage ecosystems sustainably. Scientists have as yet not been able to develop predictive whole ecosystem simulation models and, to our knowledge, no single method of food web analysis has been subjected to extensive validation of its predictive capabilities. Analytical methods targeting the smaller scale of populations or single species, on the other hand, have collected more credibility throughout their longer application history.

In the subsequent sections of this chapter, we discuss several areas, which, whilst they may not be peculiar to the aquatic environment, seem pivotal for future progress in understanding food webs. We deal first with factors that act at the level of individuals and populations (life history and storage effects, chemicals released by individuals and consumer-driven nutrient cycling) but whose influence on the food web dynamics and ecosystem functioning is as yet poorly understood. We suggest that the neglect of these factors, amongst others to be discovered and explored, contributes to the uncertainty of predictive ecosystem simulation models. We then emphasize the use of larger-scale network approaches and the need for quantitative descriptors of food webs, and conclude with challenges in application of food web theory to management issues in aquatic systems.

INDIVIDUAL AND POPULATION-BASED FACTORS THAT SHAPE FOOD WEB DYNAMICS

Influence of Life History on Aquatic Food Web Dynamics

Food webs are strongly influenced by life-history variation among populations (Polis et al., 1996b), and this is perhaps most apparent in aquatic systems. A fundamental aspect of life history is generation time and its influence on potential population growth rate and the storage effect (a sort of inertia in the biomass dynamics of the adult populations of long-lived organisms with relatively low vulnerability to mortality factors). Demographic and physiological tradeoffs that dictate limited suites of life-history attributes characterise three strategies, which represent endpoints in a triangular model of evolution of primary strategies (Winemiller and Rose, 1992). The *opportunistic strategy* characterizes populations with shorter generation times, which tend to have higher intrinsic rates of increase (r) and greater capacities to maintain positive population growth rates under density-independent environmental settings (e.g., following major disturbances). This strategy tends to be associated with small body size, early maturation, and continuous high reproductive effort, such as algae, decomposers, zooplankton, meiofauna, some aquatic macroinvertebrates (e.g., chironomids) and small planktivorous fishes. These organisms often occupy positions low within the trophic continuum of aquatic systems, but this is not always the case (e.g., parasites and small piscivores).

Species with longer generation times and life spans may be associated with either of two suites of life-history attributes. The *equilibrium endpoint,* associated with relatively low fecundity but high investment in individual offspring, presumably increases fitness (survival, growth, reproductive success, etc.) in environmental settings with frequent or chronic density-dependent population regulation via resource limitation, predation, or other biotic factors (Pianka, 1970), but the trade-off is a lower intrinsic rate of increase in density-independent settings (Lewontin, 1965). The equilibrium strategy, more common in tropical than temperate aquatic systems, is extremely rare among pelagic fishes, but predominant among marine birds and mammals. It is also observed among certain fishes and invertebrates inhabiting caves.

The second slow life-cycle strategy, associated with high fecundity but low investment per offspring, is the *periodic strategy* (Winemiller and Rose, 1992). This strategy is extremely common among aquatic macroinvertebrates and fishes and may promote fitness when environmental variation is predictable in space and/or time at relatively large scales (Winemiller and Rose, 1992, 1993). As a consequence, periodic strategists tend to have high inter-annual and regional variation in recruitment in which strong cohorts are produced in favorable locations or during favorable periods (Persson and Johansson, 1992; Scharf, 2000). In the absence of appropriate conditions for subsequent recruitment, strong cohorts of long-lived species may dominate the population for several years, yielding the phenomenon often referred to as the storage effect (Chesson and Huntley, 1989).

The periodic strategy tends to be associated with high dispersal in the egg and larval stages and can have major effects on the structure and dynamics of food webs. Seasonal mass movements of larval organisms into estuaries, that often are nurseries for marine crustaceans and fishes, provides pulses of food resources for predators (Deegan, 1993); the return of juvenile organisms to coastal marine waters results in a major reduction in consumers in the estuarine food web. Given the prevalence of the periodic strategy among aquatic/marine invertebrates and fishes, donor control, or supply-side, ecological dynamics should be common in aquatic ecosystems (Roughgarden et al., 1987; Menge, 2004). Thus, in addition to estimating effects of slow versus fast life cycles and concomitant intrinsic rates of increase and storage effects associated with life-history strategies, aquatic food web models will also have to include important migrations and subsidies (Polis et al., 1997) that connect local and regional food webs.

Life-history responses to variable environmental factors (e.g., productivity) have profound influence on food web dynamics. For example, phytoplankton in pelagic systems are opportunistic populations with fast life cycles. These populations respond relatively fast to changes in resources and sources of mortality. Moving up pelagic food chains, life cycles tend to become progressively slower but with a greater storage effect. As a consequence, species high in the food web tend to be relatively less responsive to short-term variation in food resources, however time lags in the responses of physiological, behavioral, and demographic parameters could yield complex population dynamics. Whereas opportunistic species should be more variable and responsive to short-term changes in resources, the storage effect should allow equilibrium and periodic strategists to remain more stable through periods of resource fluctuation. Because of the strong tendency for life cycles to become slower in successively higher levels of aquatic and marine food webs, models that employ simple linear Lotka-Volterra consumer-resource dynamics probably will be poor predictors of dynamics in these systems. To some degree, food web models must capture the essential ecological features that yield time lags and non-linear responses among populations with different life history strategies. For instance, Persson et al. (2003) analyzed shifts in trophic cascades caused by intrinsically driven population dynamics. Trophic cascades were explored experimentally in aquatic systems at the ecosystem level (Polis, 1999) and have been a central paradigm in explaining the structure of ecological communities. Yet, trophic cascades may alternate over time between two different configurations. By using empirical data and a size-structured population model, Persson et al. (2003) showed how according to the energy extracted from their prey, which constrains the reproduction, the size structure of a fish population may shift from a dominance of dwarfs to a dominance of giants. This intrinsically driven population dynamics limits the whole community and is the driving force behind distinct abundance switches observed in zooplankton and phytoplankton (see also De Roos and Persson, Chapter 3.2; Koen-Alonso and Yodzis, Chapter 7.3; Persson and De Roos, Chapter 4.2).

The Chemical Matrix and Food Web Dynamics

Whilst life history shapes food web dynamics on a population level, chemicals released by plants and animals modulate both trophic and competitive interactions with consequences to the entire natural food web. Dicke and Vet (1999) coined the term information web, or information network, for the combined effects of infochemicals on interactions in multitrophic communities (also see Vet, 1999; Vos et al., 2001). Here we extend this perspective and propose that food webs are embedded within a chemical matrix that includes both infochemicals and allelopathic substances. The effects of this matrix manifest themselves at the individual level in modified behaviors or rates of feeding, growth, and mortality, inducible defenses or shifts in competitive interactions. At the community level, plant- and animal-released chemicals affect both direct and indirect interactions in food webs, often through trait-mediated effects (Werner and Peacor, 2003). The study of these processes in aquatic systems requires specific approaches because the physical aspects of the medium are different from terrestrial ecosystems. For instance, the diffusion of chemicals in the medium depends on their solubility but also on the viscosity of the medium. Here we choose to focus on infochemical effects on trophic interactions and allelopathic effects on competitive and trophic interactions.

Infochemical Signals and Trophic Interactions

Infochemical signals affect trophic interactions in a variety of ways. Prey use predator-released chemicals as cues to tune refuge use and other induced defenses to actual predation risk. Such defenses affect functional responses and, as a consequence, interaction strengths (Jeschke and Tollrian, 2000; Vos et al., 2001; Vos et al., 2004a; Vos et al., 2004b). Predators, on the other hand, may use prey-released waste products and pheromones to locate resource species. This implies that trophic interactions may have an information aspect. Empirical ecologists have accumulated an impressive body of knowledge on the ecology and evolution of such interactions (Kats and Dill, 1998; Tollrian and Harvell, 1999). Kats and Dill (1998) reviewed prey responses to predator-released chemical cues and described changes in prey behavior, life history, and morphology in more than 200 species. Despite this wealth of information at the level of individuals, much less light has been shed on the implications for food web dynamics. Theoretical work predicts that infochemical use as manifested in induced defenses affects community persistence (Vos et al., 2001; Vos et al., 2002), stability (Edelstein-Keshet and Rausher, 1989; Abrams and Walters, 1996; Underwood, 1999; Vos et al.,

2004a), resilience (see Vos et al., Chapter 3.4) and trophic structure (Abrams and Vos, 2003; Vos et al., 2004b). This suggests that infochemical-mediated effects may play an important role in the functioning of food webs in nature (also see Bolker et al., 2003; Werner and Peacor, 2003). In terrestrial plant-insect systems, extensive experimental work has focused on infochemical-mediated mutualisms between plants and the predators that attack their herbivores. Such research on the scope for infochemical-mediated multitrophic interactions is relatively underdeveloped in aquatic systems and deserves more attention (also see Vos et al., Chapter 3.4).

Allelopathic Interactions

We define allelopathy here as the inhibitory effects of chemicals released by one species on a sensitive species. For instance, chemicals of some cyanobacteria species have a targeted toxic effect by inhibiting the photosystem II of other primary producers (Mason et al., 1982; Gleason and Paulson, 1984; Gross et al., 1991). Allelopathy has also been reported in zooplankton species, with effects including reduced feeding and growth and increased mortality (Halbach, 1969; Folt and Goldman, 1981; Matveev, 1993; Lürling et al., 2003). As a consequence, allelopathy can modulate competitive interactions. Allelochemicals produced by macrophytes affect the composition of algal communities, both in marine and freshwater systems (Gross, 2003; Mulderij et al., 2003), whilst inhibitory effects reported between zooplankton species suggest a regulatory role for allelopathy in zooplankton communities.

When released by a prey (see examples in Landsberg, 2002), extracellular compounds that are toxic for a predator act as a defense. Such inhibitory effects on higher trophic levels are very similar to allelopathic effects on competitors. The production of compounds that are toxic against competitors and predators alleviates the producer from controls, which implies a potential for strong impacts on community dynamics (Hulot and Huisman, 2004). For instance, a variety of factors may explain shifts between turbid and clear water states in shallow lakes or be involved in the maintenance of such alternative states (Scheffer, 1998). Toxic compounds released by macrophytes may be among these factors, although their importance relative to other factors is poorly understood. Similarly, whether allelopathy against predators or competitors plays an important role in the development of toxic phytoplankton blooms is an open question (Hulot and Huisman, 2004).

There is great potential for the chemical matrix to affect both trophic and competitive interactions in food webs. We expect that incorporation of the chemical matrix in food web theory will result in a more complete understanding of how food webs function. One particular challenge is to explore to which extent infochemicals and allelopathic substances, which manifest themselves at the individual level, may affect the ecosystem dynamics.

Consumer-Driven Nutrient Recycling (CNR)

Consumer-driven nutrient recycling (CNR) is an important component of aquatic ecosystem dynamics that has recently gained considerable attention among food web ecologists (DeAngelis, 1992a; Vanni, 1996; Sterner, 1997; Elser and Urabe, 1999; Vanni, 2002). Consumptive interactions within food webs are inextricably linked to the excretion or egestion of nutrients that exceed an organism's metabolic or reproductive requirements (Elser and Urabe, 1999). These recycled dissolved or particulate nutrients are released into the aquatic environment and may be reincorporated into the food web at all levels from basal species to top predators. Such dynamic recycling of essential elements may have strong implications for the structure, productivity, and stability of aquatic food webs (Kitchell et al., 1979).

Recent studies of consumers in lakes and streams (Elser and Urabe, 1999; Vanni, 2002) reveal a large degree of interspecific variation in dissolved nutrient recycling which is often correlated with the nutrient content of consumers. For example, Vanni et al. (2002) demonstrated that the N:P ratio of excreted nutrients by stream fish and amphibians was negatively related to organism body N:P ratios. Similar variation in nutrient recycling ratios has been discovered among species of zooplankton (Sterner et al., 1992; Elser et al., 1998) that differ largely in their body nutrient content. Such relationships between species identity and nutrient cycling imply that community composition, or even the presence of one dominant species, can determine the availability of essential nutrients and productivity of the system. Understanding the importance of CNR to whole-system nutrient dynamics will likely require a combination of detailed empirical studies of individual interactions in addition to analysing the system as a whole. Ulanowicz and Baird (1999) describe a promising whole-ecosystem approach which uses information on nutrient content, metabolic and recycling rates for all species or trophospecies in the system for various nutrients simultaneously (e.g., C, N, P etc.) to identify the limiting nutrient and the limiting source for that nutrient for each species in the food web. Network analysis techniques underscore and incorporate the importance of species connectedness via nutrient recycling, in addition to the more commonly recognized connectedness via direct consumption.

Although most empirical and theoretical studies of CNR focus on dissolved nutrients (i.e., excretion), consumers can potentially have large effects on nutrient availability through recycling of particulate nutrients (i.e., egestion). If large elemental imbalances exist between consumers and resources, the nutrient content of egested material will be lower than that of ingested food items because of nutrient sequestration by consumers. Alternatively, consumers may increase the nutrient content of egested material by altering its size-structure (i.e., surface:volume ratio) and susceptibility to nutrient-rich microbial colonization (Turner and Ferrante, 1979; see Cross et al., Chapter 5.4). In aquatic systems, coprophagy is relatively common (Wotton and Malmqvist, 2001; Pilati et al., 2004), and particulate nutrient recycling may have important ramifications for the dynamics of these food webs. Although food web linkages via egestion are well recognized in the analysis of trophic flow networks, empirical studies of this process are few (Short and Maslin, 1977) and considerable research is needed to identify when and where food web flows are dependent upon nutrient cycling through egestion.

Understanding the importance of CNR in aquatic food webs will require a great deal of empirical research in a wide variety of systems. Investigation of CNR in non-pelagic communities (e.g., benthic lake, stream, or marine communities; Frost et al., 2002; Sterner and Elser, 2002) will undoubtedly provide valuable insight into the generality of current theory.

A MACROSCOPIC APPROACH TO ECOSYSTEM STUDIES

The factors influencing the structure and dynamics of food webs often differ in scale in both signal and response. We previously considered selected issues such as life-history strategies, chemical interactions, behavioral responses, and nutrient stoichiometry. They all describe interactions on the relatively small scale of species and populations. Frequently a broader scale to view the ecosystem is required to understand interactions between species within an ecosystem context, as well as ecosystem level descriptors, which result from such direct and indirect species interactions.

Such a macroscopic or holistic approach, which can reveal descriptors of food web structure and processes, can be applied to quantified trophic flow networks. In these networks, the biomass of each node and the trophic exchanges between species and the surrounding environment are quantified using currencies of energy or material (e.g., C, N, P, S;

Wulff et al., 1989). The magnitude of material transfer between species is a result of a multitude of factors such as new nutrient availability and nutrient recycling, detritus production, interaction patterns between species (e.g., predator–prey interactions), recruitment patterns, habitat availability and exchanges with the environment (detritus and migration). They constitute as such an integration of environmental and biotic factors. Stable isotope ratios, which are now routinely employed for tracing trophic connections (see Layman et al., Chapter 7.4), also provide valuable information to build such quantified flow networks. They are used to estimate basal energy sources and overall patterns of flow of material through webs as a time-integrated measure of feeding relationships, reflecting all pathways of material flow from the base of the food web to an organism of interest. Especially where stomach content analyses prove futile (e.g., nutrients, primary producers, small organisms, etc.), stable isotopes provide valuable information of feeding interactions.

The magnitudes of transfers between species so measured by and calculated from, amongst others, above observations shape a number of species-, community-, and ecosystem-level properties, which can be calculated from a transfer matrix of material (i.e., species j (column) consuming species i (row)). Ecological Network analysis (ENA) is a tool used to analyze such quantified flow networks and comprises several types of analyses that are briefly mentioned here. The quantitative data describe direct and indirect effects between species. By applying Input-Output Analysis to weighted networks (Szyrmer and Ulanowicz, 1987), the combined direct and indirect effects reveal dependencies of any one species on any other species in the network, which can be quantified in terms of material transfer. It is thus possible to track the dependencies (also called 'extended diets') of, for example, a commercially exploited species, or a 'keystone' species. The dependencies are expressed as the fraction of the total ingestion by j which has passed through compartment i on its way to j, over all direct and indirect pathways (Szyrmer and Ulanowicz, 1987). In addition to transfers between any two species, cycles can be enumerated and the amount of material flowing through cycles of various lengths computed with ENA. The amount of material flowing through cycles (or recycling) can be quantified as opposed to straight through flows (material entering the system and leaving it without being cycled once), and gives, together with the cycle structure, a clue about the developmental state of the system. In the light of human exploitation and the urgent question of ecosystem resilience, information inherent in the flow structure (including the magnitudes of flows) can been calculated to express the development capacity, organization, and resilience of ecosystems as information theoretical indices (Ulanowicz, 2004).

The importance of quantitative data in food web studies has more recently been highlighted by, for example, Bersier et al. (2002) who compared several qualitative and quantitative food web descriptors to show that there is more information inherent in quantitative data than in a qualitative approach. Another recent example for a quantitative descriptor of food webs includes the assignment of a 'loop weight' to cycles (Neutel et al., 2002), which has allowed interaction strength and cycle length to be related to the stability of ecosystems.

The analysis of quantified networks is used to compare ecosystem attributes on local and global scales (Baird et al., 1991; Monaco and Ulanowicz, 1997) and to illustrate changes in system behavior over time (Baird and Ulanowicz, 1989; Baird et al., 1998). Other studies highlight the effects of environmental impacts such as the construction of a dam (Baird and Heymans, 1996), nitrogen overloading in an estuary (Christian and Thomas, 2003), or a change in freshwater inflow into estuaries (Scharler and Baird, 2005). In such circumstances, the network approach gives valuable insight into food web attributes through (1) changes in the topology of the food web and (2) changes in the magnitude of energy or material transfers. The latter constitutes a more subtle response of ecosystems to perturbations compared to changes in food web topology (i.e., a loss of links).

ENA as a whole ecosystem approach describes effects of abiotic and biotic interrelationships through numerous parameters. The advantage of such an approach lies in the inclusion of several species interactions, including feedback mechanisms, in an ecosystem context as opposed to single species interactions and dynamics. As such it describes patterns arising from the structure and magnitudes of flows, which is valuable information required for the management of ecosystems. Such an analysis can be complementary to dynamic ecosystem simulation models.

THE APPLICATION OF FOOD WEB THEORY IN MANAGEMENT

Food web theory has broad application in relation to management issues, yet there is often reluctance on the part of modellers and theoreticians to put their models to the test by addressing the concerns of regulators and the public. Of course, the contrary argument can be made that managers themselves can often be suspicious of models and distrustful of their predictions, and models are seen as either too complex to convey meaning to the end-user or too simplistic to apply in situations which are economically or politically sensitive (in the sense that they might not stand up to close public scrutiny). To date, the use of food web models in environmental management has largely been restricted to two areas: the trophic transfer of contaminants (Camphens and Mackay, 1997) and the management of marine fisheries (see Koen-Alonso and Yodzis, Chapter 7.3).

Trophic transfer models have generally been applied in the study of persistent organic pollutants (e.g., PCBs), where it is possible to predict the movement of substances through food webs by a combined knowledge of their chemical properties, and the structure and functioning of the food web itself. Such models remain highly simplistic, however, and it is perhaps surprising that little progress has been made in refining these models beyond the pioneering work of the Canadian Environmental Modelling Centre (http://www.trentu.ca/cemc/). There is a need here to introduce more ecological thinking into the fugacity approach (Camphens and Mackay, 1997), and this should be a fruitful area for future collaborations between ecologists and environmental chemists.

The use of food web models in fisheries management should constitute the greatest success story for the use of such models in environmental management. However this is not the case: there has been some reluctance on the part of regulators and operators to fully engage in the application of these models in regulation of fishing activity. This is further exacerbated by the unwillingness on the part of the fishing industry to accept the social and economic consequences of the management actions suggested by the model predictions. Even in marine fisheries management, decisions continue to be taken on a species by species basis, even though this is clearly failing to address the consequences of fishing, which inevitably involves exploitation of food webs rather than individual species (Pauly et al., 2003).

Food web models clearly have much to offer the field of environmental management: they can be used for forecasting the consequences of management actions, or the impacts of environmental pressures. Alternatively, they can be employed to hindcast the historical state of existing, degraded systems and to provide target images for ecological restoration schemes. They have particular value in the field of environmental risk assessment, where they can form the basis of a diagnostic methodology to tease apart the effects of multiple stressors (Baird et al., 2001; see Culp et al., Chapter 7.1), and it is in this complex area where adoption of a food web approach can perhaps offer the greatest immediate advantages to the environmental manager.

DISCUSSION AND CONCLUSIONS

The issues previously discussed are examples of study areas that provide insights into the structure, function, and dynamics of aquatic food webs, and which the authors feel deserve increased attention. They also illustrate how new perspectives may emerge from their combination. Importantly, these study areas range in scale from the population and species level (e.g., life history strategies, chemical cues and defenses, nutrient consumption and sequestration), to the community and ecosystem level (e.g., nutrient cycling, nutrient transfers, indirect effects, organization, development and resilience, management). However, the degree of influence on food web dynamics and ecosystem functioning of the processes described at these different scales are still poorly assessed. For instance, processes linked to specific organisms such as long-lived species with a great storage effect or seasonal mass dispersal of egg and larval stages may drive, at different time and spatial scales, the dynamics of food webs. However, the extent of their effects is poorly known and opens questions as to which temporal and spatial scales to consider in such studies. The issues described at the population and species level are also examples of processes, which may contribute to the uncertainty of food web model predictions. One important challenge is therefore to reduce this uncertainty by identifying the non-trophic processes influencing food web dynamics, and including them in models of predictive capability.

Despite the increasing number of food webs documented in the literature and on the World Wide Web, there is a need for the identification of pathways and the quantification of matter transfer in ecosystems to facilitate analyses of entire ecosystems and not only parts thereof. This is even more relevant if we aim to extend the analysis to interaction networks by including information transfer and allelopathy, or the size-structure of populations as a driving factor of aquatic communities.

The area of food web research has great importance for the management of ecosystems to ensure performance under circumstances of continuous exploitation by humans. Ecosystem approaches facilitate taking a step towards 'whole ecosystem management' and away from reductionist approaches such as single species management. Although the predictive capability of ecosystem models has a poor record of validation, the analysis of a given structure of ecosystems can already provide valuable information on several issues important to their management. A major challenge of food web research is to reduce modellers' and managers' reluctance to respectively test and apply food web theory.

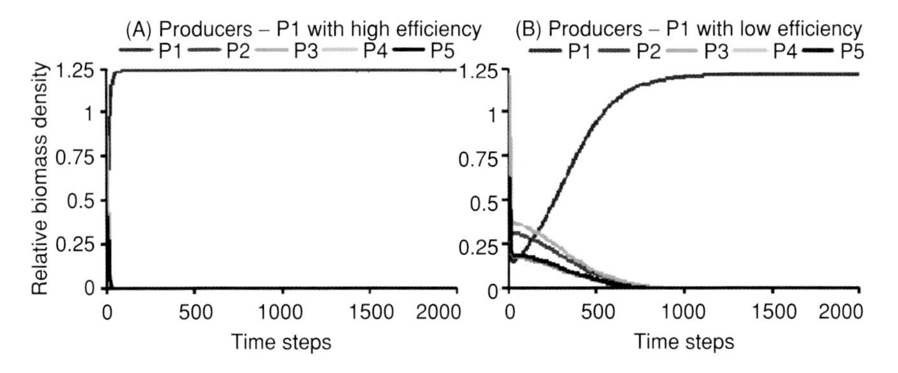

CHAPTER 2.1 FIGURE 1 | Biomass evolution in producer communities. The guzzler (P1) is strong **(A)** or weak **(B)**.

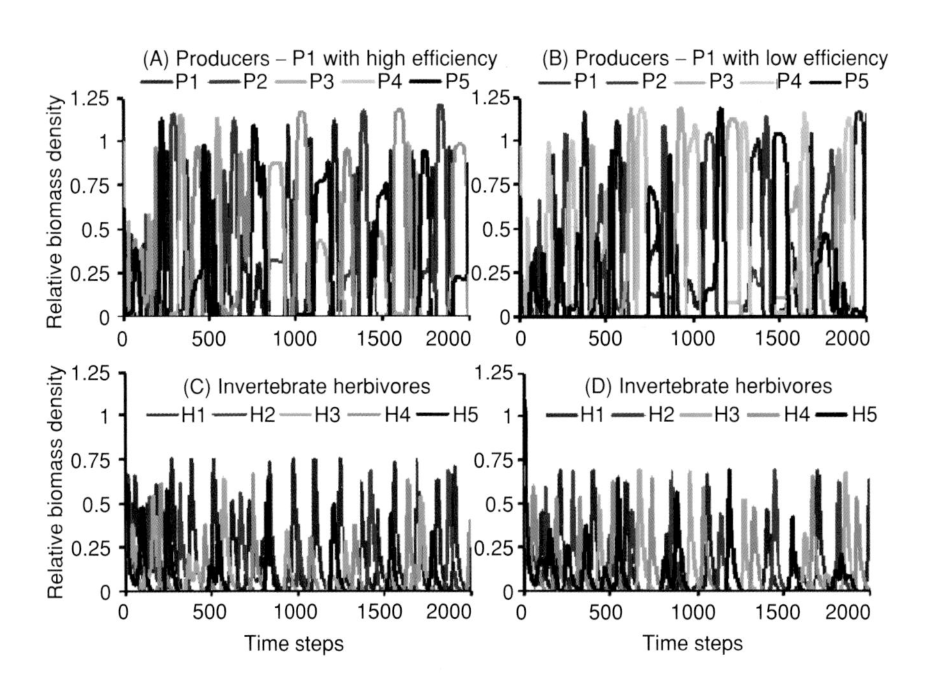

CHAPTER 2.1 FIGURE 2 | Biomass evolution in producer-invertebrate herbivore communities. The guzzler (P1) is strong **(A, C)** or weak **(B, D)**.

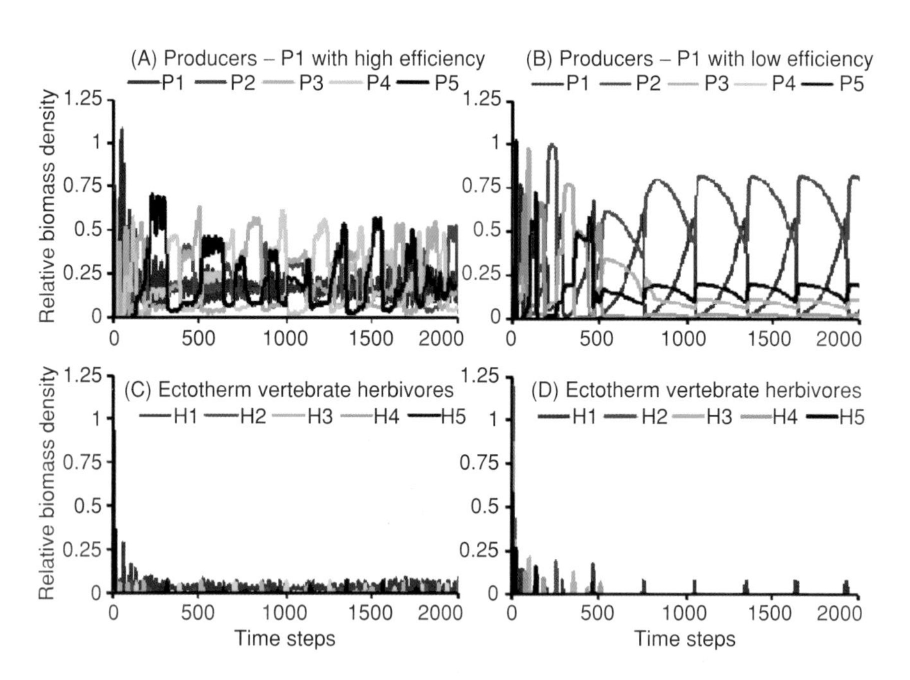

CHAPTER 2.1 FIGURE 3 | Biomass evolution in producer-ectotherm vertebrate herbivore communities. The guzzler (P1) is strong **(A, C)** or weak **(B, D)**.

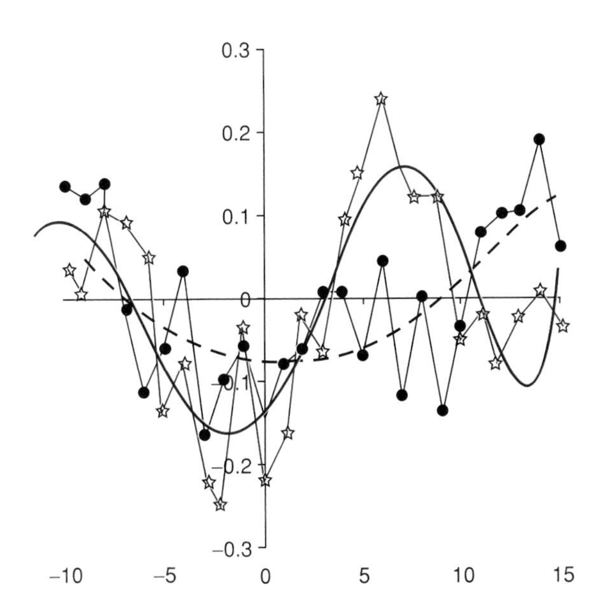

CHAPTER 2.2 FIGURE 3 | Time series cross-correlation analysis of log-biomass of edible algae against zooplankton. Black dots show data (averaged across 4 replicates of each treatment each with 20 data points) for the cross correlation function for cases where *Daphnia* are the only grazers present, red stars indicates cases where there are three species (*Daphnia, Ceriodaphnia* and *Chydorus*). The X-axis is time lag (days), the Y-axis is the correlation coefficient. The graph shows lines that connect the data and smoothed estimators for each function (dashed black line for *Daphnia* only, solid red for three species together).

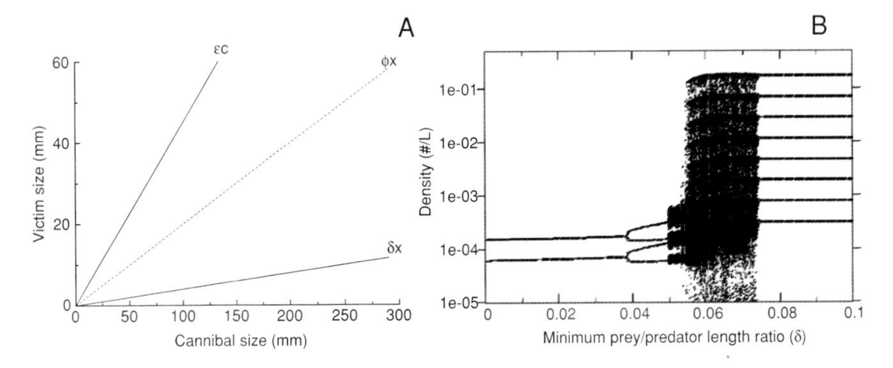

CHAPTER 4.2 FIGURE 3 | **A,** Two-dimensional plot of the cannibalistic window showing the minimum size ratio (δ), the optimum size ratio (φ) and the maximum size ratio (ε) of victims to cannibals for which cannibalistic interactions take place. **B,** Bifurcation plot showing the effects of the lower size limit (δ) on cannibalistic population dynamics for β = 200 and ε = 0.45. Densities reflect all cannibals excluding young-of-the-year individuals.

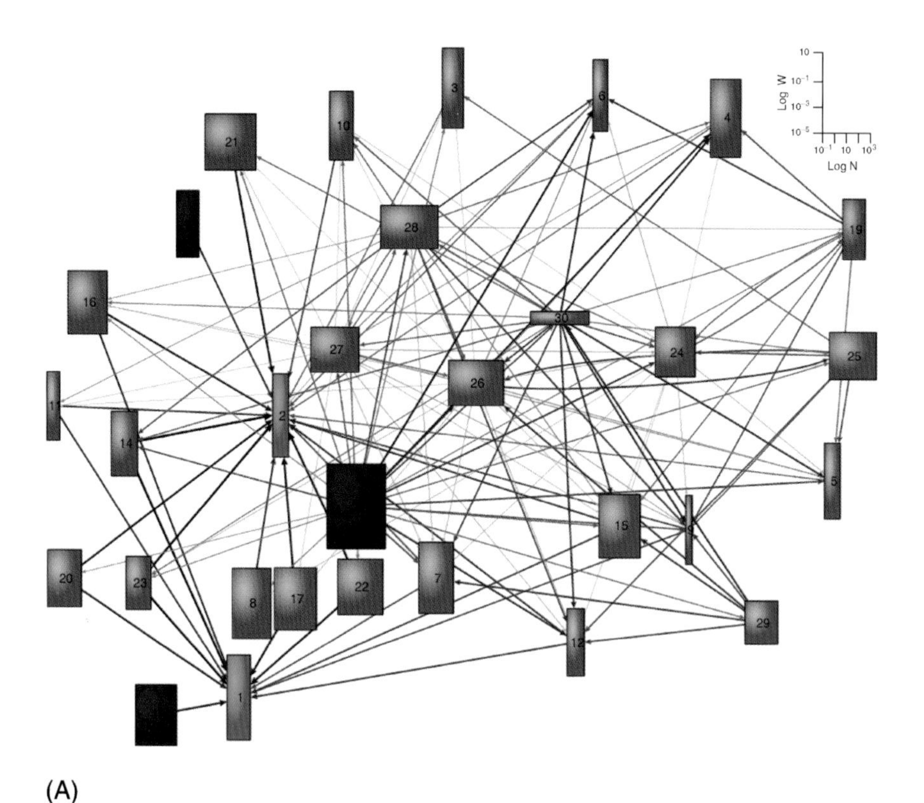

(A)

CHAPTER 7.2 FIGURE 1 | Representation of Broadstone **(A)** and Ythan **(B)** food webs. Arrows point from predators to prey, and line thickness and darkness correspond to predator per capita interaction strengths (a_{ij}) in the community matrix. Rectangles correspond to species (most of them taxonomic species), and green, blue and red correspond to basal, intermediate and top species respectively. The base of the rectangle corresponds to species' numerical abundance (number/m^2), and its height to body size (mg of mean individual or mean adult, for the Broadstone and Ythan respectively).

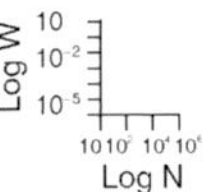

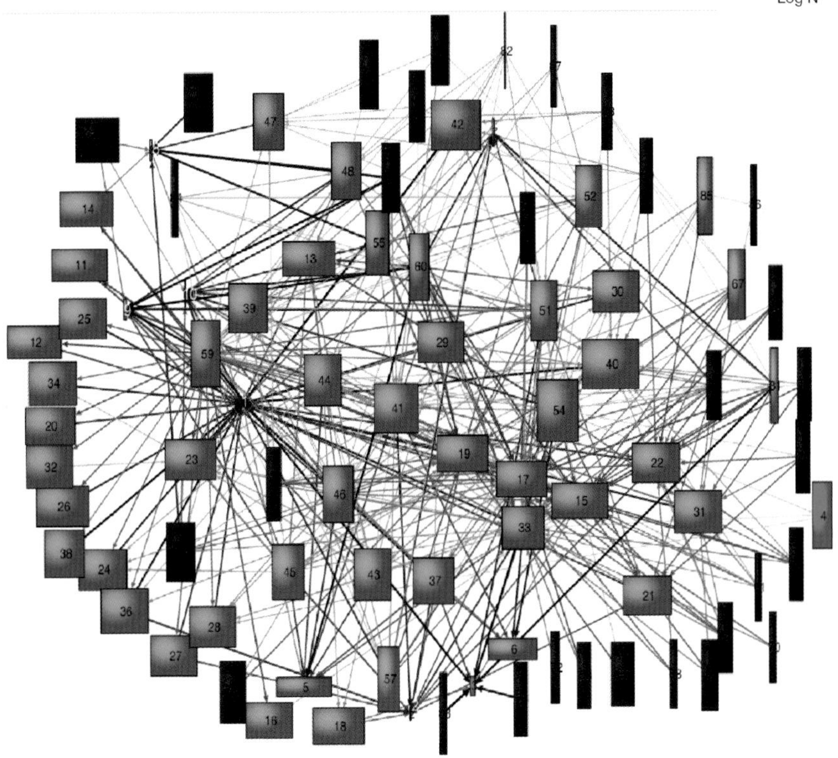

(B)

CHAPTER 7.2 FIGURE 1 | *Cont'd* All data are log-transformed, and the scales for abundance and body size are indicated. For the Ythan web, we do not have direct measurements of some species abundances, so we have estimated them from their body size using observed allometric relationship between abundance (X) and body size (W) for the species whose abundance was already known: $X = 177.51^{*}W^{-1.185}$ (r^2=0,84; $p<0.0001$) (after Leaper and Raffaelli, 1999; Emmerson and Raffaelli, 2004b). This type of representation might be very helpful for visualizing the temporal variability in food webs (e.g., exploring how abundances, body sizes, or interaction strengths change through time), or the effects of the introduction and extinctions of species in complex food webs. Figures have been generated using the program Pajek.

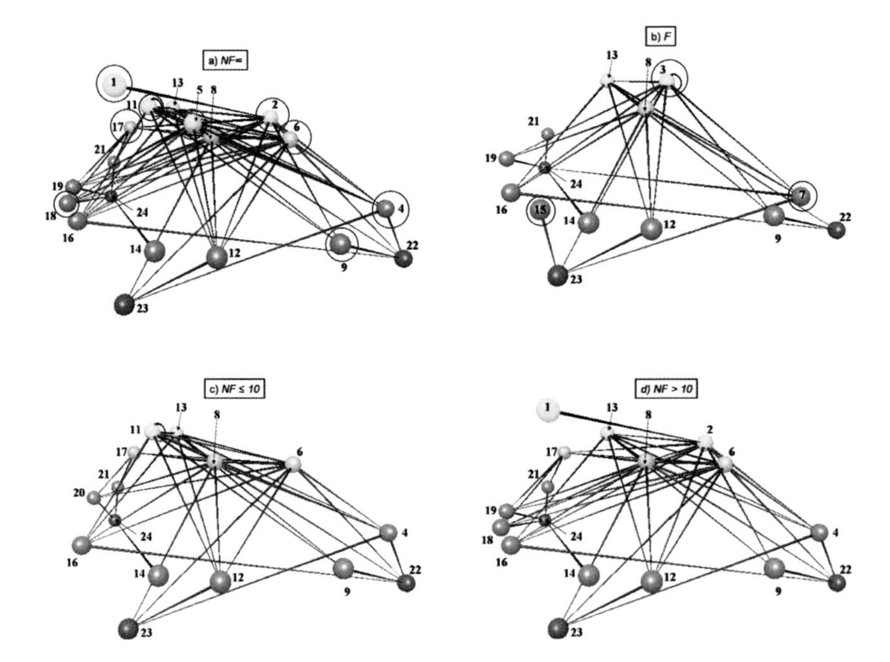

CHAPTER 7.5 FIGURE 1 | Food webs for the four lake categories: (a) “*NF*∞” = Never any Fish, (b) “*F*” = Fish present, (c) “*NF* ≤ *10*” = No Fish ≤ 10 years, (d) “*NF* > *10*” = No Fish > 10 years. Each node is a separate trophic species which may consist of one or more taxa that share all the same predators and prey. Webs are oriented so that the basal species are at the bottom in dark orange and the top predators are at the top in yellow. Link shape indicates the direction of the feeding relation, with the wider end being the predator and the narrow end the prey. If two species eat each other, the link is narrowest in the middle. Cannibalistic links are indicated by a loop returning to the same node. In the *NF*∞ web, red circles highlight the trophic species that are lost in the *F* web. In the *F* web, blue circles indicate new trophic species relative to the ‘Never Stocked’ web. For a list of taxa and trophic species, see Table 2.

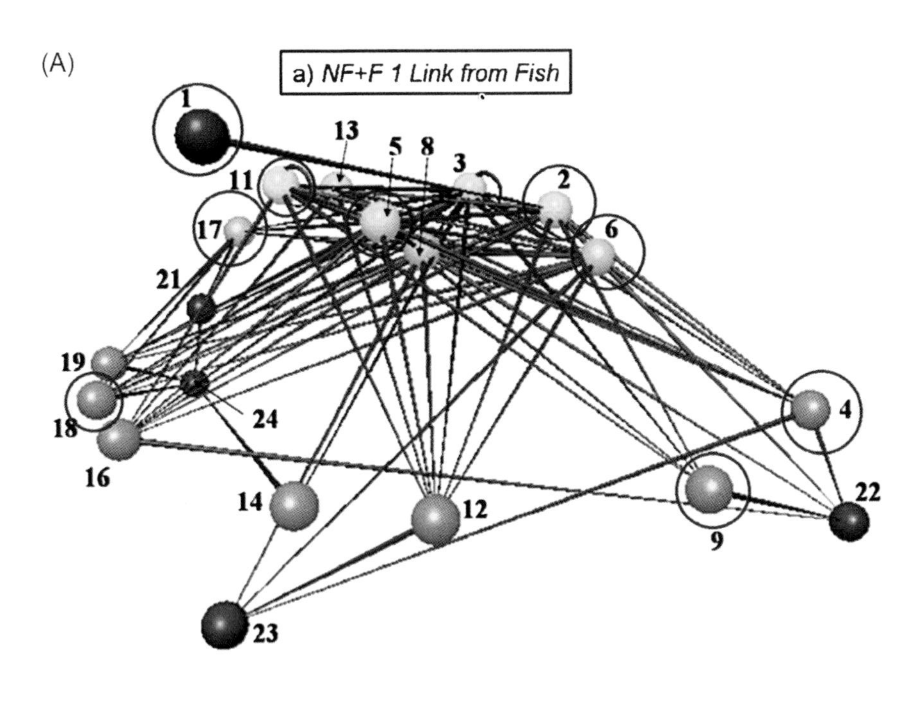

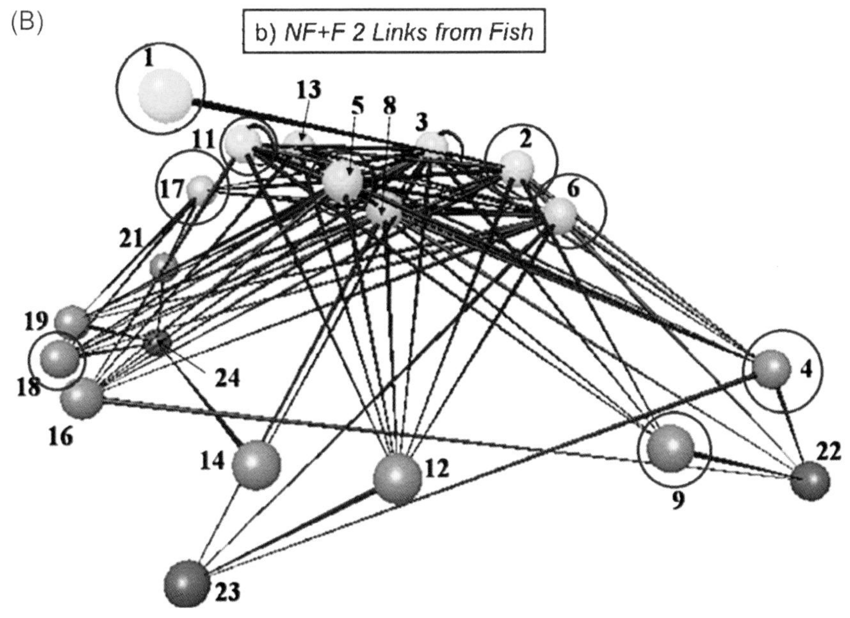

CHAPTER 7.5 FIGURE 2 | A hypothetical *NF+F* web showing the *NF*∞ condition with fish added. **A,** All direct feeding links from trout are highlighted in color, with all other links and nodes blackened. **B,** All nodes and links two feeding links away from trout are shown in color, which includes the entire web. Red circles indicate trophic species that are lost in lakes where fish are present. For a list of taxa and trophic species, see Table 2.

8.4 | SPATIAL ASPECTS OF FOOD WEBS

Ulrich Brose, Mitchell Pavao-Zuckerman, Anna Eklöf, Janne Bengtsson, Matty P. Berg, Steven H. Cousins, Christian Mulder, Herman A. Verhoef, and Volkmar Wolters

Aspects of spatial scale have until recently been largely ignored in empirical and theoretical food web studies (Cohen and Briand, 1984; Martinez, 1992; but see Bengtsson et al., 2002; Bengtsson and Berg, Chapter 5.1). Most ecologists tend to conceptualize and represent food webs as static representations of communities, depicting a community assemblage as sampled at a particular point in time, or highly aggregated trophic group composites over broader scales of time and space (Polis et al., 1996a). Moreover, most researchers depict potential food webs, which contain all species sampled and all potential trophic links based on literature reviews, several sampling events, or laboratory feeding trials. In reality, however, not all these potential feeding links are realized as not all species co-occur, and not all samples in space or time can contain all species (Schoenly and Cohen, 1991), hence, yielding a variance of food web architecture in space (Brose et al., 2004). In recent years, food web ecologists have recognized that food webs are open systems—that are influenced by processes in adjacent systems—and spatially heterogeneous (Polis et al., 1996a). This influence of adjacent systems can be bottom-up, due to allochthonous inputs of resources (Polis and Strong, 1996; Huxel and McCann, 1998; Mulder and De Zwart, 2003), or top-down due to the regular or irregular presence of top predators (Post et al., 2000a; Scheu, 2001). However, without a clear understanding of the size of a system and a definition of its boundaries it is not possible to judge if flows are internal or driven by adjacent systems. Similarly, the importance of allochthony is only assessable when the balance of inputs and outputs are known relative to the scale and throughputs within the system itself. At the largest scale of the food web—the home range of a predator such as wolf, lion, shark, or eagle of roughly 50 to 300 square kilometers—the balance of inputs and outputs caused by wind and movement of water may be small compared to the total trophic flows within the home range of the large predator (Cousins, 1990).

Acknowledging these issues of space, Polis et al. (1996a) argued that progress toward the next phase of food web studies would require addressing spatial and temporal processes. Here, we present a conceptual framework with some nuclei about the role of space in food web ecology. Although we primarily address spatial aspects, this framework is linked to a more general concept of spatio-temporal scales of ecological research.

DEFINING SPATIAL SCALES

Given that not every species occurs everywhere, spatial distributions of trophic interactions define (1) the spatial heterogeneity of food webs and (2) differences between local and macroecological food webs (Brose et al., 2004). More specifically, even food webs within the same metacommunity differ to some extent when studied at different locations. In such local community food webs, all species co-occur and all potential trophic links—based on the physiological feeding capacities of the species—are realized. This concept of co-occurrence has to be given up when larger spatial scales are considered that integrate different local community food webs into a metacommunity food web. In the following sections, we will give two examples for this. First, some large-bodied predators are too low in numerical abundance to invade all local community food webs simultaneously. Second, not all potential resource species in a metacommunity can persist under strong top-down pressure by their consumer species and thus avoid co-existence in the same local communities. Therefore, in macroecological metacommunity food webs not all species co-occur and not all potential links are realized due to spatial and temporal separation (Brose et al., 2004). The variance in local food web structure and the difference between food webs studied within local communities versus metacommunities are important challenges for future research. In the following, we will discuss some consequences of these spatial considerations for theory on food web structure and dynamics.

SPECIES OPERATE AT DIFFERENT SPATIAL SCALES

Food webs consist of organisms that vary in their taxonomic identity, body size, trophic interactions, and their trophic position and thus might have very different spatial scales of interactions. In general, species' home ranges increase with their body sizes (see Kelt and Van Vuren, 2001, for detailed explanations), and in many food webs, mean body size increases with trophic level (Warren and Lawton, 1987; Cohen et al., 1993a; Cohen et al., 2003). However, frequent exceptions to this pattern such as parasitoid food webs with constant predator–prey body size ratios (Memmott et al., 2000) and food webs with mobile prey but sessile consumers (Östman et al., 2001) suggest a cautious use of this relationship. In general, mean body size is inversely related to numerical abundance (Brown et al., 2000; Gaston and Blackburn, 2000; Cohen et al., 2003) and positively related to territory size (cf. Ghilarov, 1944; Hendriks, 1999; Jennings and Mackinson, 2003). Accordingly, a metacommunity is likely to contain small and abundant species that occur more frequently in local community food webs than larger and less abundant species. These differently sized species operate at different spatial scales. As a consequence, large predators of low abundance are likely to be distributed unevenly within the metacommunity thus having a spatio-temporally heterogeneous impact on other species in local communities. Furthermore, in consequence of the abiotic heterogeneity of most habitats not all small species at the base of the food web occur everywhere in the metacommunity.

LARGE PREDATORS INTEGRATE LOCAL WEBS

We illustrate a metacommunity example of seven species that are not evenly distributed in space (Figure 1A) and trophically linked in a potential food web module (see Figure 1B). The top predator has the largest home range, whereas the other species with smaller home ranges occur locally (see Figure 1A, species 1–6). This yields a turnover of species and trophic interactions across space (see Figure 1C), and hence a beta-diversity. Exemplary samples at different locations yield a three species food chain (see Figure 1C, *I*), a four-species module (see Figure 1C, *II*), or a four-species food chain (see Figure 1C, *III*). Although species and trophic interactions vary, this exemplary food web module is linked by the predator with the largest home range that occurs in all three samples. Documenting the coupling of the local webs by the largest predator

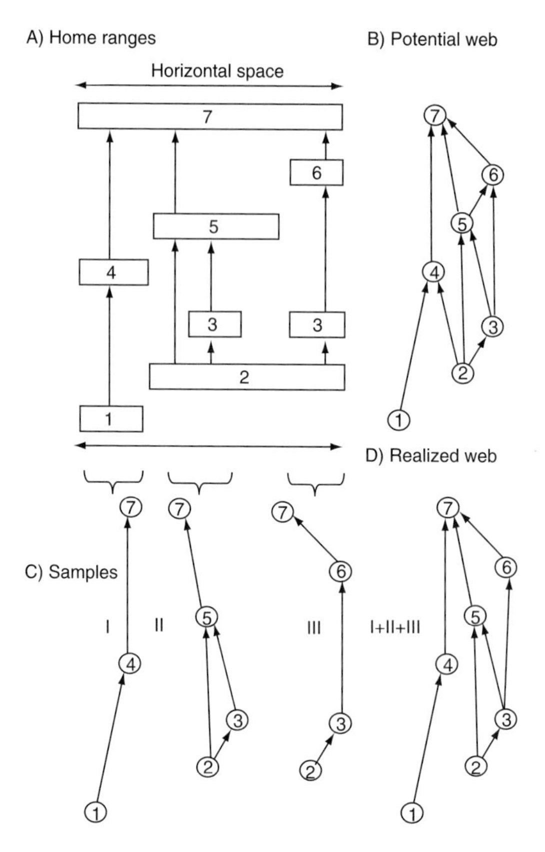

FIGURE 1 | A spatially explicit food web of seven species (1–7) that is connected by a single top predator. **A,** Distribution of the seven species' home ranges (boxes) in horizontal space. **B,** The potential food web showing physiologically possible feeding interactions (arrows) between the seven species (nodes). **C,** Three exemplary samples at different points in space with different local food webs. **D,** Realized food web showing the feeding interactions that are realized in the spatial food web.

requires studies on a time scale long enough for the predator to visit all sites in its home range. Thus, large predators may couple local webs, but the extent to which they do so depends on time. This example illustrates the impact of spatio-temporal scaling on food web studies: the more local webs are sampled and combined, and the more samples in time, the more the food web architecture approaches the metacommunity food web configuration (see Figure 1C, D). Empirical examples for this spatio-temporal concept are nested microbial compartments that are integrated by the predation activities of larger soil fauna (Pokarzhevskii et al., 2003).

The largest predator's top down influence on the local food webs is temporally heterogeneous could be conceptualized as catastrophic disturbance events that drastically reduce the biomass of the predator's prey species. Such catastrophic predation events are followed by recovery periods during which the predators move to other local webs because of their higher profitability after the local web has been exploited. This concept treats pulsed predation as a disturbance and links it conceptually to abiotic disturbance events that maintain species' coexistence and biodiversity in ecosystems (Huston, 1979).

MOVING BETWEEN LOCAL WEBS STABILIZES PREDATOR POPULATIONS

The existence of local webs or food web compartments has been discussed in context of its potential stabilizing influence on complex food webs (May, 1973; Moore and Hunt, 1988). In particular, top predators gain a higher dynamic stability when moving between two separate food chains (Post et al., 2000a; Hunt and Wall, 2002; Mulder and De Zwart, 2003). However, whether the results obtained for food chains can be transferred to more complex food webs (i.e., if top predators of complex food webs gain dynamic stability from moving between local food webs) remains to be demonstrated.

Moreover, some trophic links might occur only temporally in local habitats. For instance, the predator moving between local webs (see Figure 1, species 7) creates temporally varying links in the sub-webs (see Figure 1, links between species 7/4, 7/5, 7/6). Such temporal links have previously been modeled as energetic subsidies or allochthonous inputs for a predator, which is based on reasoning of resource species in adjacent habitats (Polis and Strong, 1996; Huxel and McCann, 1998). Interestingly, our example of an occasionally occurring predator suggests that the opposite effect of energetic efflux might also exist.

PREY SPECIES ESCAPE THEIR PREDATORS

Some links of potential metacommunity webs are not realized because predator and prey species occur in different local webs (e.g., Figure 1, links between species 2/4 and 5/6). For instance, some prey species might effectively escape their potential predators in space. Accordingly, if there is a spatial turnover in the community composition—measured as beta-diversity—it is most likely that the number of realized links does not equal the number of potential links (compare Figure 1B,D). For any given species-area relationship, the difference between potential and realized webs depends on the area studied and the efficiency of predators tracking their prey species (Holt et al., 1999; Brose et al., 2004).

SPATIAL ASPECTS OF FOOD WEB DYNAMICS

Many studies have shown how patch density, patch size, and species' dispersal rates affect metacommunity population dynamics, but surprisingly few studies have considered food web structure (Melian and Bascompte, 2002; Ellner and Fussmann, 2003; Brose et al., 2004). Brose et al. (2004) demonstrated that variances in patch size systematically affect food web diversity and complexity. Accordingly, habitat destruction may yield qualitatively new consequences when considering species that are embedded in a web of ecological relationships (Ellner and Fussmann, 2003). Species' abundances and extinction thresholds vary with food web structure (Melian and Bascompte, 2002). Accordingly, the consequences of habitat loss for a specific species do not only depend on the species' own traits but also on community characteristics such as food web architecture and diversity. Furthermore, depending on food web structure primary extinctions within a local habitat patch can result in a cascade of secondary extinctions (Borrvall et al., 2000). Such cascading extinctions can lead to community closure that prevents re-colonization by the originally lost species (Lundberg et al., 2000). The result of species extinctions—secondary extinctions or re-colonization by the originally lost species—depends on the balance between relaxation time and dispersal rate within the metacommunity. Relaxation time is the time elapsed from initial species loss until the new species composition is reached, and long relaxation times means that there will be time available for dispersal of species between local communities and hence time for rescue effects to operate and re-colonization to take place. Recent evidence suggests that relaxation times are longer in diverse communities than in communities with few species. Together, all these studies indicate that metacommunity dynamics differ substantially depending on community diversity and food web structure. Clearly, it will be an important future task to study metapopulation dynamics and the consequences of habitat loss in more complex and thus more realistic communities under varying food web structure.

DISCUSSION AND CONCLUSIONS

Current food web models generally assume that all species within the food web occur everywhere in the local habitat under study (Cohen and Briand, 1984; Martinez, 1992). This causes all potential links—who is able to eat whom—to be realized links—who eats whom. The above spatial considerations, however, suggest that not all species occur everywhere at any time and that some potential links are not realized at any single point in time. As shown in this chapter, a more thorough consideration of space in food webs extends the simple one-way concept of "spatial subsidies" (Polis and Strong, 1996) considerably. The dynamical links between habitats are more generally two-way rather that one-way. Recognition of the importance of spatial scale in food web studies has several implications for food web theory. In particular, the potential food webs that are frequently described by ecologists will often differ from how food webs are realized in actual space and time. Clearly, choosing the right spatio-temporal scale for a food web study depends on the species studied (e.g., their home ranges and locomotive behavior) and the study objective (e.g., local interaction structure versus metacommunity patterns). Although suggestions on spatio-temporal scales and boundary conditions are hard to generalize, we anticipate that our definitions of local versus metacommunity food webs will foster a healthy debate on how to deal with scaling in empirical food web research. Moreover, integrating spatial processes such as extinction and colonization by dispersal in food web models is an important step towards understanding population dynamics in complex communities, and understanding the consequences of habitat loss for the community structure and food web dynamics. Such important spatial characteristics of food webs need to be treated as such if the merging of food web and landscape ecology that Gary Polis and others envisioned (Cousins, 1993; Polis et al., 1996a; Power and Rainey, 2000; Chust et al., 2003) is to be realized.

REFERENCES

Aarssen, L.W. 1997. High productivity in grassland ecosystems: effected by species diversity or productive species? *Oikos* 80, 183–184.

Abrahamsen, G., and Thompson, W.N. 1979. A long-term study of the enchytraeid (Oligochaeta) fauna of a mixed coniferous forest and the effects of urea fertilization. *Oikos* 32, 318–327.

Abrams, P.A. 1982. Functional responses of optimal foragers. *Am. Nat.* 120, 382–390.

Abrams, P.A. 1984. Foraging time optimization and interactions in food webs. *Am. Nat.* 124, 80–96.

Abrams, P.A. 1992. Predators that benefit prey and prey that harm predators: unusual effects of interacting foraging adaptations. *Am. Nat.* 140, 573–600.

Abrams, P.A. 1993. Effects of increased productivity on the abundances of trophic levels. *Am. Nat.* 141, 351–371.

Abrams, P.A. 1995. Implications of dynamically variable traits for identifying, classifying, and measuring direct and indirect effects in ecological communities. *Am. Nat.* 146, 112–134.

Abrams, P.A. 1996. Dynamics and interactions in food webs with adaptive foragers. In: G.A. Polis and K.O. Winemiller (Editors), *Food Webs: Integration of Pattern & Dynamics*. Kluwer Academic Publishers, Norwell, pp 113–121.

Abrams, P.A. 2001. Describing and quantifying interspecific interactions: a commentary on recent approaches. *Oikos* 94, 209–218.

Abrams, P.A., and Roth, J. 1994. The responses of unstable food chains to enrichment. *Evol. Ecol.* 8, 150–171.

Abrams, P.A., and Walters, C.J. 1996. Invulnerable prey and the paradox of enrichment. *Ecology* 77, 1125–1133.

Abrams, P.A., and Matsuda, H. 1996. Positive indirect effects between prey species that share predators. *Ecology* 77, 610–616.

Abrams, P.A., and Vos, M. 2003. Adaptation, density dependence and the responses of trophic level abundances to mortality. *Evol. Ecol. Res.* 5, 1113–1132.

Abrams, P.A., Menge, B.A., Mittelbach, G.G., Spiller, D., and Yodzis, P. 1996. The role of indirect effects in food webs. In: G.A. Polis and K.O. Winemiller, (Editors), *Food Webs: Integration of Patterns and Dynamics.* Chapman and Hall, New York, pp 371–395.

Açacaya, H.R., Arditi, R., and Guinzburg, L.R. 1995. Ratio-dependent predation: an abstraction that works. *Ecology* 76, 995–1004.

Adams, J. 1985. The definition and innterpretation of guild structure in ecological communities. *J. Anim. Ecol.* 54, 43–59.

Adams, T.S., and Sterner, R.W. 2000. The effect of dietary nitrogen content on trophic level d15N enrichment. *Limno. Oceanogr.* 45, 601–607.

Ågren, G.I., and Bosatta, E. 1996. *Theoretical Ecosystem Ecology. Understanding Element Cycles.* Cambridge Univiversity Press, Cambridge.

Ågren, G.I., Bosatta, E., and Magill, A.H. 2001. Combining theory and experiment to understand effects of inorganic nitrogen on litter decomposition. *Oecologia* 128, 94–98.

Akcakaya, H.R., Arditi, R., and Ginzburg, L.R. 1995. Ratio-dependent predation: an abstraction that works. *Ecology* 76, 995–1004.

Albert, R., Jeong, H., and Barabasi, A.L. 2000. Error and attack tolerance of complex networks. *Nature* 406, 378–382.

Alcamo, J. 2002. Introduction to special issue on regional air pollution and climate change in Europe. *Environ. Sci. Policy* 5, 255.

Allen, M.F. 1991. *The Ecology of Mycorrhizae.* Cambridge University Press, Cambridge.

Allesina, S., and Bodini, A.J. 2004. Who dominates whom in the ecosystem? Energy flow bottlenecks and cascading extinctions. *J. Theor. Biol.* 230, 351–358.

Alongi, D.M. 2002. Present state and future of the world's mangrove forests. *Environ. Cons.* 29, 331–349.

Alphei, J., Bonkowski, M., and Scheu, S. 1996. Protozoa, Nematoda and Lumbricidae in the rhizosphere of Hordelymus europaeus (Poaceae): faunal interactions, response of microorganisms and effects on plant growth. *Oecologia* 106, 111–126.

Anderies, J.M., and Beisner, B.E. 2000. Fluctuating environments and phytoplankton community structure: a stochastic model. *Am. Nat.* 155, 556–569.

Andersen, T. 1997. *Pelagic Nutrient Cycles: Herbivores as Sources and Sinks.* Springer-Verlag, New York, NY.

Anderson, J.M. 1975. Sucession, diversity and trophic relationships of some soil animals in decomposing leaf litter. *J. Anim. Ecol.* 44.

Anderson, J.M. 1978. Inter- and intra-habitat relationships between woodland Cryptostigmata species diversity and the diversity of soil and litter microhabitats. *Oecologia* 32, 341–348.

Anderson, J.M. 1988. Spatiotemporal effects of invertebrates on soil processes. *Biol. Fert. Soils* 6, 216–227.

Anderson, J.M. 1995. Soil organisms as engineers: microsite modulation of macroscale processes. In: C.G. Jones and J.H. Lawton (Editors), *Linking Species and Ecosystems.* Chapman & Hall, New York, pp 94–106.

Anderson, J.M., and Ineson, P. 1984. Interactions between microorganisms and soil invertebrates in nutrient flux pathways of forest ecosystems. In: J.M. Anderson, A.D.M.Rayner, and D.W.S. Watson (Editors), *Invertebrate-Microbial Interactions*. Cambridge University Press, Cambridge, pp 89–132.

Anderson, R.V., Coleman, D.C., and Cole, C.V. 1981. Effects of saprophytic grazing on net mineralization. *Ecol. Bull.* 33, 201–216.

Andrzejewska, L. 1979. Herbivorous fauna and its role in the economy of grassland ecosystems. II. The role of herbivores in trophic relationships. *Pol. Ecol. Stud.* 5, 45–76.

Angelescu, V. 1982. Ecología trófica de la anchoita del Mar Argentino (Engraulidae, *Engraulis anchoita*). Parte II. Alimentación, comportamiento y relaciones tróficas en el ecosistema. Contribuciones del Instituto Nacional de Investigación y Desarrollo Pesquero (INIDEP), 409.

Angelescu, V., and Prenski, L.B. 1987. Ecología trófica de la merluza común del Mar Argentino (Merlucciidae, Merluccius hubbsi). Parte 2. Dinámica de la alimentación analizada sobre la base de las condiciones ambientales, la estructura y las evaluaciones de los efectivos en su área de distribución. Contribuciones del Instituto Nacional de Investigación y Desarrollo Pesquero (INIDEP) 561, 205.

Arditi, R., and Ginzburg, L.R. 1989. Coupling in predator prey dynamics-ratio dependence. *J. Theor. Biol.* 139, 311–326.

Arii, K., and Parrott, L. 2004. Emergence of non-random structure in local food webs generated from randomly structured regional webs. *J. Theor. Biol.* 227, 327–333.

Armstrong, R.A., and McGehee, R. 1976. Coexistence of species competing for shared resources. *Theor. Popul. Biol.* 9, 317–328.

Arrington, D.A. 2002. Evaluation of the Relationship Between Habitat Structure, Community Structure, and Community Assembly in a Neotropical Blackwater River. PhD Dissertation. Texas A&M University, College Station, Texas, USA.

Arrington, D.A., and Winemiller, K.O. 2002. Use of preserved biological specimens for stable isotope analysis. *Trans. Am. Fish. Soc.* 131, 337–342.

Arrington, D.A., and Winemiller, K.O. 2003. Diel changeover in sandbank fish assemblages in a Neotropical floodplain river. *J. Fish Biol.* 63, 442–459.

Attree, K. 1998. The development of a technique for measuring the change in energy availability caused by a food chain step in natural systems. MSc. Cranfield University, Cranfield.

Aubone, A., Bezzi, S., Castrucci, R., Dato, C., Ibañez, P., Irusta, G., Pérez, M., Renzi, M., Santos, B., Scarlato, N., Simonazzi, M., L., T., and Villarino, F. 1999. Merluza (*Merluccius hubbsi*). In: J.L. Cajal and L.B. Prenski, (Editors), *Diagnóstico de los recursos pesqueros de la República Argentina*, pp 27–35.

Aumen, N.G., Bottomley, P.J., Ward, G.M., and Gregory, S.V. 1983. Microbial deomposition of wood in streams -Distribution of microflora and factors affecting [C-14] lignocellulose mineralization. *Appl. Environ. Microb.* 46, 1409–1416.

Bahls, P.F. 1992. The status of fish populations and management of high mountain lakes in the western United States. *Northwest Sci.* 66, 183–193.

Baird, D., and Ulanowicz, R.E. 1989. The seasonal dynamics of the Chesapeake Bay ecosystem. *Ecol. Monogr.* 59, 329–364.

Baird, D., and Heymans, J.J. 1996. Assessment of the ecosystem changes in response to freshwater inflow of the Kromme River Estuary, St. Francis Bay, South Africa: a network analysis approach. *Water SA* 22, 307–318.

Baird, D., McGlade, J.M., and Ulanowicz, R.E. 1991. The comparative ecology of six marine ecosystems. *Philos. T. Roy. Soc.* B 333, 15–29.

Baird, D., Luczkovich, J., and Christian, R.R. 1998. Assessment of spatial and temporal variability in ecosystem attributes of the St. Marks National Wildlife Refuge, Apalachee Bay, Florida. Estuar. *Coast. Shelf. Sci.* 47, 329–349.

Baird, D.J., Brock, T.C.M., de Ruiter, P.C., Boxall, A.B.A., Culp, J., M., Eldridge, P., Hommen, U., Jak, R.G., Kidd, K.A., and Dewitt, T. 2001. The food web approach in the environmental management of toxic substances. In: D.J. Baird and G.A. Burton (Editors), *Ecological Variability: Separating Anthropogenic from Natural Causes of Ecosystem Impairment*. SETAC Press, Pensacola, USA, pp 83–122.

Balvanera, P., Daily, G.C., Ehrlich, P.R., Ricketts, T.H., Bailey, S.A., Kark, S., Kremen, C., and Pereira, H. 2001. Conserving biodiversity and ecosystem services. *Science* 291, 2047.

Banasek-Richter, C., Cattin, M.F., and Bersier, L.F. 2004. Sampling effects and the robustness of quantitative and qualitative food-web descriptors. *J. Theor. Biol.* 226, 23–32.

Barabási, A.L., and Albert, R. 1999. Emergence of scaling in random networks. *Science* 286, 509–512.

Barahona, F., Evans, S.E., Mateo, J.A., García-Márquez, M., and López-Jurado, L.F. 2000. Endemism, gigantism and extinction in island lizards: the genus Gallotia on the Canary Islands. *J. Zool.* 250, 373–388.

Barbour, M.T., Gerritsen, J., Snyder, B.D., and Stribling, J.B. 1999. Rapid Bioassessment Protocols for Use in Streams and Wadeable Rivers: Periphyton, Benthic Macroinvertebrates and Fish, Second Edition. EPA 841-B-99-002, U.S. Environmental Protection Agency; Office of Water, Washington, D.C.

Bardgett, R.D., and Wardle, D.A. 2003. Herbivore-mediated linkages between aboveground and belowground communities. *Ecology* 84, 2258–2268.

Barkai, A., and McQuaid, C. 1988. Predator-prey reversal in a marine benthic ecosystem. *Science* 242, 62–64.

Barman, C., and Mayer, C. 1994. An analysis of high school students' concepts and textbook presentations of food chains and food webs. *Amer. Biol. Teacher* 56, 160–163.

Bauer, G., Persson, H., Persson, T., Mund, M., Hein, M., Kummetz, E., Matteucci, G., van Oene, H., Scarascia-Mugnozza, G., and Schulze, E.D. 2000. Linking plant nutrition and ecosystem processes. In: E.D. Schulze (Editor), *Carbon and Nitrogen Cycling in Forest Ecosystems*. Springer, Heidelberg, pp 63–98.

Beare, M.H., Parmelee, R.W., Hendrix, P.F., Cheng, W.X., Coleman, D.C., and Crossley, D.A. 1992. Microbial and faunal interactions and effects on litter nitrogen and decomposition in agroecosystems. *Ecol. Monogr.* 62, 569–591.

Beddington, J. 1975. Mutual interference between parasites or predators and its effect on searching efficiency. *J. Anim. Ecol.* 51, 597–624.

Beeton, A.M. 2002. Large freshwater lakes: present state, trends and future. *Environ. Cons.* 29, 21–38.

Begon, M., Harper, J.L., and Townsend, C.R. 1990. *Ecology: Individuals, Populations, and Communities, 2nd edition.* Blackwell Scientific Publications, London.

Behan-Pelletier, V.M. 1999. Oribatid mite biodiversity in agroecosystems: role of bioindication. *Agr. Ecosyst. Envir.* 74, 411–423.

Beisner, B.E., Ives, A.R., and Carpenter, S.R. 2003. The effects of an exotic fish invasion on the prey communities of two lakes. *J. Anim. Ecol.* 72, 331–342.

Bender, E.A., Case, T.J., and Gilpin, M.E. 1984. Perturbation experiments in community ecology: theory and practice. *Ecology* 64, 1–13.

Bengtsson, J. 1994a. Confounding variables and independent observations in comparative analyses of food webs. *Ecology* 75, 1282–1288.

Bengtsson, J. 1994b. Temporal predictability in forest soil communities. *J. Anim. Ecol.* 63, 653–665.

Bengtsson, J. 1998. Which species? What kind of diversity? Which ecosystem function? Some problems in studies of relations between biodiversity and ecosystem function. *Appl. Soil Ecol.* 10, 191–199.

Bengtsson, J., Setälä, H., and Zheng, D.W. 1996. Food-webs and nutrient cycling in soils: interactions and positive feedbacks. In: G.A. Polis and K.O. Winemiller, (Editors), *Food Webs: Integration of Pattern and Dynamics.* Chapman and Hall, New York, pp 30–38.

Bengtsson, J., Baillie, S., and Lawton, J.H. 1997. Community variability increases with time. *Oikos* 78, 249–256.

Bengtsson, J., Zheng, D.W., G.I., Å., and Persson, T. 1995. Food webs in soil: an interface between population and ecosystem ecology. In: C.G. Jones and J. Lawton (Editors), *Linking Species and Ecosystems.* Chapman and Hall, London, pp 159–165.

Bengtsson, J., Engelhardt, K., Giller, P., Hobbie, S., Lawrence, D., Levine, J., Vilà, M., and Wolters, V. 2002. Slippin' and slidin' between the scales: the scaling components of biodiversity-ecosystem functioning relations. In: M. Loreau, S. Naeem, and P. Inchausti (Editors), *Biodiversity and Ecosystem Functioning: Synthesis and Perspectives.* Cambridge University Press, Cambridge, pp 209–220.

Benke, A.C. 1979. A modification of the Hynes method for estimating secondary production with particular significance for multivoltine populations. *Limno. Oceanogr.* 24, 168–171.

Benke, A.C. 1993. Concepts and patterns of invertebrate production in running waters. *Verh. Internat. Verein. Limnol.* 25, 1–24.

Benke, A.C., and Wallace, J.B. 1990. Wood dynamics in coastal-plain blackwater streams. *Can. J. Fish. Aquat. Sci.* 47, 92–99.

Benke, A.C., and Wallace, J.B. 1997. Trophic basics of production among riverine caddisflies: implications for food web analysis. *Ecology* 78, 1132–1145.

Berg, B., and Matzner, E. 1997. Effect of N deposition on decomposition of plant litter and soil organic matter in forest systems. *Environ. Rev.* 5, 1–25.

Berg, B., Hannus, K., Popoff, T., and Theander, O. 1982. Changes in organic chemical components of needle litter during decomposition Long-term decomposition in Scots pine forest. *I. Can. J. Bot.* 60, 1310–1319.

Berg, M., De Ruiter, P.C., Didden, W.A.M., Janssen, M.P.M., Schouten, A.J., and Verhoef, H.A. 2001. Community food web, decomposition and nitrogen mineralisation in a stratified Scots pine forest soil. *Oikos* 94, 130–142.

Berg, M.P., Kniese, J.P., and Verhoef, H.A. 1998a. Dynamics and stratification of bacteria and fungi in the organic layers of a Scots pine forest soil. *Biol. Fert. Soils* 26, 313–322.

Berg, M.P., Kniese, J.P., Zoomer, R., and Verhoef, H.A. 1998b. Long-term decomposition of successive organic strata in a nitrogen saturated Scots pine forest soil. *For. Ecol. Manag.* 107, 159–172.

Berlow, E.L. 1999. Strong effects of weak interactions in ecological communities. *Nature* 398, 330–334.

Berlow, E.L., Navarrete, A.A., Briggs, C.J., Power, M.E., and Menge, B.A. 1999. Quantifying variation in the strengths of species interactions. *Ecology* 80, 2206–2224.

Berlow, E.L., Neutel, A.M., Cohel, J.E., de Ruiter, P.C., Ebenman, B., Emmerson, M., Fox, J.W., Jansen, V.A.A., Jones, J.I., Kokkoris, G.D., Logofet, D.O., McKane, A.J., Montoya, J.M., and Petchey, O. 2004. Interaction strengths in food webs: issues and opportunities. *J. Anim. Ecol.* in press.

Bernhardt, E.S., Likens, G.E., Buso, D.C., and Driscoll, C.T. 2003. In-stream uptake dampens effects of major forest disturbance on watershed nitrogen export. *P. Natl. Acad. Sci. USA* 100, 10304–10308.

Berryman, A.A. 1992. The origins and evolution of predator prey theory. *Ecology* 73, 1530–1535.

Bersier, L.F., and Sugihara, G. 1997. Scaling regions for food-web properties. *P. Natl. Acad. Sci. USA* 94, 1247–1251.

Bersier, L.-F., Dixon, P., and Sugihara, G. 1999. Scale-invariant of scale-dependent behavior of the link density property in food webs: a matter of sampling effort? *Am. Nat.* 153, 676–682.

Bersier, L.-F., Banasek-Richter, C., and Cattin, M.F. 2002. Quantitative descriptors of food-web matrices. *Ecology* 83, 2394–2407.

Bjornstad, O.N., and Grenfell, B.T. 2001. Noisy clockwork: Time series analysis of population fluctuations in animals. *Science* 293, 638–643.

Blackburn, T.M., and Gaston, K.J. 1997. A critical assessment of the form of the interspecific relationship between abundance and body size in animals. *J. Anim. Ecol.* 66, 233–249.

Blackburn, T.M., and Gaston, K.J. 1998. Some methodological issues in macroecology. *Am. Nat.* 151, 68–83.

Blackburn, T.M., and Gaston, K.J. 1999. The relationship between animal abundance and body size: a review of the mechanisms. *Adv. Ecol. Res.* 28, 181–210.

Blair, J.M., Parmelee, R.W., and Wyman, R.L. 1994. A comparison of the forest floor invertebrate communities of 4 forest types in the Northeast United States. *Pedobiologia* 38, 146–160.

Blondel, J. 2003. Guilds or functional groups: does it matter? *Oikos* 100, 223–231.

Bobbink, R., Hornung, M., and Roelofs, J.G.M. 1998. The effects of air-borne nitrogen pollutants on species diversity in natural and semi-natural European vegetation. *J. Ecol.* 86, 717–738.

Bohannan, B.J.M., and Lenski, R.E. 2000. The relative importance of competition and predation varies with productivity in a model community. *Am. Nat.* 156, 329–340.

Bøhn, T., and Amundsen, P.A. 2004. Ecological interactions and evolution: Forgotten parts of biodiversity? *Bioscience* 54, 804–805.

Bolker, B., Holyoak, M., Krivan, V., Rowe, L., and Schmitz, O. 2003. Connecting theoretical and empirical studies of trait-mediated interactions. *Ecology* 84, 1101–1114.

Bonkowski, M., and Brandt, F. 2002. Do soil protozoa enhance plant growth by hormonal effects? *Soil Biol. Biochem.* 34, 1709–1715.

Bonkowski, M., Geoghegan, I.E., Birch, A.N.E., and Griffiths, B.S. 2001. Effects of soil decomposer invertebrates (protozoa and earthworms) on an above-ground phytophagous insect (cereal aphid) mediated through changes in the host plant. *Oikos* 95, 441–450.

Borer, E.T., Anderson, K., Blanchette, C.A., Broitman, B., Cooper, S.D., and Halpern, B.S. 2002. Topological approaches to food web analyses: a few modifications may improve our insights. *Oikos* 99, 397–401.

Borvall, C., Ebenman, B., and Jonsson, T. 2000. Biodiversity lessens the risk of cascading extinction in model food webs. *Ecol. Lett.* 3, 131–136.

Boulière, F., 1975. Mammals, small and large: the ecological implications of size. In: F.B. Golley, K. Petrusewicz, and L. Ryszkowski (Editors), *Small Mammals: Their Productivity and Population Dynamics*. Cambridge University Press, Cambridge, pp 1–8.

Bowen, S.H. 1983. Detritivory in neotropical fish communities. *Environ. Biol. Fishes* 9, 137–144.

Bowen, S.H., Bonetto, A.A., and Alhlgren, M.O. 1984. Microorganisms and detritus in the diet of a typical neotropical riverine detritivore, Prochilodus platensis (Pisces: Prochilodontidae). *Limno. Oceanogr.* 29, 1120–1122.

Bracewell, K.V., 2000. An exploration of new methods to assess energy availability in the English oak (Quercus robur). Phd. Cranfield University, Cranfield.

Bradford, D.F. 1989. Allotopic distribution of native frogs and introduced fishes in high Sierra Nevada lakes of California: implication of the negative effect of fish introductions. *Copeia* 3, 775–778.

Bradford, D.F., Graber, D.M., and Tabatabai, F. 1994. Population declines of the native frog, Rana muscosa, in Sequoia and Kings Canyon National Parks, California. *Southwest. Nat.* 39, 323–327.

Brett, M.T., and Goldman, C.R. 1997. Consumer versus resource control in freshwater pelagic food webs. *Science* 275, 384–386.

Briand, F., and Cohen, J.E. 1984. Community food webs have scale-invariant structure. *Nature* 307, 264–267.

Briand, F., and Cohen, J.E. 1987. Environmental correlates of food-chain length. *Science* 238, 956–960.

Briand, F., and Cohen, J.E. 1989. Habitat compartmentation and environmental correlates of food-chain length response. *Science* 243, 239–240.

Briggs, C.J. 1993. Competition among parasitoid species on a stage-structured host and its effect on host suppression. *Am. Nat.* 141, 372–397.

Briones, M.J.I., Ineson, P., and Poskitt, J. 1998. Climate change and Cognettia sphagnetorum: effects on carbon dynamics in organic soils. *Func. Ecol.* 12, 528–535.

Brönmark, C., Pettersson, L.B., and Nilsson, P.A. 1999. Predator-induced defense in crucian carp. In: R. Tollrian and C.D. Harvell, (Editors), *The Ecology and Evolution of Inducible Defenses*. Princeton University Press, Princeton New Jersey, pp 89–103.

Brönmark, C., Paszkowksi, C.A., Tonn, W.M., and Hargeby, A. 1995. Predation as a determinant of size structure in populations of crucian carp (*Carassius carassius*) and tench (*Tinca tinca*). *Ecol. Freshw. Fish* 4, 85–92.

Brooks, J.L., and Dodson, S.I. 1965. Predation, body size, and composition of plankton. *Science* 150, 28–35.

Brose, U., Williams, R.J., and Martinez, N.D. 2003. Comment on "Foraging Adaptation and the Relationship Between Food-Web Complexity and Stability". *Science* 301, 918b.

Brose, U., Ostling, A., Harrison, K., and Martinez, N.D. 2004. Unified spatial scaling of species and their trophic interactions. *Nature* 428, 167–171.

Brown, D.E. 1994. *Biotic Communities: Southwestern United States and Northwestern Mexico*. University of Utah Press, Salt Lake City.

Brown, J.H. 1995a. *Macroecology*. University of Chicago Press, Chicago,

Brown, J.H. 1995b. Organisms and species as complex adaptive systems: linking the biology of populations with the physics of ecosystems. In: C.G. Jones and J.H. Lawton (Editors), *Linking Species and Ecosystems*. Chapman and Hall, New York, pp 6–29.

Brown, J.H., and Gillooly, J.F. 2003. Ecological food webs: high-quality data facilitate theoretical unification. *P. Natl. Acad. Sci. USA* 100, 1467–1468.

Brown, J.H., West, G.B., and Enquist, B.J. 2000. Scaling in biology: patterns and processes, causes and consequences. In: J.H. Brown and G.B. West (Editors), *Scaling in Biology*. Oxford University Press, New York, pp 1–24.

Brown, J.H., Gillooly, J.F., Allen, A.P., Svage, V.M., and West, G.B. 2004. Toward a metabolic theory of ecology. *Ecology* 85, 1771–1789.

Brown, V.K., and Gange, A. 2002. Tritrophic below- and above-ground interactions in succession. In: T. Tscharntke and B.A. Hawkins (Editors), *Multitrophic Interactions in Terrestrial Systems*. Cambridge University Press, Cambridge, pp 197–222.

Bruno, J.F., Stachowicz, J.J., and Bertness, M.D. 2003. Inclusion of facilitation into ecological theory. *Trends Ecol. Evol.* 18, 119–125.

Brussaard, L., Behan-Pelletier, V.M., Bignell, D.E., Brown, V.K., Didden, W., Folgarait, P., Fragoso, C., Freckman, D.W., Gupta, V.V.S.R., Hattori, T., Hawksworth, D.L., Klopatek, C., Lavelle, P., Malloch, D.W., Rusek, J., Soderstrom, B., Tiedje, J.M., and Virginia, R.A. 1997. Biodiversity and ecosystem functioning in soil. *Ambio* 26, 563–570.

Bunn, S.E., Davies, P.M., and Winning, M. 2003. Sources of organic carbon supporting the food web of an arid zone floodplain river. *Freshwat. Biol.* 4, 619–635.

Burnham, K.P., and Anderson, D.R., 2002. *Model Selection and Multimodel Inference. A Practical Information-Theoretic Approach*. Springer, New York, p 488.

Burton, A.J., Pregitzer, K.S., and Hendrick, R.L. 2001. Relationships between fine root dynamics and nitrogen availability in Michigan northern hardwood forests. *Oecologia* 125, 389–399.

Byström, P., and Gàrcia-Berthóu, E. 1999. Density dependent growth and stage-specific competitive interactions in young fish. *Oikos* 86.

Cabana, G., and Rasmussen, J.B. 1996. Comparison of aquatic food chains using nitrogen isotopes. *P. Natl. Acad. Sci.* USA 93, 10844–10847.

Cadisch, G., and Giller, K.E. 1997. *Driven by Nature: Plant Litter Quality and Decomposition*. CAB International, Wallingford.

Cairns, J.J., and Pratt, J.R. 1993. A history of biological monitoring using benthic macroinvertebrates. In: D.M. Rosenberg and V.H. Resh (Editors), *Biomonitoring and Benthic Invertebrates*. Chapman and Hall, New York, pp 10–27.

Caldarelli, G., Higgs, P.G., and McKane, A.J. 1998. Modelling coevolution in multispecies communities. *J. Theor. Biol.* 193, 345–358.

Calder, W.A. 1984. *Size, Function and Life History*. Harvard University Press, Cambridge, Massachusets.

California Division of Mines and Geology, 1958. *Geologic Atlas of California*. California Department of Conservation, Sacramento, California, USA.

Callaway, R.M., Brooker, R.W., Choler, P., Kikvidze, Z., Lortie, C.J., Michalet, R., Paolini, L., Pugnaire, F.I., Newingham, B., Aschehoug, E.T., Armas, C., Kikodze, D., and Cook, B.J. 2002. Positive interactions among alpine plants increase with stress. *Nature* 417, 844–848.

Camacho, J., Guimerà, R., and Amaral, L.A.N. 2002a. Robust patterns in food web structure. *Phys. Rev. Lett.* 88.

Camacho, J., Guimerà, R., and Amaral, L.A.N. 2002b. Analytical solution of a model for complex food webs. *Phys. Rev. E* 65, 030901.

Camerano, L. 1880. Dell'equilibrio dei viventi mercè la reciproca distruzione. *Atti. Accad. Sci.* 15, 393–414.

Camphens, J., and Mackay, D. 1997. Fugacity-based model of PCB bioaccumulation in complex aquatic food webs. *Environ. Sci. Technol.* 31, 577–583.

Cardinale, B.J., Palmer, M.A., and Collins, S.L. 2002. Species diversity enhances ecosystem functioning through interspecific facilitation. *Nature* 415, 426–429.

Carpenter, S.J., and Kitchell, J.F. 1993a. *The Trophic Cascade in Lakes*. Cambridge Univ. Press, Cambridge, pp 385.

Carpenter, S.R. 1992. Destabilization of planktonic ecosystems and blooms of blue-green algae. In: J.F. Kitchell (Editor). *Food Web Management*. Springer-Verlag, New York, pp 461–481.

Carpenter, S.R., and Kitchell, J.F. 1988. Consumer control of lake productivity. *Bioscience* 38, 764–769.

Carpenter, S.R., and Kitchell, J.F. 1993b. Plankton community structure and limnetic primary production. *Am. Nat.* 124, 159–172.

Carpenter, S.R., Kitchell, J.F., and Hodgson, J.R. 1985. Cascading trophic interactions and lake productivity. *Bioscience* 35, 634–639.

Carpenter, S.R., Frost, T.M., Heisey, D., and Kratz, T.K. 1989. Randomized intervention analysis and the interpretation of whole-ecosystem experiments. *Ecology* 70, 1142–1152.

Carpenter, S.R., Kraft, C.E., Wright, R., He, X., Soranno, P.A., and Hodgson, J.R. 1992. Resilience and resistance of a lake phosphorus cycle before and after food web manipulation. *Am. Nat.* 140, 781–798.

Carscadden, J.E., Frank, K.T., and Leggett, W.C. 2001. Ecosystem changes and the effects on capelin (*Mallotus villosus*), a major forage species. *Can. J. Fish. Aquat. Sci.* 58, 73–85.

Case, T.J. 1990. Invasion resistance arises in strongly interacting species-rich model competition communities. *P. Natl. Acad. Sci. USA* 87, 9610–9614.

Case, T.J. 2000. *An Illustrated Guide to Theoretical Ecology.* Oxford University Press, Oxford.

Cattin, M.F., Blandenier, G., Banašek-Richter, C., and Bersier, L.F. 2003. Effects of mowing on the spider (Araneae) community as management practice in wet meadows. *Biol. Con.* 113, 179–188.

Cattin, M.F., Bersier, L.F., Banasek-Richter, C., Baltensberger, R., and Gabriel, J.P. 2004. Phylogenetic constraints and adaptation explain food-web structure. *Nature* 427, 835–839.

Chalcraft, D.R., and Resetarits, W.J. 2003. Predator identity and ecological impacts: Functional redundancy or functional diversity? *Ecology* 84, 2407–2418.

Chapin, F.S.I., Bret-Harte, M.S., Hobbie, S.E., and Hailan, Z. 1996. Plant functional types as predictors of transient responses of arctic vegetation to global change. *J. Veg. Sci* 7, 347–358.

Chapin, F.S.I., Sala, O.E., Burke, I.C., Grime, J.P., Hooper, D.U., Lauenroth, W.K., Lombard, A., Mooney, H.A., Mosier, A.R., Naeem, S., Pacala, S.W., Roy, J., Steffen, W.L., and Tilman, D. 1998. Ecosystem consequences of changing biodiversity: experimental evidence and a research agenda for the future. *Bioscience* 48, 45–52.

Chapin III, F.S., Zavaleta, E.S., Eviner, V.T., Naylor, R.L., Vitousek, P.M., Reynolds, H.H., Hooper, D.U., avorel, S., Sala, O.E., Hobbie, S.E., Mack, M.C., and Diaz, S. 2000. Consequences of changing biodiversity. *Nature* 405, 234–242.

Charnov, E.L. 1976. Optimal foraging: Attack strategy of a Mantid. *Am. Nat.* 110, 141–151.

Chase, J.M. 2000. Are there real differences among aquatic and terrestrial food webs? *Trends Ecol. Evol.* 15, 408–412.

Chase, J.M., Leibold, M.A., and Simms, E. 2000a. Plant tolerance and resistance in food webs: community-level predictions and evolutionary implications. *Evol. Ecol.* 14, 289–314.

Chase, J.M., Leibold, M.A., Downing, A.L., and Shurin, J.B. 2000b. The effects of productivity, herbivory, and plant species turnover in grassland food webs. *Ecology* 81, 2485–2497.

Chase, J.M., Abrams, P., Grover, J., Diehl, S., Chesson, P., Holt, R., Richards, S.A., Nisbet, R.M., and Case, T.J. 2002. The interaction between predation and competition: a review and synthesis. *Ecol. Lett.* 5, 302–315.

Chen, X., and Cohen, J.E. 2001a. Global stability, local stability and permanence in model food webs. *J. Theor. Biol.* 212, 223–235.

Chen, X., and Cohen, J.E. 2001b. Transient dynamics and food-web complexity in the Lotka Volterra cascade model. *P. Roy. Soc. Lond. B Biol.* 268, 869–877.

Chesson, J. 1983. The estimation and analysis of preference and its relationship to foraging models. *Ecology* 64, 1297–1304.

Chesson, P. 2000. Mechanisms and maintenance of diversity. *Ann. Rev. Ecol. Syst.* 31, 343–366.

Chesson, P., and Huntly, N. 1997. The role of harsh and fluctuating conditions in the dynamics of ecological communities. *Am. Nat.* 150, 519–553.

Chesson, P.L., and Huntley, N. 1989. Short term instabilities and long-term community dynamics. *Trends Ecol. Evol.* 4, 293–298.

Chmielewski, K. 1995. Microorganisms and enzymatic activity of soil under the influence of phytophagous insects. *Acta Zool. Fenn.* 196, 146–149.

Christensen, V., and Pauly, D. 1992. ECOPATH -a software for balancing steady-state ecosystem models and calculating network characteristics. *Ecol. Model.* 61, 169–185.

Christian, R.R., and Thomas, C.R. 2003. Network analysis of nitrogen inputs and cycling in the Neuse River Estuary, North Carolina, USA. *Estuaries* 26, 815–828.

Chust, G., Pretus, J.L., Ducrot, D., Bedòs, A., and Deharveng, L. 2003. Identification of landscape units from an insect perspective. *Ecography* 26, 257–268.

Claessen, D., and De Roos, A.M. 2003. Bistability in a size-structured population model of cannibalistic fish -a continuation study. *Theor. Popul. Biol.* 64, 49–65.

Claessen, D., De Roos, A.M., and Persson, L. 2000. Dwarfs and giants—cannibalism and competition in size-structured populations. *Am. Nat.* 155, 219–237.

Claessen, D., De Roos, A.M., and Person, L. 2004. Population dynamic theory of size dependent cannibalism. *P. Roy. Soc. Lond. B Biol.* 271, 333–340.

Claessen, D., van Oss, C., De Roos, A.M., and Person, L. 2002. The impact of size-dependent predation on population dynamics and individual life history. *Ecology* 83, 1660–1675.

Clarholm, M. 1985. Interactions of bacteria, Protozoa and plants leading mineralization of soil nitrogen. *Soil Biol. Biochem.* 17, 181–187.

Clays-Josserand, A., Lensi, R., and Gourbiére, F. 1988. Vertical distribution of nitrification potential in an acid forest soil. *Soil Biol. Biochem.* 20, 405–406.

Clements, W.H., and Kiffney, P.M., 1996. Validation of whole effluent toxicity tests: Integrated studies using field assessments, microcosms, and mesocosms. In: D.R. Grothe, K.L. Dickson, and D.K. Reed-Judkins (Editors), *Whole Effluent Toxicity Testing: An Evaluation of Methods and Prediction of Receiving System Impacts.* SETAC Press, Pensacola, Florida, pp 229–244.

Closs, G.P., and Lake, P.S. 1994. Spatial and temporal variation in the structure of an intermittent-stream food web. *Ecol. Monogr.* 64, 1–21.

Clutton-Brock, M. 1967. Likelihood distributions for estimating functions when both variables are subject to error. *Technometrics* 9, 261–269.

Cochran, V.L., Horton, K.A., and Cole, C.V. 1988. An estimation of microbial death rate and limitations of N or C during wheat straw decomposition. *Soil Biol. Biochem.* 20, 293–298.

Cochran-Stafira, D.L., and von Ende, C.N. 1998. Integrating bacteria into food webs: studies with Sarracenia purpurea inquilines. *Ecology* 79.

Cohen, J.E. 1977a. Food webs and dimensionality of trophic niche space. *P. Natl. Acad. Sci. USA* 74, 4533–4536.

Cohen, J.E. 1977b. Ratio of prey to predators in community food webs. *Nature* 270, 165–166.

Cohen, J.E. 1978. *Food Webs in Niche Space.* Princeton University Press, Princeton, NJ,

Cohen, J.E. 1989. Food webs and community structure. In: J. Roughgarden, R.M. May, and S.A. Levin, (Editors), *Perspectives on ecological theory.* Princeton University Press, Princeton, New Jersey, USA, pp 181–202.

Cohen, J.E. 1994. Marine and continental food webs: three paradoxes. *Philos. T. Roy. Soc. B.* 343, 57–69.

Cohen, J.E., and Briand, F. 1984. Trophic links of community food web. *P. Natl. Acad. Sci. USA* 81, 4105–4109.

Cohen, J.E., and Newman, C.M. 1984. The stability of large random matrices and their products. *Ann. Probab.* 12, 283–310.

Cohen, J.E., and Newman, C.M. 1985a. When will a large complex system be stable? *J. Theor. Biol.* 113, 153–156.

Cohen, J.E., and Newman, C.M. 1985b. A stochastic theory of community food webs. I. Models and aggregated data. *P. Roy. Soc. Lond. B Biol.* 224, 421–448.

Cohen, J.E., and Newman, C.M. 1988. Dynamic basis of food web organization. *Ecology* 69, 1655–1664.

Cohen, J.E., and Luczak, T. 1992. Trophic levels in community food webs. *Evol. Ecol.* 6, 73–89.

Cohen, J.E., Briand, F., and Newman, C.M. 1990a. *Community Food Webs: Data and Theory.* Springer-Verlag, New York.

Cohen, J.E., Jonsson, T., and Carpenter, S.R. 2003. Ecological community description using food web, species abundance, and body-size. *P. Natl. Acad. Sci. USA* 100, 1781–1786.

Cohen, J.E., Luczak, T., Newman, C.M., and Zhou, Z.M. 1990b. Stochastic structure and nonlinear dynamics of food webs: qualitative stability in a Lotka-Volterra cascade model. *P. Roy. Soc. Lond. B Biol.* 240, 607–627.

Cohen, J.E., Pimm, S.L., Yodzis, P., and Saldaña, J. 1993a. Body sizes of animal predators and animal prey in food webs. *J. Anim. Ecol.* 62, 67–78.

Cohen, J.E., Jonsson, T., Muller, C.B., Godfray, H.C.J., and Savage, V.M. 2005. Body sizes of hosts and parasitoids in individual feeding relationships. *P. Natl. Acad. Sci. USA* 102, 684–689.

Cohen, J.E., Beaver, R.A., Cousins, S.H., DeAngelis, D.L., Goldwasser, L., Heong, K.L., Holt, R., Kohn, A.J., Lawton, J.H., Martinez, N., Omalley, R., Page, L.M., Patten, B.C., Pimm, S.L., Polis, G.A., Rejmánek, M., Schoener, T.W., Schoenly, K., Sprules, W.G., Teal, J.M., Ulanowicz, R.E., Warren, P.H., Wilbur, H.M., and Yodzis, P. 1993b. Improving food webs. *Ecology* 74, 252–258.

Coleman, D.C. 1996. Energetics of detritivory and microbivory in soil in theory and practice. In: G.A.Polis and K.O. Winemiller (Editors), *Food Webs: Integration of Pattern and Dynamics.* Chapman and Hall, New York, pp 39–50.

Coleman, D.C., and Crossley, D.A.J. 1996. *Fundamentals of Soil Ecology*. Academic Press, San Diego.

Coleman, D.C., Reid, C.P.P., and Cole, C.V. 1983. Biological strategies of nutrient cycling in soil systems. *Adv. Ecol. Res.* 13, 1–55.

Colinvaux, P. 1978. *Why Big Fierce Animals Are Rare: An Ecologist's Perspective*. Princeton University Press, Princeton.

Collins, S.L. 2000. Disturbance frequency and community stability in native tallgrass prairie. *Am. Nat.* 155, 311–325.

Connell, J.H. 1983. On the prevalence and relative importance of interspecific competition: evidence from field experiments. *Am. Nat.* 122, 661–696.

Costantino, R.F., and Desharnais, R.A. 1991. *Population Dynamics and the Tribolium Model: Genetics and Demography*. Springer-Verlag, New York.

Costantino, R.F., Desharnais, R.A., Cushing, J.M., and Dennis, B. 1997. Chaotic dynamics in an insect population. *Science* 275, 389–391.

Costanza, R., d'Arge, R., Groot, R.d., Grasso, M., Hannon, B., Limburg, K., Naeem, S., O'Neill, R.V., Paruelo, J., Raskin, R.G., Sutton, P., and Belt, M.V.D. 1997. The value of the world's ecosystem services and natural capital. *Nature* 387, 253–260.

Cotgrave, P. 1993. The relationship between body size and population abundance in animals. *Trends Ecol. Evol.* 8, 244–248.

Courchamp, F., Clutton-Brock, T., and Grenfell, B. 1999. Inverse density dependence and the Allee effect. *Trends Ecol. Evol.* 14, 405–410.

Courchamp, F., Chapuis, J.L., and Pascal, M. 2003. Mammal invaders on islands: impact, control and control impact. *Biol. Rev.* 78, 347–838.

Cousins, S.H. 1980. A trophic continuum derived from plant structure, animal size and a detritus cascade. *J. Theor. Biol.* 82, 607–618.

Cousins, S.H. 1985. The trophic continuum in marine ecosystems: structure and equations for a predictive model. *Can. Bull. Fish. Aquat. Sci.* 213, 76–93.

Cousins, S.H. 1987. The decline of the trophic level concept. *Trends Ecol. Evol.* 2, 312–316.

Cousins, S.H. 1990. Countable ecosystems deriving from a new food web entity. *Oikos* 57, 270–275.

Cousins, S.H. 1993. Hierarchy in ecology: its relevance to landscape ecology and geographical information systems. In: R. Haines-Young, D.R. Green, and S.H. Cousins (Editors), *Landscape Ecology and GIS*. Taylor & Francis Ltd, London, pp 75–86.

Cousins, S.H. 1995. Ecologists build new pyramids again. *New Sci.* 4, 50–54.

Cousins, S.H. 1996. Food webs: from the Lindeman paradigm to a taxonomic general theory of ecology. In: G.A. Polis and K.O. Winemiller (Editors), *Food Webs: Integration of Pattern and Dynamics*. Chapman and Hall, New York, pp 243–251.

Crespo, E.A., and Pedraza, S.N. 1991. Estado actual y tendencia de la población de lobos marinos de un pelo (*Otaria flavescens*) en el litoral norpatagónico. *Ecologia Austral.* 1, 87–95.

Critchlow, R.E., and Stearns, S.C. 1982. The structure of food webs. *Am. Nat.* 120, 478–499.

Cross, W.F., Benstead, J.P., Rosemond, A.D., and Wallace, J.B. 2003. Consumer-resource stoichiometry in detritus-based streams. *Ecol. Lett.* 6, 721–732.

Crowder, L.B., Reagan, D.P., and Freckman, D.W. 1996. Food web dynamics and applied problems. In: G.A. Polis and K.O. Winemiller (Editors), *Food Webs: Integration of Patterns and Dynamics*. Chapman and Hall, New York, pp 327–336.

Crowl, T.A., McDowell, W.H., Covich, A.P., and Johnson, S.L. 2001. Freshwater shrimp effects on detrital processing and nutrients in a tropical headwater stream. *Ecology* 82, 775–783.

Cryer, M., Peirson, G., and Townsend, C.R. 1986. Reciprocal interactions between roach, Rutilus rutilus, and zooplankton in a small lake: Prey dynamics and fish growth and recruitment. *Limno. Oceanogr.* 31, 1022–1038.

Cuddington, K.M., and Yodzis, P. 1999. Black noise and population persistence. *P. Roy. Soc. Lond. B Biol.* 266, 2551.

Cuffney, T.F., Wallace, J.B., and Lugthart, G.J. 1990. Experimental evidence quantifying the role of benthic invertebrates in organic matter dynamics of headwater streams. *Freshwat. Biol.* 23.

Culp, J.M., Lowell, R.B., and Cash, K.J. 2000a. Integrating in situ community experiments with field studies to generate weight-of-evidence risk assessments for large rivers. *Environmental Toxicology and Chemistry* 19, 1167–1173.

Culp, J.M., Podemski, C.L., and Cash, K.J. 2000b. Interactive effects of nutrients and contaminants from pulp mill effluents on riverine benthos. *J. Aquatic Ecosyst. Stress Recov.* 8, 67–75.

Culp, J.M., Cash, K.J., Glozier, N.E., and Brua, R.B. 2003a. Effects of pulp mill effluent on benthic assemblages in mesocosms along the St. John River, Canada. Environmental Toxicology and Chemistry 12, 2916–2925.

Culp, J.M., Cash, K.J., Dubé, M.G., Glozier, N.E., and MacLatchy, D.L. 2003b. Alternative approaches for cumulative effects bioassessment. *Tox. Subst. Res. Init.* Cumulative effects of toxic substances.

Culp, J.M., Glozier, N.E., Cash, K.J., Dubé, M.G., MacLatchy, D.L., and Brua, R.B. 2004. Cumulative effects investigation of pulp mill and sewage effluent impacts on benthic food webs: a mesocosm example. In: D.L. Borton, T.J. Hall, R.P. Fisher, and J.F. Thomas, (Editors), *Pulp and Paper Mill Effluent Environmental Fate and Effects*. DEStech Publications, Inc., Lancaster, PA, USA, pp 464–472.

Cummins, K.W., and Klug, M.J. 1979. Feeding ecology of stream invertebrates. *Ann. Rev. Ecol. Syst.* 10.

Currin, C.A., Wainwright, S.C., Able, K.W., Weinstein, M.P., and Fuller, C.M. 2003. Determination of food web support and trophic position of the mummichog, Fundulus heteroclitus, in New Jersey smooth cordgrass (*Spartina alterniflora*), common reed (*Phragmites australis*), and restored marshes. *Estuaries* 26, 495–510.

Cushing, J.M., Henson, S.M., Desharnais, R.A., Dennis, B., Costantino, R.F., and King, A. 2001. A chaotic attractor in ecology: theory and experimental data. *Chaos Soliton. Frac.* 12, 219–234.

Cyr, H. 2000. Individual energy use and the allometry of population density. In: J.H. Brown and G.B. West (Editors), *Scaling in Biology*. Oxford University Press, New York, pp 267–298.

Cyr, H., Downing, J.A., and Peters, R.H. 1997a. Density-body size relationships in local aquatic communities. *Oikos* 79, 333–346.

Cyr, H., Peters, R.H., and Downing, J.A. 1997b. Population density and community size structure: comparison of aquatic and terrestrial systems. *Oikos* 80, 139–149.

Daily, G.C., Editors. 1997. *Nature's Services: Societal Dependence on Natural Ecosystems.* Island Press, Washington D.C.

Dame, R., and Libes, S. 1993. Oyster reefs and nutrient retention in tidal creeks. *J. Exp. Mar. Biol. Ecol.* 171, 251–258.

Damuth, J. 1981. Population density and body size in mammals. *Nature* 290, 699–700.

Dans, S.L., Crespo, E.A., Pedraza, S.N., and Koen-Alonso, M. 2004. Recovery of the South American sea lion (*Otaria flavescens*) population in northern Patagonia. *Can. J. Fish. Aquat. Sci.* 61, 1681–1690.

Dans, S.L., Koen-Alonso, M., Crespo, E.A., Pedraza, S.N., and García, N.A. 2003. Interactions between marine mammals and high seas fisheries in Patagonia under an integrated approach. In: N. Gales, M. Hindell, and R. Kirkwood, (Editors), *Marine Mammals and Humans: Fisheries, Tourism, and Management.* CSIRO Publishing, Melbourne, Australia, pp 100–115.

Darwin, C. 1859. *The Origin of Species by Means of Natural Selection.* John Murray, London.

Daugherty, M.P., and Juliano, S.A. 2002. Testing for context-dependence in a processing chain interaction among detritus-feeding aquatic insects. *Ecol. Entomol.* 27, 541–553.

David, J.F., Malet, N., Couteaux, M.M., and Roy, J. 2001. Feeding rates of the woodlouse *Armadillidium vulgare* on herb litters produced at two levels of atmospheric CO2. *Oecologia* 127, 343–349.

Davis, M.A., and Pelsor, M. 2001. Experimental support for a resource-based mechanistic model of invisibility. *Ecol. Lett.* 4, 421–428.

Davis, M.A., Grime, J.P., and Thompson, K. 2000. Fluctuating resources in plant communities: a general theory of invasibility. *J. Ecol.* 88, 528–534.

Dawson, T.E., and Ehleringer, J.R. 1991. Streamside trees that do not use stream water. *Nature* 350, 335–337.

de Haas, E.M., Reuvers, B., Moermond, C.T.A., Koelmans, A.A., and Kraak, M.H.S. 2002. Responses of benthic invertebrates to combined toxicant and food input in floodplain lake sediments. *Environmental Toxicology and Chemistry* 21, 2165–2171.

de Mazancourt, C., and Loreau, M. 2000. Effect of herbivory and plant species replacement on primary production. *Am. Nat.* 155, 735–754.

de Mazancourt, C., Loreau, M., and Abbadie, L. 1998. Grazing optimization and nutrient cycling: when do herbivores enhance plant production? *Ecology* 79, 2242–2252.

De Roos, A.M. 1997. A gentle introduction to physiologically structured population models. In: S. Tuljapurkar and H. Caswell, (Editors), *Structured-Population Models in Marine, Terrestrial, and Freshwater Systems.* Chapman & Hall, New York, pp 119–204.

De Roos, A.M., and Persson, L. 2001. Physiologically structured models—from versatile technique to ecological theory. *Oikos* 94, 51–71.

de Roos, A.M., and Persson, L. 2002. Size-dependent life-history traits promote catastrophic collapses of top predators. *P. Natl. Acad. Sci. USA* 99, 12907–12912.

De Roos, A.M., Persson, L., and Thieme, H.R. 2003a. Emergent Allee effects in top predators feeding on structured prey populations. *P. Roy. Soc. Lond. B Biol.* 270, 611–618.

De Roos, A.M., Persson, L., and McCauley, E. 2003b. The influence of size-dependent life-history traits on the structure and dynamics of populations and communities. *Ecol. Lett.* 6, 473–487.

De Roos, A.M., Metz, J.A.J., Evers, E., and Leipoldt, A. 1990. A size dependent predator–prey interaction: who pursues whom? *J. Math. Biol.* 28, 609–643.

de Ruiter, P., Neutel, A.M., and Moore, J.C. 1995. Energetics, patterns of interaction strengths, and stability in real ecosystems. *Science* 269, 1257–1260.

de Ruiter, P.C., and van Faassen, H.G. 1994. A comparison between an organic matter dynamics model and a food web model simulating nitrogen mineralization in agro-ecosystem. *Eur. J. Agron.* 3, 347–354.

de Ruiter, P.C., Neutel, A.M., and Moore, J.C. 1994. Modelling food webs and nutrient cycling in agro-ecosystems. *Trends Ecol. Evol.* 9, 378–383.

de Ruiter, P.C., Neutel, A.M., and Moore, J.C. 1998. Biodiversity in soil ecosystems: the role of energy flow and community stability. *Appl. Soil Ecol.* 10, 217–228.

de Ruiter, P.C., Wolters, V., Moore, J.C., and Winemiller, K.O. 2005. Food web ecology: playing Jenga and beyond. *Science* 309, 68–71.

de Ruiter, P.C., Van Veen, J.A., Moore, J.C., Brussaard, L., and Hunt, H.W. 1993a. Calculation of nitrogen mineralization in soil food webs. *Plant Soil* 157, 263–273.

de Ruiter, P.C., Moore, J.C., Bloem, J., Zwart, K.B., Bouwman, L.A., Hassink, J., De Vos, J.A., Marinissen, J.C.Y., Didden, W.A.M., Lebbink, G., and Brussaard, L. 1993b. Simulation of nitrogen dynamics in the belowground food webs of two winter-wheat fields. *J. Ap. Ecol.* 30, 95–106.

De Valpine, P. 2003. Better inferences from population-dynamics experiments using Monte Carlo state-space likelihood methods. *Ecology* 84, 3064–3077.

De Valpine, P. 2004. Monte Carlo state-space likelihoods by weighted posterior kernel density estimation. *J. Amer. Stat. Assoc.* 99, 523–536.

De Valpine, P., and Hastings, A. 2002. Fitting population models incorporating process noise and observation error. *Ecol. Monogr.* 72, 57–76.

DeAngelis, D.L. 1975. Stability and connectance in food web models. *Ecology* 56, 238–243.

DeAngelis, D.L. 1980. Energy flow, nutrient cycling, and ecosystem resilience. *Ecology* 61, 764–771.

DeAngelis, D.L. 1992a. *Dynamics of nutrient cycling and food webs.* Chapman & Hall, London.

DeAngelis, D.L. 1992b. *Dynamics in Food Webs and Nutrient Cycling.* Chapman & Hall, London, UK.

DeAngelis, D.L., Goldstein, R.L., and O'Neill, R.V. 1975. A model for trophic interaction. *Ecology* 56, 881–892.

DeAngelis, D.L., Post, W., and Sugihara, G. 1982. *Current Trends in Food Web Theory.* 5983, Oak Ridge National Laboratory, Oak Ridge, TN.

DeAngelis, D.L., Post, W., and Sugihara, G. 1983. *Current Trends in Food Web Theory.* 5983, Oak Ridge National Laboratory, Oak Ridge, TN.

DeAngelis, D.L., Bartell, S.M., and Brenkert, A.L. 1989a. Effects of nutrient recycling and food chain length on resilience. *Am. Nat.* 134, 778–805.

DeAngelis, D.L., Gardner, R.H., Mankin, J.B., Post, W.M., and Carney, J.H. 1978. Energy-flow and number of trophic levels in ecological communities. *Nature* 273, 406–407.

DeAngelis, D.L., Mulholland, P.J., Palumbo, A.V., Steinman, A.D., Huston, M.A., and Elwood, J.W. 1989b. Nutrient dynamics and food web stability. *Ann. Rev. Ecol. Syst.* 20, 71–95.

Deb, D. 1995. Scale-dependence of food web structures: Tropical ponds a paradigm. *Oikos* 72, 245–262.

Deb, D. 1997. Trophic uncertainty vs parsimony in food web research. *Oikos* 78, 191–194.

Deegan, L.A. 1993. Nutrient and energy transport between estuaries and coastal marine ecosystems by fish migration. *Can. J. Fish. Aquat. Sci.* 50, 74–79.

Den Boer, P.J. 1977. Dispersal power and survival. Carabids in a cultivated country-side. In: *Miscellaneous Papers,* Landbouwhogeschool, Wageningen, The Netherlands, pp 1–189.

DeNiro, M.J., and Epstein, S. 1981. Influence of diet on the distribution of nitrogen isotopes in animals. *Geochim. Cosmochim. Ac.* 45, 341–351.

Dennis, B., Desharnais, R.A., Cushing, J.M., Henson, S.M., and Costantino, R.F. 2001. Estimating chaos and complex dynamics in an insect population. *Ecol. Monogr.* 71, 277–303.

Dethier, M.N., and Duggins, D.O. 1984. An indirect commensalisms between marine herbivores and the importance of competitive hierarchies. *Am. Nat.* 124, 205–219.

deYoung, B., Heat, M., Werner, F., Chai, F., Megrey, B., and Monfray, P. 2004. Challenges of modeling ocean basin ecosystems. *Science* 304, 1463–1466.

Díaz, S., and Cabido, M. 1997. Plant functional types and ecosystem function in relation to global change. *J. Veg. Sci* 8, 463–474.

Díaz, S., and Cabido, M. 2001. Vive la différence: plant functional diversity matters to ecosystem processes. *Trends Ecol. Evol.* 16, 646–655.

Dicke, M., and Vet, L.E.M. 1999. Plant-carnivore interactions: evolutionary and ecological consequences for plant, herbivore and carnivore. In: H. Olff (Editor). *Herbivores: Between Plants and Predators.* Blackwell Science, Oxford, pp 483–520.

Dickie, L.M., Kerr, S.R., and Boudreau, P.R. 1987. Size-dependent processes underlying regularities in ecosystem structure. *Ecol. Monogr.* 57, 233–250.

Diehl, S. 1992. Fish predation and benthic community structure—the role of omnivory and habitat complexity. *Ecology* 73, 1464–1661.

Diehl, S. 1995. Direct and indirect effects of omnivory in a littoral lake community. *Ecology* 76, 1727–1740.

Dighton, J. 1978. Effects of synthetic lime aphid honeydew on populations of soil organisms. *Soil Biol. Biochem.* 10, 369–376.

Doak, D.F., Bigger, D., Harding, E.K., Marvier, M.A., O'Malley, R.E., and Thomson, D. 1998. The statistical inevitability of stability-diversity relationships in community ecology. *Am. Nat.* 151, 264–276.

Doedel, E.J., Keller, H.B., and Kernévez, J.P. 1991a. Numerical analysis and control of bifurcation problems: (I) Bifurcation in finite dimensions. *Int. J. Bifurcat. Chaos* 1, 493–520.

Doedel, E.J., Keller, H.B., and Kernévez, J.P. 1991b. Numerical analysis and control of bifurcation problems: (II) Bifurcation in infinite dimensions. *Int. J. Bifurcat. Chaos* 1, 745–772.

Doedel, E.J., Champneys, A.R., Fairgrieve, T.F., Kuznetsov, Y.A., Sandstede, B., and Wang, X. 1998. AUTO 97: Continuation and Bifurcation Software for Ordinary Differential Equations. *in.*

Downing, A.L. 2001. The Role of Biological Diversity for the Functioning and Stability of Pond Ecosystems. Dissertation. University of Chicago, Chicago, Illinois.

Downing, A.L. 2005. Relative effects of species composition and richness on ecosystem properties in ponds. *Ecology* 86, 701–715.

Downing, A.L., and Leibold, M.A. 2002. Ecosystem consequences of species richness and composition in pond food webs. *Nature* 416, 837–841.

Drake, J.A. 1990. The mechanics of community assembly and succession. *J. Theor. Biol.* 147, 213–233.

Drossel, B., and McKane, A.J. 2003. Modelling food webs. In: S. Bornholdt and H.G. Schuster, (Editors), *Handbook of Graphs and Networks*. Wiley-VCH, Berlin, pp 218–247.

Drossel, B., Higgs, P.G., and McKane, A.J. 2001. The influence of predator-prey population dynamics on the long-term evolution of food web structure. *J. Theor. Biol.* 208, 91–107.

Drossel, B., McKane, A.J., and Quince, C. 2004. The impact of non-linear functional responses on the long-term evolution of food web structure. *J. Theor. Biol.* 229, 539–548.

Drost, C.A., and Fellers, G.M. 1996. Collapse of a regional frog fauna in the Yosemite area of the California Sierra Nevada, USA. *Cons. Biol.* 10, 414–425.

Dubé, M.G., Culp, J.M., Cash, K.J., Glozier, N.E., MacLatchy, D.L., Podemski, C.L., and Lowell, R.B. 2002. Artificial streams for environmental effects monitoring (EEM): Development and application in Canada over the past decade. *Water Qual. Res. J. Can.* 37, 155–180.

Duffy, J.E. 2002. Biodiversity and ecosystem function: the consumer connection. *Oikos* 99, 201–219.

Duffy, J.E. 2003. Biodiversity loss, trophic skew, and ecosystem functioning. *Ecol. Lett.* 6, 680–687.

Duffy, J.E., Richardson, J.P., and Canuel, E.A. 2003. Grazer diversity effects on ecosystem functioning in seagrass beds. *Ecol. Lett.* 6, 637–645.

Duffy, J.E., Macdonald, K.S., Rhode, J.M., and Parker, J.D. 2001. Grazer diversity, functional redundancy, and productivity in seagrass beds: an experimental test. *Ecology* 82, 2417–2434.

Dufresne, J.L., Fairhead, L., Le Treut, H., Berthelot, M., Bopp, L., Ciais, P., Friedlingstein, P., and Monfray, P. 2002. On the magnitude of positive feedback between future climate change and the carbon cycle. *Geophys. Res. Lett.* 29, 43/41–43/44.

Dungan, M.L. 1986. Three-way interactions barnacles limpets and algae in a Sonoran desert rocky intertidal zone. *Am. Nat.* 127, 292–316.

Dunne, J.A., Williams, R.J., and Martinez, N.D. 2002a. Network structure and biodiversity loss in food webs: robustness increases with connectance. *Ecol. Lett.* 5, 558–567.

Dunne, J.A., Williams, R.J., and Martinez, N.D. 2002b. Food-web structure and network theory: the role of connectance and size. *P. Natl. Acad. Sci.* USA 99, 12917–12922.

Dunne, J.A., Brose, U., Williams, R.J., and Martinez, N.D. 2004. Network structure and robustness of marine food webs. *Mar. Ecol.-Prog. Ser.* 273, 291–302.

Duplisea, D.P., and Kerr, S.R. 1995. Application of a biomass size spectrum model to demersal fish data from the Scotian Shelf. *J. Theor. Biol.* 177, 263–269.

Ebenhard, T. 1988. Introduced birds and mammals and their ecological effects. *Swed. Wildlife Res.* 13, 1–107.

Ebenman, B. 1992. Evolution in organisms that change their niches during the life cycle. *Am. Nat.* 139, 990–1021.

Ebenman, B., and Persson, L. 1988. *Size-Structured Populations: Ecology and Evolution.* Springer, Berlin.

Ebenman, B., Law, R., and Borrvall, C. 2004. Community viability analysis: the response of ecological communities to species loss. *Ecology* 85, 2591–2600.

Ebise, S., and Inoue, T. 1991. Change in C-N-P ratios during passage of water areas from rivers to a lake. *Water Res.* 25, 95–100.

Edelstein-Keshet, L., and Rausher, M.D. 1989. The effects of inducible plant defenses on herbivore populations. 1. Mobile herbivores in continuous time. *Am. Nat.* 133, 787–810.

Edmondson, W.T. (Editor). 1959. *Freshwater Biology, 2nd edition.* John Wiley, New York.

Ehrlich, P.R., and Raven, P.H. 1969. Differentiation of populations. *Science* 165, 1228–1232.

Ek, H., Sjögren, M., Arnebrant, K., and Söderström, B. 1994. Extramatrical mycelial growth, biomass allocation and nitrogen uptake in ectomycorrhizal systems in response to collembolan grazing. *Appl. Soil Ecol.* 1, 155–169.

Ellner, S.P., and Fussmann, G. 2003. Effects of successional dynamics on metapopulation persistence. *Ecology* 84, 882–889.

Elser, J.J., and Hassett, R.P. 1994. A stoichiometric analysis of the zooplankton-phytoplankton interaction in marine and freshwater ecosystems. *Nature* 370, 211–213.

Elser, J.J., and Urabe, J. 1999. The stoichiometry of consumer-driven nutrient recycling: theory, observations, and consequences. *Ecology* 80, 735–751.

Elser, J.J., Elser, M.M., MacKay, N.A., and Carpenter, S.R. 1998. Zooplankton mediated transitions between N and P limited algal growth. Limno. *Oceanogr.* 33, 1–14.

Elser, J.J., Fagan, W.F., Denno, R.F., Dobberfuhl, D.R., Folarin, A., Hubery, A., Interlandi, S., Kilham, S.S., McCauley, E., Schulz, K.L., Siemann, E.H., and Sterner, R.W. 2000. Nutritional constraints in terrestrial and freshwater food webs. *Nature* 408, 578–580.

Elton, C.S. 1927. *Animal Ecology.* Sidgwick and Jackson, London.

Elton, C.S. 1958. *Ecology of Invasions by Animals and Plants.* Chapman and Hall, London.

Elton, C.S., and Miller, R.S. 1954. The ecological survey of animal communities: with a practical system of classifying habitats by structural characters. *J. Ecol.* 42, 460–496.

Emlen, J.M. 1966. The role of time and energy in food preference. *Am. Nat.* 100, 611–617.

Emmerson, M.C., and Raffaelli, D.G. 2004a. Body size, patterns of interaction strength and the stability of a real food web. *J. Anim. Ecol.* 73, 399–409.

Emmerson, M.C., and Raffaelli, D. 2004b. Predator-prey body size, interaction strength and the stability of a real food web. *J. Anim. Ecol.* 73, 399–409.

Emmerson, M.C., Solan, M., Emes, C., Paterson, D.M., and Raffaelli, D. 2001. Consistent patterns and the idiosyncratic effects of biodiversity in marine ecosystems. *Nature* 411, 73–77.

Emmons, L.H. 1987. Comparative feeding ecology of felids in a neotropical rainforest. *Behav. Ecol. Sociobiol.* 20, 271–283.

Engelhardt, K. 2001. Effects of macrophyte species richness of wetland ecosystem functioning & services. *Nature* 405, 687–689.

Englund, G., and Moen, J. 2003. Testing models of trophic dynamics: The problem of translating from model to nature. *Aust. J. Ecol.* 28, 61–69.

Enquist, B.J., Brown, J.H., and West, G.B. 1998. Allometric scaling of plant energetics and population density. *Nature* 395, 163–165.

Estes, J.A. 1995. Top-level carnivores and ecosystem effects: questions and approaches. In: C.G. Jones and J.H. Lawton (Editors), *Linking Species and Ecosystems.* Chapman and Hall, New York, pp 151–158

Estes, J.A., and Palmisano, J.F. 1974. Sea otters: their role in structuring nearshore communities. *Science* 185, 1058–1060.

Estes, J.A., Smith, N.S., and Palmisano, J.F. 1978. Sea otter predation and community organization in Western Aleutian Islands, Alaska. *Ecology* 59, 822–833.

Ettema, C.H., and Wardle, D.A. 2002. Spatial soil ecology. *Trends Ecol. Evol.* 17, 177–183.

Evans, G.T., and Garcon, V.C. 1997. *One-Dimensional Models of Water Column Biogeochemistry.* 23/97, Joint Global Ocean Flux Study (JGOFS), JGOFS Bergen, Norway.

Everett, R.A., and Ruiz, G.M. 1993. Coarse woody debris as a refuge from predation in aquatic communities-an experimental test. *Oecologia* 93.

Faber, J.H. 1991. Functional classification of soil fauna: a new approach. *Oikos* 62, 110–117.

Fagan, W.F. 1997. Omnivory as a stabilizing feature of natural communities. *Am. Nat.* 150, 554–567.

Fagan, W.F., and Hurd, L.E. 1994. Hatch density variation of a generalist arthropod predator-population consequences and community impact. *Ecology* 75, 2022–2032.

Farlow, J.O. 1976. A consideration of the trophic dynamics of a late Cretaceous large-dinosaur community (Oldman Formation). *Ecology* 57, 841–857.

Fath, B.D., and Patten, B.C. 1999. Review of the foundations of network environ analysis. *Ecosystems* 2, 167–179.

Fausch, K.D., and Northcote, T.G. 1992. Large woody debris and salmonid habitat in a small coastal British Columbia stream. *Can. J. Fish. Aquat. Sci.* 49.

Fauth, J.E. 1999. Identifying potential keystones species from field data -an example from temporary ponds. *Ecol. Lett.* 2, 36–43.

Federal Register 2003. 12-month finding for a petition to list the Sierra Nevada distinct population segment of the mountain yellow-legged frog (*Rana muscosa*). *Federal Register* 68, 2283–2303.

Findlay, S., Pace, M., and Fisher, D. 1996. Spatial and temporal variability in the lower food web of the tidal freshwater Hudson River. *Estuaries* 19, 866–873.

Finke, D.L., and R.F., D. 2004. Predator diversity dampens trophic cascades. *Nature* 429, 407–410.

Finlay, R.D. 1985. Interaction between soil micro-arthropods and endomycorrhizal associations of higher plants. In: A.H. Fitter, (Editor). *Ecological Interactions in Soil.* Blackwell Science, Oxford, pp 319–331.

Fisher, S.G., and Likens, G.E. 1973. Energy flow in Bear Brook, New Hampshire: integrative approach to stream ecosystem metabolism. *Ecol. Monogr.* 43, 421–439.

Flannery, T. 1994. *The Future Eaters.* Grove Press, New York, 423.

Flannery, T. 2001. *The Eternal Frontier.* Grove Press, New York, 390.

Flecker, A.S. 1984. The effects of predation and detritus on the structure of a stream insect community-a field test. *Oecologia* 64, 300–305.

Flecker, A.S. 1992. Fish trophic guilds and the structure of a tropical stream: weak direct vs. strong indirect effects. *Ecology* 73, 927–940.

Flecker, A.S., Taylor, B.W., Berhardt, E.S., Hood, J.M., Cornwell, W.K., Cassatt, S.R., Vanni, M.J., and Altman, N.S. 2002. Interactions between herbivorous fishes and limiting nutrients in a tropical stream ecosystem. *Ecology* 83, 1831–1844.

Fogel, R. 1980. Mycorrhizae and nutrient cycling in natural forest ecosystems. *New Phytol.* 86, 199–212.

Fogel, R., and Hunt, G. 1983. Contrigution of mycorrhizae and soil fungi to nutrient cycling in a douglas fir ecosystem. *Can. J. Forest Res.* 13, 219–232.

Folt, C., and Goldman, C.R. 1981. Allelopathy between zooplankton: A mechanism for interference competition. *Science* 213, 1133–1135.

Forbes, V.E., and Forbes, T.L. 1994. *Ecotoxicology in Theory and Practice.* Chapman and Hall, New York, USA.

Ford, B., and Smith, B.M. 1994. Food webs and environmental disturbance. What's the connection? *Amer. Biol. Teacher* 56, 247–249.

Forge, T.A., and Simard, S.W. 2001. Short-term effects of nitrogen and phosphorus fertilizers on nitrogen mineralisation and trophic structure of the soil ecosystem in forest clearcuts in the southern interior of British Columbia. *Can. J. Soil. Sci.* 81, 11–20.

Forsberg, B.R., Araujo-Lima, C.A.R.M., Martinelli, L.A., Victoria, R.L., and Bonassi, J.A. 1993. Autotrophic carbon sources for fish of the central Amazon. *Ecology* 74, 643–652.

Fowler, H.G., and Whitford, W.G. 1980. Termites, microarthropods and the decomposition of senescent and fresh creosotebush (Larrea-Tridentata) leaf litter. *J. Arid Environ.* 3, 63–68.

Fox, J.W. 2003. The long-term relationship between plant diversity and total plant biomass depends on the mechanism maintaining diversity. *Oikos* 102, 630–640.

Fox, J.W. 2004a. Effects of algal and herbivore diversity on the partitioning of biomass within and among trophic levels. *Ecology* 85, 549–559.

Fox, J.W. 2004b. Modelling the joint effects of predator and prey diversity on total prey biomass. *J. Anim. Ecol.* 73, 88–96.

Fox, J.W., and McGrady-Steed, J. 2002. Stability and complexity in microcosm communities. *J. Anim. Ecol.* 71, 749–756.

France, R., Chandler, M., and Peters, R. 1998. Mapping trophic continua of benthic food webs: body size-delta N15 relationships. *Mar. Ecol.-Prog. Ser.* 174, 301–306.

Free, C.A., Beddington, J., and Lawton, J. 1977. Inadequacy of simple-models of mutual interference for parasitism and predation. *J. Anim. Ecol.* 46, 543–554.

Fretwell, S.D. 1977. The regulation of plant communities by food chain exploring them. *Perspect Biol. Med.* 20, 169–185.

Fridley, J.D. 2002. Resource availability dominates and alters the relationship between species diversity and ecosystem productivity in experimental plant communities. *Oecologia* 132, 271–277.

Frost, P.C., Stelzer, R.S., Lamberti, G.A., and Elser, J.J. 2002. Ecological stoichiometry of trophic interactions in the benthos: understanding the role of C : N : P ratios in lentic and lotic habitats. *J. N. Am. Benthol. Soc.* 21, 515–528.

Frost, P.C., Tank, S.E., Turner, M.A., and Elser, J.J. 2003. Elemental composition of littoral invertebrates from oligotrophic and eutrophic Canadian lakes. *J. N. Am. Benthol. Soc.* 22, 51–62.

Fry, B., and Arnold, C. 1982. Rapid d13C/d12C turnover during growth of brown shrimp (*Penaeus aztecus*). *Oecologia* 54, 200–204.

Fu, S., Cabrera, M.L., Coleman, D.C., Kisselle, K.W., Garrett, C.J., Hendrix, P.F., and Crossley, J.D.A. 2000. Soil carbon dynamics of conventional tillage and no-till agroecosystems at Georgia Piedmont-HSB-C model. *Ecol. Model.* 131, 229–248.

Fulton, E.A., Smith, A.D.M., and Johnson, C.R. 2003. Effect of complexity on marine ecosystem models. *Mar. Ecol.-Prog. Ser.* 253, 1–16.

Fussmann, G.F., and Heber, G. 2002. Food web complexity and chaotic population dynamics. *Ecol. Lett.* 5, 394–401.

Gardner, M.R., and Ashby, W.R. 1970. Connectance of large (cybernetic) systems: critical values for stability. *Nature* 228, 784.

Garlaschelli, D., Caldarelli, G., and Pietronero, L. 2003. Universal scaling relations in food webs. *Nature* 423, 165–168.

Garrity, S.D., and Levings, S.C. 1981. A predator-prey interaction between two physically and biologically constrained tropical rocky shore gastropods: direct, indirect and community effects. *Ecol. Monogr.* 51, 267–286.

Gaston, K.J., and Lawton, J.H. 1988. Patterns in the abundance and distribution of insect populations. *Nature* 331.

Gaston, K.J., and Lawton, J. 1990. Effects of scale and habitat on the relationship between regional distribution and local abundance. *Oikos* 58, 329–335.

Gaston, K.J., and Blackburn, T.M. 2000. *Pattern and Process in Macroecology*. Blackwell Science, Oxford.

Gause, G.F. 1934. *The Struggle for Existence*. Williams & Wilkins Co., Baltimore.

Gelman, A., Carlin, J.B., Stern, H.S., and Rubin, D.B. 2004. *Bayesian Data Analysis, 2nd edition*. Chapman and Hall/CRC, Boca Raton, p 668.

Gende, S.M., Edwards, R.T., Wilson, M.F., and Wipfli, M.S. 2002. Pacific salmon in aquatic and terrestrial ecosystems. *Bioscience* 52, 917–928.

Gentleman, W., Leising, A., Frost, B., Strom, S., and Murray, J. 2003. Functional responses for zooplankton feeding on multiple resources: a review of assumptions and biological dynamics. *Deep-Sea Research II* 50, 2847–2875.

Getz, W.M. 1984. Population-dynamics: a per-capita resource approach. *J. Theor. Biol.* 108, 623–643.

Ghilarov, M.S. 1944. Correlations between size and number of soil animals. *P. Acad. Sci. USSR* 43, 267–269.

Gibbons, W.N., Munkittrick, K.R., and Taylor, W.D. 1998. Monitoring aquatic environments receiving industrial effluents using small fish species 1: response of spoonhead sculpin (*Cottus ricei*) downstream of a bleached-kraft pulp mill. *Environmental Toxicology and Chemistry* 17, 2227–2237.

Gilbert, L.E. 1980. Food web organization and the conservation of Neotropical diversity. In: M.E. Soulé and B.A. Wilcox (Editors), *Conservation Biology: An Evolutionary Perspective*. Sinauer, Sunderland, Massachusetts, pp 11–33.

Giller, P.S. 1996. The diversity of soil communities, the 'poor man's tropical rainforest. *Biodivers. Conserv.* 5, 135–168.

Gillooly, J.F., Brown, J.H., West, G.B., Savage, V.M., and Charbov, E.L. 2001. Effects of size and temperature on metabolic rate. *Science* 293, 2248–2251.

Gilpin, M.E. 1975. Stability of feasible predator-prey systems. *Nature* 254, 137–139.

Gittleman, J.L. 1985. Carnivore body size: Ecological and taxonomic correlates. *Oecologia* 67, 540–554.

Gleason, F.K., and Paulson, J.L. 1984. Site of action of the natural algicide, cyanobacterin, in the blue-green alga, Synechococcus sp. *Arch. Microbiol.* 138, 273–277.

Glozier, N.E., Culp, J.M., Reynoldson, T.B., Bailey, R.C., Lowell, R.B., and Trudel, L. 2002. Assessing metal mine effects using benthic invertebrates for Canada's Environmental Effects Program. *Water Qual. Res. J. Can.* 37, 251–278.

Godfray, H.C.J., and Hassell, M.P. 1989. Discrete and continuous insect populations in tropical environments. *J. Anim. Ecol.* 58, 153–174.

Goldwasser, L., and Roughgarden, J. 1993. Construction and analysis of a large Caribbean food web. *Ecology* 74, 1216–1233.

Goldwasser, L., and Roughgarden, J. 1997. Sampling effects and the estimation of food-web properties. *Ecology* 78, 41–54.

Gottsberger, G. 1978. Seed dispersal by fish in the inundated regions of Humaitá, Amazonia. *Biotropica* 10, 170–183.

Goulding, M. 1980. *The Fishes and the Forest.* University of California Press, Berkeley, California, 280.

Goulding, M., Carvalho, M.L., and Ferreira, E.G. 1988. *Rio Negro: Rich Life in Poor Water.* SPB Academic Publishing, The Hague, The Netherlands.

Gragnani, A., Scheffer, M., and Rinaldi, S. 1999. Top-down control of cyanobacteria: A theoretical analysis. *Am. Nat.* 153, 59–72.

Griffiths, B.S., Ritz, K., Bardgett, R.D., Cook, R., Christensen, S., Ekelund, F., Sørensen, S.J., Bååth, E., Bloem, J., de Ruiter, P.C., Dolfing, J., and Nicolardot, B. 2000. Ecosystem response of pasture soil communities to fumigation-induced microbial diversity reductions: an examination of the biodiversity-ecosystem function relationship. *Oikos* 90, 279–294.

Griffiths, D. 1992. Size, abundance, and energy use in communities. *J. Anim. Ecol.* 61, 307–316.

Griffiths, D. 1998. Sampling effort, regression method, and the shape and slope of size-abundance relations. *J. Anim. Ecol.* 67, 795–804.

Grimm, N.B. 1988. Role of macroinvertebrates in nitrogen dynamics of a desert stream. *Ecology* 69, 1884–1893.

Grimm, V., and Wissel, C. 1997. Babel, or the ecological stability discussions: an inventory and analysis of terminology and a guide for avoiding confusion. *Oecologia* 109, 323–334.

Gross, E.M. 2003. Allelopathy of aquatic autotrophs. *Crit. Rev. Plant Sci.* 22, 313–339.

Gross, E.M., Wolk, C.P., and Jüttner, F. 1991. Fischerellin, a new allelochemical from the freshwater cyanobacterium *Fischerella muscicola. J. Phycol.* 27, 686–692.

Grover, J. 1994. Assembly rules for communities of nutrient-limited plants and specialist herbivors. *Am. Nat.* 143, 258–282.

Grover, J. 1995. Competition, herbivory, and enrichment: nutrient-based models for edible and inedible plants. *Am. Nat.* 147, 746–774.

Grover, J. 1997. *Resource competition.* Chapman and Hall, London.

Gulis, V., and Suberkropp, K. 2003. Leaf litter decomposition and microbial activity in nutrient-enriched and unaltered reaches of a headwater stream. *Freshwat. Biol.* 48, 123–134.

Gurevitch, J., Morrison, J.A., and Hedges, L.V. 2000. The interaction between competition and predation: A meta-analysis of field experiments. *Am. Nat.* 155, 435–453.

Hafen, E., and Stocker, H. 2003. How are the sizes of cells, organ, and bodies controlled? *PLoS Biol.* 1, 319–323.

Hagen, N.T., Andersen, A., and Stabell, O.B. 2002. Alarm responses of the green sea urchin, Strongylocentrotus droebachiensis, induced by chemically labelled durophagous predators and simulated acts of predation. *Mar. Biol.* 140, 365–374.

Haimi, J., Huhta, V., and Boucelham, M. 1992. Growth increase of birch seedlings under the influence of earthworms: a laboratory study. *Soil Biol. Biochem.* 24, 1525–1528.

Hairston, N.G., Smith, F.E., and Slobodkin, L.B. 1960. Community structure, population control, and competition. *Am. Nat.* 94, 421–425.

Hairston, N.G.J., and N.G. Hairston, S. 1993. Cause-effect relationships in energy flow, trophic structure, and interspecific interactions. *Am. Nat.* 142, 379–411.

Halbach, U. 1969. Das Zusammenwirken von Konkurrenz und Räuber-Beutebeziehungen bei Rädertieren. *Verh. Zool. Ges., Zool. Anz. Suppl.-Bd.* 33, 72–79.

Hall, R.O.J., and Meyer, J.L. 1998. The trophic significance of bacteria in a detritus-based stream food web. *Ecology* 79, 1995–2012.

Hall, R.O.J., Wallace, J.B., and Eggert, S.L. 2000. Organic matter flow in stream food webs with reduced detrital resource base. *Ecology* 81, 3445–3463.

Hall, R.O.J., Tank, J.L., and Dybdahl, M.F. 2003. Exotic snails dominate nitrogen and carbon cycling in a highly productive stream. *Front. Ecol.* 8, 407–411.

Hall, S.J., 1999. *The Effects of Fishing on Marine Ecosystems and Communities. Fish and Aquatic Resources Series 1.* Blackwell Science, Oxford, U.K.

Hall, S.J. 2002. The continental shelf benthic ecosystem: current status, agents for change and future prospects. *Environ. Cons.* 29, 350–374.

Hall, S.J., and Raffaelli, D.G. 1991. Food web patterns: lessons from a species rich web. *J. Anim. Ecol.* 60, 823–841.

Hall, S.J., and Raffaelli, D.G. 1993. Food webs: theory and reality. *Adv. Ecol. Res.* 24, 187–237.

Hall, S.J., and Raffaelli, D. 1997. Food Web Patterns: What do we Really Know? In: A. Gange and A.C. Brown, (Editors), *Multitrophic Interactions in Terrestrial Systems.* Blackwell University Press,

Hall, S.R. 2004. Stoichiometrically explicit competition between grazers: Species replacement, coexistence, and priority effects along resource supply gradients. *Am. Nat.* 164, 157–172.

Hall, S.R., Leibold, M.A., Lytle, D.A., and Smith. 2004. Stoichiometry and planktonic grazer composition over gradients of light, nutrients, and predation risk. *Ecology* 85, 2291–2301.

Hamilton, A.L. 1969. On estimating annual production. Limno. *Oceanogr.* 14, 771–782.

Hamilton, S.K., and Lewis, W.M.J. 1987. Causes of seasonality in the chemistry of a lake of the Orinoco River floodplain, Venezuela. Limno. *Oceanogr.* 32, 1277–1290.

Hamilton, S.K., and Lewis, W.M.J. 1992. Stable carbon and nitrogen isotopes in algae and detritus from the Orinoco River floodplain, Venezuela. Geochim. *Cosmochim. Ac.* 56, 4237–4246.

Hamrin, S.F., and Persson, L. 1986. Asymmetrical competition between age classes as a factor causing population oscillations in an obligate planktivorous fish. *Oikos* 47, 223–232.

Hardy, A.C. 1924. *The herring in relation to its animate environment. Part 1. The food and feeding habits of the herring with special reference to the east coast of England.* Ministry of Agriculture and Fisheries, Fisheries Investigations, Series II 7, 1–45.

Harmon, M.E., Franklin, J.F., Swanson, F.J., Sollins, P., Gregory, S.V., Lattin, J.D., Anderson, N.H., Cline, S.P., Aumen, N.G., Sedell, J.R., Lienkaemper, G.W., Cromack,

K., and Cummins, K.W. 1986. Ecology of coarse wood debris in temperate ecosystems. *Adv. Ecol. Res.* 15, 133–302.

Harte, J., and Kinzig, A.P. 1993. Mutualism and competition between plants and decomposers: implications for nutrient allocation in ecosystems. *Am. Nat.* 141, 829–846.

Harwood, J., and Stokes, K. 2003. Coping with uncertainty in ecological advice: lessons from fisheries. *Trends Ecol. Evol.* 18, 617–622.

Hassett, R.P., Cardinale, B., Stabler, L.B., and Elser, J.J. 1997. Ecological stoichiometry of N and P in pelagic ecosystems: comparison of lakes and oceans with emphasis on the zooplankton-phytoplankton interaction. Limno. *Oceanogr.* 42, 648–662.

Hastings, A. 1996. What equilibrium behavior of Lotka-Volterra models does not tell us about food webs. In: G.A. Polis and K.O. Winemiller, (Editors), *Food Webs: Integration of Pattern and Dynamics.* Chapman and Hall, New York, pp 211–217.

Hastings, A. 2004. Transients: the key to long-term ecological understanding? *TREE* 19, 39–45.

Hastings, H.M., and Conrad, M. 1979. Length and the evolutionary stability of food chains. *Nature* 282, 838–839.

Hattenschwiler, S., and Bretscher, D. 2001. Isopod effects on decomposition of litter produced under elevated CO2, N eposition and different soil types. *Glob. Chang. Biol.* 7, 565–579.

Haukioja, E. 1980. On the role of plant defences in the fluctuation of herbivore populations. *Oikos* 35, 202–213.

Havens, K.E. 1992. Scale and Structure in Natural Food Webs. *Science* 257, 1107–1109.

Havens, K.E. 1997. Unique structural properties of pelagic food webs. *Oikos* 78, 75–80.

Haydon, D.T. 2000. Maximally stable model ecosystems can be highly connected. *Ecology* 81, 2631–2636.

Heal, O.W., MacLean Jr. S.F. 1975. Comparative productivity in ecosystems: secondary productivity. In: W.H. Van Dobben, Lowe-Mcconnell, R.H., (Editor). *Unifying concepts in ecology.* Junk, The Hague.

Hector, A. 1998. The effect of diversity on productivity: detecting the role of species complementarity. *Oikos* 82, 597–599.

Hector, A., Bazeley-White, E., Loreau, M., Otway, S., and Schmid, B. 2002. Overyielding in grassland communities: testing the sampling effect hypothesis with replicated biodiversity experiments. *Ecol. Lett.* 5, 502–511.

Hector, A., Schmid, B., Beierkuhnlein, C., Caldeira, M.C., Diemer, M., Dimitrakopoulos, P.G., Finn, J.A., Freitas, H., Giller, P.S., Good, J., Harris, R., Högberg, P., Huss-Danell, K., Joshi, J., Jumpponen, A., Körner, C., Leadley, P.W., Loreau, M., Minns, A., Mulder, C.P.H., O'Donovan, G., Otway, S.J., Pereira, J.S., Prinz, A., Read, D.J., Scherer-Lorenzen, M., Schulze, E.D., Siamantziouras, A.S.D., Spehn, E.M., Terry, A.C., Troumbis, A.Y., Woodward, F.I., Yachi, S., and Lawton, J.H. 1999. Plant diversity and productivity experiments in European grasslands. *Science* 286, 1123–1127.

Hector, A., Schmid, B., Beierkuhnlein, C., Caldeira, M.C., Diemer, M., Dimitrakopoulos, P.G., Finn, J.A., Freitas, H., Giller, P.S., Good, J., Harris, R., Högberg, P., Huss-Danell, K., Joshi, J., Jumpponen, A., Körner, C., Leadley, P.W.,

Loreau, M., Minns, A., Mulder, C.P.H., O'Donovan, G., Otway, S.J., Pereira, J.S., Prinz, A., Read, D.J., Scherer-Lorenzen, M., Schulze, E.D., Siamantziouras, A.S.D., Spehn, E.M., Terry, A.C., Troumbis, A.Y., Woodward, F.I., Yachi, S., and Lawton, J.H. 2000. No consistent effect of plant diversity on productivity: response. *Science* 289, 1255a.

Hendriks, A.J. 1999. Allometric scaling of rate, age, and density parameters in ecological models. *Oikos* 86, 293–310.

Hessen, D.O., Andersen, T., and Faafeng, B. 1992. Zooplankton contribution to particulate phosphorus and nitrogen in lakes. *J. Plankton Res.* 14, 937–947.

Hesslein, R.H., Hallard, K.A., and Ramlal, P. 1993. Replacement of sulfur, carbon, and nitrogen, in tissue of growing broad whitefish (*Coregonus nasus*) in response to a change in diet traced by d34S, d13C, and d15N. *Can. J. Fish. Aquat. Sci.* 50, 2071–2076.

Higashi, M., and Nakajima, H. 1995. Indirect effects in ecological interaction networks. I. The chain rule approach. *Math. Biosci.* 130, 99–128.

Higgins, K.A., Hastings, A., Sarvela, J.N., and Botsford, L.W. 1997. Stochastic dynamics and deterministic skeletons: Population behavior of Dungeness crab. *Science* 276, 1431–1435.

Hilborn, R., and Mangel, M. 1997. The ecological detective. Confronting models with data. In: *Monographs in Population Biology*. Princeton University Press, Princeton, pp 315.

Hildebrand, S.F., and Schroeder, W.C. 1972. *Fishes of Chesapeake Bay, T.F.H.* Neptune, New Jersey, 388.

Hilderbrand, G.V., Hanley, T.A., Robbins, C.T., and Schwartz, C.C. 1999. Role of brown bears (*Ursus arctos*) in the flow of marine nitrogen into a terrestrial ecosystem. *Oecologia* 121, 546–550.

Hildrew, A.G. 1992. Food webs and species interactions. In: P. Calow and G.E. Petts, (Editors), *The Rivers Handbook*. Blackwell Sciences, Oxford, pp 309–330.

Hildrew, A.G., Townsend, C.R., and Hasham, A. 1985. The predatory Chironomidae of an iron-rich stream: feeding ecology and food web structure. *Ecol. Entomol.* 10, 403–413.

Hirsch, M.W., and Smale, S. 1974. *Differential Equations, Dynamical Systems, and Linear Algebra*. Academic Press, Boston.

Hjelm, J., and Persson, L. 2001. Size-dependent attack rate and handling capacity-inter-cohort competition in a zooplanktivorous fish. *Oikos* 95, 520–532.

Hobson, K.A. 1993. Trophic relationships among High Artic sea-birds: insights from tissue-dependent stable-isotope models. *Mar. Ecol.-Prog. Ser.* 95, 7–18.

Hobson, K.A., and Clark, R.G. 1992. Assessing avian diets using stable isotopes II: factors influencing diet-tissue fractionation. *Condor* 94, 189–197.

Hobson, K.A., Alisauskas, R.T., and Clark, R.G. 1993. Stable-nitrogen isotope enrichment in avian tissues due to fasting and nutritional stress: implications for isotopic analysis of diet. *Condor* 95, 388–394.

Hoeinghaus, D.J., Layman, C.A., Arrington, D.A., and Winemiller, K.O. 2003a. Spatiotemporal variation in fish assemblage structure in tropical floodplain creeks. *Environ. Biol. Fishes* 67, 379–387.

Hoeinghaus, D.J., Layman, C.A., Arrington, D.A., and Winemiller, K.O. 2003b. Movement of Cichla spp. (Cichlidae) in a Venezuelan floodplain river. *Neo. Ichthyol.* 1, 121–126.

Hoffmeister, D.F. 1986. *Mammals of Arizona.* University of Arizona Press, Tucson, Arizona.

Hogervorst, R.F., Dijkhuis, M.A.J., van der Schaar, M.A., Berg, M.P., and Verhoef, H.A. 2003. Indications for the tracking of elevated nitrogen levels through the fungal route in a soil food web. *Environ. Pollut.* 126, 257–266.

Holling, C.S. 1959a. Some characteristics of simple types of predation and parasitism. *Can. Entomol.* 91, 385–398.

Holling, C.S. 1959b. The components of predation as revealed by a study of small mammal predation on the European pine sawfly. *Can. Entomol.* 91, 293–320.

Holling, C.S. 1965. The functional response of predators to prey density and its role in mimicry and population regulation. *Mem. Entomol. Soc. Can.* 45, 5–60.

Holt, R., and Lawton, J. 1994. The ecological consequences of shared natural enimies. *Ann. Rev. Ecol. Syst.* 25, 495–520.

Holt, R., Grover, J., and Tilman, D. 1994. Simple rules for interspecific dominance in systems with exploitative and apparent competition. *Am. Nat.* 144, 741–771.

Holt, R.B. 2002. Food webs in space: On the interplay of dynamic instability and spatial processes. *Ecological Research* 17, 261–273.

Holt, R.D. 1977. Predation, apparent competition, and the structure of prey communities. *Theor. Popul. Biol.* 12, 197–229.

Holt, R.D. 1983. Optimal foraging and the form of the predator isocline. *Am. Nat.* 122, 521–541.

Holt, R.D. 1997. Community modules. In: A.C. Gange and V.K. Brown, (Editors), *Multitrophic Interactions in Terrestrial Systems.* Blackwell Science, Oxford, pp 333–351.

Holt, R.D., and Polis, G.A. 1997. A theoretical framework for intraguild predation. *Am. Nat.* 149, 745–754.

Holt, R.D., and Loreau, M. 2002. Biodiversity and ecosystem functioning, the role of trophic interactions, and the importance of system openness. In: A.P. Kinzig, S. Pacala, and D. Tilman, (Editors), *The Functional Consequences of Biodiversity: Empirical Progress and Theoretical Extensions.* Princeton University Press, Princeton, pp 246–262.

Holt, R.D., Lawton, J.H., Polis, G.A., and Martinez, N.D. 1999. Trophic rank and the species-area relationship. *Ecology* 80, 1495–1504.

Hooper, D.U. 1998. The role of complementarity and competition in ecosystem responses to variation in plant diversity. *Ecology* 79, 704–719.

Hooper, D.U., and Vitousek, P.M. 1998. Effects of plant composition and diversity on nutrient cycling. *Ecol. Monogr.* 68.

Hooper, D.U., Solan, M., Symstad, A.J., Díaz, S., Gessner, M.O., Buchmann, N., Degrande, V., Grime, P., Hulot, F.D., Mermillod-Blondin, F., Roy, J., Spehn, E.M., and Van Peer, L. 2002. Species Diversity, Functional Diversity and Ecosystem Functioning. In: M. Loreau, S. Naeem, and P. Inchausti (Editors), *Biodiversity and*

Ecosystem Functioning: Syntheses and Perspectives. Oxford University Press, Oxford, pp 195–208.

Hsieh, H.L., Chen, C.P., Chen, Y.G., and Yang, H.H. 2002. Diversity and benthic organic matter flows through polychaetes and crabs in a mangrove estuary: d13C and d34S signals. *Mar. Ecol.-Prog.* Ser. 227, 145–155.

Hubbell, S.P. 2001. *The Unified Neutral Theory of Biodiversity and Biogeography*. Princeton University Press, Princeton.

Hubbell, S.P., and Foster, R.B. 1986. Commonness and rarity in a Neotropical forest: implications for tropical tree conservation. In: M. Soulé (Editor), *Conservation Biology: Science of Scarcity and Diversity*. Sinauer Associates, Sunderland, Massachusetts, pp 205–231.

Huenneke, L.F., Hamburg, S.P., Koide, R.T., Mooney, H.A., and Vitousek, P.M. 1990. Effects of soil resources on plant invasion and community structure in Californian serpentine grasslands. *Ecology* 71, 478–491.

Hughes, R.N. 1990. *Behavioural Mechanisms of Food Selection*, NATO-ASI Series, vol. G20. Springler-Verlag, Berlin.

Huhta, V., Persson, T., and Setälä, H. 1998. Functional implications of soil fauna diversity in boreal forests. *Appl. Soil Ecol.* 10, 277–288.

Huhta, V., Karppinen, E., Nurminen, M., and Valmas, A. 1967. Effect of silvicultural practices on arthropod, annelid and nematode populations in coniferous forest soil. *Ann. Zool. Fennici* 4, 87–145.

Huisman, J., and Weissing, F.J. 1999. Biodiversity of plankton by species oscillations and chaos. *Nature* 402, 407–410.

Huisman, J., and Weissing, F.J. 2000. Coexistence and resource competition. *Nature* 407, 694.

Huisman, J., and .Weissing, F.J. 2001. Biological conditions for oscillations and chaos generated by multispecies competition. *Ecology* 82, 2682–2695.

Huisman, J., Jonker, R.R., Zonneveld, C., and Weissing, F.J. 1999. Competition for light between phytoplankton species: experimental test of mechanistic theory. *Ecology* 80, 211–222.

Hulot, F.D., and Huisman, J. 2004. Allelopathic interactions between phytoplankton species: the role of mixing intensity and heterotrophic bacteria. *Limno. Oceanogr.* 49, 1424-1434.

Hulot, F.D., Lacroix, G., Lescher-Moutoue, F.O., and Loreau, M. 2000. Functional diversity governs ecosystem response to nutrient enrichment. *Nature* 405, 340–344.

Humborg, C., Ittekkot, V., Cociasu, A., and von Bodungen, B. 1997. Effect of Danube River dam on Black Sea biogeochemistry and ecosystem structure. *Nature* 386, 385–388.

Humphries, M.M., Umbanhowar, J., and McCann, K.S. 2004. Bioenergetic prediction of climate change impacts on northern mammals. *Integr. Comp. Biol.* 44, 152–162.

Hunt, H.W., and Wall, D.H. 2002. Modelling the effects of loss of soil biodiversity on ecosystem function. *Glob. Chang. Biol.* 8, 33–50.

Hunt, H.W., Coleman, D.C., Ingham, E.R., Ingham, R.E., Elliott, E.T., Moore, J.C., Rose, S.L., Reid, C.P.P., and Morley, C.R. 1987. The detrital food web in a shortgrass prairie. *Biol. Fert. Soils* 3, 57–68.

Hunt, R.L. 1975. Food relations and behavior of salmonid fishes. In: J.J.E. Al., (Editor). In *Coupling of Land and Water Systems*. Springer-Verlag, New York.

Hurlbert, S.H. 1997. Functional importance vs. keystoneness: reformulating some questions in theoretical biocenology. *Aust. J. Ecol.* 22, 369–382.

Huryn, A.D. 1996. An appraisal of the Allen Paradox in a New Zealand trout stream. *Limno. Oceanogr.* 41, 243–252.

Huryn, A.D. 1990. Growth and voltinism of lotic midge larvae: patterns across an Appalachian mountain basin. *Limno. Oceanogr.* 35, 339–351.

Huryn, A.D., and Wallace, J.B. 1987. Local geomorphology as a determinant of macrofaunal production in a mountain stream. *Ecology* 68, 1932–1942.

Huston, M. 1979. A general hypothesis of species diversity. *Am. Nat.* 113, 81–101.

Huston, M.A. 1997. Hidden treatment in ecological experiments: re-evaluating the ecosystem function of biodiversity. *Oecologia* 110, 449–460.

Huston, M.A., Aarsen, L.W., Austin, M.P., Cade, B.S., Fridley, J.D., Garnier, E., Grime, J.P., Hodgson, J., Lauenroth, W.K., Thompson, K., Vandermeer, J.H., and Wardle, D.A. 2000. No consistent effect of plant diversity on productivity. *Science* 289, 1255a.

Hutchens, J.J., Chung, K., and Wallace, J.B. 1998. Temporal variability of stream macroinvertebrate abundance and biomass following pesticide disturbance. *J. N. Am. Benthol. Soc.* 17, 518–534.

Hutchinson, G.E. 1957. Concluding Remarks. Cold Spring Harbour Symposium on Quantitative *Biology* 22, 415–427.

Hutchinson, G.E. 1959. Homage to Santa Rosalia or why are there so many kinds of animals? *Am. Nat.* 93, 145–159.

Huxel, G.R., and McCann, K. 1998. Food web stability: the influence of trophic flows across habitats. *Am. Nat.* 152, 460–469.

Huxham, M., Beaney, S., and Raffaelli, D. 1996. Do parasites reduce the chances of triangulation in a real food web? *Oikos* 76, 284–300.

Ingham, E.R., Coleman, D.C., and Moore, J.C. 1989. An analysis of food web structure and function in a shortgrass prairie, a mountain meadow, and a lodgepole pine forest. *Biol. Fert. Soils* 8, 29–37.

Ingham, E.R., Trofymow, J.A., Ames, R.N., Hunt, H.W., Morley, C.R., Moore, J.C., and Coleman, D.C. 1986. Trophic Interactions and Nitrogen Cycling in a Semi-Arid Grassland Soil. II. System Responses to Removal of Different Groups of Soil Microbes or Fauna. *J. Ap. Ecol.* 23, 615–630.

Ingham, R.E., Trofymow, J.A., Ingham, E.R., and Coleman, D.C. 1985. Interactions of bacteria, fungi, and their nematode grazers: effects on nutrient cycling and plant growth. *Ecol. Monogr.* 55, 119–140.

Ivanovic, M.L., and Brunetti, N.E. 1994. Food and feeding of *Illex argentinus*. *Antarct. Sci.* 6, 185–193.

Ives, A.R. 1995. Measureing resilience in stochastic-systems. *Ecol. Monogr.* 65, 217–233.

Ives, A.R., and Jansen, V.A.A. 1998. Complex dynamics in stochastic tritrophic models. *Ecology* 79, 1039–1052.

Ives, A.R., and Hughes, J.B. 2002. General relationships between species diversity and stability in competitive systems. *Am. Nat.* 159, 388–395.

Ives, A.R., Klug, J.L., and Gross, K. 2000. Stability and species richness in complex communities. *Ecol. Lett.* 3, 399–411.

Ives, A.R., Dennis, B., Cottingham, K.L., and Carpenter, S.R. 2003. Estimating community stability and ecological interactions from time-series data. *Ecol. Monogr.* 73, 301–330.

Jackson, J.B.C., Kirby, M.X., Berger, W.H., et al. 2001. Historical overfishing and the recent collapse of coastal ecosystems. *Science* 293, 629–638.

Jain, S., and Krishna, S. 2002. Large extinctions in an evolutionary model: The role of innovation and keystone species. *P. Natl. Acad. Sci. USA* 99, 2055–2060.

Jansen, V.A.A., and Kokkoris, G.D. 2003. Complexity and stability revisited. *Ecol. Lett.* 6, 498–502.

Janzen, D.H., and Martin, P.S. 1982. Neotropical anachronisms: the fruits the gomphotheres ate. *Science* 215, 19–27.

Janzen, D.H., and Hallwachs, W. 1994. *All Taxa Biodiversity Inventory (ATBI) of Terrestrial Systems. A Generic Protocol for Preparing Wildland Biodiversity for Non-Damaging Use.* Report of a NSF Workshop, Philadelphia, Pennsylvania.

Jenkins, D.G., and Buikema, J., A.L. 1998. Do similar communities develop in similar sites? A test with zooplankton structure and function. *Ecol. Monogr.* 68, 421–443.

Jennings, M.R. 1996. *Status of Amphibians Sierra Nevada ecosystem project: final report to Congress. Volume II.* Centers for Water and Wildland Resources, University of California, Davis, California, USA.

Jennings, S., and Mackinson, S. 2003. Abundance-body mass relationships in size-structured food webs. *Ecol. Lett.* 6, 971–974.

Jennings, S., and Warr, K.J. 2003. Smaller predator-prey body size ratios in longer food chains. *P. Roy. Soc. Lond. B Biol.* 270, 1413–1417.

Jennings, S., Warr, K.J., and Mackinson, S. 2002. Use of size-based production and stable isotope analyses to predict trophic transfer efficiencies and predator-prey body mass ratios in food webs. *Mar. Ecol.-Prog. Ser.* 240, 11–20.

Jennings, S., Pinnegar, J.K., Polunin, N.V.C., and Boon, T.W. 2001. Weak cross-species relationships between body-size and trophic level belie powerful size-based trophic structuring in fish communities. *J. Anim. Ecol.* 70, 934–944.

Jepsen, D.B. 1999. Analysis of Trophic Pathways in Freshwater Ecosystems Using Stable Isotope Signatures. Dissertation. Texas A&M University, College Station, Texas, USA.

Jepsen, D.B., and Winemiller, K.O. 2002. Structure of tropical river food webs revealed by stable isotope ratios. *Oikos* 96, 46–55.

Jepsen, D.B., Winemiller, K.O., and Taphorn, D.C. 1997. Temporal patterns of resource partitioning among Cichla species in a Venezuelan black-water river. *J. Fish Biol.* 51, 1085–1108.

Jeschke, J.M., and Tollrian, R. 2000. Density-dependent effects of prey defences. *Oecologia* 123, 391–396.

Jeschke, J.M., Kopp, M., and Tollrian, R. 2002. Predator functional responses: discriminating between handling and digesting prey. *Ecol. Monogr*. 72, 95–112.

Jones, C.G., and Lawton, J.H.1995. *Linking Species and Ecosystems*. Chapman & Hall, New York; London.

Jones, C.G., Lawton, J.H., and Shachak, M. 1994. Organisms as ecosystem engineers. *Oikos* 69, 373–386.

Jones, J.I., and Sayer, C.D. 2003. Does the fish-invertebrate-periphyton cascade precipitate plant loss in shallow lakes? *Ecology* 84, 2155–2167.

Jonsson, T., and Ebenman, B. 1998. Effects of predator-prey body size ratios on the stability of food chains. *J. Theor. Biol.* 193, 407–417.

Jonsson, T., Cohen, J.E., and Carpenter, S.R. 2004. Food webs, body-size and species abundance in ecological community description. *Adv. Ecol. Res.* 36, 1–84.

Jordan, F., and Molnar, I. 1999. Reliable flows and preferred patterns in food webs. *Evol. Ecol. Res.* 1, 591–609.

Jordán, F., and Scheuring, I. 2002. Searching for keystones in ecological networks. *Oikos* 99, 607–612.

Jordano, P. 1987. Patterns of mutualistic interactions in pollination and seed dispersal: Connectance, dependence, and coevolution. *Am. Nat.* 129, 657–677.

Jordano, P., Bascompte, J., and Olesen, J.M. 2003. Invariant properties in coevolutionary networks of plant-animal interactions. *Ecol. Lett.* 6, 69–81.

Jordano, P., Bascompte, J., and Olesen, J.M. 2004. The ecological consequences of complex topology and nested structure in pollination webs. In: N.M. Waser and J. Ollerton, (Editors), *Specialisation and Generalisation in Plant-Pollinator Interactions*. Univ. Chicago Press.

Jorgensen, S.E. 1998. *A Pattern of Ecosystem Theories*. Kluwer, Dordecht, 320.

Jost, C., and Arditi, R. 2000. Identifying predator-prey processes from time-series. *Theor. Popul. Biol.* 57, 325–337.

Justic, D., Rabalais, N.N., and Turner, R.E. 1995. Stoichiometric nutrient balance and origin of coastal eutrophication. *Mar. Pollut. Bull.* 30, 41–46.

Kajak, A. 1995. The role of soil predators in decomposition processes. *Eur. J. Entomol.* 92, 573–580.

Kajak, A., and Jakubczyk, H. 1977. Experimental studies on predation in the soil-litter interface. *Ecol. Bull.* 25, 493–496.

Kajak, A., Chmielewski, K., Kaczmarek, M., and Rembialkowska, E. 1993. Experimental studies on the effect of epigeic predators on matter decomposition processes in managed peat grasslands. *Pol. Ecol. Stud.* 17, 289–310.

Kajak, A., Kusińska, A., Stanuszek, S., Stefaniak, O., Szanser, M., Wasylik, A., and Wasilewska, L. 2000. Effects of macroarthropods on organic matter content in soil. *Pol. J. Ecol.* 48, 339–354.

Kalff, J. 2002. *Limnology*. Prentice Hall, New York.

Karban, R., and Baldwin, I.T. 1997. *Induced Responses to Herbivory*. The University of Chicago Press, Chicago.

Kats, L.B., and Dill, M. 1998. The scent of death: Chemosensory assessment of predation risk by prey animals. *Ecoscience* 5, 361–394.

Kaunzinger, C.M.K., and Morin, P.J. 1998. Productivity controls food chain properties in microbial communities. *Nature* 395, 495–497.

Kaye, J.P., and Hart, S.C. 1997. Competition for nitrogen between plants and soil microorganisms. *Trends Ecol. Evol.* 12, 139–143.

Keeley, E.R. 2003. An experimental analysis of self-thinning in juvenile steelhead trout. *Oikos* 102, 543–550.

Kelt, D.A., and Van Vuren, D.H. 2001. The ecology and macroecology of mammalian home range area. *Am. Nat.* 157, 637–645.

Kendall, B.E., Pendergast, J., and Bjornstad, O.N. 1998. The macroecology of population dynamics: taxonomic and biogeographic patterns of population cycles. *Ecol. Lett.* 1, 160–164.

Kendall, B.E., Bjornstad, O.N., Bascompte, J., Keitt, T.H., and Fagan, W.F. 2000. Dispersal, environmental correlation, and spatial synchrony in population dynamics. *Am. Nat.* 155, 628–636.

Kendall, B.E., Briggs, C.J., Murdoch, W.W., Turchin, P., Ellner, S.P., McCauley, E., Nisbet, R.M., and Wood, S.N. 1999. Why do populations cycle? A synthesis of statistical and mechanistic modeling approaches. *Ecology* 80, 1789–1805.

Kennedy, T.A., Naeem, S., Howe, K.M., Knops, J.M.H., and Tilman, D. 2002. Biodiversity as a barrier to ecological invasion. *Nature* 417, 636–638.

Kennish, M.J., Raffaelli, D.G., Reise, K., and Livingston, R.J. 2005. Environmental threats and environmental futures of estuaries. In: N.V.C. Polunin (Editor), *Environmental Future of Aquatic Ecosystems*, in press.

Kenny, D., and Loehle, C. 1991. Are Food Webs Randomly Connected? *Ecology* 72, 1794–1799.

Kerr, S.R. 1974. Theory of size distribution in ecological communities. *J. Fish. Res. Bd. Can.* 31, 1859–1862.

Kerr, S.R., and Dickie, L.M. 2001. *Biomass Spectrum: A Predator-Prey Theory of Aquatic Production.* Columbia University Press, New York.

Kilpatrick, A.M., and Ives, A.R. 2003. Species interactions can explain Taylor's power law for ecological time series. *Nature* 422, 65–68.

Kimball, K.D., and Levin, S.A. 1985. Limitations of laboratory bioassays: the need for ecosystem-level testing. *Bioscience* 35, 165–171.

Kinzig, A.P., Pacala, S., and Tilman, D. (Editors). 2002. *The Functional Consequences of Biodiversity: Empirical Progress and Theoretical Extensions.* Princeton University Press, Princeton, New Jersey.

Kitchell, J.F., Oneill, R.V., Webb, D., Gallepp, G.W., Bartell, S.M., Koonce, J.F., and Ausmus, B.S. 1979. Consumer regulation of nutrient cycling. *Bioscience* 29, 28–34.

Kitching, R.L. 1987. Spatial and temporal variation in food webs in water-filled treeholes. *Oikos* 48, 280–288.

Kleiber, M. 1947. Body size and metabolic rate. *Physiol. Rev.* 27, 511–541.

Kleiber, M. 1961. *The Fire of Life. An Introduction of Animal Energetics.* Wiley, New York.

Klironomos, J.N., Bednarczuk, E.M., and Neville, J. 1999. Reproductive significance of feeding on saprobic and arbuscular mycorrhizal fungi by the collembolan, *Folsomia candida. Func. Ecol.* 13, 756–761.

Knapp, E.B., Elliot, L.F., and Campbell, G.S. 1983. Carbon, nitrogen and microbial biomass interrelationships during the decomposition of wheat straw: a mechanistic simulation model. *Soil Biol. Biochem.* 15, 455–461.

Knapp, R.A. 1996. *Non-Native Trout in Natural Lakes of the Sierra Nevada: An Analysis of Their Distribution and Impacts on Native Aquatic Biota. Sierra Nevada Ecosystem Project: Final report to Congress. Volume III,* University of California, Centers for Water and Wildland Resources, Davis.

Knapp, R.A. 2003. *Mountain Yellow-Legged Frog,* http://www.mylfrog.com.

Knapp, R.A., and Matthews, K.R. 1998. Eradication of nonnative fish by gill-netting from a small mountain lake in California. *Restor. Ecol.* 6, 207–213.

Knapp, R.A., and Matthews, K.R. 2000. Non-native fish introductions and the decline of the mountain yellow-legged frog from within protected areas. *Cons. Biol.* 14, 428–438.

Knapp, R.A., Matthews, K.R., and Sarnelle, O. 2001. Resistance and resilience of alpine lake fauna to fish introductions. *Ecol. Monogr.* 71, 410–421.

Koen-Alonso, M., and Yodzis, P. 2005. Multispecies modeling of some components of the northern and central Patagonia marine community, Argentina. *Can. J. Fish. Aquat. Sci.* 62, in press.

Koen-Alonso, M., Crespo, E.A., Pedraza, S.N., García, N.A., and Coscarella, M. 2000. Food habits of the South American sea lion, *Otaria flavescens,* off Patagonia, Argentina. *Fish. Bull.* 98, 250–263.

Kokkoris, G.D., Troumbis, A.Y., and Lawton, J.H. 1999. Patterns of species interaction strength in assembled theoretical competition communities. *Ecol. Lett.* 2, 70–74.

Kokkoris, G.D., Jansen, V.A.A., Loreau, M., and Troumbis, A.Y. 2002. Variability in interaction strength and implications for biodiversity. *J. Anim. Ecol.* 71, 362–371.

Kondoh, M. 2003a. Foraging adaptation and the relationship between food-web complexity and stability. *Science* 299, 1388–1391.

Kondoh, M. 2003b. Response to comment on "foraging adaptation and the relationship between food-web complexity and stability". *Science* 301, 918c.

Kondoh, M. 2006. Does foraging adaptation create the positive complexity-stability relationship in realistic food web structure? J. *of Theor. Biol.,* in press.

Kooi, B.W. 2003. Numerical bifurcation analysis of ecosystems in a spatially homogeneous environment. *Acta Biotheor.* 51, 189–222.

Kooi, B.W., Kuijper, L.D.J., Boer, M.P., and Kooijman, S.A.L.M. 2002. Numerical bifurcation analysis of a tri-trophic food web with omnivory. *Math. Biosci.* 177, 201–228.

Kooijman, S.A.L.M. 2000. *Dynamic Energy and Mass Budgets in Biological Systems, 2nd edition.* Cambridge University Press, Cambridge, UK.

Krause, A.E., Frank, K.A., Mason, D.M., Ulanowicz, R.E., and Taylor, W.W. 2003. Compartments revealed in food-web structure. *Nature* 426, 282–285.

Krebs, C.J. 1999. *Ecological Methodology.* Addison Wesley Longman, Inc, Don Mills Ontario.

Kretzschmar, M., Nisbet, R.M., and McCauley, E. 1993. A predator-prey model for zooplankton grazing on competing algal populations. *Theor. Popul. Biol.* 44, 32–66.

Krivan, V. 2000. Optimal intraguild foraging and population stability. *Theor. Popul. Biol.* 58, 79–94.

Krivan, V. 2002. *Adaptive Behavior & Food Web Dynamics.* University Presses of California, Princeton, New Jersey.

Kusinska, A., and Kajak, A. 2000. Mineralization and humification of Dactylis glomerata litter in a field experiment excluding macroarthropods. *Pol. J. Ecol.* 48, 299–310.

Kuznetsov, Y.A. 1998. *Elements of Applied Bifurcation Theory, Volume 112 of Applied Mathematical Sciences, 2nd edition.* Springer-Verlag, New York.

Kuznetsov, Y.A., and Levitin, V.V. 1997. *CONTENT: Integrated environment for the analysis of dynamical systems., 1.5 edition.* Centrum voor Wiskunde en Informatica (CWI), Kruislaan 413, 1098 SJ Amsterdam, The Netherlands.

La Rivers, I. 1962. *Fish and Fisheries of Nevada.* Nevada State Fish and Game Commission.

Laakso, J. 1999. Short-term effects of wood ants (*Formica aquilonia Yarr.*) on soil animal community structure. *Soil Biol. Biochem.* 31, 337–343.

Laakso, J., and Setälä, H. 1999a. Sensitivity of primary production to changes in the architecture of belowground food webs. *Oikos* 87, 57–64.

Laakso, J., and Setälä, H. 1999b. Population- and ecosystem-level effects of predation on microbial-feeding nematodes. *Oecologia* 120, 279–286.

Laakso, J., Setälä, H., and Palojärvi, A. 2000. Control of primary production by decomposer food web structure in relation to nitrogen availability. *Plant Soil* 225, 153–165.

Lamberti, G.A., and Steinman, A.D. 1993. Research in artificial streams: Applications, uses, and abuses. *J. N. Am. Benthol. Soc.* 12, 313–384.

Lammens, E.H.R.R., Van New, E.H., and Mooij, W.M. 2002. Differences in exploitation of bream in three shallow lake systems and their relation to water quality. *Freshwat. Biol.* 47, 2435–2442.

Lancaster, J., and Robertson, A.L. 1995. Microcrustacean prey and macroinvertebrate predators in a stream food-web. *Freshwat. Biol.* 34, 123–134.

Landsberg, J.H. 2002. The effects of harmful algal blooms on aquatic organisms. *Rev. Fisher. Sci.* 10, 113–390.

Laska, M.S., and Wootton, J.T. 1998. Theoretical concepts and empirical approaches to measuring interaction strength. *Ecology* 79, 461–476.

Lässig, M., Bastolla, U., Manrubia, S.C., and Valleriani, A. 2001. Shape of ecological networks. *Phys. Rev. Lett.* 86, 4418–4421.

Lathrop, R.C., and Carpenter, S.R. 1992. Zooplankton and their relationship to phytoplankton. In: J.F. Kitchell (Editor), *Food Web Management.* Springer-Verlag, New York.

Law, R. 1999. Theoretical aspects of community assembly. In: J. Mcglade (Editor), *Advanced Ecological Theory: Principles and Applications.* Blackwell, Oxford, pp 143–171.

Lawler, S.P., and Morin, P.J. 1993. Food web architecture and population dynamics in laboratory microcosms of protists. *Am. Nat.* 141, 675–686.

Lawlor, L.R. 1978. A comment on randomly constructed model ecosystems. *Am. Nat.* 112, 445–447.

Lawrence, K.L., and Wise, D.H. 2000. Spider predation on forest-floor Collembola and evidence for indirect effects on decomposition. *Pedobiologia* 44, 33–39.

Lawton, J., and Pimm, S.L. 1978. Population-dynamics and length of food-chains: Reply. *Nature* 272, 190.

Lawton, J.H. 1990. Species richness and population dynamics of animal assemblages. Patterns in body size abundance space. *Philos. T. Roy. Soc.* B 330, 230–291.

Lawton, J.H. 1994. What do species do in ecosystems? *Oikos* 71, 367–374.

Lawton, J.H. 1996. The Ecotron Facility at Silwood Park: The value of "Big Bottle" experiments. *Ecology* 77, 665–669.

Lawton, J.H., and Jones, C.G. 1995. Linking species and ecosystems: organisms as ecosystem engineers. In: C.G. Jones and J.H. Lawton (Editors), *Linking Species and Ecosystems*. Chapmann & Hall, New York, pp 141–150.

Layman, C.A., and Winemiller, K.O. 2004. Size-based response of multiple species to piscivore exclusion in a Neotropical river. *Ecology* 85, 1311–1320.

Layman, C.A., Winemiller, K.O., and Arrington, D.A. 2005. Patterns of habitat segregation among large fishes in Venezuelan floodplain river. *Neo. Ichthyol.* 3, 111–117.

Leake, J.R., and Read, D.J. 1997. Mycorrhizal Fungi in Terrestrial Habitats. In: D.T. Wicklow and B.E. Söderström (Editors), *Environmental and Microbial Relationships*. Springer-Verlag, Berlin, pp 281–301.

Leaper, R., and Raffaelli, D. 1999. Defining the abundance body-size constraint space: data from a real food web. *Ecol. Lett.* 2, 191–199.

Lee, K.E., and Pankhurst, C.E. 1992. Soil organisms and sustainable productivity. *Aust. J. Soil Res.* 30, 855–892.

Legendre, P., and Legendre, L. 1998. *Numerical Ecology. Developments in Environmental Modelling 20*. Elsevier, Amsterdam.

Lehman, C.L., and Tilman, D. 2000. Biodiversity, stability, and productivity in competitive communities. *Am. Nat.* 156, 534–552.

Leibold, M. 1996. A graphical model of keystone predators in food webs: trophic regulation of abundance, incindence, and diversity patterns in communites. *Am. Nat.* 147, 784–812.

Leibold, M.A. 1989. Resource edibility and the effects of predators and productivity on the outcome of trophic interactions. *Am. Nat.* 134, 922–949.

Leibold, M.A., and Wilbur, H.M. 1992. Interactions between food web structure and nutrients on pond organisms. *Nature* 360, 341–343.

Leibold, M.A., and Wootton, T.J. 2001. Introduction to the Reprinted Edition of: Elton, C.S. (1927). *Animal Ecology*. Chicago University Press, Chicago.

Leibold, M.A., Chase, J.M., Shurin, J.B., and Downing, A.L. 1997. Species turnover and the regulation of trophic structure. *Ann. Rev. Ecol. Syst.* 28, 467–494.

León, J.A., and Tumpson, D.B. 1975. Competition between two species for two complementary or subsitutable resources. *J. Theor. Biol.* 50, 185–201.

Levin, S. 1992. The problem of pattern and scale in ecology. *Ecology* 73, 1943–1967.

Levine, J.M. 2000. Species diversity and biological invasions: relating local processes to community pattern. *Science* 288, 852–854.

Levins, R. 1974. The qualitative analysis of partially specified systems. *Ann. NY Acad. Sci.* 231, 123–138.

Lewis, W.M.J., Hamilton, S.K., Rodríguez, M.A., Saunders, J.F.I., and Lasi, M.A. 2001. Food web analysis of the Orinoco floodplain based on production estimates and stable isotope data. *J. N. Am. Benthol. Soc.* 20, 241–254.

Lewontin, R.C. 1965. Selection for colonizing ability. In: H.G. Baker and G.L. Stebbins, (Editors), *The Genetics of Colonizing Species*. Academic Press, New York, pp 79–94.

Liiri, M., Setälä, H., and Ilmarinen, K. 2002a. The significance of Cognettia sphagnetorum (Enchytraeidae) on nitrogen availability in wood ash-treted humus soil. *Plant Soil* 246, 31–39.

Liiri, M., Setälä, H., Haimi, J., Pennanen, T., and Fritze, H. 2002b. Relationship between soil microarthropod species diversity and plant growth does not change when the system is disturbed. *Oikos* 96, 137–149.

Lima, S.L., and Dill, L.M. 1996. Behavioral decisions made under the risk of predation: a review and prospectus. *Can. J. Zoolog.* 68, 619–640.

Lindahl, B.O., Taylor, A.F.S., and Finlay, R.D. 2002. Defining nutritional constraints on carbon cycling in boreal forests—towards a less 'phytocentric' perspective. *Plant Soil* 242, 123–135.

Lindberg, A.B., and Olesen, J.M. 2001. The fragility of extreme specialisation: *Passiflora mixta* and its pollinating hummingbird *Ensifera ensifera*. *J. Trop. Ecol.* 17, 1–7.

Lindeman, R.L. 1942. The trophic-dynamic aspect of ecology. *Ecology* 23, 399–418.

Litvin, S.Y., and Weinstein, M.P. 2003. Life history strategies of estuarine nekton: the role of marsh macrophytes, benthic microalgae, and phytoplankton in the trophic spectrum. *Estuaries* 26, 552–562.

Lodge, D.M., Kershner, M.W., Aloi, J.E., and Covich, A.P. 1994. Effects of an omnivorous crayfish (*Orconectes rusticus*) on a freshwater littoral food web. *Ecology* 75, 1265–1281.

Loeuille, N., and Loreau, M. 2005. Evolutionary emergence of size-structured food webs. *P. Natl. Acad. Sci. USA* 102, 5761–5766.

Loladze, I., Kuang, Y., Elser, J.J., and Fagan, W.F. 2004. Competition and stoichiometry: coexistence of two predators on one prey. *Theor. Popul. Biol.* 65, 1–15.

Loreau, M. 1996. Coexistence of multiple food chains in a heterogeneous environment: interactions among community structure, ecosystem functioning, and nutrient dynamics. *Math. Biosci.* 134, 153–188.

Loreau, M. 1998a. Biodiversity and ecosystem functioning: a mechanistic model. *P. Natl. Acad. Sci. USA* 95, 5632–5636.

Loreau, M. 1998b. Separating sampling and other effects in biodiversity experiments. *Oikos* 82, 600–602.

Loreau, M. 2000. Biodiversity and ecosystem functioning: recent theoretical advances. *Oikos* 91, 3–17.

Loreau, M. 2001. Microbial diversity, producer-decomposer interactions and ecosystem processes: a theoretical model. *P. Roy. Soc. Lond. B Biol.* 268, 303–309.

Loreau, M., and Hector, A. 2001. Partitioning selection and complementarity in biodiversity experiments. *Nature* 412, 72–76.

Loreau, M., and Holt, R.D. 2004. Spatial flows and the regulation of ecosystems. *Am. Nat.* 163, 606–615.

Loreau, M., Naeem, S., and Inchausti, P., Editors. 2002. *Biodiversity and Ecosystem Functioning: Synthesis and Perspectives.* Oxford University Press, Oxford.

Loreau, M., Naeem, S., Inchausti, P., Bengtsson, J., Grime, J.P., Hector, A., Hooper, D.U., Huston, M.A., Raffaelli, D., Schmid, B., Tilman, D., and Wardle, D.A. 2001. Biodiversity and ecosystem functioning: Current knowledge and future challenges. *Science* 294, 804–808.

Lotka, A.J. 1925. *Elements of Physical Biology.* Williams and Williams.

Lovett, G.M., Christenson, L.M., Groffman, P.M., Jones, C.G., Hart, J.E., and Mitchell, M.J. 2002. Insect defoliation and nitrogen cycling in forests. *Bioscience* 52, 335–341.

Lowell, R.B., Culp, J.M., and Dubé, M.G. 2000. A weight-of evidence approach for northern river risk assessment: integrating the effects of multiple stressors. *Environmental Toxicology and Chemistry* 19, 1182–1190.

Lowe-McConnell, R.H. 1987. *Ecological Studies in Tropical Fish Communities.* Cambridge University Press, London, UK, 382.

Lubchenco, J. 1978. Plant species diversity in a marine intertidal community: importance of herbivore food preferences and algal competitive abilities. *Am. Nat.* 112, 23–39.

Luczkovich, J.J., Borgatti, S.P., Johnson, J.C., and Everett, M.G. 2003. Defining and measuring trophic role similarity in food webs using regular equivalence. *J. Theor. Biol.* 220, 303–321.

Lugthart, G.J., and Wallace, J.B. 1992. Effects of disturbance on benthic functional structure and production in mountain streams. *J. N. Am. Benthol. Soc.* 11, 138–164.

Lundberg, P., Ranta, E., and Kaitala, V. 2000. Species loss leads to community closure. *Ecol. Lett.* 3, 465–468.

Lundberg, S., and Persson, L. 1993. Optimal body size and resource density. *J. Theor. Biol.* 164, 163–180.

Lundberg, S., Järemo, J., and Nilsson, P. 1994. Herbivory, inducible defence and population oscillations: a preliminary theoretical analysis. *Oikos* 71, 537–540.

Lürling, M., Roozen, F., Donk, E.V., and Goser, B. 2003. Response of Daphnia to substances released from crowded congeners and conspecifics. *J. Plankton Res.* 25, 967–978.

Lussenhop, J. 1992. Mechanisms of microarthropod-microbial interactions in soil. *Adv. Ecol. Res.* 23, 1–33.

MacArthur, R. 1955. Fluctuations of animal populations, and a measure of community stability. *Ecology* 36, 533–536.

MacArthur, R.H., and Levins, R. 1964. Competition, habitat selection and character displacement in a patchy environment. *P. Natl. Acad. Sci. USA* 51, 1207–1210.

MacArthur, R.H., and Pianka, E.R. 1966. On optimal use of a patchy environment. *Am. Nat.* 100, 603–609.

Macko, S.A., Lee, W.Y., and Parker, P.L. 1982. Nitrogen and carbon isotope fractionation by two species of marine amphipods: laboratory and field studies. *J. Exp. Mar. Biol. Ecol.* 63, 145–149.

MacNamara, J.M., and Houston, A.I. 1987. Starvation and predation as factors limiting population size. *Ecology* 68, 1515–1519.

Malmqvist, B., and Rundle, S. 2002. Threats to the running water ecosystems of the world. *Environ. Cons.* 29, 134–153.

Mankowski, M.E., Schowalter, T.D., Morrell, J.J., and Lyons, B. 1998. Feeding habits and gut fauna of *Zootermopsis angusticollis* (Isoptera: Termopsidae) in response to wood species and fungal associates. *Environ. Entomol.* 27, 1315–1322.

Maraun, M., Martens, H., Migge, S., Theenhaus, A., and Scheu, S. 2003. Adding to 'the enigma of soil animal diversity': fungal feeders and saprophagous soil invertebrates prefer similar food substrates. *Eur. J. Soil Sci.* 39, 85–95.

Margalef, R. 1963. Certain unifying principles in ecology. *Am. Nat.* 97, 357.

Margalef, R. 1968. *Perspectives in Ecological Theory*. The University of Chicago Press, Chicago.

Marks, J.C., Power, M.E., and Parker, M.S. 2000. Flood disturbance, algal productivity, and interannual variation in food chain length. *Oikos* 90, 20–27.

Marquet, P.A., Navarrete, S.A., and Castilla, J.C. 1990. Scaling population density to body size in rocky intertidal communities. *Science* 250, 1125–1127.

Martinez, N., and Dunne, J.A. 2004. Response to comment on Virtual Ecosystems. *Conserv. Pract.* 5, 39.

Martinez, N.D. 1991. Artifacts of Attributes? Effects of resolution the Little Rock Lake food web. *Ecol. Monogr.* 61, 367–392.

Martinez, N.D. 1992. Constant connectance in community food webs. *Am. Nat.* 139, 1208–1218.

Martinez, N.D. 1993a. Effects of resolution on food web structure. *Oikos* 66, 403–412.

Martinez, N.D. 1993b. Effect of scale on food web structure. *Science* 260, 242–243.

Martinez, N.D. 1994. Scale-Dependent Constraints on Food-Web Structure. *Am. Nat.* 144, 935–953.

Martinez, N.D., Hawkins, B.A., Dawah, H.A., and Feifarek, B.P. 1999. Effects of sampling effort on characterization of food-web structure. *Ecology* 80, 1044–1055.

Mason, C.P., Edwards, K.R., Carlson, R.E., Pignatello, J., Gleason, F.K., and Wood, J.M. 1982. Isolation of chlorine-containing antibiotic from the freshwater cyanobacterium Scytonema hofmanni. *Science* 215, 400–402.

Mason, N.W.H., MacGillivray, K., Steel, J.B., and Wilson, J.B. 2003. An index of functional diversity. *J. Veg. Sci* 14, 571–578.

Matson, P., Lohse, K.A., and Hall, S.J. 2002. The globalization of nitrogen deposition: consequences for terrestrial ecosystems. *Ambio* 31, 113–119.

Matsuda, H., and Namba, T. 1991. Food web graph of a coevolutionarily stable community. *Ecology* 72, 267–276.

Matsuda, H., and Abrams, P.A. 2004. Effects of predator-prey interactions and adaptive change of sustainable yield. *Can. J. Fish. Aquat. Sci.* 61, 175–184.

Matsuda, H., Abrams, P.A., and Hori, M. 1993. The effect of adaptive antipredator behavior on exploitative competition and mutualism between predators. *Oikos* 68, 549–559.

Matsuda, H., Hori, M., and Abrams, P.A. 1996. Effects of predator-specific defense on biodiversity and community complexity in two-trophic-level communities. *Evol. Ecol.* 10, 13–28.

Matveev, V. 1993. An investigation of allelopathic effects of Daphnia. *Freshwat. Biol.* 29, 99–105.

Mautz, W.J., and Nagy, K.A. 1987. Ontogenetic changes in diet, field metabolic rate, and water flux in the herbivorous lizard *Dipsosaurus dorsalis*. *Physiol. Zool.* 60, 640–658.

May, R.M. 1972. Will a large complex system be stable? *Nature* 238, 413–414.

May, R.M. 1973. *Stability and Complexity in Model Ecosystems, 2nd edition*. Princeton University Press, Princeton, USA.

May, R.M. 1975. Patterns of species abundance and diversity. In: M. Cody and J. Diamond (Editors), *Ecology and Evolution of Communities*. Harvard University Press, Cambridge, pp 81–120.

May, R.M. 1983. The structure of food webs. *Nature* 301, 566–568.

May, R.M. 2004. Uses and abuses of mathematics in biology. *Science* 303, 790–793.

Maynard Smith, J. 1974. *Models in Ecology*. Cambridge University Press, Cambridge.

McCann, K. 2000. The diversity-stability debate. *Nature* 405, 228–233.

McCann, K., and Yodzis, P. 1994. Biological conditions for chaos in a three-species food chain. *Ecology* 75, 561–564.

McCann, K., and Hastings, A. 1997. Re-evaluating the omnivory-stability relationship in food webs. *P. Roy. Soc. Lond. B Biol.* 264, 1249–1254.

McCann, K., Hastings, A., and Huxel, G.R. 1998. Weak trophic interactions and the balance of nature. *Nature* 395, 794–798.

McCann, K.S., Rasmussen, J.R., and Umbanhowar, J. 2005. The dynamics of spatially coupled food webs. *Ecol. Lett.* 8, 513–523.

McCauley, E., Nisbet, R.M., Murdoch, W., de Roos, A.M., and Gurney, W.S.C. 1999. Large-amplitude cycles of Daphnia and its algal prey in enriched environments. *Nature* 402, 653–656.

McClanahan, T.R. 2002. The near future of coral reefs. *Environ. Cons.* 29, 460–483.

McGrady-Steed, J., and Morin, P.J. 2000. Biodiversity, density compensation, and the dynamics of populations and functional groups. *Ecology* 81, 361–373.

McGrady-Steed, J., Harris, P.M., and Morin, P.J. 1997. Biodiversity regulates ecosystem predictability. *Nature* 390, 162–165.

McIntosh, R.P. 1985. *The Background of Ecology-Concept and Theory*. Cambridge University Press, London.

McNaughton, S.J. 1977. Diversity and stability of ecological communities: a comment on the role of empiricism in ecology. *Am. Nat.* 111, 515–525.

McNaughton, S.J. 1979. Grazing as an optimization process: grass ungulate relationships in the Serengeti. *Am. Nat.* 113, 691–703.

McQueen, D.J., Johannes, M.R.S., Post, J.R., Stewart, T.J., and Lean, D.R. 1989. Bottom-up and top-down impacts on freshwater pelagic community structure. *Ecol. Monogr.* 59, 289–309.

Megraw, S., Reynoldson, T., Bailey, R., Burd, B., Corkum, L., Culp, J., Langlois, C., Porter, E., Rosenberg, D., Wildish, D., and Wrona, F. 1997. *Benthic Invertebrate Community Expert Working Group Final Report: Recommendations From Cycle 1 Review.* EEM/1997/7, Ottawa, Canada.

Melack, J.M., Stoddard, J.L., and Ochs, C.A. 1985. Major ion chemistry and sensitivity to acid precipitation of Sierra Nevada lakes. *Water Resour. Res.* 21, 27–32.

Melian, C.J., and Bascompte, J. 2002. Food web structure and habitat loss. *Ecol. Lett.* 5, 37–46.

Memmott, J., Martinez, N.D., and Cohen, J.E. 2000. Predators, parasitoids and pathogens: species richness, trophic generality and body sizes in a natural food web. *J. Anim. Ecol.* 69, 1–15.

Menge, B.A. 1995. Indirect effects in rocky intertidal interaction webs: patterns and importance. *Ecol. Monogr.* 65, 21–74.

Menge, B.A. 1997. Detection of direct versus indirect effects: were experiments long enough? *Am. Nat.* 149, 801–823.

Menge, B.A. 2004. Bottom-up:top-down determination of rocky intertidal shorescape dynamics. In: G.A. Polis, M.E. Power, and G. Huxel, (Editors), *Food Webs at the Landscape Level.* University of Chicago Press, Chicago, pp 62–81.

Menge, B.A., and Sutherland, J.P. 1987. Community regulation: variation in disturbance, competition, and predation in relation to environmental stress and recruitment. *Am. Nat.* 130, 730–757.

Menge, B.A., Berlow, E.L., Blanchette, C.A., Navarrete, S.A., and Yamada, S.B. 1994. The keystone species concept: variation in interaction strength in a rocky intertidal habitat. *Ecol. Monogr.* 64, 249–286.

Menge, B.A., Lubchenco, J., Bracken, M.E.S., Chan, F., Foley, M.M., Freidenburg, T.L., Gaines, S.D., Hudson, G., Krenz, C., Menge, D.N.L., Russell, R., and Webster, M.S. 2003. Coastal oceanography sets the pace of rocky intertidal community dynamics. *P. Natl. Acad. Sci. USA* 100, 12229–12234.

Merritt, R.W., and Cummins, K.W., Editors. 1984. *An Introduction to North American Aquatic Insects, 2nd edition.* Kendall-Hunt, Dubuque, Iowa, USA.

Merritt, R.W., and Cummins, K.W. 1996. *An Introduction to the Aquatic Insects of North America, 3rd edition.* Kendal/Hunt Publishing Company, Dubuque, Iowa, USA.

Metz, J.A.J., and Diekmann, O. 1986. *The dynamics of physiologically structured populations.* Springer, Heidelberg.

Metz, J.A.J., De Roos, A.M., and Van den Bosch, F. 1988. Population models incorporating physiological structure: A quick survey of the basic concepts and an application to size-structured population dynamics in waterfleas. In: B. Ebenman and L. Persson (Editors), *Size-Structured Populations—Ecology and Evolution.* Springer Verlag, Berlin, pp 106–126.

Meyer, J.L., and Schultz, E.T. 1985. Migrating Haemulid fishes as a source of nutrients and organic matter on coral reefs. *Limno. Oceanogr.* 30, 146–156.

Meyer, J.L., and Wallace, J.B. 2001. Lost linkages and lotic ecology: rediscovering small streams. In: N.J. Huntly and S. Levin (Editors), *Ecology: Achievement and Challenge.* M. C. Press, Blackwell Science Press.

Meyer, J.L., Benke, A.C., Edwards, R.T., and Wallace, J.B. 1997. Organic matter dynamics in the Ogeechee River, a blackwater river in Georgia, USA. *J. N. Am. Benthol. Soc.* 16, 82–87.

Mikola, J. 1998. Effects of microbivore species composition and basal resource enrichment on trophic -level biomasses in an experimental microbial-based soil food web. *Oecologia* 117, 396–403.

Mikola, J., and Setälä, H. 1998a. Relating species diversity to ecosystem functioning: mechanistic backgrounds and experimental approach with a decomposer food web. *Oikos* 83, 180–194.

Mikola, J., and Setälä, H. 1998b. No evidence of trophic cascades in an experimental microbial-based soil food-web. *Ecology* 79, 153–164.

Mikola, J., Bardgett, R.D., and Hedlund, K. 2002. Biodiversity and Ecosystem Functioning. In: M. Loreau, S. Naeem, and P. Inchausti (Editors), *Biodiversity and Ecosystem Functioning: Synthesis and Perspectives.* Oxford University Press, Oxford, pp 169–180.

Miller, T.E., and Kerfoot, W.C. 1987. Redefining indirect effects. In: W.C. Kerfoot and A. Sih, (Editors), *Predation: Direct and Indirect Effects on Aquatic Communities.* Hanover, Hanover, N.H, pp 33–37.

Milliron, C. 1999. *Aquatic Biodiversity Management Plan for the Big Pine Creek Wilderness Basin of the Sierra Nevada.* California Department of Fish and Game, Bishop, CA.

Milner, A.M., Knudssen, E.E., Soiseth, C., Robertson, A.L., Schell, D., Phillips, I.T., and Magnusson, K. 2000. Colonization and development of stream communities across a 200-year gradient in Glacier Bay National Park, Alaska, USA. *Can. J. Fish. Aquat. Sci.* 57, 2319–2335.

Minagawa, M., and Wada, E. 1984. Stepwise enrichment of d15N along food chains: further evidence and the relation between d15N and animal age. *Geochim. Cosmochim. Ac.* 48, 1135–1140.

Mitchell, C.E., and Power, A.G. 2003. Release of invasive plants from fungal and viral pathogens. *Nature* 421, 625–627.

Mitchell, J.J., and Parkinson, D. 1976. Fungal feeding of oribatid mites (Acari-Cryptostigmata) in an aspen woodland soil. *Ecology* 57, 302–312.

Mittelbach, G.G. 1981. Foraging efficiency and body size: a study of optimal diet and habitat use by bluegills. *Ecology* 62, 1370–1386.

Moen, J., and Oksanen, L. 1999. Ecosystem trends. *Nature* 353, 510.

Monaco, M.E., and Ulanowicz, R.E. 1997. Comparative ecosystem trophic structure of three U.S. mid-Atlantic estuaries. *Mar. Ecol.-Prog. Ser.* 161, 239–254.

Montoya, J.M., and Solé, R.V. 2002. Small-world patterns in food webs. *J. Theor. Biol.* 214, 405–412.

Montoya, J.M., Rodríguez, M.A., and Hawkins, B.A. 2003. Food web complexity and higher-level ecosystem services. *Ecol. Lett.* 6, 587–593.

Moore, J.C., and Hunt, H.W. 1988. Resource compartmentation and the stability of real ecosystems. *Nature* 333, 261–263.

Moore, J.C., and De Ruiter, P.C. 1991. Temporal and spatial heterogeneity of trophic interactions within below-ground food webs. *Agr. Ecosyst. Envir.* 34, 371–379.

Moore, J.C., and de Ruiter, P.C. 1997. Compartmentalization of resource utilization within soil ecosystems. In: V.K. Brown (Editor), *Multitrophic Interactions in Terrestrial Systems*. Blackwell Science, London, pp 375–393.

Moore, J.C., and de Ruiter, P.C. 2000. Invertebrates in detrital food webs along gradients of productivity. In: D.C. Coleman and P.F. Hendrix (Editors), *Invertebrates as Webmasters in Ecosystems*. CABI, Oxford, UK, pp 161–184.

Moore, J.C., St. John, T.V., and Coleman, D.C. 1985. Ingestion of vesicular-arbuscular mycorrhizal hyphae and spores by soil microarthropods. *Ecology* 66, 1979–1981.

Moore, J.C., Walter, D.E., and Hunt, H.W. 1988. Arthropod regulation of micro- and mesobiota in below-ground ditrital food webs. *Ann. Rev. Entomol.* 33, 419–439.

Moore, J.C., de Ruiter, P.C., and Hunt, W.J. 1993. Influence of productivity on the stability of real and model ecosystems. *Science* 261, 906–908.

Moore, J.C., McCann, K., Setälä, H., and de Ruiter, P. 2003. Top-down is bottom-up: does predation in the rhizosphere regulate aboveground dynamics? *Ecology* 84, 846–857.

Moore, J.C., Berlow, E.L., Coleman, D.C., de Ruiter, P., Dong, Q., Hastings, A., Johnson, N.C., McCann, K.S., Melville, K., Morin, P.J., Nadelhoffer, K., Rosemond, A.D., Post, D.M., Sabo, J.L., Scow, K.M., Vanni, M.J., and Wall, D. 2004. Detritus, trophic dynamics and biodiversity. *Ecol. Lett.* 7, 584–600.

Morin, P.J. 1995. Functional redundancy, nonadditive interactions, and supply-side dynamics in experimental pond communities. *Ecology* 76, 133–149.

Mork, M. 1996. The effect of kelp in wave damping. *Sarsia* 80, 323–327.

Mouquet, N., Moore, J.L., and Loreau, M. 2002. Plant species richness and community productivity: why the mechanism that promotes coexistence matters. *Ecol. Lett.* 5, 56–65.

Moyle, P.B., and Light, T. 1996. Biological invasions of fresh water: empirical rules and assembly theory. *Biol. Con.* 78, 149–161.

Mulder, C.P.H., Uliassi, D.D., and Doak, D.F. 2001. Physical stress and diversity-productivity relationships: the role of positive interactions. *P. Natl. Acad. Sci. USA* 98, 6704–6708.

Mulder, C., and De Zwart, D. 2003. Assessing fungal species sensitivity to environmental gradients by the Ellenberg indicator values of above-ground vegetation. *Basic Appl. Ecol.* 4, 557–568.

Mulder, C.P.H., Koricheva, J., Huss-Danell, K., Höberg, P., and Joshi, J. 1999. Insects affect relationships between plant species richness and ecosystem processes. *Ecol. Lett.* 2, 237–246.

Mulderij, G., Donk, E.V., and Roelofs, J.G.M. 2003. Differential sensitivity of green algae to allelopathic substances from Chara. *Hydrobiologia* 491, 261–271.

Mulholland, P.J., Tank, J.L., Sanzone, D.M., W.Wollheim, Peterson, B.J., Webster, J.R., and Meyer, J.L. 2000. Nitrogen cycling in a deciduous forest stream determined from a tracer 15N addition experiment in Walker Branch, Tennessee. *Ecol. Monogr.* 70, 471–493.

Muller, E.B., Nisbet, R.M., Kooijman, S.A.L.M., Elser, J.J., and McCauley, E. 2001. Stoichiometric food quality and herbivore dynamics. *Ecol. Lett.* 4, 519–529.

Murdoch, W.W. 1969. Switching in general predators: experiments on predator specificity and stability of prey populations. *Ecol. Monogr.* 39, 335–354.

Murdoch, W.W. 1973. The functional response of predators. *J. Ap. Ecol.* 10, 335–342.

Murdoch, W.W., and McCauley, E. 1985. Three distinct types of dynamics behaviour shown by a single planktonic system. *Nature* 316, 628–630.

Murdoch, W.W., Briggs, C.J., and Nisbet, R.M. 2003. *Consumer-Resource Dynamics. Monographs in Population Biology.* Princeton University Press, Princeton.

Murdoch, W.W., Nisbet, R.M., Gurney, W.S.C., and Reeve, J.D. 1987. An invulnerable age class and stability in delay-differential parasitoid-host models. *Am. Nat.* 129, 263–282.

Murdoch, W.W., Kendall, B.E., Nisbet, R.M., Briggs, C.J., McCauley, E., and Bolser, R. 2002. Single-species models for many-species food webs. *Nature* 417, 541–543.

Murtaugh, P.A., and Kollath, J.P. 1997. Variation of trophic fractions and connectance in food webs. *Ecology* 78, 1382–1387.

Muthukrishnan, J., Mathavan, S., Navarathina, J. V. 1978. Restriction of feeding duration in *Bombyx. Monnitore Zoologico Italiano* 12, 87–94.

Naeem, S. 1998. Species redundancy and ecosystem reliability. *Cons. Biol.* 12, 39–45.

Naeem, S., and Li, S. 1997. Biodiversity enhances ecosystem reliability. *Nature* 390, 507–509.

Naeem, S., and Li, S. 1998. Consumer species richness and autotrophic biovolume. *Ecology* 79.

Naeem, S., Hahn, D.R., and Schuurman, G. 2000a. Producer-decomposer co-dependency influences biodiversity effects. *Nature* 403, 762–764.

Naeem, S., Thompson, L.J., Lawler, S.P., Lawton, J.H., and Woodfin, R.M. 1994. Declining biodiversity can alter the performance of ecosystems. *Nature* 368, 743–737.

Naeem, S., Knops, J.M.H., Tilman, D., Howe, K.M., Kennedy, T., and Gale, S. 2000b. Plant diversity increases resistance to invasion in the absence of covarying extrinsic factors. *Oikos* 91, 97–108.

Nakano, S., and Murakami, M. 2001. Reciprocal subsidies: dynamic interdependence between terrestrial and aquatic food webs. *P. Natl. Acad. Sci. USA* 98, 166–170.

Nee, S., and Lawton, J.H. 1996. Body-size and biodiversity. *Nature* 380, 672–673.

Nee, S., Read, A.F., Greenwood, J.J.D., and Harvey, P. 1991. The relationship between abundance and body size in British birds. *Nature* 351, 312–313.

Neff, J.C., Townsend, A.R., Gleixner, G., Lehman, S.J., Turnbull, J., and Bowman, W.D. 2002. Variable effects of nitrogen additions on the stability and turnover of soil carbon. *Nature* 419, 915–917.

Neill, W.E. 1975. Experimental studies of microcrustacean competition, community composition and efficiency of resource utilization. *Ecology* 56, 809–826.

Nelson, W.A., McCauley, E., and Wrona, F.J. 2001. Multiple dynamics in a single predator-prey system: experimental effects of food quality. *P. Roy. Soc. Lond. B Biol.* 268, 1223–1230.

Neubert, M.G., and Caswell, H. 1997. Alternatives to resilience for measuring the responses of ecological systems to perturbations. *Ecology* 78, 653–665.

Neutel, A.M. 2001. *Stability of Complex Food Webs: Pyramids of Biomass, Interaction Strengths and the Weight of Trophic Loops.* University Utrecht, Utrecht.

Neutel, A.M., Heesterbeek, J.A.P., and de Ruiter, P.C. 2002. Stability in real food webs: weak links in long loops. *Science* 296, 1120–1123.

Niyogi, D.K., Lewis, W.M., and McKnight, D.M. 2002. Effects of stress from mine drainage on diversity, biomass, and function of primary producers in mountain streams. *Ecosystems* 54, 554–567.

Norberg, J. 2000. Resource-niche complementarity and autotrophic compensation determines ecosystem-level responses to increased cladoceran species richness. *Oecologia* 122, 264–272.

Numerics, V. 1997. *IMSL: Fortran Subroutines for Mathematical Applications. Math/Library Volume 1 and 2. in.* Visual Numerics Inc., Houston.

Oaten, A., and Murdoch, W.W. 1975a. Switching, functional response and stability in predator-prey populations. *Am. Nat.* 109, 299–318.

Oaten, A., and Murdoch, W. 1975b. Functional response and stability in predator-prey systems. *Am. Nat.* 109, 289–298.

Odum, E.P. 1953. *Fundamentals of Ecology, 3rd edition.* Saunders, Philidelphia.

Odum, E.P. 1963. *Ecology.* Holt, Rinehard, and Winston, New York.

Odum, E.P. 1969. The strategy of ecosystem development. *Science* 164, 262–279.

Odum, E.P. 1973. *Fundamentals of Ecology, 3rd edition.* Saunders, Philadelphia.

Odum, H.T. 1957. Trophic structure and productivity of Silver Springs, Florida. *Ecol. Monogr.* 27, 55–112.

Oksanen, L. 1988. Ecosystem organization-mutualism and cybernetics or plain Darwinian struggle for existence. *Am. Nat.* 131, 424–444.

Oksanen, L., Fretwell, S.D., Arruda, J., and Niemelä, P. 1981. Exploitation ecosystems in gradients of primary productivity. *Am. Nat.* 118, 240–261.

Olesen, J.M., and Jordano, P. 2002. Geographic patterns in plant/pollinator mutualistic networks. *Ecology* 83, 2416–2424.

O'Neill, R.V. 1969. Indirect Estimation of Energy Fluxes in Animal Food Webs. *J. Theor. Biol.* 22, 284–290.

Östman, Ö., Ekbom, B., and Bengtsson, J. 2001. Natural enemy impacts on a pest aphid varies with landscape structure and farming practice. *Basic Appl. Ecol.* 2, 365–371.

Pacala, S., and Roughgarden, J. 1984. Control of arthropod abundance by anolis lizards on St. Eustatius (Neth Antilles). *Oecologia* 64, 160–162.

Paine, R. 2004. Comment on virtual ecosystems. *Conserv. Pract.* 5, 39.

Paine, R.T. 1966. Food web complexity and species diversity. *Am. Nat.* 100, 65–75.

Paine, R.T. 1969. A note on trophic complexity and community stability. *Am. Nat.* 103, 91–93.

Paine, R.T. 1974. Intertidal community structure. Experimental studies on the relationship between a dominant competitor and its principal predator. *Oecologia* 15, 93–120.

Paine, R.T. 1980. Food webs: linkage, interaction strength and community infrastructure. *J. Anim. Ecol.* 49, 667–685.

Paine, R.T. 1988. Food webs: Road maps of interactions or grist for theoretical development? *Ecology* 69, 1648–1654.

Paine, R.T. 1992. Food-web analysis through field measurements of per capita interaction strength. *Nature* 355, 73–75.

Paine, R.T. 2002. Trophic control of production in a rocky intertidal community. *Science* 296, 736–739.

Parkin, H., Cousins, S.H. 1981. *Towards a global model of large ecosystems: equations for the trophic continuum*. ERG 041, Open University, Milton Keynes.

Parmelee, R.W. 1995. Soil fauna: linking different levels of the ecological hierarchy. In: C.G. Jones and J. Lawton (Editors), *Linking Species and Ecosystems*. Chapman and Hall, London, pp 107–116.

Pascual, M.A., and Kareiva, P. 1996. Predicting the outcome of competition using experimental data: maximum likelihood and Bayesian approaches. *Ecology* 77, 337–349.

Patten, B.C. 1959. An introduction to the cybernetics of the ecosystem: The trophic-dynamic aspect. *Ecology* 40, 221–231.

Paul, E.A., and Clark, F.E. 1989. *Soil Microbiology and Biochemistry*. Academic Press, San Diego.

Pauly, D., Christensen, V., Dalsgaard, J., Froese, R., and Torres, F. 1998. Fishing down marine food webs. *Science* 279, 860–863.

Pauly, D., Alder, J., Bennett, E., Christensen, V., Tyedmers, P., and Watson, R. 2003. The future for fisheries. *Science* 302, 1359–1361.

Peacor, S.D., and Werner, E.E. 2004. How dependent are species-pair interaction strengths on other species in the food web? *Ecology* 85, 2754–2763.

Pelletier, J.D. 2000. Are large complex ecosystems more unstable? A theoretical reassessment with predator switching. *Math. Biosci.* 163, 91–96.

Pennak, R.W. 1989. *Freshwater Invertebrates of the United States: Protozoa to Mollusca, 3rd edition*. John Wiley, New York, New York, USA.

Pennanen, T., Liski, J., Baath, E., Kitunen, V., Uotila, J., Westman, C.J., and Fritze, H. 1999. Structure of the microbial communities in coniferous forest soils in relation to site fertility and stand development stage. *Microbial Ecol.* 38, 168–179.

Persson, A., Hansson, L.A., Bronmark, C., Lundberg, P., Pettersson, L.B., Greenberg, L., Nilsson, P.A., Nystrom, P., Romare, P., and Tranvik, L. 2001. Effects of enrichment on simple aquatic food webs. *Am. Nat.* 157, 654–669.

Persson, L. 1999. Trophic cascades: abiding heterogeneity and the trophic level concept at the end of the road. *Oikos* 85, 385–397.

Persson, L., and Johansson, L. 1992. On competition and temporal variation in temperate freshwater fish populations. *Neth. J. Zool.* 42, 304–322.

Persson, L., De Roos, A.M., and Bertolo, A. 2004. Predicting shifts in dynamical regimes in cannibalistic field populations using individual-based models. *P. Roy. Soc. Lond. B Biol.* 271, 2489–2493.

Persson, L., Diehl, S., Johansson, L., Andersson, G., and Hamrin, S.F. 1992. Trophic interactions in temperate lake ecosystems: a tes of food-chain theory. *Am. Nat.* 140, 59–84.

Persson, L., Leonardsson, K., Gyllenberg, M., De Roos, A.M., and Christensen, B. 1998. Ontogenetic scaling of foraging rates and the dynamics of a size-structured consumer-resource model. *Theor. Popul. Biol.* 54, 270–293.

Persson, L., De Roos, A.M., Claessen, D., Byström, P., Lövgren, J., Svanbäck, R., Wahlström, E., and Westman, E. 2003. Gigantic cannibals driving a whole lake trophic cascade. *P. Natl. Acad. Sci. USA* 100, 4035–4039.

Persson, T., van Oene, H., Harrison, A.F., Karlsson, P., Bauer, G., Cerny, J., Coûteaux, M.M., Dambrine, E., Högberg, P., Kjøller, A., Matteucci, G., Rudebeck, A., Schulze, E.D., and Paces, T. 2000. Experimental Sites in the NIPHYS/CANIF Project. In: E.D. Schulze, (Editor), *Carbon and Nitrogen Cycling in Forest Ecosystems.* Springer, Heidelberg, pp 14–46.

Petchey, O.L. 2000. Prey diversity, prey composition, and predator population dynamics in experimental microcosms. *J. Anim. Ecol.* 69, 874–882.

Petchey, O.L. 2004. On the statistical significance of functional diversity. *Func. Ecol.* 18, 297–303.

Petchey, O.L., and Gaston, K.J. 2002a. Extinction and the loss of functional diversity. *P. Roy. Soc. Lond. B Biol.* 269, 1721–1727.

Petchey, O.L., and Gaston, K.J. 2002b. Functional diversity (FD), species richness and community composition. *Ecol. Lett.* 5, 402–411.

Petchey, O.L., Hector, A., and Gaston, K.J. 2004a. How do different measures of functional diversity perform? *Ecology* 85, 847–857.

Petchey, O.L., McPhearson, P.T., Casey, T.M., and Morin, P.J. 1999. Environmental warming alters food-web structure and ecosystem function. *Nature* 402, 69–72.

Petchey, O.L., Downing, A.L., Mittelbach, G.G., Persson, L., Steiner, C.F., Warren, P.H., and Woodward, G. 2004b. Species loss and the structure and functioning of multitrophic aquatic systems. *Oikos* 104, 467–478.

Peters, R.H. 1983. *The Ecological Implications of Body Size.* Cambridge University Press, Cambridge.

Peters, R.H., and Raelson, J.V. 1984. Relations between individual size and mammalian population density. *Am. Nat.* 124, 498–517.

Peters, R. H., and Wassenberg, K. 1983. The effect of body size on animal abundance. *Oecologia* 60, 89–96.

Petersen, H., and Luxton, M. 1982. A comparative analysis of soil fauna populations and their role in decomposition processes. *Oikos* 39, 286–388.

Peterson, B.J., and Fry, B. 1987. Stable isotopes in ecosystem studies. *Ann. Rev. Ecol. Syst.* 18, 293–320.

Peterson, B.J., Bahr, M., and Kling, G.W. 1997. A tracer investigation of nitrogen cycling in a pristine tundra river. *Can. J. Fish. Aquat. Sci.* 54, 2361–2367.

Peterson, B.J., Wollheim, W.M., Mulholland, P.J., Webster, J.R., Meyer, J.L., Tank, J.L., Marti, E., Bowden, W.B., Valett, H.M., Hershey, A.E., McDowell, W.H., Dodds, W.K., Hamilton, S.K., S.Gregory, and Morrall, D.D. 2001. Control of nitrogen export from watersheds by headwater streams. *Science* 292, 86–90.

Petrusewitz, K., MacFadyen, A. 1970. *Productivity of terrestrial animals.* Blackwell, Oxford.

Phillips, D.L., and Koch, P.L. 2002. Incorporating concentration dependence in stable isotope mixing models. *Oecologia* 130, 114–125.

Phillips, O.M. 1974. The equilibrium and stability of simple marine systems. II. Herbivores. *Arch. Hydrobiol.* 73, 310–333.

Phillipson, J. 1966. *Ecological Energetics.* Edward Arnold, London.

Pianka, E.R. 1970. On r- and K-selection. *Am. Nat.* 104, 592–597.

Pilati, A., Wurtsbaugh, W.A., and Brindza, N.R. 2004. Evidence of coprophagy in freshwater zooplankton. *Freshwat. Biol.* 48, 913–918.

Pimentel, D. 1971. Evolution of niche width. *Am. Nat.* 106, 683–718.

Pimm, S.L. 1979. The structure of food webs. *Theor. Popul. Biol.* 16, 144–158.

Pimm, S.L. 1980a. Bounds of food web connectance. *Nature* 285, 591.

Pimm, S.L. 1980b. Properties of food webs. *Ecology* 61, 219–225.

Pimm, S.L. 1982. *Food Webs.* Chapman and Hall, London.

Pimm, S.L. 1991. *The Balance of Nature?: Ecological Issues in the Conservation of Species and Communities.* University of Chicago Press, Chicago.

Pimm, S.L. 1993. Understand indirect effects: is it possible? In: H. Kawanabe, J.E. Cohen, and K. Iwasaki, (Editors), *Mutualism and Community Organization.* Oxford University Press, New York, pp 199–209.

Pimm, S.L., and Lawton, J.H. 1977. The number of trophic levels in ecological communities. *Nature* 268, 329–331.

Pimm, S.L., and Lawton, J.H. 1978. On feeding on more than one trophic level. *Nature* 275, 542–544.

Pimm, S.L., and Lawton, J.H. 1980. Are food webs divided into compartments. *J. Anim. Ecol.* 49, 879–898.

Pimm, S.L., and Rice, J.C. 1987. The dynamics of multispecies, multi-life-stage models of aquatic food webs. *Theor. Popul. Biol.* 32, 303–325.

Pimm, S.L., and Kitching, R.L. 1987. The determinants of food chain lengths. *Oikos* 50, 302–307.

Pimm, S.L., Lawton, J.H., and Cohen, J.E. 1991. Food web patterns and their consequences. *Nature* 350, 669–674.

Platt, T., and Denman, K. 1977. Organisation in the pelagic ecosystem. *Helgolander. Meeresun.* 30, 575–581.

Platt, T., and Denman, K. 1978. *The structure of pelagic marine ecosystems.* Rapports et Procés-Verbaux des Réunions, Conseil International pour l'Exploration de la Mer 173, 60–65.

Poff, N.L., Wellnitz, T., and Monroe, J.B. 2003. Redundancy among three herbivorous insects across an experimental current velocity gradient. *Oecologia* 134, 262–269.

Pokarzhevskii, A.D., van Straalen, N.M., Zaboev, D.P., and Zaitsev, A.S. 2003. Microbial links and elemental flows in nested detrital food-webs. *Pedobiologia* 47, 213–224.

Polis, G.A. 1991. Complex trophic interactions in deserts: an empirical critique of food web theory. *Am. Nat.* 138, 123–155.

Polis, G.A. 1994. Food webs, trophic cascades and community structure. *Aust. J. Ecol.* 19, 121–136.

Polis, G.A. 1998. Stability is woven by complex food webs. *Nature* 395, 744.

Polis, G.A. 1999. Why are parts of the world green? Multiple factors control productivity and the distribution of biomass. *Oikos* 86, 3–15.

Polis, G.A., and Holt, R.D. 1992. Intraguild predation: the dynamics of complex trophic interactions. *Trends Ecol. Evol.* 7, 151–155.

Polis, G.A., and Winemiller, K.O. Editors. 1996. *Food Webs: Integration of Patterns and Dynamics.* Chapman & Hall, New York.

Polis, G.A., and Hurd, S.D. 1996. Linking marine and terrestrial food webs: Allochthonous input from the ocean supports high secondary productivity on small islands and coastal land communities. *Am. Nat.* 147, 396–423.

Polis, G.A., and Strong, D.R. 1996. Food web complexity and community dynamics. *Am. Nat.* 147, 813–846.

Polis, G.A., Myers, C.A., and Holt, R.D. 1989. The ecology and evolution of intraguild predation: potential competitors that eat each other. *Ann. Rev. Ecol. Syst.* 20, 297–330.

Polis, G.A., Anderson, W.B., and Holt, R.D. 1997. Towards an integration of landscape and food web ecology. *Ann. Rev. Ecol. Syst.* 28, 289–316.

Polis, G.A., Holt, R.D., Menge, B.A., and Winemiller, K.O. 1996a. Time, space, and life history: influences on food webs. In: G.A. Polis and K.O. Winemiller, (Editors), *Food Webs: Integration of Patterns and Dynamics.* Chapman and Hall, New York, pp 435–460.

Polis, G.A., Holt, R.D., Menge, B.A., and Winemiller, K.O. 1996b. Temporal and spatial components of food webs. In: G.A. Polis and K.O. Winemiller (Editors), *Food Webs: Integration of Patterns and Dynamics.* Chapman and Hall, New York, pp 435–460.

Ponge, J.F. 1991. Food resources and diets of soil animals in a small area of Scots pine litter. *Geoderma* 49, 33–62.

Ponge, J.F. 2000. Vertical distribution of Collembola (Hexapoda) and their food resources in organic horizons of beech forests. *Biol. Fert. Soils* 32, 508–522.

Ponsard, S., and Arditi, R. 2000. What can stable isotopes (delta15N and delta13C) tell about the food web of soil macro-invertebrates? *Ecology* 81, 852–864.

Pope, D.P. 1984. Methods of vegetative regeneration and moisture requirements of selected southwest riparian species. MS Thesis. Arizona State University, Tempe.

Post, D.M. 2002a. The long and short of food-chain length. *Trends Ecol. Evol.* 17, 269–277.

Post, D.M. 2002b. Using stable isotopes to estimate trophic position: Models, methods, and assumptions. *Ecology* 83, 703–718.

Post, D.M. 2003. Individual variation in the timing of ontogenetic niche shifts in largemouth bass. *Ecology* 84, 1298–1310.

Post, D.M., Conners, M.E., and Goldberg, D.S. 2000a. Prey preference by a top predator and the stability of linked food chains. *Ecology* 81, 8–14.

Post, D.M., Pace, M.L., and Jr., N.G.H. 2000b. Ecosystem size determines food-chain length in lakes. *Nature* 405, 1047–1049.

Power, M.E. 1990. Effects of fish in river food webs. *Science* 250, 811–814.

Power, M.E. 1992. Top-down and bottom-up forces in food webs: Do plants have primacy? *Ecology* 73, 733–746.

Power, M.E., and Rainey, W.E. 2000. Food webs and resource sheds: towards spatially delimiting trophic interactions. In: M.J. Hutchings, E.A. John, and A.J.A. Stewart (Editors), *The Ecological Consequencies of Environmental Heterogeneity*. Blackwell Science Ltd, Oxford, pp 291–314.

Power, M.E., Parker, M.S., and Wootton, J.T. 1996a. Disturbance and food chain length in rivers. In: G.A. Polis and K.O. Winemiller (Editors), *Food Webs: Integration of Pattern and Dynamics*. Chapmann & Hall, New York, pp 286–297.

Power, M.E., Tilman, D., Estes, J.A., Menge, B.A., Bond, W.J., Mills, L.S., Daily, G., Castilla, J.C., Lubchenco, J., and Paine, R.T. 1996b. Challenges in the quest for keystones. *Bioscience* 46, 609–620.

Prenski, L.B., and Angelescu, V. 1993. Ecología trófica de la merluza común (*Merluccius hubbsi*) del Mar Argentino. Parte 3. Consumo anual de alimento a nivel poblacional y su relación con la explotación de las pesquerías multiespecíficas. INIDEP Documento Científico 1, 1–118.

Pringle, C.M., Hemphill, N., McDowell, W.H., Bednarek, A., and March, J.G. 1999. Linking species and ecosystems: different biotic assemblages cause interstream differences in organic matter. *Ecology* 80, 1860–1872.

Proulx, M., and Mazunder, A. 1998. Reversal of grazing impact on plant species reichness in nutrient-poor vs. nutrient-rich ecosystems. *Ecology* 79, 2581–2592.

Pulliam, H.R. 1974. On the theory of optimal diets. *American Naturalist* 109, 765–768.

Pulliam, H.R. 1988. Sources, sinks, and population regulation. *Am. Nat.* 132, 652–661.

Pulliam, H.R., and Danielson, B.J. 1991. Sources, sinks, and habitat selection: A landscape perspective on population-dynamics. *Am. Nat.* 137, S50–S60.

Quammen, D. 1996. *The Song of the Dodo: Island Biogeography in an Age of Extinctions*. Simon and Schuster, New York.

Quince, C., Higgs, P.G., and McKane, A.J. 2002. Food web Structure and the Evolution of Ecological Comminities. In: M. Lässig and A. Valleriani, (Editors), *Biological Evolution and Statistical Physics*. Springer-Verlag, Berlin.

Raffaelli, D. 2002. Ecology - From Elton to mathematics and back again. *Science* 296, 1035.

Raffaelli, D., and Hall, S.J. 1992. Compartmentation and predation in an estuarine food web. *J. Anim. Ecol.* 61.

Raffaelli, D., Emmerson, M., Solan, M., Biles, C., and Paterson, D. 2003. Biodiversity and ecosystem processes in shallow coastal waters: an experimental approach. *J. Sea Res.* 49, 133–141.

Raffaelli, D., van der Putten, W.H., Persson, L., Wardle, D.A., Petchey, O.L., Koricheva, J., van der Heijden, M.G.A., Mikola, J., and Kennedy, T. 2002. Multi-Trophic Processes and Ecosystem Functioning. In: M. Loreau, S. Naeem, and P. Inchausti (Editors), *Biodiversity and Ecosystem Functioning: Syntheses and Perspectives*. Oxford University Press, Oxford, pp 147–154.

Raffaelli, D.G. 2000. Trends in marine food research. *J. Exp. Mar. Biol. Ecol.* 250, 223–232.

Raffaelli, D.G., and Hall, S.J. 1996. Assessing the relative importance of trophic links in food webs. In: G.A. Polis and K.O. Winemiller (Editors), *Food Webs: Integration of Pattern and Dynamics*. Chapman and Hall, New York, pp 185–191.

Raffaelli, D.G., and Moller, H. 2000. Manipulative experiments in animal ecology - do they promise more than they can deliver? *Adv. Ecol. Res.* 30, 299–330.

Rahel, F.J. 2002. Homogenization of freshwater fish faunas. *Ann. Rev. Ecol. Syst.* 33, 291–315.

Ramos-Jiliberto, R. 2003. Population dynamics of prey exhibiting inducible defenses: the role of associated costs and density-dependence. *Theor. Popul. Biol.* 64, 221–231.

Read, D.J. 1991. Mycorrhizas in ecosystems. *Experientia* 47, 376–391.

Reagan, D.P., and Waide, R.B. Editors. 1996. *The Food Web of a Tropical Rain Forest*. The University of Chicago Press, Chicago.

Redfield, A.C. 1958. The biological control of chemical factors in the environment. *Am. Nat.* 46, 205–221.

Reich, P.B., Tilman, D., Naeem, S., Ellsworth, D.S., Knops, J., Craine, J., Wedin, D., and Trost, J. 2004. Species and functional group diversity independently influence biomass accumulation and its response to CO2 and N. *P. Natl. Acad. Sci. USA* 101, 10101–10106.

Reid, C.P.P., and Woods, F.W. 1969. Translocation of C14-labeled compounds in mycorrhizae and its implications in interplant nutrient cycling. *Ecology* 50, 179.

Reilly, P.M., and Patino-Leal, H. 1981. A Bayesian study of the error-in-variables model. *Technometrics* 23, 221–231.

Reiners, W.A. 1986. Complementary models for ecosystems. *Am. Nat.* 127, 59–73.

Rejmánek, M., and Stary, P. 1979. Connectance in real biotic communities and critical values for stability of model ecosystems. *Nature* 280, 311–313.

Relyea, R.A., and Yurewicz, K.L. 2002. Predicting community outcomes from pairwise interactions: integrating density- and trait-mediated effects. *Oecologia* 131, 569–579.

Reuman, D., and Cohen, J.E. 2004a. Estimating relative energy fluxes using the food web, species abundance, and body size. *Adv. Ecol. Res.* 36, 137–182.

Reuman, D.C., and Cohen, J.E. 2004b. Trophic links' length and slope in the Tuesday Lake food web with species' body mass and numerical abundance. *J. Anim. Ecol.* 73, 852–866.

Reuman, D.C., and Cohen, J.E. 2005. Estimating relative energy fluxes using the food web, species abundance, and body size. *Adv. Ecol. Res.* 36, 137–182.

Reyes, L.M., Crespo, E.A., and Szapkievich, V. 1999. Distribution and population size of the southern sea lion (Otaria flavescens) in central and southern Chubut, Argentina. *Mar. Mammal Sci.* 15, 478–493.

Riisgård, H.U. 1998. No foundation of a "3/4 power scaling law" for respiration in biology. *Ecol. Lett.* 1, 71–73.

Rinaldo, A., Maritan, A., Cavender-Bares, K.K., and Chisholm, S.W. 2002. Cross-scale ecological dynamics and microbial size spectra in marine ecosystems. *P. Roy. Soc. Lond. B Biol.* 269, 2051–2059.

Ripa, J., Lundberg, P., and Kaitala, V. 1998. A general theory of environmental noise in ecological food webs. *Am. Nat.* 151, 256–263.

Ritchie, M., and Olff, H. 1999. Spatial scaling laws yield a synthetic theory of biodiversity. *Nature* 400, 557–560.

Robinson, J.G., and Redford, K.H. 1986. Body size, diet and population density of Neotropic mammals. *Am. Nat.* 128, 665–680.

Rodríguez, J.P. 2000. Impact of the Venezuelan economic crisis on wild populations of animals and plants. *Biol. Con.* 96, 151–159.

Roff, D.A. 1986. Predicting body size with life history models. *Bioscience* 36, 316–323.

Romanuk, T.N., and Kolasa, J. 2001. Simplifying the complexity of temporal diversity dynamics: a differentiation approach. *Ecoscience* 8, 259–263.

Romanuk, T.N., and Kolasa, J. 2002a. Environmental variability alters the relationship between richness and variability of community abundances in aquatic rock pool microcosms. *Ecoscience* 9, 55–62.

Romanuk, T.N., and Kolasa, J. 2002b. Abundance and species richness in natural aquatic microcosms: a test and a refinement of the niche-limitation hypothesis. *Community Ecol.* 3, 87–94.

Romanuk, T.N., and Kolasa, J. 2004. Resource limitation, biodiversity, and competitive effects interact to determine the invasibility of rock pool microcosms. *Biol. Invas.*, in press.

Root, R.B. 1967. The niche exploitation pattern of the Blue-gray Gnatcatcher. *Ecol. Monogr.* 37, 317–350.

Root, R.B. 2001. Guilds. In: S.A. Levin (Editor), *Encyclopedia of Biodiversity*. Academic Press, San Diego, pp 295–302.

Rose, M.D., and Polis, G.A. 1998. The distribution and abundance of coyotes: the effects of allochthonous food subsidies from the sea. *Ecology* 79, 998–1007.

Rosemond, A.D., Pringle, C.M., Ramirez, A., and Paul, M.J. 2001. A test of top-down and bottom-up control in a dtritus-based food web. *Ecology* 82, 2279–2293.

Rosenfeld, J. 2002. Functional redundancy in ecology and conservation. *Oikos* 98, 156–162.

Rosenzweig, M.L. 1969. Why prey curve has a hump. *Am. Nat.* 103, 81.

Rosenzweig, M.L. 1971. Paradox of enrichment: destabilization of exploitation ecosystems in ecological time. *Science* 171, 385–387.

Rosenzweig, M.L. 1973. Exploitation in 3 trophic levels. *Am. Nat.* 107, 275–294.

Rosenzweig, M.L., and MacArthur, R.M. 1963. Graphical representation and stability conditions of predator-prey interactions. *Am. Nat.* 97, 209–223.

Roughgarden, J. 1979. *Theory of Population Genetics and Evolutionary Ecology: An Introduction*. MacMillan, New York.

Roughgarden, J., Gaines, S., and Pacala, S. 1987. Supply side ecology: the role of physical transport processes. In: J.H.R. Gee and P.S. Giller (Editors), *The 27th Symposium of the The British Ecological Society: Organization of Communities, Past and Present*. Blackwell Scientific Publications, pp 491–518.

Rounick, J.S., and Winterbourn, M.J. 1986. Stable carbon isotopes and carbon flow in ecosystems. *Bioscience* 36, 171–177.

Russo, S.E., Robinson, S.K., and Terborgh, J. 2003. Size-abundance relationships in an Amazonian bird community: implications for the energetic equivalence rule. *Am. Nat.* 161, 267–283.

Rypstra, A.L., Carter, P.E., Balfour, R.A., and Marshall, S.D. 1999. Architectural features of agricultural habitats and their impact on the spider inhabitants. *J. Arachnol.* 27, 371–377.

Sabo, J.L., Bastow, J.L., and Power, M.E. 2002. Length-mass relationships for adult aquatic and terrestrial invertebrates in a California watershed. *J. N. Am. Benthol. Soc.* 21, 336–343.

Saetre, P., and Bååth, E. 2000. Spatial variation and patterns of the soil microbial community structure in a mixed spruce-birch stand. *Soil Biol. Biochem.* 32, 909–917.

Sala, E., and Graham, M.H. 2002. Community-wide distribution of predator-prey interaction strength in kelp forests. *P. Natl. Acad. Sci. USA* 99, 3678–3683.

Sanderson, B.L., Hrabik, T.R., Magnuson, J.J., and Post, D.M. 1999. Cyclic dynamics of a yellow perch (*Perca flavescens*) population in an oligotrophic lake: evidence for the role of intraspecfic interactions. *Can. J. Fish. Aquat. Sci.* 56, 1534–1542.

Sankaran, M., and McNaughton, S.J. 1999. Determinants of biodiversity regulate compositional stability of communities. *Nature* 401, 691–693.

Santos, P.F., Phillips, J., and Whitford, W.G. 1981. The role of mites and nematodes in early stages of buried litter decomposition in a desert. *Ecology* 62, 664–669.

Scarascia-Mugnozza, G., Bauer, G.A., Persson, H., Matteucci, G., and Masci, A. 2000. Tree biomass, growth and nutrient pools. In: E.D. Schulze (Editor), *Carbon and Nitrogen Cycling in Forest Ecosystems*. Springer, Heidelberg, pp 49–62.

Schaffer, W.M. 1981. Ecological abstraction: the consequences of reduced dimensionality in ecological models. *Ecol. Monogr.* 51, 383–401.

Scharf, F.S. 2000. Patterns in abundance, growth, and mortality of juvenile red drum across estuaries of the Texas coast with implications for recruitment and stock enhancement. *Trans. Am. Fish. Soc.* 129, 1207–1222.

Scharler, U.M., and Baird, D. 2005. A comparison of selected ecosystem attributes of three Sounth African estuaries with different freshwater inflow regimes, using network analysis. J. Mar. Syst. 56, 283–308.

Scheffer, M. 1998. *Ecology of Shallow Lakes: Population and Community Biology*. Chapman & Hall, London.

Scheffer, M., and Carpenter, S.R. 2003. Catastrophic regime shifts ecosystems: linking theory and observation. *TREE* 18, 648–656.

Scheu, S. 2001. Plants and generalist predators as links between the below-ground and above-ground system. *Basic Appl. Ecol.* 2, 3–13.

Scheu, S., and Falca, M. 2000. The soil food web of two beech forests (*Fagus sylvatica*) of contrasting humus type: stable isotope analysis of a macro- and a mesofauna-dominated community. *Oecologia* 123, 285–296.

Scheu, S., and Setälä, H. 2002. Multitrophic interactions in decomposer communities. In: T. Tscharntke and B.A. Hawkins (Editors), *Multitrophic Interactions in Terrestrial Systems*. Cambridge University Press, Cambridge, pp 223–264.

Scheu, S., Theenhaus, A., and Jones, T.H. 1999. Links between the detritivore and the herbivore system: effects of earthworms and Collembola on plant growth and aphid development. *Oecologia* 119, 541–551.

Schindler, D.E.R., Knapp, A., and Leavitt, P. 2001. Alteration of nutrient cycles and algal production resulting from fish introductions into mountain lakes. *Ecosystems* 4, 308–321.

Schindler, D.W. 1998. Replication versus realism: the need for ecosystem-scale experiments. *Ecosystems* 1, 323–334.

Schmid, B., Joshi, J., and Schläpfer, F. 2002. Empirical evidence for biodiversity-ecosystem functioning relationships. In: A. Kinzig, S. Pacala, and D. Tilman, (Editors), *Linking Biodiversity and Ecosystem Functioning*. Princeton University Press, Princeton, pp 120–150.

Schmid, P.E., Tokeshi, M., and Schmid-Araya, J.M. 2000. Relation between population density and body-size in stream communities. *Science* 289, 1557–1560.

Schmid-Araya, J.M., Hildrew, A.G., Robertson, A., Schmid, P.E., and Winterbottom, J. 2002a. The importance of meiofauna in food webs: Evidence from an acid stream. *Ecology* 83, 1271–1285.

Schmid-Araya, J.M., Schmid, P.E., Robertson, A., Winterbottom, J.H., Gjerløv, C., and Hildrew, A.G. 2002b. Connectance in stream food webs. *J. Anim. Ecol.*

Schmidt, I.K., Michelsen, A., and Jonasson, S. 1997. Effects of labile soil carbon on nutrient partitioning between an arctic graminoid and microbes. *Oecologia* 112, 557–565.

Schmidt, I.K., Ruess, L., Baath, E., Michelsen, A., Ekelund, F., and Jonasson, S. 2000. Long-term manipulation of the microbes and microfauna of two subarctic heaths by addition of fungicide, bactericide, carbon and fertilizer. *Soil Biol. Biochem.* 32, 707–720.

Schmidt, K., Atkinson, A., Stübing, D., McClelland, J.W., Montoya, J.P., and Voss, M. 2003. Trophic relationships among Southern Ocean copepods and krill: some uses and limitations of a stable isotope approach. *Limno. Oceanogr.* 48, 277–289.

Schmitt, R.J. 1987. Indirect interactions between prey: Apparent competition, predator aggregation, and habitat segregation. *Ecology* 68, 161–183.

Schmitz, O. 1993. Trophic exploitation in grassland food-cahins: Simple-models and a field experiment. *Oecologia* 93, 327–335.

Schmitz, O.J. 1997. Press perturbations and the predictability of ecological interaction in a food web. *Ecology* 78, 55–69.

Schmitz, O.J. 2001. From interesting detail to dynamic relevance: toward more effective use of empirical insights in theory construction. *Oikos* 94, 39–50.

Schmitz, O.J. 2003. Top predator control of plant biodiversity and productivity in an old field ecosystem. *Ecol. Lett.* 6, 156–163.

Schoener, T.H. 1983. Field experiments on interspecific competition. *Am. Nat.* 122, 240–285.

Schoener, T.W. 1973. Population-growth regulated by intraspecific competition for energy or time: some simple representations. *Theor. Popul. Biol.* 4, 56–84.

Schoener, T.W. 1989. Food webs from the small to the large. *Ecology* 70, 1559–1589.

Schoener, T.W. 1993. On the relative importance of direct versus indirect effects in ecological communities. In: H. Kawanabe, J.E. Cohen, and K. Iwasaki (Editors), *Multitrophic Interactions in Terrestrial Systems.* Oxford University Press, New York, pp 365–411.

Schoener, T.W., and Spiller, D. 1987. Effect of lizards on spider populations: manipulative reconstruction of a natural experiment. *Science* 236, 949–952.

Schoenly, K., and Cohen, J.E. 1991. Temporal variation in food web structure: sixteen empirical cases. *Ecol. Monogr.* 61, 267–298.

Schoenly, K., Beaver, R.A., and Heumier, T.A. 1991. On the trophic relations of insects: A food-web approach. *Am. Nat.* 137, 597–638.

Schröter, D. 2001. *Structure and Function of the Decomposer Food Webs of Forests Along a European North-South-Transect with Special Focus on Testate Amoebae (Protozoa).* Shaker Verlag, Aachen, 138.

Schröter, D., Wolters, V., and de Ruiter, P.C. 2003. C and N mineralisation in the decomposer food webs of a European forest transect. *Oikos* 102, 294–308.

Schultz, P.A. 1991. Grazing preference of two collembolan species, *Folsomia candida* and *Proisotoma minuta,* for ectomycorrhizal fungi. *Pedobiologia* 35, 313–325.

Schulze, E.D., Editors. 2000. *Carbon and Nitrogen Cycling in Forest Ecosystems.* Springer, Heidelberg, 498.

Schulze, E.D., and Mooney, H.A. 1993. *Biodiversity and Ecosystem Function.* Springer Verlag, New York.

Schwartz, M.W., Brigham, C.A., Hoeksema, J.D., Lyons, K.G., Mills, M.H., and van Mantgem, P.J. 2000. Linking biodiversity to ecosystem function: implications for conservation ecology. *Oecologia* 122, 297–305.

Schwinghamer, P. 1981. Characteristic size distributions of integral benthic communities. *Can. J. Fish. Aquat. Sci.* 38, 1255–1263.

Seastedt, T.R., and Crossley, D.A. 1984. The influence of arthropods on ecosystems. *Bioscience* 34, 157–161.

Sebens, K.P. 1982. The limits to indeterminate growth: An optimal model applied to passive suspension feeders. *Ecology* 63, 209–222.

Sebens, K.P. 1987. The ecology of indeterminate growth in animals. *Ann. Rev. Ecol. Syst.* 18, 371–407.

Seldal, T., Andersen, K.J., and Högstedt, G. 1994. Grazing-induced proteinase inhibitors: a possible cause for lemming population cycles. *Oikos* 70, 3–11.

Setälä, H. 1995. Growth of birch and pine seedlings in relation to grazing by soil fauna on ectomycorrhizal fungi. *Ecology* 76, 1844–1851.

Setälä, H. 2000. Reciprocal interactions between Scots pine and soil food web structure in the presence and absence of ectomycorrhiza. *Oecologia* 25, 109–118.

Setälä, H. 2002. Sensitivity of ecosystem functioning to changes in trophic structure, functional group composition and species diversity in belowgroup food webs. *Ecol. Res.* 17, 207–215.

Setälä, H., and Huhta, V. 1991. Soil fauna increase Betula pendula growth: laboratory experiments with coniferous forest floor. *Ecology* 72, 665–671.

Setälä, H., and MacLean, M.A. 2003. Decomposition of organic substrates in relation to the species diversity of soil saprophytic fungi. *Oecologia* 139, 98–107.

Setälä, H., Marshall, V.G., and Trofymow, J.A. 1996. Influence of body size of soil fauna on litter decomposition an 15N uptake by poplar in a pot trial. *Soil Biol. Biochem.* 28, 1661–1675.

Setälä, H., Rissanen, J., and Markkola, A.M. 1997. Conditional outcomes in the relationship between pine and ectomycorrhizal fungi in relation to biotic and abiotic environment. *Oikos* 80, 112–122.

Setälä, H., Berg, M., and Jones, T.H. 2005. Trophic structure and functional redundancy in soil communities. In: Bardgett, R.D., Hopkins, D.W., and Usher, M, (Editors) *Biol. Div. and Function in Soils.* Cambridge University Press, Cambridge, in press.

Setälä, H., Martikainen, E., Tyynismaa, M., and Huhta, V. 1990. Effects of soil fauna on leaching of nitrogen and phosphorus from experimental systems simulating coniferous forest floor. *Biol. Fert. Soils* 10, 170–177.

Shannon, C.E. 1948. A mathematical theory of communications. *AT&T Tech. J.* 27, 379–423.

Shaus, M.H., and Vanni, M.J. 2000. Effects of gizzard shad on phytoplankton and nutrient dynamics: role of sediment feeding and fish size. *Ecology* 81, 1701–1719.

Shaw, J.L., and Kennedy, J.H. 1996. The use of aquatic field mesocosm studies in risk assessment. *Environmental Toxicology and Chemistry* 15, 605–607.

Shea, K.L., and Chesson, P. 2002. Community ecology theory as a framework for biological invasions. *Trends Ecol. Evol.* 17, 170–176.

Sheldon, R.W., and Parsons, T.R. 1967. A continuous size spectrum for particulate matter in the sea. *J. Fish. Res. Bd. Can.* 24, 909–915.

Sheldon, R.W., Prakash, A., and Sutcliffe, W.H. 1972. The size distribution of particles in the ocean. *Limno. Oceanogr.* 17, 327–339.

Shertzer, K.W., Ellner, S.P., Fussmann, G.F., and Hairston, N.G. 2002. Predator-prey cycles in an aquatic microcosm: testing hypotheses of mechanism. *J. Anim. Ecol.* 71, 802–815.

Shiomoto, A., Tadokoro, K., Nagasawa, K., and Ishida, Y. 1997. Trophic relations in the subArctic North Pacific ecosystem: Possible feeding effect from pink salmon. *Mar. Ecol.-Prog. Ser.* 150, 75–85.

Short, R.A., and Maslin, P.E. 1977. Processing of leaf litter by a stream detritivore: effect on nutrient availability to collectors. *Ecology* 58, 935–938.

Shurin, J.B. 2000. Dispersal limitation, invasion resistance, and the structure of pond zooplankton communities. *Ecology* 81, 3074–3086.

Shurin, J.B., Borer, E.T., Seabloom, E.W., Anderson, K., Blanchette, C.A., Broitman, B., Cooper, S.D., and Halpern, B.S. 2002. A cross-ecosystem comparison of the strength of trophic cascades. *Ecol. Lett.* 5, 785–791.

Siepel, H., and de Ruiter-Dijkman, E.M. 1993. Feeding guilds of oribatid mites based on their carbohydrase activities. *Soil Biol. Biochem.* 25, 1491–1497.

Sih, A. 1984. Optimal behavior and density-dependent predation. *Am. Nat.* 123, 314–326.

Sih, A., Englund, G., and Wooster, D. 1998. Emergent impacts of multiple predators on prey. *Trends Ecol. Evol.* 13, 350–355.

Sih, A., Crowley, P., McPeek, M., Petranka, J., and Strohmeir, K. 1985. Predation, competition and prey communities: a review of filed experiments. *Ann. Rev. Ecol. Syst.* 16, 269–311.

Siira-Pietikainen, A., Haimi, J., Kanninen, A., Pietikainen, J., and Fritze, H. 2001. Responses of decomposer community to root-isolation and addition of slash. *Soil Biol. Biochem.* 33, 1993–2004.

Silva, M., and Downing, J.A. 1995. The allometric scaling of density and body mass: a non-linear relationship for terrestrial mammals. *Am. Nat.* 145, 704–727.

Silvert, W., Platt, T. 1980. Dynamic energy-flow model of particle size distribution in pelagic ecosystems. In: W.C. Kerfoot (Editor), *Evolution and ecology of zooplankton communities.* University Press of New England, New Hampshire.

Simberloff, D. 2000. Extinction-proneness of island species-causes and mangement implications. *Raffles Bulletin of Zoology* 48, 1–9.

Simberloff, D., and Dayan, T. 1991. The guild concept and the structure of ecological communities. *Ann. Rev. Ecol. Syst.* 22, 115–143.

Sinclair, A.R.E., Mduma, S., and Brashares, J.S. 2003. Patterns of predation in a diverse predator-prey system. *Nature* 425, 288–290.

Sinsabaugh, R.L., and Linkens, A.E. 1990. Enzymatic and chemical analysis of particulate organic matter from a boreal river. *Freshwat. Biol.* 23, 301–309.

Smith, O.L. 1982. Soil Microbiology: a model of decomposition and nutrient cycling. In: M.J. Bazin (Editor), *CRC Series in Mathematical Modelling in Microbiology.* CRC Press Inc., Boca Raton, Florida.

Sneath, P.H.A., and Sokal, R.R. 1973. *Numerical Taxonomy.* W.H. Freeman and Company, San Francisco.

Solé, R.V., and Montoya, J.M. 2001. Complexity and fragility in ecological networks. *P. Roy. Soc. Lond. B Biol.* 268, 2039–2045.

Solé, R.V., and Montoya, J.M. 2002. Complexity and fragility in ecological networks. *P. Roy. Soc. Lond. B Biol.* 268, 2039–2045.

Sommer, U., Sommer, F., Santer, B., Jamieson, C., Boersma, M., Becker, C., and Hansen, T. 2001. Complementary impacts of copepods and cladocerans on phytoplankton. *Ecol. Lett.* 4, 545–550.

Sousa, W.P. 1984. The role of disturbance in natural communities. *Ann. Rev. Ecol. Syst.* 15, 353–391.

Speirs, D.C., Gurney, W.S.C., Winterbottom, J.H., and Hildrew, A.G. 2000. Long-term demographic balance in the Broadstone Stream insect community. *J. Anim. Ecol.* 69, 45–58.

Spiller, D.A., and Schoener, T.W. 1988. An experimental study of the effect of lizards on web-spider communities. *Ecol. Monogr.* 58, 57–77.

Spiller, D.A., and Schoener, T.W. 1994. Effects of top and intermediate predators in the terrestrial food web. *Ecology* 75, 182–196.

Spiller, D.A., and Schoener, T.W. 1995. Long-term variation in the effect of lizards on spider density is linked to rainfall. *Oecologia* 103, 133–139.

Springer, A.M., Estes, J.A., van Vliet, G.B., Doak, D.F., Danner, E.M., Forney, K.A., and Pfister, B. 2003. Sequential megafaunal collapse in the North Pacific Ocean: an ongoing legacy of industrial whaling? *P. Natl. Acad. Sci. USA* 100, 12223–12228.

Sprules, G.W., and Munawar, M. 1986. Plankton size spectra in relation to ecosystem productivity, size and perturbation. *Can. J. Fish. Aquat. Sci.* 43, 1789–1784.

Srivastava, D.S. 2002. The role of conservation in expanding biodiversity research. *Oikos* 98, 351–360.

St. John, T.V., Coleman, D.C., and Reid, C.P.P. 1983. Association of vesicular-arbuscular mycorrhizal hyphae with soil organic particles. *Ecology* 64, 957–959.

Stabell, O.B., Ogbebo, F., and Primicerio, R. 2003. Inducible defences in Daphnia depend on latent alarm signals from conspecific prey activated in predators. *Chem. Senses* 28, 141–153.

Stachowicz, J.J., Fried, H., Whitlatch, R.B., and Osman, R.W. 2002. Biodiversity, invasion resistance and marine ecosystem function: reconciling pattern and process. *Ecology* 83, 2575–2590.

Steel, E.A., Naiman, R.J., and West, S.D. 1999. Use of woody debris piles by birds and small mammals in a riparian corridor. *Northwest Sci.* 73, 19–26.

Steinberg, P.D., Estes, J.A., and Winter, F.C. 1995. Evolutionary consequences of food chain length in kelp forest communities. *P. Natl. Acad. Sci. USA* 92, 8145–8148.

Steiner, C.F. 2001. The effects of prey heterogeneity and consumer identity on the limitation of trophic-level biovolume. *Ecology* 82, 2495–2506.

Steiner, C.F., and Leibold, M. 2004. Cyclic assembly trajectories and scale-dependent productivity-diversity relationships. *Ecology* 85, 107–113.

Stelzer, R.S., and Lamberti, G.A. 2002. Ecological stoichiometry in running waters: periphyton chemical composition and snail growth. *Ecology* 83, 1039–1051.

Stemberger, R.S. 1979. *A Guide to Rotifers of the Laurentian Great Lakes Final Report, Contract Number R-804652.* Environmental Monitoring and Support Laboratory, Office of Research and Development, U.S. Environmental Protection Agency, Cincinnati, Ohio.

Stenseth, N.C. 1985. The structure of food webs predicted from optimal food selection models: an alternative to Pimm's stability hypothesis. *Oikos* 44, 361–364.

Stephens, D.W., and Krebs, J.R. 1986. *Foraging Theory.* Princeton University Press, Princeton, New Jersey.

Stephens, P.A., and Sutherland, W.J. 1999. Consequences of the Allee effect for behaviour, ecology and conservation. *Trends Ecol. Evol.* 14, 401–405.

Sterner, R.W. Editor. 1995. *Elemental Stoichiometry of Species in Ecosystems. Linking Species and Ecosystems.* Chapman and Hall, New York.

Sterner, R.W. 1997. Modelling interactions of food quality and quantity in homeostatic consumers. *Freshwat. Biol.* 38, 473–481.

Sterner, R.W., and Elser, J.J. 2002. *Ecological Stoichiometry: The Biology of Elements from Molecules to the Biosphere.* Princeton University Press, Princeton.

Sterner, R.W., Elser, J.J., and Hessen, D.O. 1992. Stoichiometric relationships among producers, consumers, and nutrient cycling in pelagic ecosystems. *Biogeochemistry* 17, 49–67.

Sterner, R.W., Bajpai, A., and Adams, T. 1997. The enigma of food chain length: Absence of theoretical evidence for dynamic constraints. *Ecology* 78, 2258–2262.

Sterner, R.W., Clasen, J., Lampert, W., and Weisse, T. 1998. Carbon: phosphorus stoichiometry and food chain production. *Ecol. Lett.* 1, 146–150.

Stone, L., and Roberts, A. 1991. Conditions for a species to gain advantage from the presence of competitors. *Ecology* 72, 1964–1972.

Strathmann, R.R. 1990. Testing size-abundance rules in a human exclusion experiment. *Science* 250, 1091.

Strayer, D.L. 1994. Body size and abundance of benthic animals in Mirror Lake, New Hampshire. *Freshwat. Biol.* 32, 83–90.

Strayer, D.L., Caraco, N.F., Cole, J.J., Findlay, S., and Pace, M.L. 1999. Transformation of freshwater ecosystems by bivalves: a case study of zebra mussels in the Hudson River. *Bioscience* 49, 19–27.

Stromberg, J.C. 1993a. Fremont cottonwood-gooding willow riparian forests: A review of their ecology, threats, and recovery potential. *J. AZ-NV Acad. Sci.* 26, 97–110.

Stromberg, J.C. 1993b. Riparian mesquite forests: A review of their ecology, threats, and recovery potential. *J. AZ-NV Acad. Sci.* 26, 111–124.

Strong, D.R. 1992. Are trophic cascades all wet? Differentiation and donor-control in speciose ecosystems. *Ecology* 73, 747–754.

Stuart, A., and Ord, J.K. 1991. *Kendall's Advanced Theory of Statistics Vol. 1 and 2, 5th edition.* Oxford University Press, Oxford.

Suberkropp, K., and Wallace, J.B. 1992. Aquatic hyphomycetes in insecticide-treated and untreated streams. *J. N. Am. Benthol. Soc.* 11, 165–171.

Sugihara, G., Schoenly, K., and Trombla, A. 1989. Scale invariance in food web properties. *Science* 245, 48–52.

Sulkava, P., and Huhta, V. 1998. Habitat patchiness affects decompositions and faunal diversity: a microcosm experiment on forest floor. *Oecologia* 116, 390–396.

Summerhayes, and Elton, C. 1923. Contributions to the ecology of Spitsbergen and Bear Island. *J. Ecol.* 11, 214–286.

Sutherland, W. 1996. *From Individual Behaviour to Population Ecology.* Oxford University Press, Oxford.

Swank, W.T., Waide, J.B., Crossley, D.A., and Todd, R.L. 1981. Insect defoliation enhances nitrate export from forest ecosystems. *Oecologia* 51, 297–299.

Swift, M.J., Heal, O.W., and Anderson, J.M. 1979. *Decomposition in terrestrial ecosystems.* University of California Press, Berkeley, California.

Symstad, A.J., Tilman, D., Wilson, J., and Knops, J.M.H. 1998. Species loss and ecosystem functioning: effects of species identity and community composition. *Oikos* 81, 389–397.

Szanser, M. 2000a. Effect of macroarthropods patrolling soil surface on decomposition rate of grass litter (*Dactylis glomerata*) in a field experiment. *Pol. J. Ecol.* 48, 283–297.

Szanser, M. 2000b. Rozklad sciólki trawiastej modyfikowany aktywnoscia epigeicznej fauny makrostawonogów (Decomposition of grass litter modified by the activity of epigeic macroarthropod fauna). Phd. Dziekanów Lesny (Instytut Ekologii PAN, Dziekanów Lesny).

Szanser, M. 2003. Impact of shelterbelts on litter decomposition and fauna of adjacent fields: in situ experiment. *Pol. J. Ecol.* 51, 309–321.

Szyrmer, J., and Ulanowicz, R.E. 1987. Total flows in ecosystems. *Ecol. Model.* 35, 123–136.

Takeda, H. 1987. Dynamics and maintenance of collembolan community structure in a forest soil system. *Res. Popul. Ecol.* 29, 292–346.

Takimoto, G. 2003. Adaptive plasticity in ontogenetic niche shifts stabilizes consumer-resource dynamics. *Am. Nat.* 162, 93–109.

Tank, J.L., and Webster, J.R. 1998. Interaction of substrate and nutrient availability on wood biofilm processes in streams. *Ecology* 79, 2168–2179.

Tansky 1978. Switching effect in prey-predator system. *J. Theor. Biol.* 70, 263–271.

Tansley, A.G. 1935. The use and abuse of vegetational concepts and terms. *Am. Nat.* 16, 284–307.

Tansley, A.G., and Adamson, R.S. 1925. Studies of the vegetation of the English chalk. III. The chalk grasslands of the Hampshire-Sussex border. *J. Ecol.* 13, 177–223.

Tavares-Cromar, A.F., and Williams, D.D. 1996. The importance of temporal resolution in food web analysis: evidence from a detritus-based stream. *Ecol. Monogr.* 66, 91–113.

Teal, J.M. 1962. Energy flow in the salt marsh ecosytem of Georgia. *Ecology* 43, 614–624.

Teng, J., and McCann, K. 2004. The dynamics of compartmented and reticulate food webs in relation to energetic flows. *Am. Nat.* 164, 86–100.

Teramoto, E., Kawasaki, K., and Shigesada, N. 1979. Switching effect of predation on competitive prey species. *J. Theor. Biol.* 79, 305–315.

Terborgh, J. 1992. Maintenance of diversity in tropical forests. *Biotropica* 24, 283–292.

Terborgh, J., Robinson, S.K., Parker, T.A., III, Munn, C.A., and Pieront, N. 1990. Structure and organization of an Amazonian forest bird community. *Ecol. Monogr.* 60, 213–238.

Terborgh, J., Estes, J.A., Paquet, P., Ralls, K., Boyd-Herger, D., Miller, B.J., and Noss, R.F. 1999. The role of top carnivores in regulating terrestrial ecosystems. In: M.E. Soulé and J. Terborgh, (Editors), *Continental Conservation.* Island Press, Washington D.C., pp 39–64.

Thébault, E., and Loreau, M. 2003. Food-web constraints on biodiversity-ecosystem functioning relationships. *P. Natl. Acad. Sci. USA* 100, 14,949–914,954.

Thébault, E., and Loreau, M. 2005. Trophic interactions and the relationship between species diversity and ecosystem stability. *The American Naturalist,* in press.

Thompson, J.N. 1998. Rapid evolution as an ecological process. *TREE* 13, 329–332.

Thompson, R.M., and Townsend, C.R. 1999. Is resolution the solution?: the effect of taxonomic resolution on the calculated properties of three stream food webs. *Freshwat. Biol.* 44, 413–422.

Thorp, J.H., and DeLong, A.D. 2002. Dominance of autochthonous autotrophic carbon in food webs of heterotrophic rivers. *Oikos* 96, 543–550.

Thorpe, J., and Covich, A. 1991. *Ecology and Classification of North American Freshwater Invertebrates.* Academic Press, San Diego, California.

Thrush, S.F., Whitalch, R.B., Pridmore, R.D., Hewitt, J.E., Cummings, V.J., and Wilkinson, M.R. 1996. Scale-dependent recolonisation.: the role of sediment stability in a dynamic sandflat habitat. *Ecology* 77, 2472–2487.

Thrush, S.F., Schneider, D.C., Legendre, P., Whitalch, R.B., Dayton, P.K., and Hewitt, J.E. 1997. Scaling up from experiments to comples ecological systems: where to next? *J. Exp. Mar. Biol. Ecol.* 216, 243–254.

Tietema, A. 1998. Microbial carbon and nitrogen dynamics in coniferous forest floor material collected along a European nitrogen deposition gradient. *For. Ecol. Manag.* 101, 29–36.

Tilman, D. 1977. Resource competition between planktonic algae: an experimental and theoretical approach. *Ecology* 58, 338–348.

Tilman, D. 1982. *Resource Competition and Community Structure.* Princeton University Press, Princeton, New Jersey.

Tilman, D. 1996. Biodiversity: population versus ecosystem stability. *Ecology* 77, 350–363.

Tilman, D. 1997a. Distinguishing between the effects of species diversity and species composition. *Oikos* 80, 185.

Tilman, D. 1997b. Community invasibility, recruitment limitation, and grassland biodiversity. *Ecology* 78, 81–92.

Tilman, D. 2001. Functional diversity. In: S.A. Levin (Editor), *Encyclopedia of Biodiversity.* Academic Press, San Diego, pp 109–120.

Tilman, D., and Kareiva, P. 1997. *The role of space in population dynamics and interspecific interactions.* Princeton University Press, Princeton.

Tilman, D., Wedin, D., and Knops, J. 1996. Productivity and sustainability influenced by biodiversity in grassland ecosystems. *Nature* 379, 718–720.

Tilman, D., Lehman, C.L., and Thomson, K. 1997. Plant diversity and ecosystem productivity, theoretical considerations. *P. Natl. Acad. Sci. USA* 94, 1857–1861.

Tilman, D., Lehman, C., and Bristow, C.E. 1998. Diversity-stability relationships: statistical inevitability or ecological consequence? *Am. Nat.* 151, 277–282.

Tilman, D., Reich, P.B., Knops, J., Wedin, D., Mielke, T., and Lehman, C. 2001. Diversity and productivity in a long-term grassland experiment. *Science* 294, 843–845.

Tittel, J., Bissinger, V., Zippel, B., Gaedke, U., Bell, E., Lorke, A., and Kamjunke, N. 2003. Mixotrophs combine resource use to outcompete specialists: Implications for aquatic food webs. *P. Natl. Acad. Sci. USA* 100, 12776–12781.

Tokita, K., and Yasutomi, A. 2003. Emergence of a complex and stable network in a model ecosystem with extinction and mutation. *Theor. Popul. Biol.* 63, 131–146.

Tollrian, R., and Harvell, C.D. Editors. 1999. *The Ecology and Evolution of Inducible Defenses.* Princeton University Press, Princeton, New Jersey.

Tonn, W.M., Paszkowksi, C.A., and Holopainen, I.J. 1992. Piscivory and recruitment: mechanisms structuring prey populations in small lakes. *Ecology* 73, 951–958.

Torsvik, V., Salte, K., Sorheim, R., and Goksoyr, J. 1990. Comparison of phenotypic diversity and DNA heterogenity in a population of soil bacteria. *Appl. Environ. Microb.* 56, 776–781.

Townsend, C.R., Sutherland, W.J., and Perrow, M.R. 1990. A modelling investigation of population cycles in the fish *Rutilus rutilus*. *J. Anim. Ecol.* 59, 469–485.

Turchin, P. 2003. *Complex population dynamics: A theoretical/empirical synthesis.* Princeton University Press, Princeton, 450.

Turchin, P., Oksanen, L., Ekerholm, P., Oksanen, T., and Henttonen, H. 2000. Are lemmings prey or predators? *Nature* 405, 562–565.

Turner, J.T., and Ferrante, J.G. 1979. Zooplankton faecal pellets in aquatic ecosystems. *Bioscience* 29, 670–677.

Uetz, G.W. 1979. Influence of variation in litter habitats on spider communities. *Oecologia* 40, 29–42.

Ulanowicz, R.E. 1986. *Growth and Development: Ecosystems Phenomenolgy.* In. Springer-Verlag, New York, pp 232.

Ulanowicz, R.E. 1996. Trophic flows as indicators of ecosystem stress. In: G. Polis and K.O. Winemiller (Editors), *Food Webs: Integration of Pattern and Dynamics.* Chapman and Hall, New York.

Ulanowicz, R.E. 2004. Quantitative methods for ecological network analysis. *Comp. Biol. Chem.* 28, 321–339.

Ulanowicz, R.E., and Wolff, W.F. 1991. Ecosystem flow networks: loaded dice? *Math. Biosci.* 103, 45–68.

Ulanowicz, R.E., and Baird, D. 1999. Nutrient controls in ecosystem dynamics: the Chesapeake mesohaline community. *J. Mar. Syst.* 19, 159–172.

Underwood, N. 1999. The influence of plant and herbivore characteristics on the interaction between induced resistance and herbivore population dynamics. *Am. Nat.* 153, 282–294.

Usher, M.B., Davis, P., Harris, J., and Longstaff, B. 1979. A profusion of species? Approaches towards understanding the dynamics of the populations of microarthropods in decomposer communities. In: R.M. Anderson, B.D. Turner, and L.R. Taylor (Editors), *Population Dynamics.* Blackwell Scientific Publications, Oxford, pp 359–384.

Usinger, R.L. Editors. 1963. *Aquatic Insects of North America.* University of California Press, Los Angeles.

Vadeboncoeur, Y., and Lodge, D.M. 2000. Periphyton production on wood and sediment: substratum-specific response to laboratory and whoel-lake nutrient manipulations. *J. N. Am. Benthol. Soc.* 19, 68–81.

Valett, H.M., Crenshaw, C.L., and Wagner, P.F. 2002. Stream nutrient uptake, forest succession, and biogeochemical theory. *Ecology* 83, 2888–2901.

Valido, A., Nogaels, M., and Medina, F.M. 2003. Fleshy fruits in the diet of Canarian lizards *Gallotia galloti* (Lacertidae) in a xeric habitat of the Island of Tenerife. *J. Herpetol.* 37, 741–747.

van Baalen, M., Krivan, V., van Rijn, P.C.J., and Sabelis, M.W. 2001. Alternative food, switching predators, and the persistence of predator-prey systems. *Am. Nat.* 157, 512–524.

van der Heijden, M.G.A. 1998. Mycorrhizal fungal diversity determines plant biodiversity, ecosystem variability and productivity. *Nature* 396, 69–72.

Van Donk, E., Lürling, M., and Lampert, W. 1999. Consumer-induced changes in phytoplankton: inducibility, costs, benefits, and the impact on grazers. In: R. Tollrian and C.D. Harvell (Editors), *The Ecology and Evolution of Inducible Defenses.* Princeton University Press, Princeton, New Jersey, pp 89–103.

Van Kooten, T. 2004. On the interplay of llife-history and population dynamics: Emergent consequences of individual variability and specialization. PhD. University of Amsterdam.

Van Ruijven, J., and Berendse, F. 2003. Positive effects of plant species diversity on productivity in the absence of legumes. *Ecol. Lett.* 6, 170–175.

Vander Zanden, J.M., Chandra, S., Allen, B.C., Reuter, J.E., and Goldman, C.R. 2003. Historical food web structure and restoration of native aquatic communities in Lake Tahoe (California-Nevada) basin. *Ecosystems* 6, 274–288.

Vander Zanden, M.J., and Rasmussen, J.B. 1999. Primary consumer delta C-13 and delta N-15 and the trophic position of aquatic consumers. *Ecology* 80, 1395–1404.

Vander Zanden, M.J., and Rasmussen, J.B. 2001. Variation in delta N-15 and delta C 13 trophic fractionation: Implications for aquatic food web studies. *Limno. Oceanogr.* 46, 2061–2066.

Vander Zanden, M.J., Casselman, J.M., and Rasmussen, J.B. 2000. Stable isotope evidence for the foodweb consequences of species invasions in lakes. *Nature* 401, 464–467.

Vanderklift, M.A., and Ponsard, S. 2003. Sources of isotopic variation in consumer-diet d15N enrichment: a meta-analysis. *Oecologia* 136, 169–182.

Vandermeer, J., Evans, M.A., Foster, P., Hook, T., Reiskind, M., and Wund, M. 2002. Increased competition may promote species coexistence. *P. Natl. Acad. Sci. USA* 99, 8731–8736.

Vanni, M.J. 1996. Nutrient transport and recycling by consumers in lake food webs: implications for algal communities. In: G.A. Polis and K.O. Winemiller (Editors), *Food Webs: Integration of Patterns and Dynamics.* Chapman and Hall, New York, pp 81–91.

Vanni, M.J. 2002. Nutrient recycling by animals in freshwater ecosystems. *Ann. Rev. Ecol. Syst.* 33, 341–370.

Vanni, M.J., and Layne, C.D. 1997. Nutrient recycling and herbivory as mechanisms in the "top-down" effect of fish on algae in lakes. *Ecology* 78, 21–40.

Vanni, M.J., Layne, C.D., and Arnott, S.E. 1997. "Top-down" trophic interactions in lakes: Effects of fish on nutrient dynamics. *Ecology* 78, 1–20.

Vanni, M.J., Flecker, A.S., Hood, J.M., and Headworth, J.L. 2002. Stoichiometry of nutrient recycling by vertebrates in a tropical stream: linking species identity and ecosystem processes. *Ecol. Lett.* 5, 285–293.

Vannote, R.L., Minshall, G.W., Cummins, K.W., Sedell, J.R., and Cushing, C.E. 1980. River continuum concept. *Can. J. Fish. Aquat. Sci.* 37, 130–137.

Vargas, A.J. 2000. Effects of fertilizer addition and debris removal on leaf-litter spider communities at two elevations. *J. Arachnol.* 28, 79–89.

Verhoef, H.A., and Brussaard, L. 1990. Decomposition and nitrogen mineralization in natural and agroecosystems: the contribution of soil animals. *Biogeochemistry* 11, 175–211.

Verity, P.G., Smetacek, V., and Smayda, T.J. 2002. Status, trends and the future of the marine pelagic ecosystem. *Environ. Cons.* 29, 207–237.

Vermeij, G.J. 1991. When biotas meet: understanding biotic interchange. *Science* 253, 1099–1104.

Verschoor, A.M., Vos, M., and Stap, I.v.d. 2004a. Inducible defences prevent strong population fluctuations in bi- and tritrophic food chains. *Ecol. Lett.* 7, 1143–1148.

Verschoor, A.M., Stap, I.v.d., Helmsing, N.R., Lürling, M., and Donk, E.v. 2004b. Inducible colony formation within the cenedesmaceae: Adaptive responses to infochemicals from two different herbivore taxa. *J. Phycol.* 40, 808–814.

Vet, L.E.M. 1999. From chemical to population ecology: Infochemical use in an evolutionary context. *J. Chem. Ecol.* 25, 31–49.

Vet, L.E.M., and Dicke, M. 1992. Ecology of infochemical use by natural enemies in a tritrophic context. *Ann. Rev. Entomol.* 37, 141–172.

Vezina, A.F. 1985. Empirical relationships between predator and prey size among terrestrial vertebrate predators. *Oecologia* 67, 555–565.

Vitousek, P.M., and Hooper, D.U. 1993. Biological Diversity and Terrestrial Ecosystem Biogeochemistry. In: E.D. Schulze and H.A. Mooney (Editors), *Biodiversity and Ecosystem Function.* Springer-Verlag, Berlin, pp 3–14.

Volterra, V. 1926. Variazioni e fluttuazioni delnumero d'indivdui in specie animalia conviventi. *Mem. R. Accad. Naz. dei Lincei* 6.

Vos, M., Kooi, B.W., DeAngelis, D.L., and Mooij, W.M. 2004a. Inducible defences and the paradox of enrichment. *Oikos* 105, 471–480.

Vos, M., Berrocal, S.M., Karamaouna, F., Hemerik, L., and Vet, L.E.M. 2001. Plant-mediated indirect effects and the persistence of parasitoid-herbivore communities. *Ecol. Lett.* 4, 38–45.

Vos, M., Flik, B.J.G., Vijverberg, J., Ringelberg, J., and Mooij, W.M. 2002. From inducible defences to population dynamics: modelling refuge use and life history changes in Daphnia. *Oikos* 99, 386–396.

Vos, M., Verschoor, A.M., Kooi, B.W., Wäckers, F.L., DeAngelis, D.L., and Mooij, W.M. 2004b. Inducible defenses and trophic structure. *Ecology* 85, 2783–2794.

Vredenburg, V.T. 2004. Reversing introduced species effects: Experimental removal of introduced fish leads to rapid recovery of a declining frog. *P. Natl. Acad. Sci. USA* 101, 7646–7650.

Wäckers, F.L., and Bonifay, C. 2004. How to be sweet? Extrafloral nectar allocation by *Gossypium hirsutum* fits optimal defense theory predictions. *Ecology* 85, 1512–1518.

Wäckers, F.L., Zuber, D., Wunderlin, R., and Keller, F. 2001. The effect of herbivory on temporal and spatial dynamics of extrafloral nectar production in cotton and castor. *Ann. Bot.-London* 87, 365–370.

Wagner, J.D., Toft, S., and Wise, D.H. 2003. Spatial stratification in litter depth by forest-floor spiders. *J. Arachnol.* 31, 28–39.

Wainwright, S.C., Weinstein, M.P., Able, K.W., and Currin, C.A. 2000. Relative importance of benthic microalgae, phytoplankton and the detritus of smooth cordgrass Spartina alterniflora and the common reed Phragmites australis to brackish-marsh food webs. *Mar. Ecol.-Prog. Ser.* 200, 77–91.

Walker, B., Kinzig, A., and Langridge, J. 1999. Plant attribute diversity, resilience, and ecosystem function: the nature and significance of dominant and minor species. *Ecosystems* 2, 95–113.

Wallace, J.B. 1988. Aquatic invertebrate research. In: W.T. Swank and D.A.C. Jr. (Editors), *Ecological Studies, Vol. 66: Forest Hydrology and Ecology at Coweeta,* Springer-Verlag, Berlin.

Wallace, J.B., and Benke, A.C. 1984. Quantification of wood habitat in sub-tropical coastal-plain streams. *Can. J. Fish. Aquat. Sci.* 41, 1643–1652.

Wallace, J.B., and Hutchens, J.J. 2000. Effects of invertebrates in lotic ecosystem processes. In: D.C. Coleman and P.F. Hendrix, (Editors), *Invertebrates as Webmasters in Ecosystems,* CABI Publishing.

Wallace, J.B., Webster, J.R., and Cuffney, T.F. 1982. Stream detritus dynamics: regulation by invertebrate consumers. *Oecologia* 53, 197–200.

Wallace, J.B., Lugthart, G.J., Cuffney, T.F., and Schurr, G.A. 1989. The impact of repeated insecticidal treatments on drift and benthos of a headwater stream. *Hydrobiologia* 179, 135–147.

Wallace, J.B., Eggert, S.L., Meyer, J.L., and Webster, J.R. 1997. Multiple trophic levels of a forest stream linked to terrestrial litter inputs. *Science* 277, 102–104.

Wallace, J.B., Eggert, S.L., Meyer, J.L., and Webster, J.R. 1999. Effects of resource limitation on a detrital-based ecosystem. *Ecol. Monogr.* 69, 409–442.

Wallace, J.B., Cuffney, T.F., Webster, J.R., Lugthart, G.J., Chung, K., and Goldowitz, B.S. 1991. Export of fine particles from headwater streams: effects of season, extreme discharges, and invertebrate manipulation. Limno. *Oceanogr.* 36, 670–682.

Wallace, J.B., Whiles, M.R., Eggert, S.L., Cuffney, T.F., Lugthart, G.J., and Chung, K. 1995. Long-term dynamics of coarse particulate organic matter in three Appalachian Mountain streams. *J. N. Am. Benthol. Soc.* 14.

Walters, C., and Kitchell, J.F. 2001. Cultivation/depensation effects on juvenile survival and recruitment: implications for the theory of fishing. *Can. J. Fish. Aquat. Sci.* 58, 39–50.

Walters, C., Christensen, V., and Pauly, D. 1997. Structuring dynamic models of exploited ecosystems from trophic mass-balance assessments. *Rev. Fish Biol. Fisher.* 7, 139–172.

Wardle, C.A. 1999a. How soil food-webs make plants grow. *Trends Ecol. Evol.* 14, 418–420.

Wardle, D.A. 1995. Impact of disturbance on detritus food-webs in agroecosystems of contrasting tillage and weed management practices. *Adv. Ecol. Res.* 26, 105–185.

Wardle, D.A. 1999b. Is "sampling effect" a problem for experiments investigating biodiversity-ecosystem function relationships? *Oikos* 87, 403–407.

Wardle, D.A. 2002a. *Communities and Ecosystems: Linking the Aboveground and Belowground Components*. Princeton University Press, Princeton.

Wardle, D.A. 2002b. *Linking the Aboveground and Belowground Components*. Princeton University Press, Princeton.

Wardle, D.A., and Nicholson, K.S. 1996. Synergistic effects of grassland plant species on soil microbial biomass and activity: implications for ecosystem-level effects of enriched plant diversity. *Func. Ecol.* 10, 410–416.

Wardle, D.A., and Lavelle, P. 1997. Linkages between soil biota, plant litter quality and decomposition. In: G. Cadisch and K.E. Giller (Editors), *Driven by Nature: Plant Litter Quality and Decomposition*. CAB International, Wallingford, pp 107–124.

Wardle, D.A., Bonner, K.I., and Nicholson, K.S. 1997a. Biodiversity and plant litter: experimental evidence which does not support the view that enhanced species richness improves ecosystem function. *Oikos* 79, 247–258.

Wardle, D.A., Zackrisson, O., Hörnberg, G., and Gallet, C. 1997b. The Influence of Island Area on Ecosystem Properties. *Science* 277, 1296–1299.

Warren, P.H. 1989. Spatial and temporal variation in the structure of a freshwater food web. *Oikos* 55, 299–311.

Warren, P.H. 1990. Variation in food-web structure: the determinants of connectance. *Am. Nat.* 136, 689–700.

Warren, P.H. 1994. Making connections in food webs. *Trends Ecol. Evol.* 9, 136–141.

Warren, P.H. 1995. Estimating morphologically determined connectance and structure of food webs of freshwater invertebrates. *Freshwater Biology* 33, 213–221.

Warren, P.H. 1996. Structural constraints on food web assembly. In: M.E. Hochberg, J. Clobert, and R. Barbault (Editors), *Aspects of the Genesis and Maintenance of Biological Diversity*. Oxford University Press, Oxford., pp 142–161.

Warren, P.H., and Lawton, J.H. 1987. Invertebrate predator-prey body size relationships: an explanation for upper triangular food webs and patterns in food web structure? *Oecologia* 74, 231–235.

Watts, D.J., and Strogatz, S.H. 1998. Collective dynamics of 'small-world' networks. *Nature* 393, 440–442.

Webb, S.C., Hedges, R.E.M., and Simpson, S.J. 1998. Diet quality influences the d13C and d15N of locusts and their biochemical components. *J. Exper. Biol.* 201, 2903–2911.

Webster, J.R. 1983. The role of benthic macroinvertebrates in detritus dynamics of streams: a computer simulation. *Ecol. Monogr.* 53, 383–404.

Webster, J.R., Meyer, J.L., Wallace, J.B., and Benfield, E.F. 1997. Organic matter dynamics in Hugh White Creek, Coweeta Hydrologic Laboratory, North Carolina, USA. *J. N. Am. Benthol. Soc.* 16, 74–78.

Webster, J.R., Mulholland, P.J., Tank, J.L., Valett, H.M., Dodds, W.K., Peterson, B.J., Bowden, W.B., Dahm, C.N., Findlay, S., Gregory, S.V., Grimm, N.B., Hamilton, S.K., Johnson, S.L., Marti, E., McDowell, W.H., Meyer, J.L., Morrall, D.D., Thomas, S.A., and Wollheim, W.M. 2003. Factors affecting ammonium uptake in streams -an inter-biome perspective. *Freshwat. Biol.* 48, 1329–1352.

Werner, E.E. 1988. Size, scaling and the evolution of complex life cycles. In: B. Ebenman and L. Persson (Editors), *Size-Structured Populations—Ecology and Evolution*. Springer Verlag, Berlin, pp 60–81.

Werner, E.E. 1994. Ontogenetic scaling of competitive relations: size-dependent effects and responses in two Anuran larvae. *Ecology* 75, 197–231.

Werner, E.E. 1998. Ecological experiments and a research program in community ecology. In: J. W.J. Restarits and J. Bernardo (Editors), *Experimental Ecology: Issues and Perspectives*. Oxford Univ. Press, Oxford, pp 3–26.

Werner, E.E., and Gilliam, J.F. 1984. The ontogenetic niche and species interactions in size-structured populations. *Ann. Rev. Ecol. Syst.* 15, 393–425.

Werner, E.E., and Peacor, S.D. 2003. A review of trait-mediated indirect interactions in ecological communities. *Ecology* 84, 1083–1100.

West, C. 1985. Factors underlying the late seasonal appearance of the lepidopterous leaf-miner guild on oak. *Ecol. Entomol.* 10, 111–120.

West, G.B., Brown, J.H., and Enquist, B.J. 1997. A general model for the origin of allometric scaling laws in biology. *Science* 276, 122–126.

West, G.B., Brown, J.H., and Enquist, B.J. 2000. The origin of universal scaling laws in biology. In: J.H. Brown and G.B. West (Editors), *Scaling in Biology*. Oxford University Press, Oxford, pp 87–112.

Wheelwright, N. 1985. Fruit size, gape width, and the diets of fruit-eating birds. *Ecology* 66, 808–818.

Whiles, M.R., and Wallace, J.B. 1992. First-year recovery of a headwater stream following a 3-year insecticide-induced disturbance. *Freshwat. Biol.* 28.

Whiles, M.R., and Wallace, J.B. 1995. Macroinvertebrate production in a headwater stream during recovery from anthropogenic disturbance and hydrologic extremes. *Can. J. Fish. Aquat. Sci.* 52, 2402–2422.

Whitford, W.G. 1989. Abiotic controls on the functional structure on soil food webs. *Biol. Fert. Soils* 8, 1–6.

Whittaker, R.J., Willis, K.J., and Field, R. 2001. Scale and species reichness: towards a general, hierchical theory of species diversity. *J. Biogeog.* 28, 453–470.

Williams, R.J., and Martinez, N.D. 2000. Simple rules yield complex food webs. *Nature* 404, 180–183.

Williams, R.J., and Martinez, N.D. 2002. *Trophic Levels in Complex Food Webs: Theory and Data*. 02-10-056, Santa Fe Inst. Working Paper.

Williams, R.J., and Martinez, N.D. 2004a. Stabilization of chaotic and non-permanent food web dynamics. *Eur. Phys. J. B* 38, 297–203.

Williams, R.J., and Martinez, N.D. 2004b. Limits to trophic levels and omnivory in complex food web: Theory and data. *Am. Nat.* 163, 458–468.

Williams, R.J., and Martinez, N.D. 2004c. *Diversity, complexity, and persistence in large model ecosystems*. Santa Fe Institute Working Paper 2004-07-022.

Williams, R.J., Martinez, N.D., Berlow, E.L., Dunne, J.A., and Barabàsi, A.L. 2002. Two degrees of separation in complex food webs. *P. Natl. Acad. Sci. USA* 99, 12913–12916.

Williams, R.J., Martinez, N.D. 2000. Simple rules yield complex food webs. *Nature* 404, 180–183.

Williamson, M.H., and Lawton, J.H. 1991. Fractal geometry of ecological habitats. In: S.S. Bell, E.D. Mccoy, and H.R. Mushinsky (Editors), *Habitat Structure: The Physical Arrangement of Objects in Space*. Chapman and Hall, London, pp 69–81.

Wilson, D.S. 1975. The adequacy of body size as a niche difference. *Am. Nat.* 109, 769–784.

Wilson, D.S., and Yoshimura, J. 1994. On the coexistence of specialists and generalists. *Am. Nat.* 144, 692–707.

Winemiller, K.O. 1989a. Patterns of variation in life history among South American fishes in seasonal environments. *Oecologia* 81, 225–241.

Winemiller, K.O. 1989b. Must connectance decrease with species richness? *Am. Nat.* 34, 960–968.

Winemiller, K.O. 1989c. Ontogenetic diet shifts and resource partitioning among piscivorous fishes in the Venezuelan llanos. *Environ. Biol. Fishes* 26, 177–199.

Winemiller, K.O. 1990. Spatial and temporal variation in tropical fish trophic networks. *Ecol. Monogr.* 60, 331–367.

Winemiller, K.O. 1996. Factors driving spatial and temporal variation in aquatic floodplain food webs. In: G.A. Polis and K.O. Winemiller, (Editors), *Food Webs: Integration of Patterns and Dynamics.* Chapman and Hall, New York, pp 298–312.

Winemiller, K.O., and Rose, K.A. 1992. Patterns of life-history diversification in North American fishes: implications for population regulation. *Can. J. Fish. Aquat. Sci.* 49, 2196–2218.

Winemiller, K.O., and Rose, K.A. 1993. Why do most fish produce so many tiny offspring? Evidence from a size-based model. *Am. Nat.* 142, 585–603.

Winemiller, K.O., and Polis, G.A. 1996. Food webs: what can they tell us about the world? In: G.A. Polis and K.O. Winemiller, (Editors), *Food Webs: Integration of Pattern and Dynamics.* Chapman and Hall, New York, USA.

Winemiller, K.O., and Jepsen, D.B. 1998. Effects of seasonality and fish movement on tropical river food webs. *J. Fish Biol.* 53, 267–296.

Winemiller, K.O., and Jepsen, D.B. 2004. Migratory neotropical fish subsidize food webs of oligotrophic blackwater rivers. In: G.A. Polis, M.E. Power, and G. Huxel, (Editors), *Food Webs at the Landscape Level.* University of Chicago Press, Chicago, pp 115–132.

Winemiller, K.O., Taphorn, D.C., and Barbarino-Duque, A. 1997. Ecology of Cichla (Cichlidae) in two blackwater rivers of southern Venezuela. *Copeia* 1997, 690–696.

Winemiller, K.O., Montoya, J.V., Layman, C.A., Roelke, D.L., and Cotner, J.B. 2006. Experimental demonstrations of seasonal fish effects on benthic ecology of a Neotropical floodplain river. *J. of the N. Am. Benthol. Soc.,* in press.

Wipfli, M.S., and Gregovich, D.P. 2002. Export of invertebrates and detritus from fishless headwater streams in southeastern Alaska: implications for downstream salmonid production. *Freshwat. Biol.* 47.

Wipfli, M.S., Chaloner, D.T., and Caouette, J.P. 1999. Influence of salmon spawner densities on stream productivity in Southeastern Alaska. *Can. J. Fish. Aquat. Sci.* 56, 1600–1611.

Wise, D.H. 1995. *Spiders in Ecological Webs.* Cambridge University Press, Cambridge.

Wise, D.H., and Chen, B.R. 1999. Impact of intraguild predators on survival of a forest-floor wolf spider. *Oecologia* 121, 129–137.

Wolters, V. 1998. Long-term dynamics of a collembolan community. *Appl. Soil Ecol.* 9, 221–227.

Woodward, G., and Hildrew, A.G. 2001. Invasion of a stream food web by a new top predator. *J. Anim. Ecol.* 70, 273–288.

Woodward, G., and Hildrew, A.G. 2002a. Body-size determinants of niche overlap and intraguild predation within a complex food web. *J. Anim. Ecol.* 71, 1063–1074.

Woodward, G., and Hildrew, A.G. 2002b. Food web structure in riverine landscapes. *Freshwat. Biol.* 47, 777–798.

Woodward, G., Jones, J.I., and Hildrew, A.G. 2002. Community persistence in Broadstone Stream (U.K.) over three decades. *Freshwat. Biol.* 47, 1419–1435.

Woodward, G., Speirs, D.C., and Hildrew, A.G. 2005. Quantification and resolution of a complex, size-structured food web. *Adv. Ecol. Res.* 36, 85–135.

Wootton, J. 1994a. Putting the pieces together: testing the independence of interactions among organisms. *Ecology* 75, 1544–1551.

Wootton, J., and Power, M.E. 1993. Productivity, consumers, and the structure of a river food-chain. *P. Natl. Acad. Sci. USA* 90, 1384–1387.

Wootton, J.T. 1992. Indirect effects, prey susceptibility and habitat selection: impacts of birds on limpets and algae. *Ecology* 73, 981–991.

Wootton, J.T. 1993. Indirect effects and habitat use in an intertidal community: interaction chains and interaction modifications. *Am. Nat.* 141, 71–89.

Wootton, J.T. 1994b. The nature and consequences of indirect effects in ecological communities. *Ann. Rev. Ecol. Syst.* 25, 443–466.

Wootton, J.T. 1994c. Predicting direct and indirect effects: an integrated approach using experiments and path analysis. *Ecology* 75, 151–165.

Wootton, J.T. 1997. Estimates and tests of per capita interaction strength: diet, abundance, and impact of intertidally foraging birds. *Ecol. Monogr.* 67, 45–64.

Wootton, J.T. 2001. Prediction in complex communities: analysis of empirically derived Markov models. *Ecology* 82, 580–598.

Wootton, J.T. 2002. Indirect effects in complex ecosystems: recent progress and future challenges. *J. Sea Res.* 48, 157–172.

Wootton, J.T., and Downing, A.L. 2003. Understanding the effects of reduced biodiversity: a comparison of two approaches. In: P.M. Kareiva and S.A. Levin (Editors), *The Importance of Species; Perspective on Expendability and Triage.* Princeton University Press, Princeton, New Jersey, pp 85–104.

Wotton, R.S., and Malmqvist, B. 2001. Feces in aquatic ecosystem. *Bioscience* 51, 537–544.

Wright, J.F. 1995. Development and use of a system for predicting the macroinvertebrate fauna in flowing waters. *Aust. J. Ecol.* 20, 181–197.

Wulff, F., Field, J.G., and Mann, K.H. 1989. *Network Analysis in Marine Ecology: Methods and Applications.* Springer Verlag, Berlin Heidelberg.

Yachi, S., and Loreau, M. 1999. Biodiversity and ecosystem productivity in a fluctuating environment: the insurance hypothesis. *P. Natl. Acad. Sci. USA* 96, 1463–1468.

Yodzis, P. 1980. The connectance of real ecosystems. *Nature* 284, 544–545.

Yodzis, P. 1981a. The stability of real ecosystems. *Nature* 289, 674–676.

Yodzis, P. 1981b. The structure of assembled communites. *J. Theor. Biol.* 92, 103–117.

Yodzis, P. 1982. The compartmentation of real and assembled ecosystems. *Am. Nat.* 120, 551–570.

Yodzis, P. 1984. How rare is omnivory. *Ecology* 65, 321–323.

Yodzis, P. 1988. The indeterminacy of ecological interactions as perceived through perturbation experiments. *Ecology* 69, 508–512.

Yodzis, P. 1989. *Introduction to Theoretical Ecology*. Harper & Row, New York, 384.

Yodzis, P. 1993. Environment and trophodiversity. In: R.E. Ricklefs and D. Schluter (Editors), *Species Diversity in Ecological Communities: Historical and Geographical Perspectives*. The University of Chicago Press, Chicago, pp 26–38.

Yodzis, P. 1994. Predator-prey theory and management of multispecies fisheries. *Ecol. Appl.* 4, 51–58.

Yodzis, P. 1996. Food webs and perturbation experiments: theory and practice. In: G.A. Polis and K.O. Winemiller (Editors), *Food Webs: Integration of Pattern and Dynamics*. Chapman and Hall, New York, pp 192–200.

Yodzis, P. 1998. Local trophodynamics and the interaction of marine mammals and fisheries in the Benguela ecosystem. *J. Anim. Ecol.* 67, 635–658.

Yodzis, P. 2000. Diffuse effects in food webs. *Ecology* 81, 261–226.

Yodzis, P. 2001. Must top predators be culled for the sake of fisheries? *Trends Ecol. Evol.* 16, 78–84.

Yodzis, P., and Innes, S. 1992. Body-size and consumer-resource dynamics. *Am. Nat.* 139, 1151–1173.

Yodzis, P., and Winemiller, K.O. 1999. In search of operational trophospecies in a tropical aquatic food web. *Oikos* 87, 327–340.

Yoshida, K. 2003. Evolutionary dynamics of species diversity in an interaction web system. *Ecol. Model.* 163, 131–143.

Yoshida, T., Jones, L.E., Ellner, S.P., et al. 2003. Rapid evolution drives ecological dynamics in a predator-prey system. *Nature* 424, 303–306.

Zar, J.H. 1984. *Biostatistical Analysis, 2nd edition*. Prentice-Hall, Englewood Cliffs, NJ, 718.

Zechmeister-Boltenstern, S., Baumgarten, A., Bruckner, A., Kampichler, C., and Kandeler, E. 1998. Impact of faunal complexity on nutrient supply in field mesocosms from a spruce forest soil. *Plant Soil* 198, 45–52.

Zheng, D.W., Bengtsson, J., and Ågren, G.I. 1997. Soil food webs and ecosystem processes: decomposition in donor-control and Lotka-Volterra systems. *Am. Nat.* 149, 125–148.

Zheng, D.W., Ågren, G.I., and Bengtsson, J. 1999. How do soil organisms affect total organic nitrogen storage and substrate nitrogen to carbon ratio in soils? A theoretical analysis. *Oikos* 86, 430–442.

KEYWORDS

Layman

- Anthropogenic effects
- Experiments
- Food chain length
- Predation
- Stable isotopes
- Stomach content analysis

Scheu

- Allochthonous resources
- Aquatic systems
- Belowground system
- Decomposer system
- Decomposer-plant interactions
- Generalist predators
- Omnivory
- Prey switching
- Rhizosphere
- Stable isotope analysis
- Terrestrial systems

McKane

- Adaptive foraging
- Beddington functional response

- Competition
- Deletion stability
- Dynamical model
- Evolutionarily stable strategy (ESS)
- Evolving webs
- Food webs
- Functional response
- Holling type II functional response
- Link strength distribution
- Lotka-Volterra functional response
- Model food webs
- Population dynamics
- Ratio-dependent functional response
- Simulations
- Stability
- Weak links

Dell

- Complexity
- Connectance
- Food chain
- Interaction strength
- Non-linear
- Omnivory
- Path lengths
- Persistence
- Stability
- Topology

Persson

- Cannibalism
- Cohort cycles
- Ecosystem shifts
- Intraspecific size scaling
- Ontogeny

Petchy

- Complementarity
- Diversity
- Ecosystem functioning
- Functional diversity

- Functional groups
- Microbial community

Woodward

- Allometric scaling
- Body-size
- Seasonality
- Spatial scaling
- Stability
- Stoichiometry

Kondoh

- Adaptive foraging
- Food web complexity
- Flexibility
- Stability

De Roos

- Alternative stable states
- Bistability
- Catastrophic collapse
- Emergent facilitation
- Community dynamics
- Community structure
- Development
- Development regulation
- Emergent Allee effect
- Growth
- Physiologically structured population model
- Population size distribution
- Size-dependent life history processes
- Size-specific predation
- Size structure

Schroter

- Bacteria
- Bacterial subsystem
- Biomass
- Biomass pyramid
- Carbon
- Decomposers

- Energy channel
- Forest
- Function
- Functional group contribution
- Fungal subsystem
- Fungi
- Global change
- Interaction strength
- Intraspecific interaction strength
- Jacobian matrix
- Loop
- Microbial food web
- Microbivores
- Mineralization
- Modelling
- Nitrogen
- Nitrogen deposition
- Protozoa
- Soil
- Stability
- Structure
- Terrestrial carbon balance
- Testate amoebae

Cohen

- Abundance, numerical
- Abundance-body size relationship
- Akaike information criterion
- Allometry
- Biomass
- Biomass spectrum
- Body mass
- Body size
- Body size distribution
- Community food web
- Confidence interval
- Distribution, body size
- Energetic model
- Error-in-variables regression
- Food web
- Food web, pelagic
- General structural relation (statistical model)

- Homoscedasticity
- Independence (of residuals in linear regression)
- Least squares regression
- Linear model
- Linear regression
- Linearity (in linear regression)
- Mass, body
- Matlab
- Model, energetic
- Normality (of residuals in linear regression)
- Normality tests
- Pelagic food web
- Quantile-quantile plot
- Reduced major axis regression
- Regression, least squares
- Regression, linear
- Regression, reduced major axis
- Residual (in linear regression)
- Size, body
- Slope (linear regression)
- Spectrum, biomass
- Tuesday Lake, Michigan
- Ythan Estuary

Leibold

- Cohort cycles
- Daphnia pulex
- Population dynamics
- Plankton dynamics
- Predator–prey oscillations

Fox

- Biodiversity
- Biomass partitioning
- Coexistence
- Ecosystem function
- Selection effect
- Structural effect
- Trade-offs

Sabo

- Detritivore

- Detritus
- Field experiment
- Leaf litter
- Predator–prey interaction
- Riparian

Setala

- Biodiversity
- Functional redundancy
- Omnivore
- Nutrient dynamics
- Trophic interactions

Loreau

- Biodiversity
- Ecosystem functioning
- Environmental fluctuations
- Food webs
- Food web connectivity
- Functional complementarity
- Functional redundancy
- Generalist consumers
- Insurance
- Interaction strength
- Models
- Niche differentiation
- Specialist consumers
- Stability
- Temporal variability
- Theory
- Top-down control
- Trade-offs

McCann

- Compartments
- Generalist consumers
- Variability
- Space
- Stability
- Sub-systems

Beisner (* top 5)

- *Exotic species invasions
- *Diversity
- *Resource availability
- *Rock pool food webs
- *Species richness
- ANOVA
- *Alona davidii*
- Aquatic
- *Armases miersii*
- Assembly rules
- Camargo index
- *Candona* sp.
- *Ceriodaphnia* sp.
- Cladocera
- Community resistance
- Competitors/Competition
- Copepods
- Correspondence analysis
- *Culex* sp.
- *Cypridopsis* sp.
- *Dasyhelea* sp.
- Detritivores
- Detritus
- Discovery Bay Marine Laboratory, Jamaica
- Diversity
- Exotic species/Invaders
- Experimental
- Food web structure
- Generation time
- Herbivores
- Insect larvae
- Invertebrate
- Multi-trophic level
- Niche overlap
- Nutrients
- Facultative predators
- *Leydigia leydigia*
- Life history
- Meiofauna

- Microcosms/Container habitats
- Midge
- Multivariate
- *Nematode* sp.
- *Nitocra spinipes*
- *Oligochaete* sp.
- *Orthocyclops modestus*
- Ostracod
- *Potamocypris* sp.
- Relative abundance
- Resources/Resource limitation
- Reticulate food webs
- Species composition
- Species invasion
- Species richness
- Stable isotopes
- Stress
- Traits
- Trophic habits
- Trophic levels
- Trophic position
- Tropical freshwater rock pools/Rock pool ecosystem
- Worms
- Zooplankton

Koen-Alonso

- Argentina
- Bifurcation analysis
- Bioenergetic-allometric models
- Engraulis anchoita
- Functional response
- Illex argentinus
- Merluccius hubbsi
- Model uncertainty
- Multispecies fisheries models
- Otaria flavescens
- Parameter uncertainty
- Patagonia
- Sampling-importance-resampling algorithm
- Simulated annealing algorithm
- Structural uncertainty

- Southwestern South Atlantic
- Trophodynamic models

Downing

- Biodiversity
- Ecosystem functioning
- Extinctions
- Functional groups
- Indirect effects
- Macroinvertebrates
- Mesocosms
- Periphyton
- Phytoplankton
- Pond food webs
- Species richness
- Species composition
- Trophic position
- Zooplankton

Brose

- Allometric scaling
- Consumer resource dynamics
- Herbivory
- Predator–prey dynamics
- Resource competition

Culp

- Algae
- Benthic invertebrate
- Biomonitoring
- Community composition and structure
- Contaminant
- Detritus
- Ecological risk assessment
- Enrichment
- Fish
- Functional feeding group (FFG)
- Insect emergence
- Mayfly
- Metal mine effluent
- Midge
- Nutrient

- Pulp mill effluent
- River
- Stonefly
- Stream mesocosm
- Stressor
- Weight-of-evidence

Montoya

- Abundance
- Body size
- Broadstone Stream
- Empirical food webs
- Indirect effects
- Interaction strength
- Inverse community matrix
- Linkage density
- Link distribution
- Lotka-Volterra model
- Macroecological patterns
- Predator–prey body size ratio
- Press perturbations
- Ythan Estuary

Bengtsson & Berg

- Bray-Curtis similarity
- Composition
- Detritus
- Food web structure
- Fragmented litter
- Functional groups
- Horizontal similarity
- Humus
- Litter
- Pine forest
- Similarity in food web composition
- Small-scale variability
- Soil food web
- Spatial
- Stratification
- Taxonomical aggregation
- Temporal
- Temporal similarity

- Variability
- Variation in food webs
- Vertical similarity

Scharler

- Allelopathy
- Aquatic food webs
- Consumer-driven nutrient recycling (CNR)
- Ecological network analysis (ENA)
- Ecosystem management
- Indirect effects
- Infochemicals
- Life history traits

Vos

- Aquatic and terrestrial systems
- Bifurcation analysis
- Defenses
- Eigenvalues
- Flexible food web links
- Food chain models
- Herbivory
- Heterogeneous food web nodes
- Induced defenses
- Interaction strength
- Paradox of enrichment
- Persistence
- Predators
- Prey
- Resilience
- Stability
- Trophic structure

INDEX